Lecture Notes in Physics

Springer
Berlin
Heidelberg
New York
Barcelona
Hong Kong
London
Milan
Paris
Tokyo

Editorial Policy

The series *Lecture Notes in Physics* (LNP), founded in 1969, reports new developments in physics research and teaching -- quickly, informally but with a high quality. Manuscripts to be considered for publication are topical volumes consisting of a limited number of contributions, carefully edited and closely related to each other. Each contribution should contain at least partly original and previously unpublished material, be written in a clear, pedagogical style and aimed at a broader readership, especially graduate students and nonspecialist researchers wishing to familiarize themselves with the topic concerned. For this reason, traditional proceedings cannot be considered for this series, though volumes to appear in this series are often based on material presented at conferences, workshops and schools (in exceptional cases the original papers and/or those not included in the printed book may be added on an accompanying CD-ROM, together with the abstracts of posters and other material suitable for publication, e.g. large tables, colour pictures, program codes, etc.).

Acceptance

A project can only be accepted tentatively for publication, by both the editorial board and the publisher, following thorough examination of the material submitted. The book proposal sent to the publisher should consist of at least a preliminary table of contents outlining the structure of the book together with abstracts of all contributions to be included.
Final acceptance is issued by the series editor in charge, in consultation with the publisher, only after receiving the complete manuscript. Final acceptance, possibly requiring minor corrections, usually follows the tentative acceptance unless the final manuscript differs significantly from expectations (project outline). In particular, the series editors are entitled to reject individual contributions if they do not meet the high quality standards of this series. The final manuscript must be camera-ready, and should include both an informative introduction and a sufficiently detailed subject index.

Contractual Aspects

Publication in LNP is free of charge. There is no formal contract, no royalties are paid, and no bulk orders are required, although special discounts are offered in this case. The volume editors receive jointly 30 free copies for their personal use and are entitled, as are the contributing authors, to purchase Springer books at a reduced rate. The publisher secures the copyright for each volume. As a rule, no reprints of individual contributions can be supplied.

Manuscript Submission

The manuscript in its final and approved version must be submitted in camera-ready form. The corresponding electronic source files are also required for the production process, in particular the online version. Technical assistance in compiling the final manuscript can be provided by the publisher's production editor(s), especially with regard to the publisher's own LaTeX macro package which has been specially designed for this series.

Online Version/ LNP Homepage

LNP homepage (list of available titles, aims and scope, editorial contacts etc.):
http://www.springer.de/phys/books/lnp/

LNP online (abstracts, full-texts, subscriptions etc.):
http://link.springer.de/series/lnp/

K. W. Weiler (Ed.)

Supernovae and Gamma-Ray Bursters

Springer

Editor

Dr. Kurt W. Weiler
NRL-Code 7213
Naval Research Laboratory
Washington, DC 20375-5320, USA

Cover picture: shows SN 1998bw/GRB 980425 in ESO 184-G82, by T.J. Galama, F.M. Freeswijk, E. Pian, F. Frontera, V. Doublier and J.-F. Gonzalez

---

Cataloging-in-Publication Data applied for

A catalog record for this book is available from the Library of Congress.

Bibliographic information published by Die Deutsche Bibliothek
Die Deutsche Bibliothek lists this publication in the Deutsche Nationalbibliografie; detailed bibliographic data is available in the Internet at http://dnb.ddb.de

ISSN 0075-8450
ISBN 3-540-44053-4 Springer-Verlag Berlin Heidelberg New York

Springer-Verlag Berlin Heidelberg New York
a member of BertelsmannSpringer Science+Business Media GmbH

http://www.springer.de

Typesetting: Camera-ready by the authors/editor
Camera-data conversion by Steingraeber Satztechnik GmbH Heidelberg
Cover design: *design & production*, Heidelberg

Printed on acid-free paper
SPIN: 10880850 54/3141/du - 5 4 3 2 1 0

Preface

Since the dawn of mankind, observers of the sky have wondered at the sudden appearance of new stars on the seemingly unchanging heavens and, for at least 2000 years, have recorded these phenomena in their annals and archives. Even in more modern times, since the discovery of SN1885A in S Andromeda which figured in the important "island universe" discussions of the 1920's, the puzzle of supernovae (SNe) has played an important role in astrophysics. Only with the seminal work of Fritz Zwicky and Walter Baade in the 1930's did we begin to understand the differences between novae and SNe and the importance of SNe as the fonts of energy for the interstellar medium and as drivers of chemical evolution in galaxies. As recently as the 1940's and 1950's the early days of radio astronomy were heavily influenced by the familiar names of Cassiopeia A and Taurus A, two young supernova remnants, and two Nobel prizes have been awarded for discovery and study of a related phenomenon, pulsars.

In spite of the great age of the study of SNe, since at least the Chinese records of SN185 and probably earlier, the field is, in fact, very young having only attracted a large devoted following since the spectacular Type II SN1987A in the Large Magellanic Cloud, the first naked-eye SN in more than 400 years.

On a seemingly non-intersecting parallel path, γ-ray bursts (GRBs) discovered by the Air Force VELA satellites in the 1960's presented a mystery to researchers for 30 years. Finally, the launch of the Italian/Dutch *BeppoSAX* satellite in 1996 provided sufficiently fast and accurate positional information to allow detection and study of their "afterglows" at other wavelengths. These results then provided evidence that, at least at some level, the fields of GRB and SN studies merge through the possible connection of Type Ib/c SNe, so that one of our most recent astronomical puzzles appears to be at least partially solved by reference to our ancient interest in SNe.

Although discovery, observation, and interpretation of new examples of SNe and GRBs continues, the end of the *Compton Gamma-Ray Observatory (CGRO)* era in 2000 and of the *BeppoSAX* operations in 2002 provides a significant breakpoint for trying to summarize the current status of these extremely active areas of study. Thus, experts from many areas of SN and GRB research have agreed to contribute chapters to this monograph to assemble a coherent picture.

Because the two areas of research have still only partially merged, and may never totally merge because of the possibility that some types of GRBs originate in other physical processes, we have chosen to roughly divide this work into two parts – SN research and GRB research with bridging chapters to explore the known and likely relations between the two areas. We hope that this monograph contributes some small part to our ultimate understanding of these exciting phenomena.

Before proceeding further, however, I would like to thank all of the people and institutions which have contributed to the assembly of this volume. Foremost, I wish to thank all of the chapter authors who have contributed their knowledge, expertize, time, and effort to providing up-to-date descriptions of the many areas of supernova and GRB research and for working so willingly with me on the preparation and editing of this volume. Obviously, nothing could have been accomplished without the support of the Office of Naval Research (ONR), which provides the 6.1 funding for my research, and the Naval Research Laboratory (NRL) which provides me with the time and facilities necessary. Although too many individuals to list have been supportive of my effort, I must separately thank Dr. Lee J Rickard, who has been extremely tolerant of my disappearing for days at a time and who has shielded me from so many other demanding and time consuming tasks.

N.B.: Because the fields of supernova and GRB research tend to use somewhat different nomenclature for the same thing – radio supernova flux density light curves are often described by $S \propto \nu^{\pm\alpha} t^{\pm\beta}$ *while GRB workers tend to use* $F_\nu \propto \nu^{\pm\beta} t^{\pm\alpha}$ *– I have attempted to standardize everywhere to the format* $F_\nu \propto \nu^{+\alpha} t^{+\beta}$*. I have also attempted to make all chapters consistent with using:* α *= spectral index,* β *= decline rate,* F_ν *= flux density,* γ *= gamma-rays,* Γ *= Lorentz factor,* τ *= optical depth,* t *= time, and* T *= temperature. Although I have tried to minimize it, there may be some remaining variation in notation between chapters.*

Washington, DC, USA
January 2003

Kurt W. Weiler
Naval Research Laboratory

Table of Contents

Part II Supernovae to γ-Ray Bursters

Part III γ-Ray Bursters

List of Contributors

E. Baron
University of Oklahoma
Norman, OK 73019
USA
baron@mail.nhn.ou.edu

David Branch
University of Oklahoma
Norman, OK 73019
USA
branch@mail.nhn.ou.edu

Enrico Cappellaro
Osservatorio Astronomico di Capodimonte
via Moiariello 16
80181 Napoli
Italy
cappellaro@na.astro.it

Roger A. Chevalier
Department of Astronomy
University of Virginia
PO Box 3818
Charlottesville, VA 22903
USA
rac5x@virginia.edu

Claes Fransson
Stockholm Observatory
Department of Astronomy
SCFAB
106 91 Stockholm
Sweden
claes@astro.su.se

Filippo Frontera
Physics Department
University of Ferrara
Ferrara
Italy
and
Ist. Astrof. Spaz. e Fis. Cosm.
CNR
Bologna
Italy
frontera@fe.infn.it

Titus J. Galama
Astronomy Department
Calif. Inst. of Technology
Pasadena, CA 91125
USA
tjg@astro.caltech.edu

David A. Green
Cavendish Laboratory
University of Cambridge
Madingley Road
Cambridge CB3 0HE
United Kingdom
dag9@cam.ac.uk

Kevin Hurley
University of California
Space Sciences Laboratory
Berkeley, CA 94720-7450
USA
khurley@sunspot.ssl.berkeley.edu

Stefan Immler
Dept. of Astron. and Astrophys.
Pennsylvania State Univ.
University Park, PA 16802
USA
immler@astro.psu.edu

Koichi Iwamoto
Department of Physics
College of Science and Technology
Nihon University
Tokyo 101-8308
Japan
iwamoto@phys.cst.nihon-u.ac.jp

David J. Jeffery
NM Inst. of Mining and Technology
Socorro, NM 87801
USA
jeffery@kestrel.nmt.edu

Bruno Leibundgut
European Southern Observatory
Karl-Schwarzschild-Strasse 2
85748 Garching
Germany
bleibundgut@eso.org

Walter H.G. Lewin
Center for Space Research
and
Department of Physics
Massachusetts Inst. of Technology
Cambridge, MA 02139-4307
USA
lewin@space.mit.edu

Zhi-Yun Li
Department of Astronomy
University of Virginia
Charlottesville, VA 22903
USA
zl4h@virginia.edu

Abraham Loeb
Astronomy Department
Harvard University
Cambridge, MA 02138
USA
aloeb@cfa.harvard.edu

Keiichi Maeda
Department of Astronomy
and
Research Centr. for Early Univ.
School of Science
The University of Tokyo
Tokyo 113-0033
Japan
maeda@astron.s.u-tokyo.ac.jp

Paolo A. Mazzali
Astronomical Observatory of Trieste
Via G.B. Tiepolo 11
34131 Trieste
Italy
mazzali@ts.astro.it

Richard McCray
JILA
University of Colorado
Boulder CO 80309-0440
USA
dick@jila.colorado.edu

Marcos J. Montes
Naval Research Laboratory
Code 7212
Washington, DC 20375-5320
USA
montes@rsd.nrl.navy.mil

Takayoshi Nakamura
Department of Astronomy
and
Research Center for the Early
Universe
School of Science
University of Tokyo
Tokyo 113-0033
Japan

Ken'ichi Nomoto
Department of Astronomy
and
Research Centr. for Early Univ.
School of Science
University of Tokyo
Tokyo 113-0033
Japan
nomoto@astron.s.u-tokyo.ac.jp

Nino Panagia
STScI
and
Astrophysics Division
Space Science Department of ESA
3700 San Martin Drive
Baltimore, MD 21218
USA
panagia@stsci.edu

Saul Perlmutter
Physics Division
Lawrence Berkeley National Lab.
University of California
Berkeley, CA 94720
USA
saul@lbl.gov

Elena Pian
INAF
Astronomical Observatory of Trieste
Via G.B. Tiepolo 11
34131 Trieste
Italy
pian@ts.astro.it

Brian P. Schmidt
Research School of Astron. and Ap.
The Australian National University
via Cotter Road
Weston Creek, ACT 2611
Australia
brian@mso.anu.edu.au

Richard Sramek
NRAO
PO Box 0
Socorro, NM 87801
USA
dsramek@nrao.edu

F. Richard Stephenson
Department of Physics
University of Durham
Durham DH1 3LE
United Kingdom
F.R.Stephenson@durham.ac.uk

Nicholas Suntzeff
Cerro Tololo
Inter-American Observatory
Casilla 603
La Serena
Chile
nsuntzeff@noao.edu

Massimo Turatto
Osservatorio Astronomico di Padova
vicolo dell'Osservatorio 5
35122 Padova
Italy
turatto@pd.astro.it

Eli Waxman
Weizmann Institute of Science
Rehovot 76100
Israel
waxman@wicc.weizmann.ac.il

Kurt W. Weiler
Naval Research Laboratory
Code 7213
Washington, DC 20375-5320
USA
Kurt.Weiler@nrl.navy.mil

Introduction

Kurt W. Weiler

Naval Research Laboratory, Code 7213, Washington, DC 20375-5320, USA

Although ancient, the study of supernovae (SNe) is still developing extremely rapidly with new instruments, new measurements, new interpretations, and new models. Work in recent years has led to major modifications of our understanding of the nature of SNe and their role in the evolution of galaxies. Typical of the rapid change in our understanding was the unexpected discovery of a blue supergiant progenitor for SN1987A rather than the expected red supergiant. The ability of the *Hubble Space Telescope (HST)* to image the development of SN1987A and the interaction of its shock with the nearby circumstellar medium, particularly the inner optical ring [15,18,19], has contributed greatly to our understanding of these extremely energetic phenomena. All of this is aided by new data covering the entire wavelength range from radio to γ-ray and new theoretical interpretation of these results.

In addition, we have a much larger sample of objects for study. The discovery rate of new SNe has grown from about two per month 20 years ago to almost one per day at present and the volume of space which can be sampled has increased dramatically from relatively nearby galaxy clusters to deep surveys for objects at $z \gtrsim 1$. In particular, the study of these very distant objects, especially type Ia SNe, has impacted cosmology by yielding an intriguing and puzzling picture of the universe in which we live – apparently it is Euclidean, rather than curved as expected by many, but seems to not only be "open," and thus expanding forever, but expanding at an accelerating rate due to the presence of a so-called "dark energy" (see, e.g., [14,22]).

Increasingly detailed study of SNe has also led to the breakdown of our long-standing classification scheme of type I (no hydrogen lines detected in the optical spectra) or type II (hydrogen lines detected) into a richer range of nomenclature including type Ia, type Ib, type Ic, type IIL, type IIP, type IIb, and type IIn to name just a few.

To provide an overview of this "zoo" of SNe, a taxonomy chart has been developed which is given in Fig. 1. (See also Fig. 1 in the chapter by Turatto in this volume.)

The study of γ-ray bursts (GRBs) is much younger than that of SNe, dating only to the U.S. Air Force VELA satellites in the 1960s. NASA's *Compton Gamma-Ray Observatory (CGRO)* satellite launched in 1990 detected several thousand bursts, with an occurrence rate of approximately one per day, and proved a uniform distribution on the sky which led theoreticians to suggest that their sources were either very near, and thus uniformly distributed around the solar system, in an unexpectedly large halo around the Galaxy, or are at cosmological distances – not very restrictive proposals.

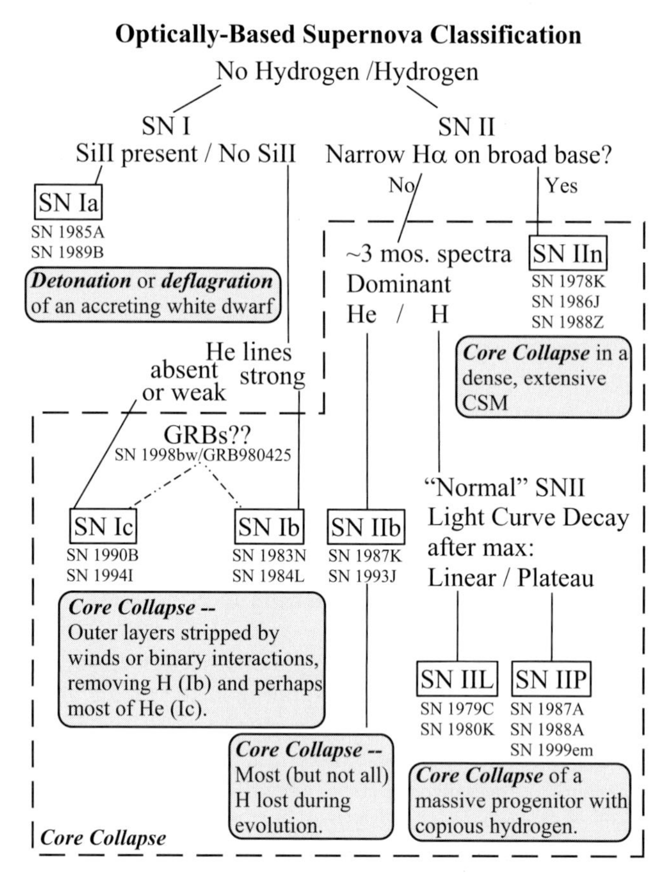

Fig. 1. Taxonomy of SNe with a likely connection to some types of GRBs. (M. J. Montes, private communication; see also Fig. 1 in the chapter by Turatto in this volume.) Expected explosion mechanisms are given in the shaded boxes with rounded corners.

However, the launch of the Italian/Dutch satellite *BeppoSAX* in 1996 made it possible to couple a quick response pointing system with a relatively high precision (few arcmin) coded mask *Wide Field Camera (WFC)* for X-rays and an all sky a *Gamma-Ray Burst Monitor (GRBM)* to finally detect and rapidly pinpoint the GRBs for follow-up at other wavelengths. These longer wavelength "afterglows" then allowed observers to probe the immediate environment of γ-ray burst sources and to assemble clues as to their nature.

The first detected optical afterglow by Groot et al. [9] for GRB970228 was followed-up with the *HST* by Sahu et al. [23]. They demonstrated that the GRB was associated with a faint (thus probably distant) late-type galaxy. A few months later Fruchter et al. [6] imaged the afterglow of GRB970508 with the *HST Space Telescope Imaging Spectrograph (STIS)* (see also, Pian et al. [20]) and proposed that the host galaxy should be a compact dwarf at a redshift of $z = 0.835$ [16]. Further *HST* imaging over one year later confirmed this pre-

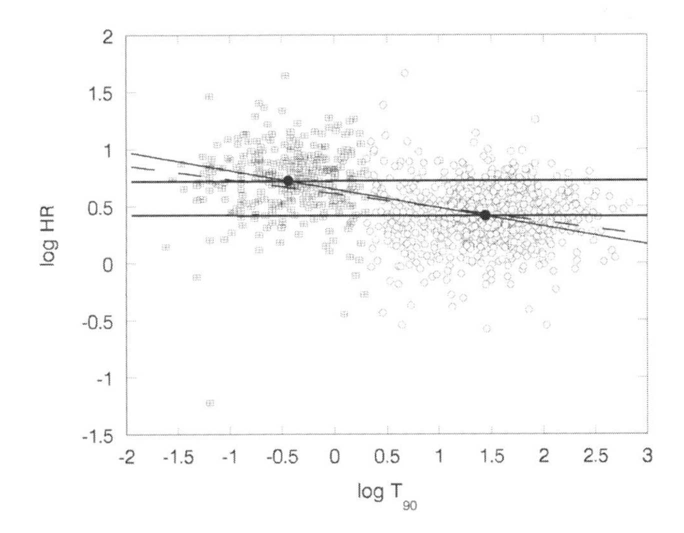

Fig. 2. Hardness Ratio-Duration relation for γ-ray bursts from Qin et al. [21].

diction by detecting a late-type spiral at the position of the optical transient [7]. Also, detection of an optical afterglow from GRB971214 [10,11,13,17] measured a redshift of $z = 3.42$ and confirmed that GRBs are at cosmological distances.

As with SNe, increased study of GRBs has led to increasingly complex taxonomy. Although the origin of the differences is not presently understood, there appear to be at least two GRB categories based on their γ-ray durations and spectral hardness ratios – succinctly characterized as "short-hard" and "long-soft." Short events (mean duration $\sim$0.2 s) with a harder γ-ray spectrum roughly separate from long events (mean duration $\sim$20 s) which tend to have a somewhat softer γ-ray spectrum. This is illustrated in Fig. 2 from Qin et al. [21]. At present, all of the GRBs with detected afterglows fall in the long-soft category, partly because their spectra imply a low flux in the X-ray band (< 30 keV), probably below the sensitivity of the wide field X-ray telescopes aboard *BeppoSAX* and partly because the trigger logic of the *GRBM* is more efficient for long duration bursts.

Although the number of detected GRB optical afterglows is very limited with only about two dozen known, most appear to occur in galaxies and a few show late-time deviations in their light curves consistent with SN-like optical contributions [1,2,8]. The remarkable type Ib/c SN1998bw, which is probably related to GRB980425, presents a possible "missing link" between SNe and GRBs. An initial attempt to place GRBs within the context of SNe is shown in the SN Taxonomy of Fig. 1.

In addition to the overview and synthesis of the SN and GRB fields which we undertake in this volume, there are a number of more specialized, recent reviews for SNe [3,4,12,14,25] and GRBs [5,24,25] which should be consulted. We hope that this monograph provides a broader and deeper understanding of both fields and helps to further the development and, perhaps, synthesis of these "greatest explosions since the Big Bang."

Acknowledgments

KWW thanks the *Office of Naval Research (ONR)* for the 6.1 funding supporting this research. Additional information and data on radio emission from SNe and GRBs can be found on *h*ttp://rsd-www.nrl.navy.mil/7214/weiler/ and linked pages.

References

1. J.S. Bloom et al.: Nature **401**, 453 (1999)
2. J.S. Bloom et al.: Astrophys. J. **572**, 45 (2002)
3. D. Branch: Ann. Rev. Astron. Astrophys. **36**, 17 (1998)
4. A.V. Filippenko: Ann. Rev. Astron. Astrophys. **35**, 309 (1997)
5. G.J. Fishman, C.A. Meegan: Ann. Rev. Astron. Astrophys. **33**, 415 (1995)
6. A. Fruchter, L. Bergeron: IAUC 6674 (1997)
7. A.S. Fruchter et al.: Astrophys. J. **545**, 664 (2000)
8. T.K. Galama et al.: Astrophys. J. **536**, 185 (2000)
9. P.J. Groot et al.: IAUC 6584 (1997)
10. J. Halpern et al.: IAUC 6788 (1997)
11. J. Heise et al.: IAUC 6787 (1997)
12. W. Hillebrandt, J.C. Niemeyer: Ann. Rev. Astron. Astrophys. **38**, 191 (2000)
13. S.R. Kulkarni, J.S. Bloom, D.A. Frail, R. Ekers, M. Wieringa, R. Wark, J.L. Higdon: IAUC 6903 (1998)
14. B. Leibundgut: Ann. Rev. Astron. Astrophys. **39**, 67 (2001)
15. R.N. Manchester, B.M. Gaensler, V.C. Wheaton, L. Staveley-Smith, A.K. Tzioumis, N.S. Bizunok, M.J. Kesteven, J.E. Reynolds: Pub. Astron. Soc. Australia **19**, 207 (2002)
16. M.R. Metzger, S.G. Djorgovski, S.R. Kulkarni, C.C. Steidel, K.L. Adelberger, D.A. Frail, E. Costa, F. Frontera: Nature **387**, 879 (1997)
17. S.C. Odewahn et al.: Astrophys. J. **509**, 5 (1998)
18. S. Park, D.N. Burrows, G.P. Garmire, J.A. Nousek, R. McCray, E. Michael, S. Zhekov: Astrophys. J. **567**, 314 (2002)
19. C.S.J. Pun et al.: Astrophys. J. **572**, 906 (2002)
20. E. Pian et al.: Astrophys. J. Lett. **492**, L103 (1998)
21. Y. Qin, G. Xie, S. Xue, E. Liang, X. Zheng, D. Mei: Pub. Astron. Soc. Japan **52**, 759 (2000)
22. A.G. Riess et al.: Astrophys. J. **560**, 49 (2001)
23. K.C. Sahu et al.: Astrophys. J. Lett. **489**, L127 (1997)
24. J. van Paradijs, C. Kouveliotou, R.A.M.J. Wijers: Ann. Rev. Astron. Astrophys. **38**, (2000) 379
25. K.W. Weiler, N. Panagia, M.J. Montes, R.A. Sramek: Ann. Rev. Astron. Astrophys. **40**, 387 (2002)

Part I

Supernovae

Historical Supernovae

David A. Green[1] and F. Richard Stephenson[2]

[1] Cavendish Laboratory, University of Cambridge, Madingley Road, Cambridge CB3 0HE, United Kingdom
[2] Department of Physics, University of Durham, Durham DH1 3LE, United Kingdom

Abstract. The available historical records of supernovae occurring in our own Galaxy over the past two thousand years are reviewed. These accounts include the well-recorded supernovae of AD1604 (Kepler's SN), AD1572 (Tycho's SN), AD1181 AD1054 (which produced the Crab Nebula) and AD1006, together with less certain events dating back to AD185. In the case of the supernovae of AD1604 and AD1572 it is European records that provide the most accurate information available, whereas for earlier supernovae records are principally from East Asian sources. Also discussed briefly are several spurious supernova candidates, and the future prospects for studies of historical supernovae.

1 Introduction

The investigation of observations of Galactic supernovae is very much an interdisciplinary exercise, far removed from the usual course of scientific endeavor. It seems well established that no supernova has been observed in our own Galaxy since AD1604. Hence for reports of Galactic supernovae it is necessary to rely entirely on observations made with the unaided eye. It is fortunate that early astronomers recorded several of these events – along with many other temporary "stars," such as comets and novae. As might be expected the quality of these early observations is very variable. In the case of the two most recent Galactic supernovae, in AD1604 and AD1572, European astronomers measured the positions with remarkable precision – to within about 1 arcmin – and the changing brightness was carefully recorded. However, in earlier centuries the available observations are of lesser quality. Nevertheless, sightings of three supernovae have been confidently identified from medieval East Asian records, while three more ancient supernovae have also been proposed.

Here we review the available historical records of supernovae occurring in our own Galaxy over the past two thousand years or so. Studies of potential supernovae from the various historical records will be concentrated on those new stars which were said to be visible for at least three months. This restriction eliminates most novae and also considerably diminishes the possibility of the object being a comet. More detailed discussions of these historical supernovae are presented in [5,21].

In Sect. 2 the available historical records – from East Asia, Europe and the Arab Dominions – of the well-recorded supernovae of AD1604, AD1572, AD1181, AD1054 and AD1006 are discussed. Other probable supernovae occurring before AD1000, from Chinese records, are discussed in Sect. 3, with less convincing possible and spurious historical supernova discussed in Sect. 4 (including the suggestion that the supernova that produced the young supernova remnant (SNR) Cas A was seen by Flamsteed in AD1680). Table 1 presents a summary of the well-recorded and probable historical supernovae seen in our Galaxy, including the sources of historical records. Finally, the future prospects for studies of historical supernovae are briefly discussed in Sect. 5.

Table 1. Summary of the Historical Supernovae

	Length of		Historical Records				
Date	Visibility	Remnant	Chinese	Japanese	Korean	Arabic	European
AD1604	12 mo.	SNR4.5 + 6.8	few	–	many	–	many
AD1572	18 mo.	SNR120.1 + 2.1	few	–	two	–	many
AD1181	6 mo.	3C58	few	few	–	–	–
AD1054	21 mo.	Crab Nebula	many	few	–	one	–
AD1006	3 yr.	SNR327.6 + 14.6	many	many	–	few	two
AD393	8 mo.	–	one	–	–	–	–
AD386?	3 mo.	–	one	–	–	–	–
AD369?	5 mo.	–	one	–	–	–	–
AD185	8 or 20 mo.	–	one	–	–	–	–

2 Well Defined Historical Supernovae

2.1 Kepler's SN of AD1604

The new star which appeared in the autumn of AD1604 was discovered in Europe on 09 October 1604, and first noticed only a day later in China, and by Korean astronomers on 13 October. The supernova, which remained visible for a whole year, was extensively observed by European astronomers, including Johannes Kepler, and this SN is often referred to as Kepler's SN. Chinese and Korean astronomers kept a regular watch on it, and valuable systematic Korean reports over many months are still preserved, as well as occasional Chinese records. There are no known Japanese or Arab accounts of this star. The European positional measurements are far superior to those from East Asia (approximately 1 arcmin precision as compared with 1 degree). Favorable circumstances assisted the discovery of the supernova, as it was only about 3 degrees to the north-west of the planets Mars and Jupiter, which were then in conjunction. This conjunction was carefully watched by European astronomers in early October of AD1604, and was also recorded in China. The peak brightness of the supernova probably did not occur until late October, so that the supernova was detected well before maximum.

Chinese observations of AD1604 are found in two approximately contemporary sources: three records in the annalistic *Ming Shenzong Shilu*, and a single record in the dynastic history of the *Mingshi*. The guest star was first detected

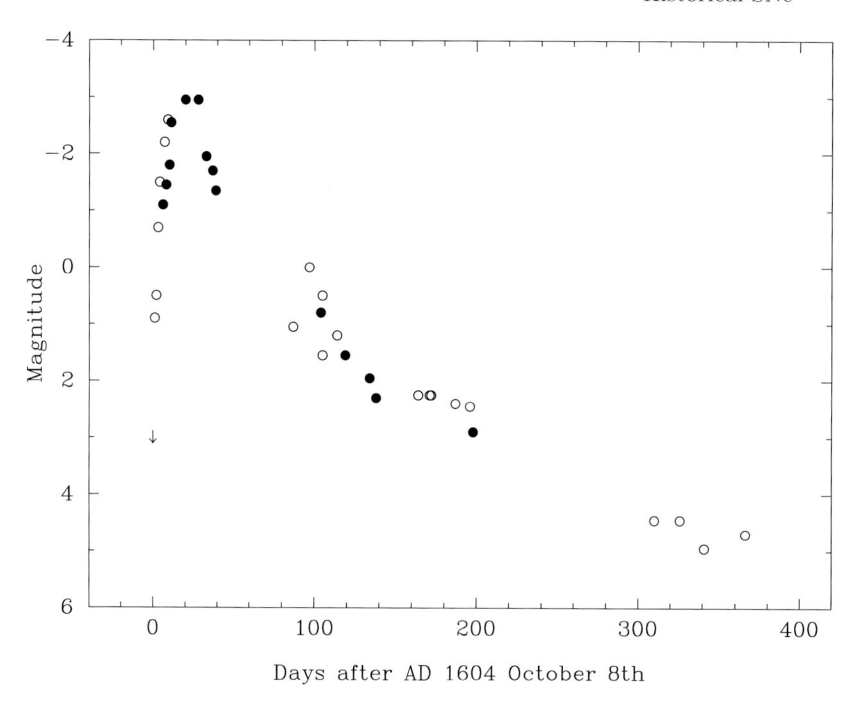

Fig. 1. Light curve of the SN of AD1604 from European (○) and Korean (•) observations, with a European upper limit on 08 October 1604.

in China on 09 October 1604 and was finally lost to sight on 07 October 1605. Although the guest star was not sighted in Korea until 13 October, it attracted considerable attention there. An almost day-to-day record of the Korean observations of the star over the first six months of visibility is available, and the regular estimates of brightness parallel the European observations – see Fig. 1. Nearly one hundred separate observations of the guest star are reported in the *Sonjo Sillok*. Several brief accounts of the new star are also to be found in the *Chungbo Munhon Pigo*, a compendium dating from AD1770.

The most important contemporary European work on the supernova is Johannes Kepler's *De Stella Nova in Pede Serpentarii* which was published in AD1606 (the more familiar name for Serpentarius is Ophiuchus). Other European accounts of the star are to be found in a wide variety of sources (see [2]). The most important aspects of the European observations are the accurate position of the star measured by Kepler and its changing brightness over the twelve months of observation.

We can be fairly confident that 09 October 1604 was the date of discovery because several European astronomers observed the conjunction of Mars and Jupiter on 08 October and did not notice anything remarkable. Due to poor weather, Kepler did not start observing the supernova until 17 October. He measured the angular distance between the new star and several planets and reference stars using a sextant. Several European astronomers estimated the brightness of the supernova in the days leading up to maximum. It can probably only be concluded that the brightness considerably exceeded that of Jupiter, with peak magnitude close to -3.0 mag. Although Kepler's SN has been identified as a type I in the past – on the basis of its light curve from the historical observations – this classification cannot be justified, as the light curves of some type I and type II SNe can be quite similar (see also [17]). With an estimated date of peak brightness of approximately 28 October 1604, the supernova was detected nearly twenty days before maximum. Using the calculated position for the new star from Kepler's observations, Baade [2] was able to locate the remnant as a patch of optical nebulosity with the 100-inch reflector at the *Mount Wilson Observatory (MWO)*. Subsequently the remnant, SNR4.5 + 6.8, has been revealed as a limb brightened "shell" supernova remnant at radio and X-ray wavelengths.

2.2 Tycho's SN of AD1572

The supernova which appeared in the constellation of Cassiopeia during the late autumn of AD1572 was compared by observers in both Europe and East Asia with Venus, and was visible in daylight. Since the most detailed observations of the supernova in Europe were made by Tycho Brahe, this supernova is often referred to as Tycho's SN.

Five Chinese records of this supernova are preserved, two in the same sources as the observations of the supernova of AD1604: the *Ming Shenzong Shilu* and in the astronomical section of the *Mingshi*. Further brief mentions occur in the biographical section of the *Mingshi* and also the *Ming Shigao*, the draft version of the *Mingshi*. A fifth account is in the *Yifeng Xianzhi*, a provincial history. Only two brief Korean reports of the AD1572 star are available. The guest star of AD1572 was discovered in Korea on 06 November 1572 and sighted two days later in China. Noticeably fading by March 1573, the star finally disappeared from sight some time between 21 April and 19 May 1574. The duration of visibility was thus about 18 months. Chinese records assert that the star was visible in daylight, while in Korea its brightness was compared with Venus. The position of the supernova is not defined very precisely by the East Asian records, but it is notable that its position is marked on at least two independent Chinese star charts.

The supernova was probably detected in Europe by Maurolycus, abbot of Messina, on 06 November 1572 (if not a day or two earlier). It was first observed by Tycho on 11 November (although he remained sceptical of the star's existence until he had questioned both the servants who were with him and passers by). Tycho immediately realized this was a new star, and "began to measure its situation and distance from the neighboring stars of Cassiopeia and to note extremely diligently those things which were visible to the eye concerning its apparent size, form, color and other aspects." Over the following year he was to make many measurements of the angular distances of the star from the nearby stars of Cassiopeia, and also determined its distance from Polaris. The accurate observations by Tycho Brahe and others, which established the fixed nature of the star, have proved of great importance to modern astronomers in a different guise: fixing the

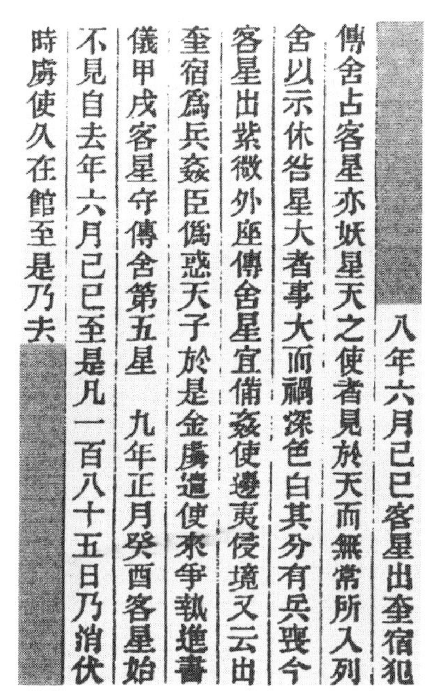
八年六月己巳客星出奎宿犯
傳舍占客星亦妖星天之使者見於天而無常所入列
舍以示休咎星大者事大而禍深色白其分有兵喪今
客星出紫微外座傳舍星宜備姦使邊夷侵境又云出
奎宿爲兵姦臣僞惑天子於是金虜遣使來爭執進書
儀甲戌客星守傳舍第五星　九年正月癸酉客星始
不見自去年六月己巳至是凡一百八十五日乃消伏
時虜使久在館至是乃去

Fig. 2. Detailed account of the supernova of AD1181 from the *Wenxian Tongkao*, which was compiled about a century after the supernova.

precise location of the supernova to within a few arcminutes. Tycho concluded that the new star was situated far beyond the Moon and among the fixed stars. Hence the supernova contravened the widely accepted Aristotelian doctrine that change could only occur in the sub-lunar region. Virtually all the important European observations of the supernova are contained in Tycho's *Astronomiae Instaurate Progymnasmata* ("Essays on the New of Astronomy"), published in AD1602. From Korean and European comparisons of the supernova with Venus, it seems that the peak magnitude of the supernova was around -4.0 mag. Tycho was evidently the only astronomer to carefully watch the decline in brightness of the new star. In the months after discovery he successively compared it with Jupiter, then with stars of fainter magnitudes.

The remnant of Tycho's SN was first tentatively identified in 1952 [8], when a radio source was found near the then available position for the SN. This was subsequently confirmed by later radio observations, which also led to the identification of the faint optical nebulosity associated with the radio source (Minkowski, private communication in [3]). At radio and X-ray wavelengths this remnant – often called either 3C10 or SNR120.1 + 2.1 – shows a limb-brightened shell ~ 8 arcmin in diameter.

2.3 The SN of AD1181

The new star of AD1181, which was extensively observed in both China and Japan, was seen for fully six months. Such a lengthy duration of visibility in the various historical records is indicative of a supernova. There are three Chinese records of the new star of AD1181, from both the North (Jin) and South (Song) Chinese empires in existence at that time, and five Japanese accounts. None of these sources report any motion of the star. The most detailed surviving Chinese account of the guest star is found in the *Wenxian Tongkao* ("Comprehensive Study of Civilization"), an extensive work compiled around AD1280 (see Fig. 2). This account clearly states that the star was first seen on 06 August 1181, and was visible for 185 days in total. There is also some valuable positional information in the record, giving the approximate location of the guest star, and describing it as guarding the fifth star of the *Chuanshe* asterism. The Japanese records of this guest star come from a variety of sources, including a retrospective account of the star (and also the new stars of AD1054 and AD1006) written in AD1230. The date of discovery of the guest star in Japan is one day after its discovery in South China. Other Japanese accounts of the star occur in various histories, and in diaries of imperial courtiers. Unlike the Chinese accounts, there are no estimates of the full duration of visibility of the guest star in the available Japanese accounts, although one states it was still visible two months after discovery.

The guest star was said to "invade" the *Chuanshe* asterism, and to guard the fifth star of that asterism. The identification of the fifth star with SAO12076 was proposed by Liu Jinyu [14], in which case the position of the SN can be deduced as within about 1 degree of SAO12076; equivalent galactic coordinates are $l \approx 130°$, $b \approx +3°$. This position is close to the radio source 3C58 (SNR130.7+3.1), which was first proposed as the remnant of this guest star in 1971 [18]. 3C58 is a "filled-center" or plerionic supernova remnant, in which a pulsar with a period of ≈ 65.58 ms has recently been identified [16].

2.4 The SN of AD1054 That Produced the Crab Nebula

The Crab Nebula, which has been known optically since the early 18th century, was first suggested as the remains of the Chinese guest star of AD1054 in the 1920s [10,15]. A substantial number of records of the guest star, from both China and Japan, were later assembled by the Dutch sinologue Duyvendak in 1942 [6], since when the Crab Nebula has generally been recognized as the remnant of the oriental guest star of AD1054. The Crab Nebula is one of few Galactic SNRs which are known to contain a pulsar – which is extremely important for explaining the energetics and structure of the whole SNR – and is the prototype of the class of "filled-center" SNRs or plerions. The nebula was first identified as a source of radio waves in 1963 and in X-rays in 1964, while the discovery of a pulsar within it in 1968 attracted huge interest internationally. No other supernova remnant has achieved such notoriety, or been the subject of so many research papers.

There are many Chinese records of the guest star of AD1054. It was first sighted in China at daybreak in the eastern sky on 04 July 1054 and remained visible until 06 April 1056. Three records of the guest star from Japan are known, apparently from two independent sources. As noted above, one Japanese record, from AD1230, also includes discussions of the guest stars of AD1006 and AD1181. Both Chinese and Japanese sources agree that the guest star appeared close to *Tianguan*, which is identified with ζ Tau. There is no hint of any motion; on the available evidence, the guest star remained fixed for the whole of the very long period of visibility. In noting that the star was visible in daylight (probably for 23 days), Chinese astronomers compared it with Venus. Japanese observers compared the brightness with Jupiter, again implying a brilliant object.

It is also likely that this supernova was recorded in Constantinople. Ibn Buṭlān, a Christian physician, provides a brief record of a new star seen at this time. Although there have been some suggestions that there are European records of the supernova of AD1054, there appears to be no definite report of it from Europe. It has also been suggested that this supernova, which was close to the ecliptic, is recorded in cave paintings in the American south-west which depict a crescent close to a circle or star symbol. However, only a very approximate date range can be deduced for the paintings (tenth to twelfth centuries AD), while only one of the pictures shows the correct orientation of the crescent relative to the new star. If the paintings are indeed astronomical, which is open to speculation, they might possibly represent one or more close approaches of the Moon to Venus over the estimated date range of some two centuries.

2.5 The Bright SN of AD1006

The new star which appeared in AD1006 was extensively observed in China and Japan, and was also recorded in Europe and the Arab dominions. The various records indicate that the star was of extreme brilliance and had an exceptionally long duration of visibility – several years.

The reports from China are by far the most detailed, giving not only a fairly accurate position, but also demonstrating that the new star was certainly seen for at least three years. Chinese observations are preserved in a wide variety of sources, including dynastic histories, chronicles and biographies. The new star was independently sighted in Japan, where it was consistently described as a *kexing* or "guest star" in several independent reports. Discovery in both China and Japan took place on 01 May 1006. Chinese records indicate that the star remained visible until some time during the lunar month between 27 August and 24 September 1006, when it set heliacally. However, Japanese reports may imply visibility to 21 September. After recovery in China on 26 November 1006, the star was seen until the following autumn (between 14 September and 13 October 1007), when it was lost to view in the evening twilight. The star was evidently sighted again at dawn some time toward the end of AD1007, or the beginning of AD1008, and – after a further conjunction with the Sun near the end of AD1008 – was apparently still visible well into the year AD1009. In China, the extreme brilliance of the star was emphasized in several ways: it was "huge... like a golden disk", "its appearance was like the half Moon and it had pointed rays;" "it was so brilliant that one could really see things clearly (by its light)." In Japan, the only direct brightness comparison was with the planet Mars, although the fact that the new star made such a profound impression on the imperial court suggests that it was a remarkable sight.

Brief Arab reports of the new star are preserved in chronicles from several regions: Egypt, Iraq, north-west Africa or Spain, and Yemen. The most likely date for the discovery of the new star in the Arab dominions is 30 April 1006, one day earlier than in China and Japan. Furthermore several Arab records suggest that the star disappeared around 01 September, a few weeks before it ceased to be reported in Japan. Two accounts from Europe – in the chronicles of the monasteries at St. Gallen in Switzerland and at Benevento in Italy – clearly refer to the new star, the former source indicating that it was visible for three months. Several other annals note the occurrence of a "comet" in or around AD1006. Since there is no notice of a comet in Chinese history around this time, it may be presumed that the European chroniclers also referred to the new star but had no separate term to describe a brilliant star-like object. The St. Gallen record mentions frequent interruptions in visibility, which imply that the star no more than skimmed the southern horizon; this provides a valuable declination limit for the position of the supernova.

The identification of the likely remnant of this SN was made in 1965 [7], when a search was made of radio catalogs covering part of the region for the SN from the historical observations. The likely associated radio source is PKS1459 – 51 – also known as MSH14 – *415*, or SNR327.4 + 14.6 from its Galactic coordinates. Subsequent improved observations confirmed this as the remnant, when its structure was revealed to be a limb-brightened "shell" supernova remnant about half a degree in diameter.

3 Probable Historical Supernovae Before AD1000

Other long duration guest stars recorded before AD1000 in China that are possibly records of supernovae before AD1000 are from AD393, AD386, AD369 and AD185, and are briefly discussed here.

The new stars of AD393, AD386 and AD369 appeared towards the end of the Jin dynasty in China. All three objects are recorded in the astronomical treatises of histories of both the Song and Jin dynasties (the *Songshu* and *Jinshu*). However, the two records of each of these stars clearly show a common origin. The guest star of AD393 was visible for some 8 months, and so is likely to have been a supernova. The position of this star is within the *Wei* asterism, which lies near the Galactic equator. There are several SNRs within this region, and it is not possible to identify the remnant of this SN unambiguously. The new star of AD386 was visible for anything from about 60 to 115 days. Since the duration of visibility may have been rather short, the possibility that this was a nova cannot be ruled out. Nevertheless, a supernova interpretation is also plausible. The position of the guest star is not well defined – it could have appeared near the star group *Nandou* (which lies close to the Galactic equator), or was merely somewhere in the range of Right Ascension (RA) defined by *Nandou*. In the latter case, although the RA would be fairly well defined the declination would

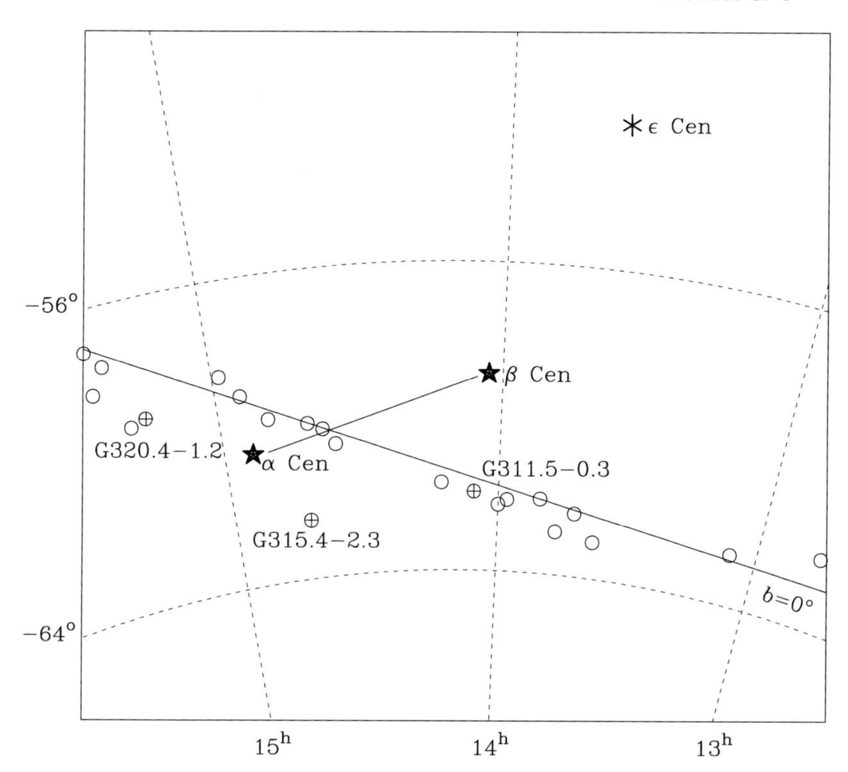

Fig. 3. The region of the *Nanmen* asterism, near which the SN of AD185 occurred. The positions of the bright stars α and β Cen, which are thought to comprise *Nanmen*, are shown, along with the fainter ϵ Cen. The circles indicate the centroids of cataloged Galactic SNRs, some with additional crosses and labels. The solid line shows the Galactic equator.

not. In the former case there are several possible identifications for the remnant of this guest star, with SNR11.2 − 0.3 perhaps being the prime candidate. The guest star of AD369 is described with only limited details. It was reported to be visible for 5 months, but only a poorly localized position is reported. If the star was near the Galactic equator a supernova interpretation would be plausible, but if it were far from the Galactic plane it was more likely a slow nova.

The earliest new star which merits investigation as a possible supernova was seen in China in AD185. This event is reported only in a single early source, the *Hou Hanshu*, which was composed towards the end of the third century AD. The new star was recorded as being visible for at least 8 months, or possibly even 20 months (depending on interpretation of part of the record to mean "next year" or "year after next"). The star was reported to be within the *Nanmen* asterism. Although some authors have questioned the identification of this asterism, comparison with contemporary records and star charts supports the usual identification of α and β Cen with *Nanmen*, which lies close to the Galactic equator. SNR315.4 − 2.3 is the prime candidate for the remnant of the SN of AD185

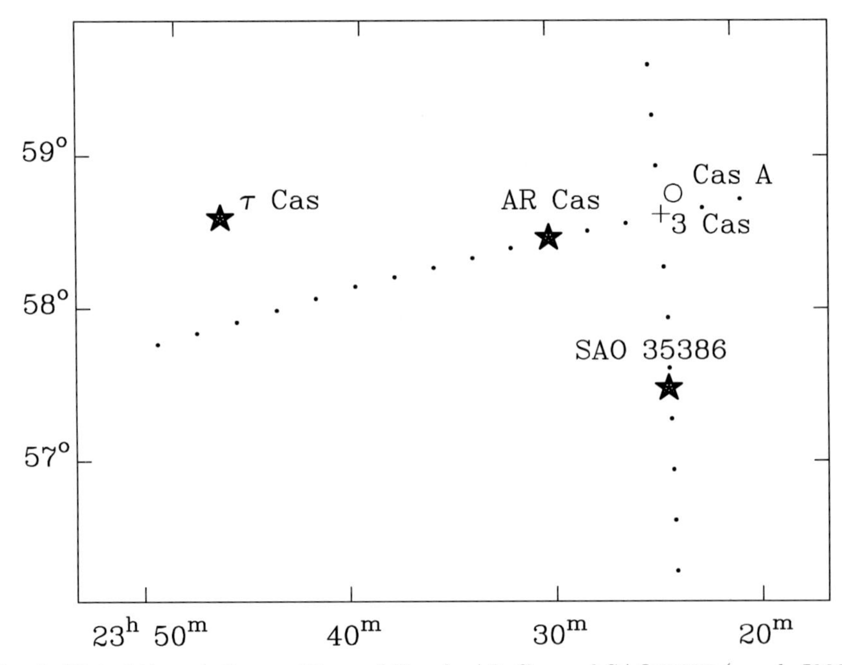

Fig. 4. Plot of the relative positions of Cas A, AR Cas and SAO35386 (epoch J2000). Also shown are τ Cas and the site of "3 Cas," and dotted lines to indicate the arcs measured by Flamsteed from β Peg and β Per.

among the SNRs in the present catalog of Galactic SNRs between α and β Cen – see Fig. 3 – although it should be noted that there are other remnants in this region which have not yet been studied in great detail.

4 Other Possible and Spurious Supernovae

4.1 Did Flamsteed See the Cas A Supernova in AD1680?

Cassiopeia A (Cas A) is an obviously young and relatively nearby SNR, which is a bright source at radio and X-ray wavelengths, showing a clumpy shell of emission. Optically Cas A shows a patchy ring of many expanding knots. The date at which the knots would converge, assuming no deceleration, is AD1671 (see [22] for a recent review). These observations are consistent with Cas A being produced by a SN in AD1671 or shortly afterwards, provided these optical knots have undergone only a very small deceleration subsequently, which is expected if these optical knots are very dense ejecta from the SN. The distance to Cas A can be found trigonometrically to be $3.4^{+0.3}_{-0.1}$ kpc, by combining the proper motion of the optical filaments in the plane of the sky with their radial velocities measured spectroscopically. Given the likely date and proximity of the supernova that produced the Cas A SNR, it has been a puzzle that no historical observations of it are available. In 1980 Ashworth [1] suggested that the supernova that produced

Cas A was indeed recorded by Flamsteed in AD1680, as he cataloged a 6th magnitude star "3 Cas," to the west of τ Cas, fairly close to the present site of Cas A, where there is no known star today. However, the discrepancy in the positions of 3 Cas and Cas A – about 10 arcmin – is very much larger than Flamsteed's typical measurement error. Alternatively – as proposed by Broughton [4] and by Kamper [12] – Flamsteed did not observe the supernova, but instead accidentally compounded his measurements of two separate stars (AR Cas and SAO35386), neither of which he actually cataloged. Since it is possible to identify the other faint stars that would have produced the erroneous 3 Cas position – see Fig. 4 – with measurement uncertainties that are plausible for Flamsteed's typical precision, it seems most unlikely that Flamsteed observed the supernova which produced Cas A.

4.2 The Korean Guest Stars of AD1592

During a period of about one month in AD1592, four separate guest stars were reported in Korea in the *Sonjo Sillok*, the official annals of the King Sonjo who reigned in Korea from AD1567 to AD1608. The first of these appeared in Cetus and was observed for 15 months. Two more guest stars (both in Cassiopeia) were seen for periods of at least three and four months, while a fourth (in Andromeda) was visible for more than a month. In each case the position remained unchanged; small refinements in the recorded locations of the three stars of longest duration only serve to emphasize their fixed nature. Surprisingly, none of these objects was reported in China or Europe, which suggests that all were by no means brilliant: probably of 2nd magnitude or fainter. The various records have been investigated in detail [20]. In summary, it would appear that as many as four novae may have occurred in AD1592, but in no instance is a supernova interpretation tenable.

4.3 The Spurious Supernovae of AD1408, AD1230 and AD837

In 1979 Li Qibin [13] assembled several Chinese records of two temporary stars observed in AD1408. Six of these accounts were from Szechuan province and described a bright star which appeared in the east, most probably on 10 September 1408. Three further reports were in official histories of China and related to a star which appeared on 24 October and "did not move." Li Qibin regarded the two objects as identical and proposed a supernova identification. Subsequently Imaeda and Kiang [11] found two further Japanese records mentioning a guest star on a date equivalent to 14 July 1408. Although no position was recorded for this object, it was inferred that the observation represented an earlier sighting of the stars seen in China on 10 September and 24 October 1408. They further concluded that the star "was quite likely to be a supernova explosion." The publications by Li Qibin and by Imaeda and Kiang led to consideration of SNR69.0+2.7 (CTB80) as the remnant of the star by various authors. However, it has been shown [19] that the "star" of 10 September was merely a meteor, and also there are insufficient grounds for linking the guest star of 14 July seen in Japan with the star appearing on 24 October as reported in China.

Wang Zhenru in 1987 suggested [23] that a “bushy star” seen for more than three months in AD1230 was a supernova, and further proposed a γ-ray source 2CG 054+01 as its remnant. (Wang Zhenru also proposed an association between a purported 14th century BC supernova record with another γ-ray source, but this is highly speculative.) The object was, in fact, a comet. Ho Peng Yoke [9] had already drawn attention to the motion of the same object as described in the astronomical treatise of the *Jinshu* – which contained records from North China.

Two guest stars appeared in AD837, which were discovered soon after Halley’s Comet had been detected in that year. Various authors have interpreted the records of one of these guest stars as evidence of a supernova, which they associate with SNR189.1 + 3.0 (IC443). Although the first star was fairly close to the galactic equator, the duration of visibility (22 days) was very short. Further, the star disappeared while still some 7 hours in RA to the east of the Sun, so that its visibility would not be impaired by the twilight glow. A supernova interpretation can thus be rejected; the star was most likely a fast nova. The second star, visible for 75 days at high galactic latitude, was evidently also a nova.

5 Future Prospects

Looking to the future, it seems unlikely that records of additional supernovae – other than those discussed above – will come to light. Most of the accessible historical sources, especially those of East Asia, have been fairly throughly examined. Many medieval Arab and European chronicles are still unpublished, but even to access a small proportion of this material, which is scattered in numerous archives, would be extremely time consuming. Furthermore, chroniclers were mainly interested in reporting only the most spectacular events. Hence, although it would seem likely that further accounts of the brilliant supernova of AD1006 might well emerge, the prospects for fainter objects – including the supernova of AD1054 – would appear to be far from promising. In particular, caution should be exercised in assessing the viability of further potential records of historical supernovae.

The remnants of the supernovae observed since AD1000 are well-established. However, improved distance measurements for the suggested remnants of the proposed supernovae of AD393, AD386, and AD185 would be valuable. These results would lead to better estimates of the physical size, and hence age, which might help distinguish between individual candidate remnants.

References

1. W.B. Ashworth: J. Hist. Astron. **11**, 1 (1980)
2. W. Baade: Astrophys. J. **97**, 119 (1943)
3. J.E. Baldwin, D.O. Edge: The Observatory **77**, 139 (1957)
4. R.P. Broughton: J. R. Astron. Soc. Canada **73**, 381 (1979)

5. D.H. Clark, F.R. Stephenson: *The Historical Supernovae* (Pergamon, Oxford 1977)
6. J.J.L. Duyvendak: Pub. Astron. Soc. Pacific **54**, 91 (1942)
7. F.F. Gardner, D.K. Milne: Astron. J. **70**, 754 (1965)
8. R. Hanbury Brown, C. Hazard: Nature **170**, 364 (1952)
9. Ho Peng Yoke: Vist. Astron. **5**, 127 (1962)
10. E. Hubble: ASP Leaflet 14 (1928)
11. K. Imaeda, T. Kiang: J. Hist. Astron. **11**, 77 (1980)
12. K.W. Kamper: The Observatory **100**, 3 (1980)
13. Li Qibin: Chinese Astron. **3**, 315 (1979)
14. Liu Jinyu: Stud. Hist. Nat. Sci. **2**, 45 (1983)
15. K. Lundmark: Pub. Astron. Soc. Pacific **33**, 234 (1921)
16. S.S. Murray, P.O. Slane, F.D. Seward, S.M. Ransom, B.M. Gaensler: Astrophys. J. **568**, 226 (2002)
17. B.E. Schaefer: Astrophys. J. **459**, 454 (1996)
18. F.R. Stephenson: Q. J. R. Astron. Soc. **12**, 10 (1971)
19. F.R. Stephenson, K.K.C. Yau: Q. J. R. Astron. Soc. **27**, 559 (1986)
20. F.R. Stephenson, K.K.C. Yau: Q. J. R. Astron. Soc. **28**, 431 (1987)
21. F.R. Stephenson, D.A. Green: *Historical Supernovae and their Remnants* (Oxford University Press, Oxford 2002)
22. J.R. Thorstensen, R.A. Fesen, S. van den Bergh: Astron. J. **122**, 297 (2001)
23. Wang Zhenru: Science **235**, 1485 (1987)

Classification of Supernovae

Massimo Turatto

Osservatorio Astronomico di Padova, vicolo dell'Osservatorio 5, I-35122, Padova, Italy

Abstract. The current classification scheme for supernovae is presented. The main observational features of the supernova types are described and the physical implications briefly addressed. Differences between the homogeneous thermonuclear type Ia and similarities among the heterogeneous core-collapse type Ib, Ic and II are highlighted. Transforming type IIb, narrow line type IIn, supernovae associated with GRBs and few peculiar objects are also discussed.

1 Introduction

An adequate and satisfactory taxonomy for a class of objects should fulfill certain elementary characteristics such as mutual exclusion, exhaustiveness, non-ambiguity, repeatability and usefulness. The taxonomy of supernovae (SNe) has been progressively developing since 1941 when Minkowskii [88] first recognized that at least two main types of SNe exist. Since then several new types of SNe have been introduced, some have been dismissed, and others have held their ground. After an early "Linnaeus stage" during which the classification was based on the recognition of the observational characteristics, the concept of SN type has been refined to include "genetic" information. Nevertheless the present day classification scheme is still not satisfactory. In particular, it is ambiguous and non-exhaustive (many objects are still classified as "peculiar") and the nomenclature of the taxonomic groups, determined by historical reasons, is confusing. However, since classification is a process which mankind naturally and instinctively carries out in order to sort out and understand vast arrays, the available SN classification scheme remains extremely important and useful.

Extensive reviews on the SN taxonomy can be found in [44,62,134]. Here we give a brief update on the subject. More details can be found in other chapters in this volume.

2 Main Supernova Types

The classification of SNe is generally performed on the optical spectra but, to some extent, also on their light curves. Since SNe are brighter near their maximum light, for obvious reasons the classification is based on the early spectra, which consist of a thermal continuum and P Cygni profiles of lines formed by resonant scattering. This means that the SN types are assigned on the basis of the chemical and physical properties of the outermost layers of the exploding stars.

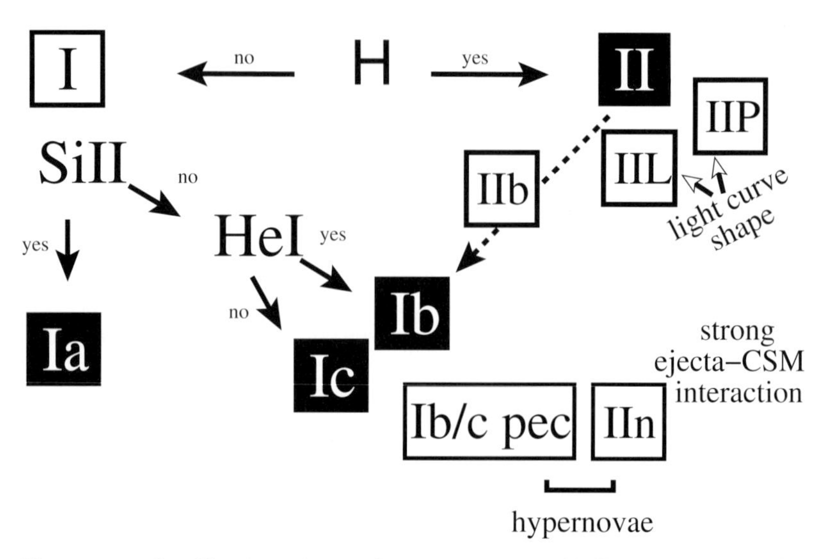

Fig. 1. The current classification scheme of supernovae. type Ia SNe are associated with the thermonuclear explosion of accreting white dwarfs. Other SN types are associated with the core-collapse of massive stars. Some type Ib/c and IIn SNe with explosion energies $E > 10^{52}$ erg are often called hypernovae.

Only in recent years have late time observations contributed to differentiating various subtypes.

The first two main classes of SNe were identified [88] on the basis of the presence or absence of hydrogen lines in their spectra: SNe of type I (SNI) did not show H lines, while those with the obvious presence of H lines were called type II (SNII). Type I SNe were also characterized by a deep absorption at 6150 Å which was not present in the spectra of some objects, therefore considered peculiar [16,17]. In 1965, Zwicky [143] introduced a schema of five classes but in recent years the scarcely populated types III, IV and V have been generally included among type II SNe.

In the mid-1980s, evidence began to accumulate that the peculiar SNI formed a class physically distinct from the others. The objects of the new class, characterized by the presence of HeI [58,63], were called type Ib (SNIb), and "classical" SNI were renamed as type Ia (SNIa). The new class further branched into another variety, SNIc, based on the absence of He I lines. Whether these are physically distinct types of objects has been long debated [62,135]. In several contexts they are referred to as SNIb/c. The current classification scheme is illustrated in Fig. 1.

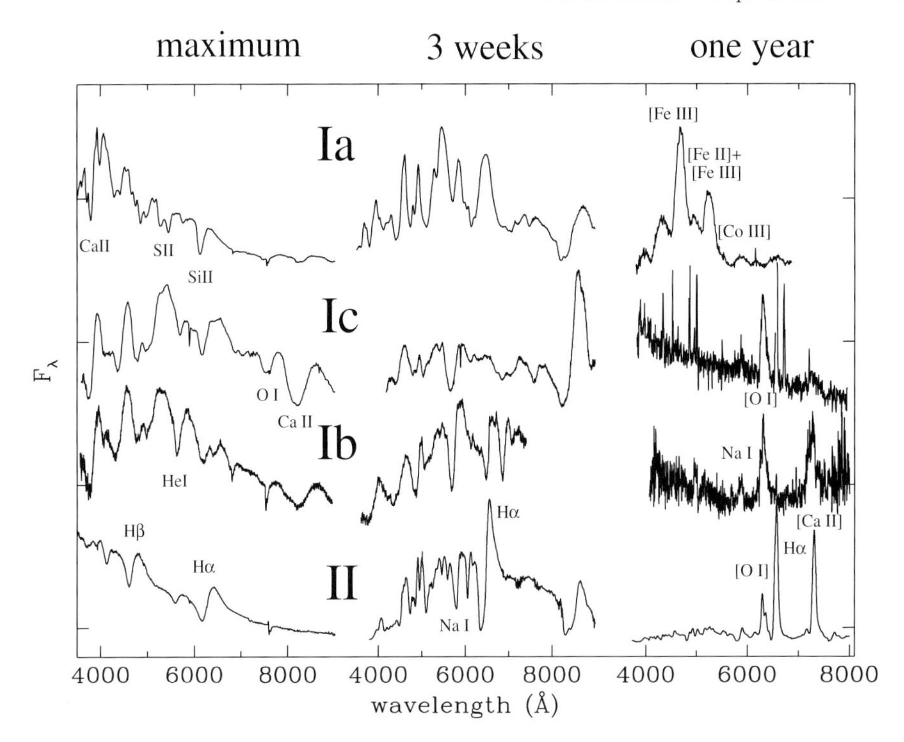

Fig. 2. The spectra of the main SN types at maximum, three weeks, and one year after maximum. The representative spectra are those of SN1996X for type Ia [109], of SN1994I (left and center) [35] and SN1997B (right) for type Ic, of SN1999dn (left and center) and SN1990I (right) for type Ib, and of SN1987A [99] for type II. At late times (especially in the case of the type Ic SN1997B the contamination from the host galaxy is evident as an underlying continuum plus unresolved emission lines. In all figures of this paper the spectra have been transformed to the parent galaxy rest frame.

3 Type Ia Supernovae

type Ia SNe have become very popular in the last decade because of their role in determining the geometry of the Universe with their high luminosity and relatively small luminosity dispersion at maximum. An extensive review of the properties of SNIa has recently appeared [69].

SNIa are discovered in all types of galaxies, also in ellipticals [8], and are not associated with the arms of spirals as strongly as other SN types [85,128]. The spectra are characterized by lines of intermediate mass elements such as calcium, oxygen, silicon and sulfur during the peak phase and by the absence of H at any time (see, e.g., Fig. 2). With age the contribution of the Fe lines increases and several months past maximum the spectra are dominated by [Fe II] and [Fe III] lines. The overall homogeneous spectroscopic and photometric behavior has led to a general consensus that they are associated with the thermonuclear explosion of a white dwarf [20].

Nevertheless, during the past decade early suggestions of significant differences among SNIa [9,102] have been confirmed by new, high signal-to-noise data. The crucial year was 1991, when the bright, slowly declining SN1991T [74,105], and the faint, intrinsically red and fast declining SN1991bg [50,71,119] were discovered. Other under- and over-luminous objects have been found since then [66].

The analysis of homogeneous sets of optical data led to the discovery of a correlation between the peak luminosity and the shape of the early light curve with brighter objects having a slower rate of decline than dimmer ones [97,98,100,104]. This correlation has been employed in restoring SNIa as useful distance indicators up to cosmological distances. A correlation between the photometric and the spectroscopic properties was also found [90].

It is known that the peak luminosity of SNIa is directly linked to the amount of radioactive ^{56}Ni produced in the explosion [3,5]. Hence SNIa, having different magnitudes at maximum, are probably the result of the synthesis of different amounts of radioactive ^{56}Ni. Moreover, there are indications of large variances (up to a factor 2) in the total mass of the ejecta [24].

Observations at other wavelengths have provided very useful information. In particular, infrared and I-band light curves have shown that the light curves of SNIa are characterized by a secondary peak 20-30 days after the B maximum [39,74,86,115]. Remarkable exceptions are the faint objects like SN1991bg.

The afore mentioned diversity of SNIa persists up to very late epochs. The light curves of faint objects are steeper than those of other SNIa, probably because of a progressive transparency to positrons from radioactive decay [24,87]. Faint SNIa also show slower expansion velocity of the emitting gas both at early [43,125] and late epochs [82].

The calibration of the absolute magnitudes of a number of type Ia SNe by Cepheids has provided the zero point for determining the Hubble constant (H_0). Average values based on limited samples and different recipes range from $M_V = -19.34$ to $M_V = -19.64$ with small dispersions [60,68,107,116]. The recent determination of the Cepheid distance to the host galaxy of SN1991T seems to confirm that it is brighter ($M_V = -19.85 \pm 0.29$) than the spectroscopically normal supernovae [108].

SN1991T was also the first SNIa to show photometric and spectroscopic evidence of a light echo from circumstellar dust [111]. So far the only other SNIa to show the same phenomenon was the slowly declining, spectroscopically normal SN1998bu [27]. In addition to the light echo, these two SNe suffered strong reddening, quite unusual among known, standard SNIa, suggesting that slow decliners may be associated to younger population objects. The fact that, on average, SNIa in late type galaxies have slower decline rates (hence are more luminous) than SNIa in early type galaxies had been already suggested [25,43,61,126].

One current issue important for the application of SNIa to cosmology is whether they evolve with redshift. Indeed, the lower metallicity of the progenitors at higher redshift might result in systematic differences in brightness.

4 Type Ib and Ic Supernovae

Type Ib and Ic appear only in spiral type galaxies [8,101] and have been associated with a parent population of massive stars, perhaps more massive than SNII progenitors [128]. SNIb/c exhibit relatively strong radio emission with steep spectral indices and fast turn-on/turn-off [132], which is thought to arise from the SN shock interaction with a dense circumstellar medium [28,29]. They are, therefore, usually thought to be associated with the core-collapse of massive stars which have been stripped of their outer H (and possibly He) envelope.

The introduction of this class of SNe is recent. As mentioned above, they were classified with other SNI until the mid-1980s, when late time observations of SN SN1983N, SN1984L and SN1985F highlighted the physical differences from SNIa [31,47,58,63,138]. The characterizing features are the absence of H and Si II lines and the presence of He I. The excited levels of He producing such lines are thought to be populated by fast electrons accelerated by γ-rays from the decay of ^{56}Ni and ^{56}Co [63,75]. It was soon recognized that some objects did not show strong He lines [136] and the class of helium poor type Ic was proposed [62,135].

In order to investigate the physical differences between these two classes, the signatures of He were searched carefully. The He I λ10830 line in type Ic was first found in the spectra of SN1994I [49] with velocity as high as 17,000 km s^{-1} [35]. It was noted that even as little as 0.1 M$_{\odot}$ of helium at such high expansion velocities implies an energy of about 3×10^{51} erg in the outer shell alone. Unless very high explosion energies are involved, the amount of He has to, therefore, be much smaller [35]. Other SNIc have shown He lines at similar (SN1987M and SN1988L) and slower (SN1990B [38]) velocities. Thus the presence of some amount of He seems to be common to several SNIc progenitors.

Helium has been unambiguously identified also in the spectra of the recent type Ic SN1999cq [76]. These lines have expansion velocities much lower than other lines, indicating that the ejecta interacts with a dense shell of almost pure He originating from a stellar wind or mass transfer to a companion. These findings support the idea that SNIc differ from SNIb by the He abundance rather than by the amount of mixing of ^{56}Ni in the helium envelope.

Absorption features attributed to Hα were first identified in the spectra of the type Ib SN1983N and SN1984L [137]. Recent analysis of the spectra of 11 objects has suggested that detached H is generally present in SNIb [22]. The optical depths of H and He are not very high, so that modest differences in the He I line optical depths might transform type Ib into type Ic objects.

In addition to the different strengths of He lines, it has been suggested [76] that permitted oxygen lines are relatively stronger in type Ic than in type Ib and the nebular emission lines broader. In general, type Ib appear more homogeneous than type Ic.

The light curves of SNIb/c have been divided in two groups depending on the luminosity decline rate [34]. However, the suggestion that type Ic include both fast and slow decliners while type Ib seem to prefer slow declines has been

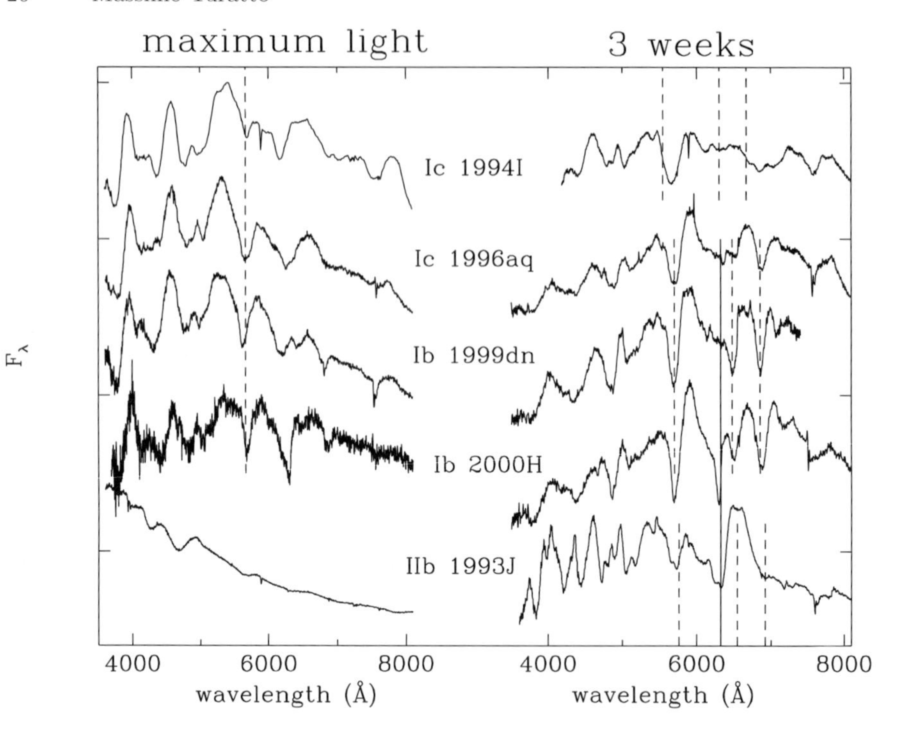

Fig. 3. Comparison among the spectra of SNe reported as type Ib/c in the Asiago SN Catalogue [8] (left: at maximum light; right: 3 weeks later). In addition to the spectra of SN1994I (SNIc) and SN1999dn (SNIb) shown in Fig. 2, the spectra of SN1996aq (SNIc), SN2000H (SNIb) [22] and SN1993J (SNIIb) [7] are displayed. With the exception of SN1993J, the spectra at maximum are rather similar. For reference the position of He I λ5876 (blueshifted by 10,500 km s^{-1}) is marked with a dashed vertical line. The dashed lines on the right indicate the positions of He I $\lambda\lambda$ 5876, 6678 and 7065 (blueshifted by 16,900 km s^{-1} for SN1994I, 6,000 km s^{-1} for SN1993J and 9,000 km s^{-1} for other objects). Hα (solid line, blueshifted by 11,000 km s^{-1}) is also shown. While He I lines in SN1994I are detectable only through a detailed analysis, they are prominent in the spectra of SN1996aq, which should be reclassified as a SNIb.

challenged by the existence of SNIb with fast light curves, e.g., SN1990I and SN1991D.

With this variety, the classification of new objects based on a single spectrum is difficult, and to some extent subjective, so that more detailed analyses of the objects often reveal discrepancies from early classifications. This is illustrated in Fig. 3 which presents the optical spectra of a number of objects reported as type Ib/c in the Asiago Supernova Catalog [8].

4.1 Type IIb

A few objects have been found to have early time spectra similar to type II (i.e. with prominent H lines) and late time spectra similar to type Ib/c SNe. For this reason they have been called type IIb. The first SNIIb was SN1987M [42], but the best studied example, and one of the best studied SNe ever, was SN1993J in M81. The analysis of seven years of observations of this object has been recently published [77,79], and will be discussed in more detail in other chapters of this volume. More examples of transition objects are SN1996cb [141] and probably SN1997dd [78].

Fig. 3 shows that while the early spectrum of SN1993J was almost featureless with a blue continuum and broad H and He I λ5876 lines typical of SNII, already three weeks later it displayed progressively stronger He I $\lambda\lambda$ 5876, 6678 and 7065 lines characteristic of SNIb.

The light curve of SN1993J was unusual with a narrow peak followed by a secondary maximum, recalling the behavior of SN1987A if the time axis were reduced by a factor of four. After another rapid luminosity decline around 50 days past the explosion, the light curve settled into an almost exponential tail with a decline rate faster than normal SNII and similar to that of SNIa, indicative of a small mass for the ejecta.

Indeed, the photometry of the progenitor of SN1993J taken before the explosion is inconsistent with the spectral energy distribution of a single star, but requires the composition of a K0Ia spectrum with a hot component [1]. The radio and X-ray emission of SN1993J [11,129,142] have been attributed to circumstellar interaction [54]. Another indication comes from the boxy shape of the emission lines of late time optical spectra [77,94]. Circumstellar gas in proximity to the exploding star was also revealed through the detection of narrow coronal lines persisting for a few days after the explosion [13].

These SNe transforming from type II to type Ib/c constitute the previously missing link between envelope retaining and envelope stripped SNe.

5 Type II Supernovae

Type II SNe are characterized by the obvious presence of H in their spectra. They avoid early type galaxies [8], are strongly associated with regions of recent star formation [81,128] and are commonly associated with the core-collapse of massive stars [140].

SNII display a wide variety of properties both in their light curves [92,93] and in their spectra [44]. Four subclasses of SNII are commonly mentioned in the literature: type IIP, type IIL, and type IIn in addition to the above mentioned type IIb. However, a number of peculiar objects do not fit into any of these categories.

SNIIP (Plateau) and SNIIL (Linear) constitute the bulk of all SNII, and are often referred as normal SNII. The subclassification is made according to the shape of the optical light curves [10]. The luminosity of SNIIP stops declining shortly after maximum forming a plateau 2-3 months long during which a recombination wave moves through the massive hydrogen envelope releasing its internal energy. SNIIL, on the other hand, show a linear, uninterrupted luminosity decline, probably because of a lower mass envelope. Indeed the two classes are not separated and there are a number of intermediate cases with short plateaus, e.g., SN1992H [36]. A quantitative criterion for the classification of the light curves has been proposed on the basis of the average decline rate of the first 100 days [93]. Starting 150 days past maximum the luminosity of both types settles into an exponential decline, consistent with complete (or constant) trapping of the energy release of the radioactive decay of ^{56}Co into ^{56}Fe.

No major spectral differences exist between SNIIP and SNIIL, although there are recurrent claims that SNIIL do not show the blueshifted absorption of the P Cygni profile evident in normal SNIIP. Indeed, a statistical analysis has shown that the presence or absence of the P Cygni absorption is correlated with the absolute brightness at maximum [93].

The progenitors of SNIIL are believed to have H envelopes of the order of 1-2 $M_\odot$, much smaller than those of SNIIP (typically 10 $M_\odot$) probably due to mass-loss during the progenitor evolution. A general scenario has been proposed in which common envelope evolution in massive binary systems with varying mass ratios and separations of the components can lead to various degrees of stripping of the envelope [89]. According to this scenario the sequence of types IIP-IIL-IIb-Ib-Ic in Fig. 1 is ordered according to a decreasing mass of the envelope.

SNIIL are often radio sources [132] and show UV excess attributed to Compton scattering of photospheric radiation by high speed electrons in shock-heated circumstellar medium (CSM) material [52,53]. A number of SNIIL, e.g., SN1994aj and SN1996L [12,14], have shown lines with double P Cygni profiles (sometimes dubbed SNIId, "d" for double) indicating the presence of strong wind episodes shortly before the explosion. These and other SNIIL show a flattening in the light curves at late stages, a broadening of the spectral lines, and Hα fluxes greater than those expected from purely radioactive models [23,41,70,124]. These features have been interpreted as signatures of the onset of interaction between the ejecta and the circumstellar material.

The epochal SN1987A in the *Large Magellanic Cloud (LMC)*, the first SN to be observed by naked eye in the last four centuries, was a "not very peculiar" type II SN with a plateau. Extensive observations at all wavelengths have explored the nature of this object in detail (see, e.g., [4] and references therein). The detection of neutrinos from SN1987A has been a spectacular confirmation of the theory of core-collapse [18,64]. The contribution of this object to our understanding of supernovae is addressed by McCray in another chapter in this volume.

Because of its closeness, SN1987A was the first SN for which it has been possible to unambiguously identify the progenitor, the B3 I star Sk-69 202. More recent, high spatial resolution prediscovery images of nearby galaxies are providing new insights into the nature of the precursor stars of other core-collapse SNe. In addition to the case of SN1993J in M81 mentioned above, tight constraints on the mass of the progenitors of SN1999em and SN1999gi have been published ($M(1999\mathrm{gi}) < 9^{+3}_{-2}\ \mathrm{M}_\odot$ and $M(1999\mathrm{em}) < 12 \pm 1\ \mathrm{M}_\odot$ [113,114]).

SN1987A was typical in both the explosion energy ($\sim 10^{51}$ erg) and the amount of ejected radioactive material ($\mathrm{M}(^{56}\mathrm{Ni}) = 0.07\ \mathrm{M}_\odot$). Recent observations of other SNe seem to indicate a considerable dispersion around such values.

High explosion energies are required to explain the high luminosities and kinetic energies of hypernovae (see, e.g., Sect. 6) and a large mass of radioactive ^{56}Ni (0.3 $M_\odot$) has been measured for SN1992am [112]. At the other extreme the faint SN1997D had a considerably smaller explosion energy ($\sim$ few $\times 10^{50}$ erg) and returned to the interstellar medium (ISM) a mass of radioactive material as small as $M(^{56}\mathrm{Ni}) = 0.002$ $M_\odot$ [122]. The discovery of similar objects (e.g., SN1999eu indicates that faint, under energetic SNII may represent a non-negligible fraction of all core-collapse SNe. These objects are the best candidates for the detection of signatures of black hole formation and for providing support to the black hole-forming SN scenario [6,15].

5.1 Type IIn

A number of peculiar SNII have been grouped into the class of SNIIn ("n" denoting narrow emission lines [110]). The spectra of these objects have a slow evolution and are dominated by strong Balmer emission lines without the characteristic broad absorptions. The early time continua are very blue, He I emission is often present and, in some cases, narrow Balmer and Na I absorptions are visible corresponding to expansion velocities of about 1,000 km s^{-1}[91], reminiscent of SNIId (see Sect. 5). Unresolved forbidden lines of [OI], [OIII], and of highly ionized elements such as [FeVII], [FeX], and [AX] are sometimes present.

One of the best cases is SN1988Z, which has been observed in the optical, radio and X-rays for over ten years [2]. Balmer lines, in particular Hα, show a well defined multiple component structure with broad (FWHM $\sim 15,000$ km s^{-1}), intermediate (2,200 km s^{-1}) and narrow (< 700 km s^{-1}) components with different time evolution [121]. SN1995N was a remarkably similar object, detected in radio and X-rays [51,132].

It is commonly believed that what we observe in these objects is the result of interaction between the ejecta and a dense CSM, which transforms the mechanical energy of the ejecta into radiation [32,117]. The interaction of the fast ejecta with the slowly expanding CSM generates a forward shock in the CSM and a reverse shock in the ejecta. The shocked material emits energetic radiation whose characteristics strongly depend on the density of both the CSM and the ejecta, and on the properties of the shock [30]. Thus the great diversity of observed SNIIn can provide clues to the different history of the mass-loss in the late evolution of progenitors.

A systematic search for radio emission from SNIIn has produced contradictory results [130]. This might be due to the high degree of heterogeneity and to a bias in the classification of the targets. In fact, a solid classification criterion for these SNe is still missing, and often objects classified as SNIIn turn out to be normal when they undergo closer scrutiny (e.g., SN1989C [120]).

A remarkable SNIIn is the recent SN1998S. Contrary to most other SNIIn, the spectrum evolved rapidly and seems to be the result of the interaction between the supernova and a two-component progenitor wind [40]. Polarimetric studies indicate significant asphericity for its continuum scattering environment [73].

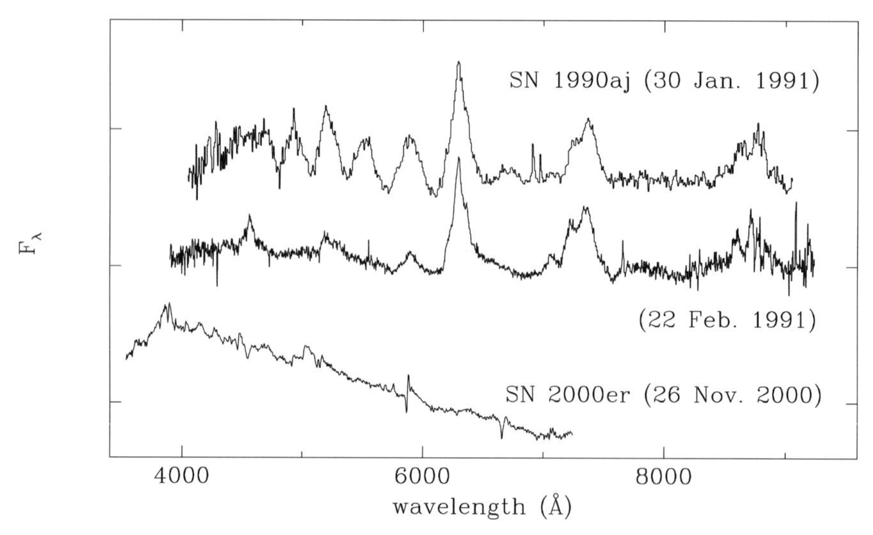

Fig. 4. Spectra of SN1990aj, a SNIb/c with exceptionally strong Fe lines at late epochs, and SN2000er, a bright, fast declining SN with evidence of narrow He I lines in the spectra at maximum.

The early spectra are well modeled by lines arising primarily in the circumstellar region while later spectra are dominated by the supernova ejecta [72].

6 SNe and GRBs

Other chapters in this volume discuss the association of SN1998bw with GRB980425 and the possibility that at least some GRBs originate from SN explosions.

In addition to its stringent spatial and temporal association to GRB980425 [55,67], SN1998bw was peculiar in many respects. The spectrum at early phases was unprecedented and led to different classifications (type Ib [106] or peculiar type Ic [46,95]). SN1998bw was as bright as an SNIa and displayed expansion velocities as high as 3×10^4 km s^{-1}, suggesting that it was the result of an extremely energetic explosion ($> 10^{52}$ erg), even if different degrees of asymmetry and beaming allow for a broad range of values [65,67,139]. Its very powerful radio emission has been interpreted as due to the presence of a mildly relativistic blast wave interacting with a clumpy, structured CSM deriving from a complex mass-loss history (see the chapter by Weiler et al. in this volume and [133]). The complete dataset of optical and IR observations spanning over one year after the explosion has recently appeared [96]. The analysis of the late time spectra [84] seems to confirm that the explosion was asymmetric as suggested by polarimetry [65]. The presence of underlying SNe similar to SN1998bw has been invoked to explain the anomalous rebrightening in light curves of other GRB afterglows at high redshift [19,103].

Other known SNe, SN1997ef [57] and SN1998ey [56], bear some spectroscopic resemblance to SN1998bw. In particular, SN1997ef was possibly associated with GRB971115 [131] and, although it was fainter than SN1998bw, its kinetic energy was close to that of SN1998bw [83]. The case of the bright SNIc SN1992ar [37], which occurred at about two σ from the position of GRB920616 [139], is also very interesting.

Additional SNe possibly associated with GRBs are SN1997cy, (GRB970514 [59,139]), observationally classified as SNIIn and possibly the brightest SN ever observed [123], and its twin SN1999E [26] (GRB980910 [118]). As in the case of the type IIn SN1988Z, they show evidence for strong ejecta-CSM interaction. Explosion energies as high as 3×10^{52} erg are required to reproduce the light curves [123].

To denote these particularly energetic objects, the term "hypernova" [67] has been used, although its meaning is not well defined or universally accepted.

7 Peculiar Supernovae

Several objects do not fit the schema described above. In many cases they are probably core-collapse SNe exploding in unusual configurations and/or conditions, even if the presence of other explosion mechanisms cannot be ruled out.

For instance, SN1993R [45] (and possibly SN1990aj, see, e.g., Fig. 4), showed hybrid spectral features of type Ia and type Ib/c at late epochs. This has been interpreted as the explosion of a SNIb/c with overproduction of ^{56}Ni or with the slow deflagration of a white dwarf.

The recent SN2000er was also classified as peculiar [33,80]. The scanty observations available show a broad maximum at $M_V = -19$ followed by a rapid decline of 5 mag in 50 days. The spectrum shows (see, e.g., Fig. 4) a strong continuum with a drop at wavelengths shorter than 4000 Å, possibly due to the presence of metals. Narrow lines of He I with P Cygni profiles are visible, indicating the presence of gas with expansion velocity ~ 900 km s^{-1} as well as several other lines possibly due to FeII, FeIII, and SiIII [21]. It might be a core-collapse SN which lost its He envelope shortly before the explosion.

A well studied peculiar object is SN1961V, Zwicky's Type V, which was probably not a genuine SN because the exploding star survived the giant eruption [48]. Similar objects might be SN1954J, SN1999bw, and the faint SN1997bs for which an extremely luminous supergiant precursor has been identified [128].

8 Conclusions

The taxonomy of supernovae is a subject which is still evolving. The experience of the last years has shown that the diversity of SNe increases with our ability to detect, study, and understand them. Even though distant SNe are disclosing new frontiers in observational cosmology, much is still to be learned from nearby objects. Indeed, when we have the chance to apply new observational techniques or to obtain high signal-to-noise observations of nearby objects, new features and phenomena are revealed. Major progress in understanding supernovae can be expected if coordinated efforts involving several teams with wide ranging expertise and access to different observational facilities are established. But SN occurrences are unpredictable: in order to really understand them, we would need their cooperation.

Acknowledgments

This work is partially based on data collected at the *European Southern Observatory (ESO)* observatory on La Silla in Chile.

References

1. G. Aldering, R.M. Humphreys, M. Richmond: Astron. J. **107**, 662 (1994)
2. I. Aretxaga, S. Benetti, R.J. Terlevich, A.C. Fabian, E. Cappellaro, M. Turatto, M. Della Valle: Mon. Not. R. Astron. Soc. **309**, 343 (1999)
3. W.D. Arnett: Astrophys. J. **253**, 785 (1982)
4. W.D. Arnett, J.N. Bahcall, R.P. Kirshner, S.E. Woosley: Ann. Rev. Astron. Astrophys. **27**, 629 (1989)
5. W.D. Arnett, D. Branch, J.C. Wheeler: Nature **314**, 337 (1985)
6. S. Balberg, L. Zampieri, S.L. Shapiro: Astrophys. J. **541**, 860 (2000)
7. R. Barbon, S. Benetti, E. Cappellaro, F. Patat, M. Turatto, T. Iijima: Astron. Astrophys. Suppl. Ser. **110**, 513 (1995)
8. R. Barbon, V. Buondì, E. Cappellaro, M. Turatto: Astron. Astrophys. Suppl. Ser. **139**, 531 (1999)
9. R. Barbon, F. Ciatti, L. Rosino: Astron. Astrophys. **25**, 241 (1973)
10. R. Barbon, F. Ciatti, L. Rosino: Astron. Astrophys. **72**, 287 (1979)
11. N. Bartel et al.: Science **287**, 112 (2000)
12. S. Benetti, E. Cappellaro, I.J. Danziger, M. Turatto, F. Patat, M. Della Valle: Mon. Not. R. Astron. Soc. **294**, 448 (1998)
13. S. Benetti, F. Patat, M. Turatto, G. Contarini, R. Gratton, E. Cappellaro: Astron. Astrophys. **285**, L13 (1994)
14. S. Benetti, M. Turatto, , P.A. Mazzali: Mon. Not. R. Astron. Soc. **305**, 811 (1999)
15. S. Benetti et al.: Mon. Not. R. Astron. Soc. **322**, 361 (2001)
16. F. Bertola: Annals Ap. **27**, 319 (1964)
17. F. Bertola, A. Mammano, M. Perinotto: Contrib. Oss. Astrofis. Padova in Asiago **174**, 51 (1965)
18. R.M. Bionta et al.: Phys. Rev. Lett. **58**, 1494 (1987)
19. J.S. Bloom et al.: Nature **401**, 453 (1999)
20. D. Branch, M. Livio, L.R. Yungelson, F.R. Boffi, E. Baron: Pub. Astron. Soc. Pacific **107**, 1019 (1995)
21. D. Branch, R. Thomas: private communication (2001)
22. D. Branch et al.: Astrophys. J. **566**, 1005 (2002)
23. E. Cappellaro, I.J. Danziger, M. Turatto: Mon. Not. R. Astron. Soc. **277**, 106 (1995)
24. E. Cappellaro, P. A. Mazzali, S. Benetti, I. J. Danziger, M. Turatto, M. Della Valle, F. Patat: Astron. Astrophys. **328**, 203 (1997)

25. E. Cappellaro, M. Turatto: In: *The Influence of Binaries on Stellar Population Studies*, ed. by D. Vanbeveren (Kluwer Academic Publishers, Dordrect 2001) p. 199
26. E. Cappellaro, M. Turatto, P. A. Mazzali: IAUC 7091 (1999)
27. E. Cappellaro et al.: Astrophys. J. Lett. **549**, L215 (2001)
28. R.A. Chevalier: Astrophys. J. **259**, 302 (1982)
29. R.A. Chevalier: Astrophys. J. Lett. **259**, L85 (1982)
30. R.A. Chevalier, C. Fransson: Astrophys. J. **420**, 268 (1994)
31. N.N. Chugai: Sov. Astron. **12**, L192 (1986)
32. N.N. Chugai, I.J. Danziger: Mon. Not. R. Astron. Soc. **268**, 173 (1994)
33. A. Clocchiatti, M. Turatto: IAUC 7528 (2000)
34. A. Clocchiatti, J.C. Wheeler: Astrophys. J. **491**, 375 (1997)
35. A. Clocchiatti, J.C. Wheeler, M.S. Brotherton, A.L. Cochran, D. Wills, E.S. Barker, M. Turatto: Astrophys. J. **462**, 462 (1996)
36. A. Clocchiatti et al.: Astron. J. **111**, 1286 (1996)
37. A. Clocchiatti et al.: Astrophys. J. **536**, 62 (2000)
38. A. Clocchiatti et al.: Astrophys. J. **553**, 886 (2001)
39. J.H. Elias, J.A. Frogel, J.A. Hackwell, S.E. Persson: Astrophys. J. Lett. **251**, L13 (1981)
40. A. Fassia et al.: Mon. Not. R. Astron. Soc. **325**, 907 (2001)
41. R.A. Fesen: Astrophys. J. Lett. **413**, L109 (1993)
42. A.V. Filippenko: Astron. J. **96**, 194 (1988)
43. A.V. Filippenko: Pub. Astron. Soc. Pacific **101**, 588 (1989)
44. A.V. Filippenko: Ann. Rev. Astron. Astrophys. **35**, 309 (1997)
45. A.V. Filippenko: In: *Thermonuclear Supernovae*, ed. by P. Ruiz-Lapuente, R. Canal, J. Isern (Kluwer Academic Publishers, Dordrect 1997) p. 795
46. A.V. Filippenko: IAUC 6969 (1998)
47. A.V. Filippenko, W.L.W. Sargent: Nature **316**, 407 (1985)
48. A.V. Filippenko, A.J. Barth, G.C. Bower, L.C. Ho, G.S. Stringfellow, R.W. Goodrich, A. Porter: Astron. J. **110**, 2261 (1995)
49. A.V. Filippenko et al.: Astrophys. J. Lett. **450**, L11 (1995)
50. A. V. Filippenko et al.: Astron. J. **104**, 1543 (1992)
51. D.W. Fox: Mon. Not. R. Astron. Soc. **319**, 1154 (2000)
52. C. Fransson: Astron. Astrophys. **111**, 140 (1982)
53. C. Fransson: Astron. Astrophys. **133**, 264 (1984)
54. C. Fransson, P. Lundqvist, R.A. Chevalier: Astrophys. J. **461**, 993 (1996)
55. T. J. Galama et al.: Nature **395**, 670 (1998)
56. P. Garnavich, S. Jha, R.P. Kirshner, P. Berlind: IAUC 7066 (1998)
57. P. Garnavich, S. Jha, R.P. Kirshner, P. Challis, D. Balam, W. Brown, C. Briceno: IAUC 6786 (1997)
58. C.M. Gaskell, E. Cappellaro, H.L. Dinerstein, D. Garnett, R.P. Harkness, J.C. Wheeler: Astrophys. J. Lett. **306**, L77 (1986)
59. L.M. Germany, D.J. Reiss, B.P. Schmidt, C.W. Stubbs, E.M. Sadler: Astrophys. J. **533**, 320 (2000)
60. B.K. Gibson et al.: Astrophys. J. **529**, 723 (2000)
61. M. Hamuy, S.C. Trager, P.A. Pinto, M.M. Phillips, R.A. Schommer, V. Ivanov, N.B. Suntzeff: Astron. J. **120**, 1479 (2000)
62. R.P. Harkness, J.C. Wheeler: In: *Supernovae*, ed. by A.G. Petshek (Springer, Berlin 1990) p. 1
63. R.P. Harkness et al.: Astrophys. J. **317**, 355 (1987)

64. K. Hirata et al.: Phys. Rev. Lett. **58**, 1490 (1987)
65. P. Höflich, J.C. Wheeler, L. Wang: Astrophys. J. **521**, 179 (1999)
66. D.A. Howell: Astrophys. J. Lett. **554**, L193 (2001)
67. K. Iwamoto et al.: Nature **395**, 672 (1998)
68. S. Jha et al.: Astrophys. J. Suppl. **125**, 73 (1999)
69. B. Leibundgut: Astron. Astrophys. Rev. **10**, 179 (2000)
70. B. Leibundgut et al.: Astrophys. J. **372**, 531 (1991)
71. B. Leibundgut et al.: Astron. J. **105**, 301 (1993)
72. E.J. Lentz et al.: Astrophys. J. **507**, 406 (2001)
73. D.C.; Leonard, A.V. Filippenko, A.J. Barth, T. Matheson: Astrophys. J. **536**, 239 (2000)
74. P. Lira et al.: Astron. J. **116**, 1006 (1998)
75. L.B. Lucy: Astrophys. J. **383**, 308 (1991)
76. T. Matheson, A.V. Filippenko, R. Chornock, D.C. Leonard, W. Li: Astron. J. **119**, 2303 (2000)
77. T. Matheson, A.V. Filippenko, L.C. Ho, A.J. Barth, D.C. Leonard: Astron. J. **120**, 1499 (2000)
78. T. Matheson, A.V. Filippenko, W. Li, D.C. Leonard, J.C. Shields: Astron. J. **121**, 1648 (2001)
79. T. Matheson et al.: Astron. J. **120**, 1487 (2000)
80. A. Maury, I. Hook, R. Gorki, F. Selman, M. Dennefeld: IAUC 7528 (2001)
81. J. Maza, S. van den Bergh: Astrophys. J. **204**, 519 (1976)
82. P.A. Mazzali, E. Cappellaro, I.J. Danziger, S. Benetti, M. Turatto: Astrophys. J. Lett. **499**, L49 (1998)
83. P.A. Mazzali, K. Iwamoto, K. Nomoto: Astrophys. J. **545**, 407 (2000)
84. P.A. Mazzali, K. Nomoto, F. Patat, K. Maeda: Astrophys. J. **559**, 1047 (2001)
85. R.J. McMillan, R. Ciardullo: Astrophys. J. **473**, 707 (1996)
86. W.P.S. Meikle: Mon. Not. R. Astron. Soc. **314**, 782 (2000)
87. P.A. Milne, L.S. The, M.D. Leising: Astrophys. J. Suppl. **124**, 503 (1999)
88. R. Minkowski: Mon. Not. R. Astron. Soc. **53**, 224 (1941)
89. K. Nomoto, K. Iwamoto, T. Suzuki: Phys. Rep. **256**, 173 (1995)
90. P. Nugent, M.M. Phillips, E. Baron, D. Branch, P. Hauschildt: Astrophys. J. Lett. **455**, L147 (1995)
91. A. Pastorello et al.: Mon. Not. R. Astron. Soc. **333**, 27 (2002)
92. F. Patat, R. Barbon, E. Cappellaro, M. Turatto: Astron. Astrophys. Suppl. Ser. **98**, 443 (1993)
93. F. Patat, R. Barbon, E. Cappellaro, M. Turatto: Astron. Astrophys. **282**, 731 (1994)
94. F. Patat, N. Chugai, P.A. Mazzali: Astron. Astrophys. **299**, 715 (1995)
95. F. Patat, A. Piemonte: IAUC 6918 (1998)
96. F. Patat et al.: Astrophys. J. **555**, 900 (2001)
97. S. Perlmutter et al.: Astrophys. J. **483**, 565 (1997)
98. M.M. Phillips: Astrophys. J. Lett. **413**, L105 (1993)
99. M.M. Phillips, S.R. Heatcote, M. Hamuy, M. Navarrete: Astron. J. **95**, 1087 (1988)
100. M.M. Phillips, P. Lira, N.B. Suntzeff, R.A. Schommer, M. Hamuy, J. Maza: Astron. J. **118**, 1766 (1999)
101. A.C. Porter, A.V. Filippenko: Astron. J. **93**, 1372 (1987)
102. Y.P. Pskovskii: Astron. Zhurnal **44**, 82 (1967)
103. D.E. Reichart: Astrophys. J. Lett. **521**, L111 (1999)

104. A.G. Riess et al.: Astron. J. **116**, 1009 (1998)
105. P. Ruiz-Lapuente, E. Cappellaro, M. Turatto, C. Gouiffes, I.J. Danziger, M. della Valle, L.B. Lucy: Astrophys. J. Lett. **387**, L33 (1992)
106. E.M. Sadler, R.A. Stathakis, B.J. Boyle, R.D. Ekers: IAUC 6901 (1998)
107. A. Saha, A. Sandage, G.A. Tammann, L. Labhardt, F.D. Macchetto, N. Panagia: Astrophys. J. **522**, 802 (1999)
108. A. Saha et al.: Astrophys. J. **551**, 973 (2001)
109. M. Salvo, E. Cappellaro, P.A. Mazzali, S. Benetti, I.J. Danziger, F. Patat, M. Turatto: Mon. Not. R. Astron. Soc. **321**, 254 (2001)
110. E.M. Schlegel: Mon. Not. R. Astron. Soc. **224**, 269 (1990)
111. B. Schmidt, R.P. Kirshner, B. Leibundgut, L. Wells, A.C. Porter, P. Ruiz-Lapuente, A.V. Filippenko: Astrophys. J. Lett. **434**, L19 (1994)
112. B. Schmidt et al.: Astron. J. **107**, 1444 (1994)
113. S.J. Smartt, G.F. Gilmore, A. Christopher, C.A. Tout, S.T. Hodgkin: Astrophys. J. **565**, 1089 (2002)
114. S.J. Smartt, G.F. Gilmore, N. Trentham, C.A. Tout, C.M. Frayn: Astrophys. J. Lett. **556**, L29 (2001)
115. N.B. Suntzeff: In: *Supernova and Supernova Remnants*, ed. by R. McCray, Z. Wang (Cambridge University Press, Cambridge 1996) p. 41
116. N.B. Suntzeff et al.: Astron. J. **117**, 1175 (1999)
117. R.J. Terlevich, G. Tenorio-Tagle, J. Franco, J. Melnick: Mon. Not. R. Astron. Soc. **255**, 713 (1992)
118. S.E. Thorsett, D.W. Hogg: GCN **197** (1999)
119. M. Turatto, S. Benetti, E. Cappellaro, I.J. Danziger, M. Della Valle, C. Gouiffes, P.A. Mazzali, F. Patat: Mon. Not. R. Astron. Soc. **283**, 1 (1996)
120. M. Turatto, E. Cappellaro, I.J. Danziger, S. Benetti: Mon. Not. R. Astron. Soc. **265**, 471 (1993)
121. M. Turatto, E. Cappellaro, I.J. Danziger, S. Benetti, C. Gouiffes, M. Della Valle: Mon. Not. R. Astron. Soc. **262**, 128 (1993)
122. M. Turatto et al.: Astrophys. J. Lett. **498**, L129 (1998)
123. M. Turatto et al.: Astrophys. J. Lett. **534**, L57 (2000)
124. A. Uomoto, R.P. Kirshner: Astrophys. J. **308**, 685 (1986)
125. S. van den Bergh, D. Branch: Astron. J. **105**, 2231 (1993)
126. S. van den Bergh, J. Pazder: Astrophys. J. **390**, 34 (1992)
127. S.D. Van Dyk, M. Hamuy, A.V. Filippenko: Astron. J. **111**, 2017 (1996)
128. S.D. Van Dyk, C.Y. Peng, A.J. Barth, A.V. Filippenko: Astron. J. **118**, 2331 (1999)
129. S.D. Van Dyk, K.W. Weiler, R.A. Sramek, M.P. Rupen, N. Panagia: Astrophys. J. Lett. **432**, L115 (1994)
130. S.D. Van Dyk, K.W. Weiler, R.A. Sramek, E.M. Schlegel, A.V. Filippenko, N. Panagia, B. Leibundgut: Astron. J. **111**, 1271 (1996)
131. L. Wang, J.C. Wheeler: Astrophys. J. Lett. **504**, L87 (1998)
132. K.W. Weiler: http://rsd-www.nrl.navy.mil/7214/weiler/kwdata/RSNtable.txt
133. K.W. Weiler, N. Panagia, M.J. Montes: Astrophys. J. **562**, 670 (2001)
134. J.C. Wheeler, S. Benetti: In: *Allen's Astrophysical Quantities*, ed. by A.N. Cox (Springer, New York 2000) p. 451
135. J.C. Wheeler, R.P. Harkness: Rep. Prog. Phys. **53**, 1467 (1990)
136. J.C. Wheeler, R.P. Harkness, E.S. Barker, A.L. Cochran, D. Wills: Astrophys. J. Lett. **313**, L69 (1987)
137. J.C. Wheeler, R.P. Harkness, A. Clocchiatti, S. Benetti, M.S. Brotherton, D.L. DePoy, J. Elias: Astrophys. J. Lett. **436**, L135 (1994)

138. J.C. Wheeler, R. Levreault: Astrophys. J. Lett. **294**, L17 (1985)
139. S.E. Woosley, R.G. Eastman, B.P. Schmidt: Astrophys. J. **516**, 788 (1999)
140. S.E. Woosley, T.A. Weaver: Ann. Rev. Astron. Astrophys. **24**, 205 (1986)
141. Q. Yulei, W. Li, Q. Qiyuan, H. Jingyao: Astron. J. **117**, 736 (1999)
142. H.U. Zimmerman et al.: Nature **367**, 621 (1994)
143. F. Zwicky: In: *Stars and Stellar Systems*, ed. by L.H. Aller, D.B. McLaughin (University of Chicago Press, Chicago 1965) p. 367

Supernova Rates

Enrico Cappellaro

Osservatorio Astronomico di Capodimonte, via Moiariello 16, I-80181, Napoli, Italy

Abstract. We review the most recent estimates of the rate of supernovae and evidence for a link with specific stellar populations. We also compare different estimates of the supernova rate in the Galaxy. The early results and the prospects of supernova searches at high redshift are reported.

1 Introduction

The rate of supernovae (SNe) is a key parameter in the ecology of stellar systems with SNII and SNIb/c, the core-collapse of massive stars, tracing the ongoing star formation and SNIa, originating in evolved binary systems, echoing the long term star formation history [33]. All give a major contribution to the Galactic chemical enrichment [11,47] and constrain the phase of the interstellar medium (ISM) [18]. Conversely, stellar evolutionary models make definite predictions for the history of the SN rates in a given stellar population which, by comparison with observations, can test on the adopted SN progenitor scenarios [67].

A comprehensive review of the history of SN rate determinations has been presented by van den Bergh and Tammann [73]. Here we focus on the most recent results and on the prospect for future research.

2 Supernova Searches

The discovery of SN1987A, the nearest supernova in 400 years, the exciting results of the study of SNIa at high redshift, and the association of some SNe with the still mysterious γ-ray burst (GRB) events have given a new boost to SN search programs. Thanks to the efforts of professional astronomers, but also to the contribution of enthusiastic amateurs, the number of SN discoveries in the last years has increased to $150-200$ SN yr^{-1} [2].

As can be seen in Fig. 1, despite an increase by an order of magnitude in the SN discovery rate, the detection rate of bright events has remained more of less constant over the past 40 years at about ten SN yr^{-1} with magnitude brighter than 15. This is because SNe are rare, of the order of one SN per century in an average galaxy, and the quest for more events requires SN searches to explore a larger volume of space and to reach fainter magnitudes. Obviously, in magnitude limited searches there is a bias against intrinsically faint objects. For this reason, even if the statistics may appear comfortably large, it is not safe to derive estimates of the rate from the general SN list. Even for an exploratory analysis,

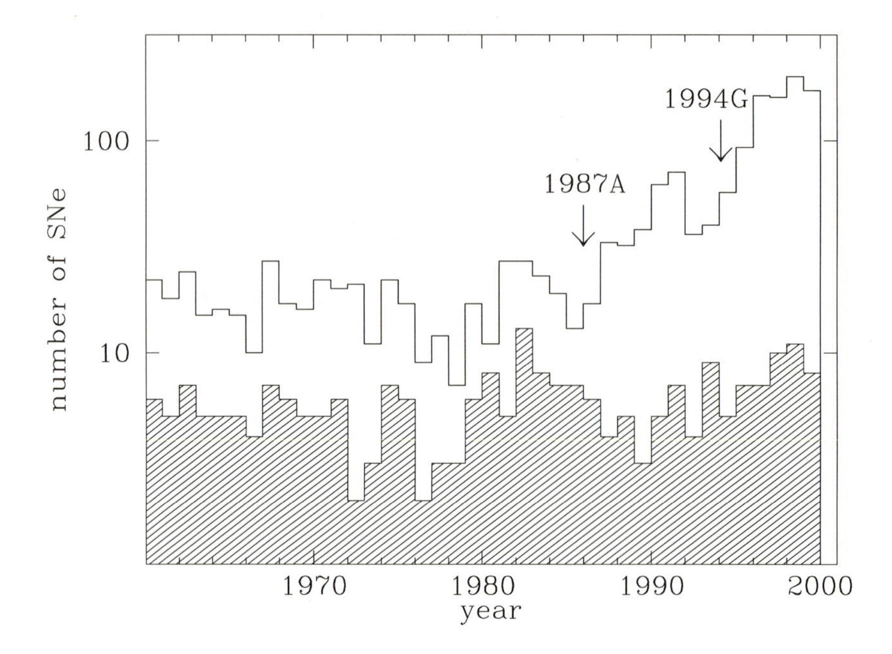

Fig. 1. Number of SN discoveries per year in the last 40 years. The shaded area indicates SNe brighter than 15 mag. The epochs of discovery of the nearby SN1987A and of SN1994G, the first SN at redshift $z > 0.4$, are indicated.

it is at least necessary to restrict the statistical sample to nearby galaxies for which it can be assumed that they have been searched uniformly for SNe. Based on this so called "fiducial sample" approach, a strong dependence of SN rates on galaxy types and a correlation between SN rate and galaxy luminosity can be proved [59–62]. This method has, however, some intrinsic limitations: it cannot be used to derive absolute SN rates and, for SN types with faint intrinsic luminosities, even relative rates may not be reliable. The fact that SN1987A, the nearest SN in modern times, was intrinsically one of the faintest, undermines the basic assumption of the fiducial sample approach.

3 Supernova Rates from Directed Searches

Reliable estimates of the SN rates can be obtained if the surveillance time is known for a sample of galaxies by a specific SN search. The method was introduced in the early days of SN searches by Zwicky [76,77] and was later applied to a number of different SN searches [10,23,41,72].

The basic ingredients of this method are [9]:

- The definition of a sample of galaxies for which detailed information is available; in particular, distance, luminosity, morphological type, and any other parameters against which the SN rate has to be tested.

- The properties of the SN types for which to derive the rates; namely absolute magnitudes, light curves, and colors.
- The characteristics of the SN search under consideration, in particular, limiting magnitude, spectral sensitivity, and spatial resolution.
- The log of observations of the search.
- The list of the SNe discovered in the search.

Knowing these properties, the first step is to compute the time interval during which a SN occurring in a given galaxy remains brighter than the limiting magnitude of the search. Each observation of the galaxy will sample just this time interval. By integrating the surveillance time of individual observations over the actual observing log, the "control time" of the galaxy sample is derived. Finally, the rate of SNe is obtained by dividing the number of SN discovered by the control time.

It must be stressed that SN searches can be affected by severe biases which need to be carefully evaluated. For instance, a simple examination of the list of discovered SNe shows that the SN detection efficiency is reduced in the central region of distant galaxies [7,57]. This bias is most severe for SN searches employing low dynamic range detectors, such as photographic plates, and/or small scale telescopes and is least severe in modern CCD searches exploiting template subtraction techniques [28,48].

Another important bias affects the detection of SNe in inclined spiral galaxies [10,60,70]. This, is due to the large extinction along the line of sight to a SN in an inclined galaxy. Because of the wavelength dependence of the extinction, the effect is expected to be less severe for SN searches with red sensitivity, but this has not yet been measured. The fact that the bias appears to quickly become important for moderately inclined spiral galaxies and then remains almost constant for greater inclinations may be an indication that the dust is not uniformly distributed in the galactic disk and that SNe explode at the bottom of chimney-like dust structures [9,70]. Actually, on the basis of a simple model for the dust and SN progenitor distributions in typical SN producing disk galaxies, it has been claimed that extinction also contributes to the deficit of SN detection in the nuclear region of disk galaxies [27].

In general, if the analysis is confined to the data of a single search, the main limitation is poor statistics. Of course, better statistics can be obtained by combining the data of different searches. This approach was adopted for Table 1 for which five SN searches have been merged to give a total sample of 137 SNe in 9,346 galaxies [10]. The combined effort is equivalent to the continuous monitoring of one galaxy for about $\sim 25,000$ yr. Due to the relatively bright limiting magnitudes of these searches, only nearby galaxies have been well searched (the average galaxy redshift, weighted according to the individual surveillance time, is 3,300 km s^{-1}). Therefore the SN rates in Table 1 refer only to the local Universe.

Although Table 1 has the smallest statistical errors of results published so far, it should not be overlooked that much of the data rely on photographic and visual searches which are severely affected by biases. It is expected that more accurate results will become available in a few years from the on-going CCD SN searches.

Since it has been demonstrated that, for normal galaxies, the SN rate is proportional to the galaxy blue luminosity [6,59,60], it is convenient to express SN rates in SN units (SNu), where 1 SNu = 1 SN $(100\text{yr})^{-1}$ $(10^{10} L_{\odot}^{\mathrm{B}})^{-1}$. Because of this normalization, the rates reported in Table 1 scale with the adopted Hubble constant (H_0) as $h_{75} = (H_0/75)$.

Table 1. SN Rates in the Local Universe [10]

Galaxy	Rate [SNu h_{75}^2]			
Type	Ia	Ib/c	II	All
E-S0	0.18 ± 0.06	< 0.01	< 0.02	0.18 ± 0.06
S0a-Sb	0.18 ± 0.07	0.11 ± 0.06	0.42 ± 0.19	0.72 ± 0.21
Scd-Sd	0.21 ± 0.08	0.14 ± 0.07	0.86 ± 0.35	1.21 ± 0.37
Other	0.40 ± 0.16	0.22 ± 0.16	0.65 ± 0.39	1.26 ± 0.45
All	0.20 ± 0.06	0.08 ± 0.04	0.40 ± 0.19	0.68 ± 0.20

As can be seen in Table 1, the rate of SNIa per unit blue luminosity is, within the errors, independent on the galaxy type while for both SNII and SNIb/c, the rate peaks in late spirals. In these late spirals, five out of six SNe are core-collapse events.

It is comforting to note that, despite the different samples and computation methods, there is a fair agreement of the numbers reported in Table 1 with the estimates of other authors [23,62,72]. However, when comparing individual values there are also noticeable differences. In particular, claims of a very high rate of SNIb/c (four time higher than SNIa [41]) and for a high rate of SNIa in elliptical galaxies (a factor of two higher than in spirals [62]) are not confirmed.

4 Supernovae and Galaxy Content

A strong correlation is found between the rate of core-collapse SNe (SNII and SNIb/c) and the galaxy colors, while no such correlation is apparent for SNIa [10]. To some extent this reflects the known correlation between galaxy types and colors. Actually, a very nice match can be obtained between the core-collapse SN rate and the star formation rate as a function of galaxy color predicted by galaxy evolutionary models [30], assuming that all stars in the mass range $8-40$ $\mathrm{M}_{\odot}$ explode as core-collapse SNe [10].

This is consistent with the finding that SNII and SNIb/c in spiral galaxies are associated with spiral arms and HII regions while SNIa show no preference for these locations [3,38,39,74]. Moreover, it seems that Ib/c are more closely associated with star formation regions than SNII. This suggests that the progenitors of SNIb/c are more massive than those of SNII and is consistent with the SNIb/c – Wolf-Rayet star connection suggested by [75].

The galaxy blue luminosity, which is used for the normalization of the SN rate, cannot be linked to a specific stellar population. Young, massive stars have a hot, blue spectrum but, since they are often embedded in dust, they are heavily affected by reddening [53]. Instead, other spectral bands may be better for tracing a stellar population. For instance, it has been argued that the near infrared H and K luminosity can be used to measure the old stellar population content in galaxies. If all SNIa result from low mass stars, we would expect the SNIa rate per unit of H and K luminosity to not change with galaxy type. Instead, it turns out that the rate normalized by galaxy H and K luminosity increases significantly when moving from ellipticals to late spirals, suggesting that a significant fraction of SNIa result from intermediate age stars [15,71]. Even if this conclusion is true, it must be stressed that the SN rate for unit H and K luminosity are not direct measurements but are obtained by a simple scaling of the SN rate per unit blue luminosity after assuming an average B-H and B-K color for each galaxy type. This is because homogeneous H and K photometry is available for only a small number of galaxies to date.

The far infrared luminosity (FIR) of a galaxy is often assumed to give a direct measurement of the massive star content, hence of the star formation rate [16]. If so, the rate of core-collapse SNe per unit FIR should change little with Hubble Type. Again, the sum of the SNII and the SNIb/c rates per unit FIR changes by almost a factor two moving from early to late spirals [10]. The most plausible interpretation is that different components are contributing to the FIR emission in galaxies, not all of which are directly related to massive stars [30].

The nuclei of a small fraction of galaxies called starburst galaxies show evidence of enhanced star formation rate. Nuclear star formation activity is found to be enhanced also in some other types of active galaxies such as Seyferts [34] and it has been suggested that nuclear activity stimulates the star formation rate in AGN host galaxies [50]. Yet, despite such expectations, no evidence has been found for an enhanced SN rate in active or starburst galaxies [10,45,49]. Most likely this is due to the fact that star formation is enhanced only in the nuclear regions of these galaxies which are inaccessible to optical SN searches because of the heavy extinction. To investigate this, SN searches could be conducted in active galaxies at longer wavelengths where extinction is reduced, but the first attempts at infrared and radio SN searches are still limited by instrumental capabilities and have not given the expected results [22,52]. However, the future looks promising [12,35,37].

One of the factors which has been suggested as a trigger for star formation in galaxies is gravitational interaction. In this scenario one would expect the SN rate to depend on the galaxy environment. However, no clear effect has been found, with the possible exception of an enhanced SN rate in strongly interacting systems [42].

5 Variance Within Basic Supernova Types

The discussion above concerns the three major types of SNe (SNIa, SNIb/c, and SNII) which cover 98% of classified events. However, there are events that, although sharing some of the properties of the main SN types, show distinct peculiarities [5,20]. It would be interesting to know if these are rare events or if they are the tip of an large, undetected population. Though much work is still needed to answer this question, there are some preliminary indications.

We know now that SNIa are far from being standard candles and that there are both faint and bright objects, the best examples being SN1991bg [19,64] and SN1991T [40,46]. In magnitude limited SN searches, faint and bright SNIa are expected to be differently biased. Because it is possible to identify these objects based on the shape of their light curves, this bias may not have a major impact on the use of SNIa as distance indicators, but it is certainly important for the modeling of galactic chemical evolution. Indeed, the mass of iron injected into the interstellar medium by extreme events may differ by one order of magnitude [8]. Although there have been claims that faint SNIa similar to SN1991bg may be the majority of SNIa events [54], current statistics suggest that they are $16-25\%$ of SNIa [9,29]. Owing to the high luminosity, SNIa similar to SN1991T may be over-represented in SN discovery lists, but their intrinsic rate is also close to 20% of SNIa events [29].

Additionally, there appears to be a correlation between the SNIa light curve shapes, hence absolute luminosities, and the morphological type of the host galaxies, with bright SNIa occurring preferentially in young stellar populations [5,24]. It is not yet clear if this can be related to a different age or metallicity of the progenitor population.

Among SNII, a very important example is SN1987A. The fact that this nearby SN was very faint at maximum luminosity raised the possibility that many similar events remain undetected in current SN searches. However, it should be noted that the slow luminosity evolution of SN1987A, with a second maximum about 3 months after the first, favors detection. Indeed, after a careful examination, it has been determined that SNe similar to SN1987A are not the most common type of SNII [56] and likely count for $10-30\%$ of all SNII [9,69]. More recently, a handful of SNII have been discovered with very low luminosity at all times. The best case is SN1997D [65] for which it has been argued that the low luminosity, low kinetic energy, and low radioactive ^{56}Ni yield are the result of a low energy explosion of a massive progenitor [4]. A direct estimate of the rate of these events is not yet available, but it has been suggested that they corresponds to the 15% of core-collapse events which lead to the formation of a black hole [1].

SNII with narrow lines in their spectra are labeled IIn [55]. They show a very slow luminosity decline which is believed to be powered by the interaction between the expanding ejecta and a dense circumstellar medium (CSM). Because they remain bright for long time, SNIIn are easily detected in SN searches. Therefore, although in recent years they represents up to $15-20\%$ of all discovered SNII, they may represent only $2-5\%$ of the actual SNII explosions [9].

No estimates are available for the occurrence rate of SNIb with respect to SNIc. This is mainly due to the poor statistics of these events and also to the fact that it is difficult to define the objects with a simple classification scheme [36]. These problems are even more acute for SNe associated with GRBs for which there are, at most, only a handful of cases and in only one case, SN1998bw [21], is this association firmly established. We should note, however, that all of the proposed GRB-SN associations, if true, would be among the intrinsically brightest SNe ever observed. Thus, a bias against detecting such objects is certainly not expected and their rate is probably low, at most a few percent of core-collapse SNe.

6 Supernova Rates in the Milky Way

Using the data in Table 1 it is possible to compute the expected rate of SNe for any galaxy (or galaxy sample) for which its morphological type and luminosity are know. In particular, the expected number of SNe in the Milky Way, assuming a luminosity of 2×10^{10} $L_\odot$ and a morphological type Sb, are: 0.4 ± 0.1 SNIa, 0.2 ± 0.1 SNIb/c and 1.2 ± 0.6 SNII per century.

Direct estimates of the Galactic SN rate can be derived from the events recorded in the last 1,000–2,000 yr or by counting SN remnants. In the former case, the uncertainties are the inaccuracy of the historical records and the incompleteness caused by Galactic reddening, whereas in the latter case there are major uncertainties in the remnant life-times. Current estimates range from $3.4 - 8.4$ events per century which are a factor $2 - 4$ higher than the numbers given above [17,58,73]. It has been argued that the disagreement indicates that the SN rate in the solar neighborhood is enhanced [17].

All core-collapse events should leave a compact remnant which, in most cases, is expected to be a neutron star. The estimates of the Galactic pulsar birth rate has decreased with time and the current estimate of ~ 0.3 events per century [26] seems significantly smaller than the estimated core-collapse SN rate.

An independent estimate of the Galactic core-collapse SN rate can be obtained from the Galactic mass of radioactive ^{26}Al which, along with other nuclear species, is injected into the CSM by SN explosions. Measurements of the diffuse galactic emission at 1.8 Mev yield a total mass of ^{26}Al in the Galaxy from 3.1 ± 0.9 to 5 ± 4 $M_\odot$ which translates into a core-collapse SN rate of 1.8 ± 0.9 to 3.8 ± 2.8 per century [31,63]. This is in very good agreement with the Galactic SN rate estimated from the rates in external galaxies.

7 Supernova Rates as a Function of Redshift

In recent years many authors have shown that, by measuring the evolution of the SN rate with redshift, it is possible to constrain the cosmic star formation history which is one of the main uncertainties in galaxy evolutionary models [14,33,51,68]. Yet, despite major efforts devoted to high-z SN searches, only some preliminary results for SNIa have been published. These are summarized

Table 2. Estimates of the rate of SNIa at different redshift[a]

$<z>$	N(SN)	SNIa Rate (SNu h^2_{75})	SNIa Rate ($Mpc^{-3}yr^{-1}$ h^3_{75})[b]	Reference
0.01	69	0.20 ± 0.06	$3.7 \pm 0.9 \times 10^{-5}$	[10]
0.1	4	$0.25^{+0.20}_{-0.12}$	$4.7^{+3.7}_{-2.2} \times 10^{-5}$ [c]	[25]
0.25	1	$0.5^{+1.9}_{-0.4}$		[66]
0.38	3	$0.46^{+0.30}_{-0.21}$	$5.8^{+5.1}_{-3.6} \times 10^{-5}$	[43]
0.54	38	$0.33^{+0.06}_{-0.05}$	$6.5^{+1.4}_{-1.3} \times 10^{-5}$	[44]
0.9	2	$0.9^{+2.8}_{-0.9}$		[66]

[a]Values are for a flat Universe with $\Omega_M = 0.3$.
[b]$h_{75} = (H_0/75)$
[c]Scaled to the value of the local luminosity density adopted here (see text).

in Table 2, where in the first column is the average redshift of the galaxy sample and in the second is the number of SNe on which the estimate is based. The rate of SNIa in the local Universe is reported for reference in the first row; the SN rate per unit volume (Mpc^{-3}) has been derived from the SN rate in SNu, assuming for the local luminosity density the value of $1.87 \times 10^8 h_{75}$ $L_\odot$ Mpc^{-3} [13].

It appears that the SNIa rate rises with redshift already at $z = 0.3 - 0.5$. Taken at face value this requires a peak in the star formation rates at relatively low z [14] and seems to exclude models where the SNIa rate remains almost constant up to $z = 2$ [32]. However, it is fair to say that, at the moment, the uncertainties are still large enough that none of the models can be rejected. Note in particular than only in one case are the statistics for high-z SNIa comparable to those of the local sample.

The emphasis on SNIa as cosmological probes has the consequence that most high-z SN searches are designed to detect un-reddened SNIa. This, combined with the fact that the average core-collapse SN is fainter than the average SNIa, precludes, for the moment, an observational estimate of the core-collapse SN rates at high z. That, in turn, precludes a direct estimate of the star formation rate at the corresponding cosmological epoch. Hopefully the situation will improve in the near future thanks to dedicated SN search programs.

References

1. S. Balberg, S.L. Shapiro: Astrophys. J. **556**, 944 (2001)
2. R. Barbon, V. Buondí, E. Cappellaro, M. Turatto: Astron. Astrophys. Suppl. Ser. **139**, 531 (1999)
3. O.S. Bartunov, D.Yu. Tsvetkov, I.V. Filimonova: Pub. Astron. Soc. Pacific **106**, 1276 (1994)

4. S. Benetti et al.: Mon. Not. R. Astron. Soc. **322**, 361 (2001)
5. E. Cappellaro, M. Turatto: 'Supernova Types and Rates'. In: *The Influence of Binaries on Stellar Population Studies, ASSL 264*, ed. by D. Vanbeveren (Kluwer Academic Publishers, Dordrecht 2001) p. 199
6. E. Cappellaro, M. Turatto, S. Benetti, D.Yu. Tsvetkov, O. Bartunov, I.N. Makarova: Astron. Astrophys. **273**, 383 (1993)
7. E. Cappellaro, M. Turatto: In: *Thermonuclear Supernovae, J. Isern NATO ASIC Proc. 486*, ed. by P. Ruiz-Lapuente, R. Canal (Kluwer Academic Publishers, Dordrecht 1997) p. 77
8. E. Cappellaro, P.A. Mazzali, S. Benetti, I.J. Danziger, M. Turatto, M. della Valle, F. Patat: Astron. Astrophys. **328**, 203 (1997)
9. E. Cappellaro, M. Turatto, D.Yu. Tsvetkov, O.S. Bartunov, C. Pollas, R. Evans, M. Hamuy: Astron. Astrophys. **322**, 431 (1997)
10. E. Cappellaro, R. Evans, M. Turatto: Astron. Astrophys. **351**, 459 (1999)
11. C. Chiappini, F. Matteucci, D. Romano: Astrophys. J. **554**, 1044 (2001)
12. L. Colina, A. Alberdi, J.M. Torrelles, N. Panagia, A.S. Wilson: Astrophys. J. Lett. **553**, L19 (2001)
13. N. Cross et al.: Mon. Not. R. Astron. Soc. **324**, 825 (2001)
14. T. Dahlén, C. Fransson: Astron. Astrophys. **350**, 349 (1999)
15. M. Della Valle, M. Livio: Astrophys. J. Lett. **423**, L31 (1994)
16. N.A. Devereux, S. Hameed: Astron. J. **113**, 599 (1997)
17. P.M. Dragicevich, D.G. Blair, R.R. Burman: Mon. Not. R. Astron. Soc. **302**, 693 (1999)
18. G. Efstathiou: Mon. Not. R. Astron. Soc. **317**, 697 (2000)
19. A.V. Filippenko et al.: Astron. J. **104**, 1543 (1992)
20. A.V. Filippenko: Ann. Rev. Astron. Astrophys. **35**, 309 (1997)
21. T.J. Galama et al.: Nature **395**, 670 (1998)
22. B. Grossan, E. Spillar, R. Tripp, N. Pirzkal, B.M. Sutin, P. Johnson, D. Barnaby: Astron. J. **118**, 705 (1999)
23. M. Hamuy, P.A. Pinto: Astron. J. **117**, 1185 (1999)
24. M. Hamuy, S.C. Trager, P.A. Pinto, M.M. Phillips, R.A. Schommer, V. Ivanov, N.B. Suntzeff: Astron. J. **120**, 1479 (2000)
25. D. Hardin et al.: Astron. Astrophys. **362**, 419 (2000)
26. J.W. Hartman, D. Bhattacharya, R. Wijers, F. Verbunt: Astron. Astrophys. **322**, 477 (1997)
27. K. Hatano, D. Branch, J. Deaton: Astrophys. J. **502**, 177 (1998)
28. D.A. Howell, L. Wang, J.C. Wheeler: Astrophys. J. **530**, 166 (2000)
29. W. Li, A.V. Filippenko, R.R. Treffers, A.G. Riess, J. Hu, Y. Qiu: Astrophys. J. **546**, 734 (2001)
30. R.C. Kennicutt: Ann. Rev. Astron. Astrophys. **36**, 189 (1998)
31. J. Knödlseder: Astrophys. J. **510**, 915 (1999)
32. C. Kobayashi, T. Tsujimoto, K. Nomoto: Astrophys. J. **539**, 26 (2000)
33. P. Madau, M. della Valle, N. Panagia: Mon. Not. R. Astron. Soc. **297**, L17 (1998)
34. R. Maiolino, A. Alonso-Herrero, S. Anders, A. Quillen, G.H. Rieke, L.E. Tacconi-Garman: Adv. Space Res. **23**, 875 (1999)
35. R. Maiolino, M. Della Valle, L. Vanzi, F. Mannucci: IAUC 7661 (2001)
36. T. Matheson, A.V. Filippenko, W. Li, D.C. Leonard, J.C. Shields, J.C.: Astron. J. **121**, 1648 (2001)
37. S. Mattila, W.P.S. Meikle: Mon. Not. R. Astron. Soc. **324**, 325 (2001)
38. J. Maza, S. van den Bergh: Astrophys. J. **204**, 519 (1976)

39. R.J. McMillan, R. Ciardullo: Astrophys. J. **473**, 707 (1996)
40. P.A. Mazzali, I.J. Danziger, M. Turatto: Astron. Astrophys. **297**, 509 (1995)
41. R.A. Muller, H.J.M. Newberg, C.R. Pennypacker, S. Perlmutter, T.P. Sasseen, C.K. Smith: Astrophys. J. Lett. **384**, L9 (1992)
42. H. Navasardyan, A.R. Petrosian, M. Turatto, E. Cappellaro, J. Boulesteix: Mon. Not. R. Astron. Soc. **328**, 1181 (2001)
43. R. Pain et al.: Astrophys. J. **473**, 356 (1996)
44. R. Pain et al.: astro-ph 0205476 (2002)
45. A.R. Petrosian, M. Turatto: Astron. Astrophys. **239**, 63 (1990)
46. M.M. Phillips, L.A. Wells, N.B. Suntzeff, M. Hamuy, B. Leibundgut, R.P. Kirshner, C.B. Foltz: Astron. J. **103**, 1632 (1992)
47. L. Portinari, C. Chiosi: Astron. Astrophys. **355**, 929 (2000)
48. D.J. Reiss, L.M. Germany, B.P. Schmidt, C.W. Stubbs: Astron. J. **115**, 26 (1998)
49. M.W. Richmond, A.V. Filippenko, J. Galisky, J.: Pub. Astron. Soc. Pacific **110**, 553 (1998)
50. J.M. Rodriguez-Espinosa, R.M. Stanga: Astrophys. Space Sci. **157**, 173 (1989)
51. R. Sadat, A. Blanchard, B. Guiderdoni, J. Silk: Astron. Astrophys. **331**, L69 (1998)
52. E.M. Sadler, D. Campbell-Wilson: Pub. Astron. Soc. Australia **14**, 159 (1997)
53. L.J. Sage, P.M. Solomon: Astrophys. J. Lett. **342**, L15 (1989)
54. B.E. Schaefer: Astrophys. J. **464**, 404 (1996)
55. E.M. Schlegel: Mon. Not. R. Astron. Soc. **244**, 269 (1990)
56. M.F. Schmitz, C.M. Gaskel: In: *Supernova 1987A in the Large Magellanic Cloud*, ed. by M. Kafatos, A. Michalitsianos (Cambridge University Press, Cambridge and New York 1988) p. 112
57. R.L. Shaw: Astron. Astrophys. **76**, 188 (1979)
58. R.G. Strom: Astron. Astrophys. **288**, L1 (1994)
59. G.A. Tammann: Astron. Astrophys. **8**, 458 (1970)
60. G.A. Tammann: In: *Supernovae and Supernova Remnants, ASSL 45*, ed. by C.B. Cosmovici (Reidel, Dordrecht 1974) p.155
61. G.A. Tammann: In: *Supernovae, ASSL 66*, ed. by D.N. Schramm (Reidel, Dordrecht 1977) p. 95
62. G.A. Tammann, W. Loeffler, A. Schroeder: Astrophys. J. Suppl. **92**, 487 (1994)
63. F.X. Timmes, R. Diehl, D.H. Hartmann: Astrophys. J. **479**, 760 (1997)
64. M. Turatto, S. Benetti, E. Cappellaro, I.J. Danziger, M. della Valle, C. Gouiffes, P.A. Mazzali, F. Patat: Mon. Not. R. Astron. Soc. **283**, 1 (1996)
65. M. Turatto et al.: Astrophys. J. Lett. **498**, L129 (1998)
66. A. Gal-Yam, D. Maoz, K. Sharon: Mon. Not. R. Astron. Soc. **332**, 37 (2002)
67. L.R. Yungelson, M. Livio: Astrophys. J. **497**, 168 (1998)
68. L.R. Yungelson, M. Livio: Astrophys. J. **528**, 108 (2000)
69. S. van den Bergh, R.D. McClure: Astrophys. J. **347**, 29 (1989)
70. S. van den Bergh, R.D. McClure: Astrophys. J. **359**, 277 (1990)
71. S. van den Bergh: Pub. Astron. Soc. Pacific **102**, 1318 (1990)
72. S. van den Bergh, R.D. McClure: Astrophys. J. **425**, 205 (1994)
73. S. van den Bergh, G.A. Tammann: Ann. Rev. Astron. Astrophys. **29**, 363 (1991)
74. S.D. van Dyk: Astron. J. **103**, 1788 (1992)
75. S.D. Van Dyk, C.Y. Peng, A.J. Barth, A.V. Filippenko: Astron. J. **118**, 2331 (1999)
76. F. Zwicky: Astrophys. J. **88**, 529 (1938)
77. F. Zwicky: Astrophys. J. **96**, 28 (1942)

Optical Spectra of Supernovae

David Branch[1], Edward A. Baron[1], and David J. Jeffery[2]

[1] University of Oklahoma, Norman, OK 73019, USA
[2] New Mexico Institute of Mining and Technology, Socorro, NM 87801, USA

Abstract. Supernova flux and polarization spectra bring vital information on the geometry, physical conditions, and composition structure of the ejected matter. For some supernovae the circumstellar matter is also probed by the observed spectra. Some of this information can be inferred directly from the observed line profiles and fluxes but, because of the Doppler broadening and severe line blending, interpretation often involves the use of *synthetic spectra*. The emphasis in this chapter is on recent results obtained with the help of *synthetic spectra*.

1 Introduction

As discussed and illustrated in the chapter by Turatto in this volume, optical spectra serve as the basis for the classification of supernovae (SNe). This chapter is concerned with extracting information from optical spectra on the geometry, physical conditions, and composition structure (element composition versus ejection velocity) of the ejected matter, and also the circumstellar medium (CSM) for some SNe. Other chapters in this volume also discuss certain aspects of SN spectroscopy: UV spectra (see the chapter by Panagia in this volume); the circumstellar and nebular spectra of SN1987A (see the chapter by McCray in this volume); and the hypernova SN1998bw (see the chapter by Iwamoto et al. in this volume).

Some of the information carried by SN spectra can be inferred directly from the observed line profiles and fluxes, but because SN ejection velocities are a few percent of the speed of light, spectral features generally are blended and interpretation often involves comparing observed spectra with *synthetic spectra* calculated for model SNe. The spectrum calculations may be simplified and rapid, or physically self-consistent and computationally intensive. Similarly, both simple parameterized physical models and numerical hydrodynamical models are used. In this chapter we place some emphasis on recent results obtained with the help of *synthetic spectra*.

In Sect. 2 the elements of spectral line formation in SNe are discussed and some of the synthetic spectrum codes that are in current use are described. In Sects. 3–6 recent comparisons of *synthetic spectra* with observed spectra of SNIa, SNIb, SNIc, and SNII are surveyed. Space limitations prevent us from referring to the numerous papers that have been important to the development of this subject; many such can be found in 1993 and 1997 review articles [36,138]. Here most of the references are to, and all of the figures are from, papers that have appeared since 1998. A brief discussion of the prospects for the future appears in Sect. 7.

2 Line Formation and Synthetic Spectrum Codes

At the time of the supernova explosion, physical scales are relatively small. Shortly thereafter the ejected matter expands freely as from a point explosion with each matter element having a constant velocity. All structures in the ejecta then just scale linearly with t, the time since explosion. The matter is in homologous expansion, which has several convenient features:

- The radial velocity v of a matter element is a useful comoving coordinate with actual radial position of the element given by $r = vt$.
- The density at any comoving point just scales as t^{-3}.
- The photon redshift between matter elements separated by velocity Δv, $\Delta\lambda = \lambda(\Delta v/c)$, is time-independent.
- The resonance surfaces for line emission at a single Doppler-shifted line frequency are just planes perpendicular to the observer's line of sight. (Relativistic effects introduce a slight curvature [64].

If continuous opacity in the line forming region is disregarded, the profile of an unblended line can be calculated when the line optical depth $\tau_\ell(v)$ and source function $S_\ell(v)$ are specified. Because SN ejection velocities ($\sim 10,000$ km s^{-1}) are much larger than the random thermal velocities (~ 10 km s^{-1}), a photon remains in resonance with an atomic transition only within a small resonance region. The *Sobolev approximation*, that the physical conditions other than the velocity are uniform within the resonance region, usually is a good one, and it allows the optical depth of a line to be simply expressed in terms of the local number densities of atoms or ions in the lower and upper levels of the transition:

$$\tau_\ell = \frac{\pi e^2}{m_e c} f \lambda t n_\ell \left(1 - \frac{g_\ell n_u}{g_u n_\ell}\right) = 0.229 f \lambda_\mu t_d n_\ell \left(1 - \frac{g_\ell n_u}{g_u n_\ell}\right), \quad (1)$$

where f is the oscillator strength, λ_μ is the line wavelength in microns, t_d is the time since explosion in days, and n_ℓ and n_u are the populations of the lower and upper levels of the transition in cm^{-3}. The term in brackets is the correction for stimulated emission. The source function is

$$S_\ell = \frac{2hc}{\lambda^3} \left(\frac{g_u n_\ell}{g_\ell n_u} - 1\right)^{-1}. \quad (2)$$

All of the radial dependence of τ_ℓ and S_ℓ is in the level populations. The specific intensity that emerges from a resonance region is

$$I = S_\ell \left(1 - e^{-\tau_\ell}\right). \quad (3)$$

Spectroscopic evolution can be divided into a photospheric phase when the SN is optically thick in the continuum below a photospheric velocity, and a subsequent nebular phase during which the whole SN is optically thin in the continuum. In the photospheric phase line formation occurs above the photosphere and in the nebular phase throughout the ejecta. There is, of course, no sharp division between the two phases. However, spectral synthesis modeling techniques in the photospheric and nebular limits can make use of different approximations which are adequate for those limits.

In the remainder of this section we first discuss line formation and synthetic spectrum codes in the photospheric and nebular phases under the assumptions of spherical symmetry and negligible circumstellar interaction. We then consider the effects of circumstellar interaction, and SN asymmetry and polarization spectra.

2.1 The Photospheric Phase

The elements of spectrum formation in the photospheric phase have been discussed and illustrated at length elsewhere [66]. Here we will only briefly summarize those elements and then discuss spectrum synthesis codes.

During the photospheric phase a continuum radiation field is emitted by a photosphere which can be idealized as an infinitely thin layer. Above the photosphere the radiation interaction with continuous opacity is small. Line opacity on the other hand can be very large for the strongest lines. The large Doppler shifts spread line opacity over a large wavelength interval increasing the effect of strong lines compared to a static atmosphere where such strong lines saturate and can only affect radiation in a narrow wavelength interval. The cumulative effect of many lines, strong and weak, can create a quasi-continuous opacity in the Eulerian frame. This effect has been called the "expansion opacity" [68,108]; it dominates in the UV, where it effectively pushes the photosphere out to a larger radius than in the optical. Full, co-moving, line-blanketed calculations automatically account for this effect [3]; introducing the observer's frame expansion opacity into comoving frame calculations is incorrect.

In the optical, which is usually the chief focus of analysis, the spectrum is characterized by P Cygni lines superimposed on the photospheric continuum. The P Cygni profile has an emission peak near the rest wavelength of the line and a blueshifted absorption feature. The peak may be formed in part by true emission or by line scattering into the line of sight of photons emitted by the photosphere. The emission peak would tend to be symmetrical about the line center wavelength if not for the blueshifted absorption. The absorption is formed by scattering out of the line of sight of photospheric photons emitted toward the observer. Since this occurs in front of the photosphere, the absorption is blueshifted. At early times the ejecta density is high, the photosphere is at high velocity, and the line opacity is strong out to still higher velocities. As expansion proceeds, the photosphere and the region of line formation recede deeper into the ejecta. The P Cygni line profile width thus decreases with time. The minimum of the absorption feature of weak lines tends to form near the photospheric velocity, thus weak lines (e.g., weak Fe II lines) can be used to determine the photospheric velocity's time evolution. The recession of the photosphere exposes the inner ejecta and permits its analysis.

In understanding P Cygni line formation in SNe it is conceptually useful, and often in simple parameterized radiative transfer modeling a good physical approximation, to consider the line-photon interaction as being pure resonance scattering; i.e., no photon creation or destruction above the photosphere. Then the source function of an unblended line is just

$$S_\ell(v) = W(v) I_{\rm phot}, \tag{4}$$

where $W(v)$ is the usual geometrical dilution factor [100] and $I_{\rm phot}$ is the continuum specific intensity (assumed angle independent) radiated by the photosphere. Given the pure resonance scattering approximation for an unblended P Cygni line, the emission is formed just by scattering into the line of sight and the absorption just by scattering out of the line of sight.

Unfortunately, in general there is strong line blending. A photon that is scattered by one transition can redshift into resonance with another, so the influence of each transition on others of longer wavelength must be taken into account. This multiple scattering corresponds to the observer's line blending. Calculations show that in blends, absorptions trump emissions: i.e., when resolved, the absorption minima of blends tend to be blueshifted by their usual amounts, but the emission peaks do not necessarily correspond to the rest wavelengths of the lines of the blend. Thus absorption minima usually are more useful than emission peaks for making line identifications during the photospheric phase.

If one has made the resonance scattering approximation for simple modeling, it turns out that the effect of line blending on source functions for some lines is often effectively very small. From a conceptual point of view it is convenient to think of $S_\ell(v) = W(v) I_{\rm phot}$ even for strongly blended lines. In general the effect of line blending on the emergent spectrum is significant and cannot be neglected.

A special case of a P Cygni line that has become of interest is a "detached" line: i.e., a line that has a significant optical depth only above some detachment velocity that exceeds the velocity at the photosphere. A detached line consists of a flat inconspicuous emission peak and an absorption having a sharp red edge at the blueshift corresponding to the detachment velocity (Fig. 1). We shall later refer to the concept of detached lines in discussing particular observed spectra.

It is possible to extract some information from spectra by simple direct analysis. Recently a method for inverting an unblended P Cygni line profile to extract the radial dependences of the line optical depth and the source function has been developed [69]. But the complex blending of P Cygni lines (which often occurs), and the need to account for effects that cannot be treated by direct approaches make analysis by synthetic spectrum modeling essential. Rather than discuss modeling in general, we will briefly describe three synthetic spectrum codes that are frequently used, proceeding from simple to complex. Example *synthetic spectra* from these codes will appear in Sects. 3–6.

The simplest code is the fast, parameterized *SYNOW* code. Technical details of the current version are in [38]. The basic assumptions are: spherical symmetry, a sharp photosphere that emits a blackbody continuous spectrum, and the *Sobolev approximation* with a resonance scattering source function. *SYNOW*

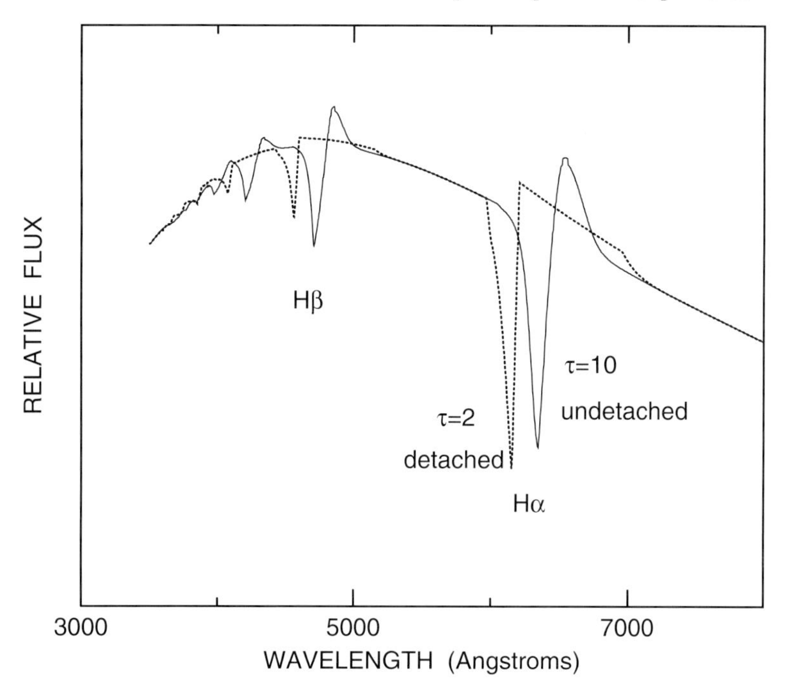

Fig. 1. A *SYNOW* synthetic spectrum (*dotted line*) that has $v_{phot} = 10,000$ km s^{-1} and hydrogen lines detached at 20,000 km s^{-1} where τ(Hα)=2, is compared with a synthetic spectrum (*solid line*) that has $v_{phot} = 10,000$ km s^{-1} and undetached hydrogen lines with $\tau(\mathrm{H}\alpha) = 10$ at the photosphere. (From [13])

does not do continuum transport; it does not solve rate equations; and it does not calculate or assume ionization ratios. Its main function is to take line multiple scattering into account so that it can be used in an empirical spirit to make line identifications and determine the velocity at the photosphere and the velocity intervals within which the presence of each ion is detected. For each ion that is introduced, the optical depth at the photosphere of a reference line is a free parameter, and the optical depths of the other lines of the ion are calculated assuming Boltzmann excitation. Reference lines are generally chosen as the strongest line in the optical for a particular species. Line optical depths are assumed to decrease radially following a power law or an exponential. When deciding which ions to introduce, use is made of the results of [50] who presented plots of local thermodynamic equilibrium (LTE) *Sobolev approximation* line optical depths versus temperature for six different compositions that might be expected to be encountered in SNe, and sample *SYNOW* optical spectra for 45 individual ions that are candidates for producing identifiable spectral features in SNe. When fitting to an observed spectrum, the important parameters are the reference line optical depths, the velocity at the photosphere, and whatever maximum and minimum (detachment) velocities may be imposed on each ion.

Somewhat more complex synthetic spectrum calculations are carried out with what we will refer to as the *ML MONTE CARLO* code (ML for Mazzali and Lucy), technical details of which are in [91] and references therein. This code also assumes a sharp photosphere, makes the *Sobolev approximation*, and does not solve rate equations. The main differences from *SYNOW* are: (1) an approximate radiative equilibrium temperature distribution and internally consistent ionization ratios are calculated for an assumed composition structure; (2) electron scattering in the atmosphere is taken into account; and (3) the original assumption of resonance scattering has been replaced by an approximate treatment of photon branching to allow an absorption in one transition to be followed by emission in another [85]. Several circumstances in which the effects of photon branching are significant are discussed in [91].

Detailed calculations are made with the multi-purpose synthetic spectrum and model atmosphere code called *PHOENIX*. Technical details of *PHOENIX* are in [53] (see also [6]). The basic assumptions are spherical symmetry and time independence. The *Sobolev approximation* is dispensed with and calculations are carried out in the comoving frame with all special relativistic effects. The aim of *PHOENIX* is to take all of the relevant physics into account as fully as possible. *PHOENIX* treats continuum transport, solves the non-local thermodynamic equilibrium (NLTE) rate equations for a large number of ions and a very large number of atomic levels with γ-ray deposition and nonthermal excitation taken into account, and determines a self-consistent radiative equilibrium temperature structure. The lower boundary condition can be chosen to be either diffusive or nebular.

One may ask why simpler codes are needed when *PHOENIX*-like codes can be implemented. *PHOENIX*-like codes demand many hours of computation and are not optimal for explorations of parameter space or for gaining insight into how simple features of modeling affect outcomes. The simpler codes are tools for rapid exploration and gaining insight. Such codes, however, cannot give definitive determinations because of their simplified physics. *PHOENIX*-like calculations must be the basis for ultimate decisions about the viability of SN models.

2.2 The Nebular Phase

Our discussion of the nebular phase will be brief. For more thorough discussions see [41,98].

In the absence of circumstellar interaction, the nebular phase is powered by radioactive decay, primarily ^{56}Co (77.23 day half-life), the daughter of explosion-synthesized ^{56}Ni (6.075 day half-life). At very late times longer lived radioactive species such as ^{57}Co and ^{44}Ti become important.

The decay γ-rays deposit their energy by Compton scattering off electrons to produce nonthermal fast electrons whose energy quickly goes into atomic ionization and excitation, and Coulomb heating of the thermal electrons [121]. Particularly for low mass SNe (SNIa, SNIb, SNIc) the unscattered escape of γ-rays (which increases as the ejecta thins) causes the kinetic energy of positrons from the ^{56}Co decays to become an important energy source: about 3 % of the

^{56}Co decay energy is in the form of positron kinetic energy. The positrons are much more strongly trapped than the γ-rays, but not completely trapped [102]. The positron kinetic energy goes into creating fast electrons leading to the same deposition processes as the γ-rays. Optical emission lines are formed by recombination, collisional excitation, and fluorescence. P Cygni profiles of permitted lines that form by scattering in the outer layers may be superimposed on the emission line spectrum.

Because of low continuous opacity and increasingly low line opacities, photons emitted by true emission processes are increasingly unlikely to be scattered again. Thus non-local radiative transfer effects in the nebular phase tend to become negligible. (Local trapping in an optical thick line can be treated easily using a *Sobolev approximation* escape probability.) Thus (absent circumstellar interaction) it is somewhat more feasible than in the photospheric phase to extract information without making synthetic spectrum calculations. This was especially true of SN1987A which, because of its proximity, was well observed in the optical and the infrared long into the nebular phase when the characteristic velocity of the line forming region was only $\sim 3,000$ km s^{-1} and blending was not too severe. Most other non-circumstellar interacting SNe are observed only for a year or so after explosion and line blending is more of a problem.

Without a photosphere, there is no continuum flux to be scattered out of the line of sight even if there are optically thick lines. Hence the characteristic line profile of the nebular phase is not a P Cygni profile but an emission line that peaks near the rest wavelength and, for several reasons, may have an extended red wing [20,43]. If the line has significant optical depth down to low velocity, the emission line has a rounded peak. If, instead, the line forms in a shell, the emission has a boxy (flat top) shape with the half width of the box corresponding to the minimum velocity of the shell.

As for the photospheric phase, nebular phase synthetic spectrum calculations of various levels of complexity prove to be useful. Relatively simple parameterized approaches that can be used as diagnostic tools are described by [10,42]. A one zone code that was developed in its original form by [111] and will be referred to here as the *RL NEBULAR* code (RL for Ruiz-Lapuente and Lucy) is frequently used (see, e.g., [92,93,110]). The input parameters are the time since explosion and the mass, composition, and outer velocity of the nebula. Heating is calculated from the ^{56}Co decay rate, and cooling is by radiative emission. The electron density, temperature, *NLTE* level populations and emission line fluxes are determined simultaneously by enforcing statistical and thermal equilibrium. Codes for calculating *synthetic spectra* of hydrodynamical models are described by [60,83] and references therein. These multi-zone codes calculate the nonthermal heating, ionization equilibrium, and the *NLTE* level populations for a large number of atomic levels. Synthetic spectra of the CO molecule have also been calculated [32,45,119]. For a description of additional nebular phase codes that have been used for SN1987A, see the chapter by McCray in this volume.

The photospheric phase forbids a direct view into the inner regions of the ejecta. The nebular phase permits this direct view but, on the other hand, the outer layers tend toward invisibility. The photospheric phase demands a powerful radiative transfer technique. The advanced nebular phase requires almost none at all, at least in the optical; the ionizing photons in the UV may need a detailed radiative transfer treatment and, of course, the γ-ray and positron energy depositions need a non-local transfer treatment probably best done by Monte Carlo. Both phases require *NLTE* treatments and extensive atomic data. The two phases offer complementary insights and challenges for analysis.

2.3 Circumstellar Interaction

The physics of the radiative and hydrodynamical interactions between SNe and their CSM (circumstellar interaction (CSI)) is discussed in the chapter by Chevalier and Fransson in this volume. No convincing detection of SNIa CSI has yet been made in any wavelength band. CSI also does not seem to affect the optical spectra of typical SNIb, SNIc (for an exception see [87]), and many SNII. Some SNII, and perhaps some hyperenergetic SNIc, are affected, and in extreme cases dominated by CSI effects.

In the idealized case, the wind of the SN progenitor star has a constant velocity and a constant mass-loss rate and, therefore, an r^{-2} density distribution. After the explosion, the CSM can be heated and accelerated by the photons emitted by the SN. Subsequent hydrodynamic interaction between high velocity SN ejecta and low velocity CSM generates photon emission that further affects the velocity distribution and the ionization and excitation of both the CSM and the SN ejecta. These interactions can produce a rich array of effects on the optical spectra. During the photospheric phase narrow emission, absorption, and P Cygni lines formed in low velocity CSM may be superimposed on the SN spectrum [8,9,113,117]. The SN resonance scattering features may be "muted" by the external illumination of the SN line forming region [14]. In extreme cases the CSM may be optically thick in the continuum so that the spectrum consists only of circumstellar features, perhaps including broad wings produced by multiple scattering off circumstellar electrons [21]. During the nebular phase CSI can power boxy emission from high velocity SN matter [42] and at very late times, even decades, CSI can allow the detection of SNe that would otherwise be unobservably faint [35].

Because the simple $r = vt$ velocity law does not apply to CSM, and blending of circumstellar lines generally is not as severe as it is for SN lines (and spherical symmetry often is not an adequate approximation for CSM), few calculations of *synthetic spectra* for CSI have been carried out so far. Line profiles for various velocity and density distributions have been calculated [23,40] and some *PHOENIX* calculations for a constant velocity circumstellar shell have been carried out [70]. Most of the analysis of optical CSI features has involved extracting information on velocities directly from individual line profiles, and on physical conditions and abundances from emission line fluxes [17,23,32,88].

2.4 Asymmetry and Polarization Spectra

It has been the heuristic and hopeful assumption that SNe are essentially spherically symmetric. But from SN polarimetry, especially spectropolarimetry, it is now clear that all types of SNe other than SNIa usually do exhibit some kind of significant asymmetry (see, e.g., [61,78,132,136]).

The polarization of SN flux in most cases arises from their atmospheres where continuous opacity is dominated by electron scattering in the optical and near infrared. Electron scattering is polarizing with polarization (for incident unpolarized radiation) varying from 100 % for 90° scattering to 0 % for forward and backward scattering according to the Rayleigh phase matrix [16]. The polarization position angle is perpendicular to the scattering plane defined by incident and scattered light. The emergent flux from a deep electron scattering atmosphere (either plane-parallel or geometrically extended) tends to exhibit a position angle aligned perpendicular to the normal to the surface and polarization increasing monotonically from 0 % at an emission angle of 0° (to the normal) to a maximum at an emission angle of 90°. For the plane-parallel case the maximum is only 11.7 % [16], but for a highly extended atmosphere it can approach 100 % [15] from the limb, if it can be resolved. SN atmospheres fall into the extended class, and so beams from their limb should be highly polarized though mostly not at 100 %. Beams from the photodisk region (the projection of the photosphere on the sky) should be much less polarized. If SNe were spherically symmetric, the integrated flux (which is all that can be observed from extragalactic SNe) would have zero polarization since all polarization alignments would contribute equally and thus cancel producing zero net polarization. Because of cancelation, the predicted net continuum polarizations are low (a few percent, see below) even for quite strong asymmetries.

Line scattering is usually depolarizing [62]. The simplest line polarization profile is an inverted P Cygni profile that is common, although not universal, in the SN1987A data [63] and is seen in some of the later spectropolarimetry [132]. In this profile a polarization maximum is associated with the P Cygni absorption feature of the flux spectrum and a polarization minimum is associated with the P Cygni flux emission feature. The inverted P Cygni profile has a natural explanation. The polarized flux at any wavelength tends to come from 90° electron scattering. At the flux absorption, unpolarized flux directly from the photosphere is scattered out of the line of sight by a line; thus, the electron polarized flux is less diluted and a polarization maximum arises. At the flux emission, the line scatters into the line of sight extra unpolarized flux diluting the polarization due to the electron scattered flux. Almost any shape asymmetry should tend to give the inverted polarization P Cygni profile. Thus, the inverted P Cygni profile is somewhat limited as a direct analysis tool for asymmetry.

Although one must be cautious in generalizing because of the paucity and inhomogeneity of published data, it seems that most core-collapse SNe that have been observed well enough to detect polarization of order a few tenths-of-percent (of order 40 to date) have shown at least that much continuum polarization. Some of this can be interstellar polarization (ISP) due to intervening dust, but intrinsic polarization seems to be present in many cases. The intrinsic polarization reveals itself through time dependence and, in spectropolarimetry, by line polarization features or by deviation from the ISP wavelength dependence [115]. Unfortunately, correcting for parent galaxy ISP can only be done from an analysis of the SN data itself because there is no other bright nearby point source. Even the Galactic ISP correction may often have to be determined using the SN data since ISP can vary rapidly with position and depth and there may be no suitable Galactic star near the angular position of the SN. Techniques for correcting for ISP have been developed [61,63,77].

Fig. 2 shows a sample of spectropolarimetry from a range of SN types overlapping the corresponding relative flux spectra. The polarization has been corrected for ISP, but in most cases only with the intention of highlighting the line polarization features; the overall polarization level is not trustworthy. Some of the line polarization features are consistent with the inverted P Cygni profile. However, line polarization features are badly distorted by ISP, and thus reliable ISP corrections are needed to determine the true line polarization profiles as well as the true continuum polarization.

For core-collapse SNe the intrinsic continuum polarization varies from probably $\sim 0.0\,\%$ up to perhaps $\sim 4\,\%$ [74,132] and a representative value would be $1\,\%$. There may be a class of core-collapse SNe (those with massive hydrogen envelopes) that have nearly zero polarization at early times [74], and so are highly spherical. Apparent trends are that polarization increases with decreasing envelope mass (i.e., SNIIP least polarized and SNIc most polarized), and that polarization increases with time until the electron optical depth becomes small [63,76,132]. The position angle of polarization tends to be nearly constant in time and wavelength [63,76,132], which is naturally accounted for by axisymmetry in the ejecta.

For SNIa most observations (~ 15 to date) have only been able to place an upper limit of about $0.3\,\%$ continuum polarization (not distinguishing intrinsic from ISP in most cases) [136]. One normal event, SN1996X, has a marginal detection of intrinsic polarization of $\sim 0.3\,\%$ [131]. Spectropolarimetry of SN1997dt also may show the signatures of intrinsic polarization (Fig. 2). So far it seems that normal SNIa are not very polarized and are quite spherically symmetric; this is consistent with the observational homogeneity that makes them such useful tools for cosmology. On the other hand, the subluminous SN1991bg-like event SN1999by had intrinsic continuum polarization reaching up to perhaps $0.8\,\%$ [61]. Based on a sample of one, subluminous SNIa may be significantly asymmetric.

Various models of asymmetry have been proposed to interpret the polarization data. The conventional model is axisymmetric ellipsoidal asymmetry, either prolate or oblate [56,62,97,116]. Even when viewed equator-on, considerable ellipsoidal asymmetry is required to account for SN-like continuum polarization. For example, Monte Carlo calculations suggest that isodensity contour axis ratios of order 1.2:1 and 2.5:1 are needed for polarizations of $1\,\%$ and $3\,\%$, respectively [58,132]. Late time imaging of SN1987A shows asymmetries suggesting

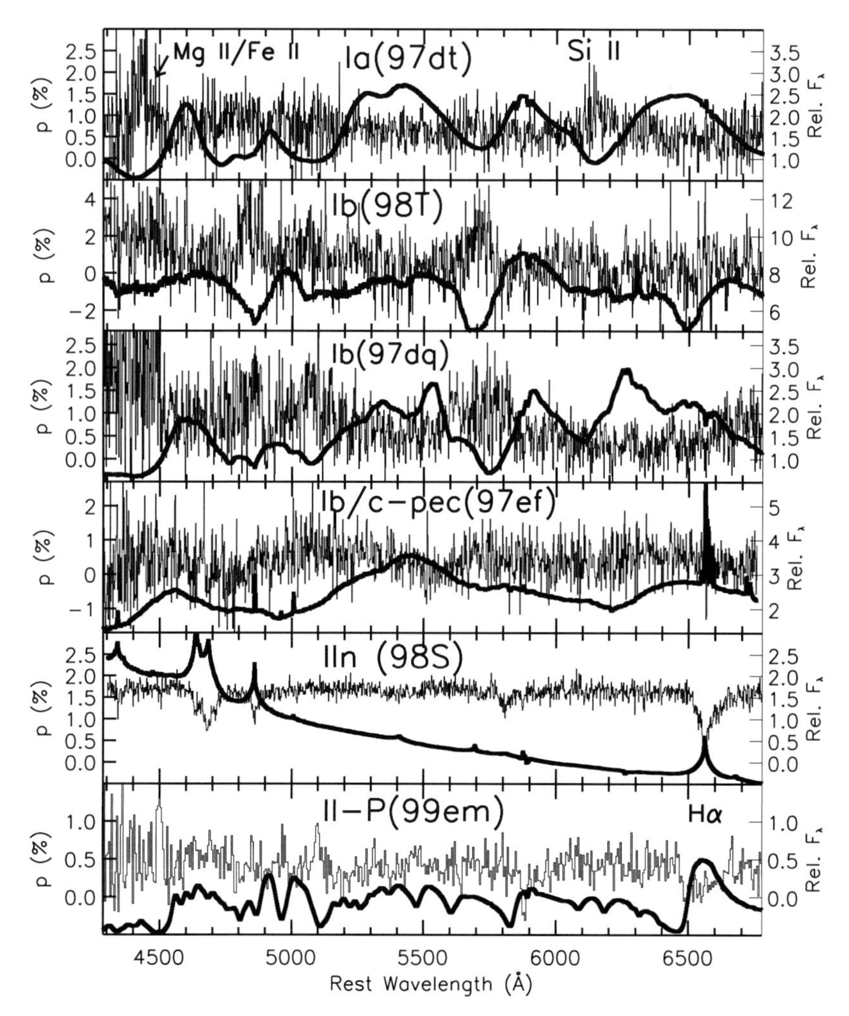

Fig. 2. Spectropolarimetry (*thin lines*) of various SN types, with the corresponding flux spectra (*thick lines*), all obtained within two months of explosion. Figure courtesy of D.C. Leonard.

axis ratios of 2:1 [57,133]. Thus large organized asymmetries something like the ellipsoidal models may account for most SN polarization. Other suggested models of polarizing large scale asymmetry are asymmetric clumpy ionization [18], or scattering of SN light off a nearby dust cloud [130], or off bipolar jets [65].

Spectropolarimetry plays an extremely important role in polarimetric observations for two reasons: (1) it is very useful for establishing intrinsic polarization and correcting for ISP, and (2) it is probably vital for understanding the SN asymmetry. Currently the most advanced synthetic polarization spectrum analyses are done with a 3-dimensional (3D) *NLTE* code that uses Monte Carlo radiative transfer in the outer region where emergent polarization is formed

[57,61]. The *NLTE* component of these calculations is necessarily more limited than in 1D calculations since 3D radiative transfer calculations are computationally very demanding. The ellipsoidal asymmetry used in the calculations of [61] is parameterized and then fitted to the flux and polarization observations. Explosion model asymmetries can also be used [57]. Because of the large parameter space to explore in investigating SN asymmetry, less elaborate codes than those of [57,61] will also continue to be useful. Such codes will mostly rely on Monte Carlo radiative transfer because of its flexibility and robustness.

In addition to the shape asymmetries discussed above, the ejected matter also may be "clumped." Both observation and theory indicate that macroscopic (not microscopic) mixing is ubiquitous in SNe. Clumping is difficult to detect via polarization (although, see [18]), but it can affect both the photospheric [22,33,34] and nebular [10,24,35,79,88,96] flux spectra. A version of *SYNOW* for calculating flux spectra with clumps (*CLUMPYSYN*) has been developed for analysis of the photospheric phase [123].

CSM may, and probably often does, have both global asymmetry and clumping (see [24,32,42,74,128,134,135] and the chapter by Sramek and Weiler in this volume).

3 Type Ia Supernovae

SNIa are thought to be thermonuclear disruptions of accreting or merging white dwarfs. They can be separated on the basis of their photospheric spectra into those that are normal and those that are peculiar. Normals, such as the recently well observed SN1996X [112] and SN1998bu [54,67], have P Cygni features due to Si II, Ca II, S II, O I, and Mg II around the time of maximum brightness, and develop strong Fe II features soon after maximum. Peculiar, weak SN1991bg-like events such as SN1998de [104], SN1997cn [126], and SN1999by [44,61,129] have, in addition, conspicuous low excitation Ti II features. Peculiar powerful SN1991T-like events such as SN1997br [80] and SN2000cx [81] have conspicuous high excitation Fe III features at maximum and the usual SNIa features develop later. Events such as SN1999aa [82] appear to link the 1991T-like events to the normals [11]. Nebular spectra of SNIa are dominated by collisionally excited forbidden emission lines of [Fe III], [Fe II], and [Co III]. SNIa ejecta basically consist of a core of iron peak isotopes (initially mostly ^{56}Ni) surrounded by a layer of intermediate-mass elements such as silicon, sulfur, and calcium. Inferring the composition of the outermost layers is difficult [39,49,92,95] because the lines are broad and line formation occurs in these layers for only a short time after explosion.

Current viable hydrodynamical models for SNIa include deflagrations and delayed detonations . The composition structure of the classic deflagration model W7 [105,122] can account reasonably well for most of the photospheric and nebular spectral features of normal SNIa (Fig. 3 and Fig. 4). On the other hand, near infrared spectra appear to contain valuable diagnostics of the composition

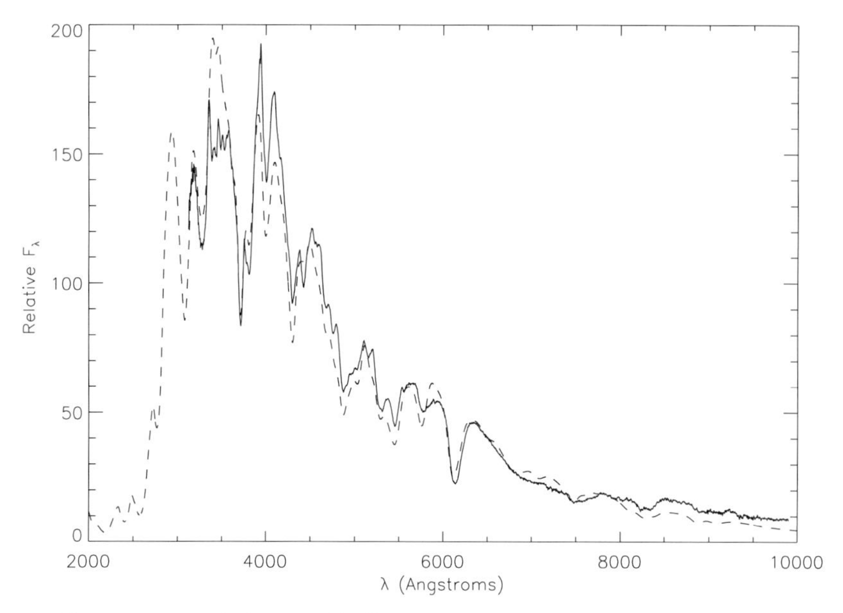

Fig. 3. The spectrum of the type Ia SN1994D three days before maximum brightness (*solid line*) and a *PHOENIX* synthetic spectrum for the deflagration model W7 17 days after explosion (*dashed line*) [72].

structure [10,54,99] and may provide evidence in favor of delayed detonation models [137].

SNIa can be arranged in a spectroscopic sequence ranging from the powerful SN1991T-like events through the normals to the weak SN1991bg-like events. The sequence usually is based on the R(Si II) parameter [106] – the ratio of the depth of an absorption feature near 5800 Å to that of the deep Si II absorption feature near 6100 Å. The ratio increases with decreasing temperature owing to the increasing contribution of numerous weak Ti II lines to the 5800 Å absorption [44] (Fig. 5). The velocity at the outer boundary of the iron peak core, as inferred from nebular spectra, also varies along the spectral sequence [90]. The principal physical variable along the sequence is thought to be the mass of ejected ^{56}Ni in a constant total ejected mass which is near the Chandrasekhar mass.

The lack of a clear correlation between R(Si II) and the blueshift of the 6100 Å feature shows that the diversity among SNIa spectra actually is at least two dimensional [51]. Blueshift differences among events that have similar values of R(Si II) appear to be caused by differences in the amount of mass that is ejected at high velocity, $\sim$ 15,000 – –20,000 km s^{-1} [71] (Fig. 6). One possibility is that the high velocity events are delayed detonations while the others are deflagrations.

The influence of metallicity on SNIa spectra is of special interest in connection with the use of high redshift SNIa as distance indicators for cosmology. The main effects are in the UV but there also are some mild effects in the optical [59,73].

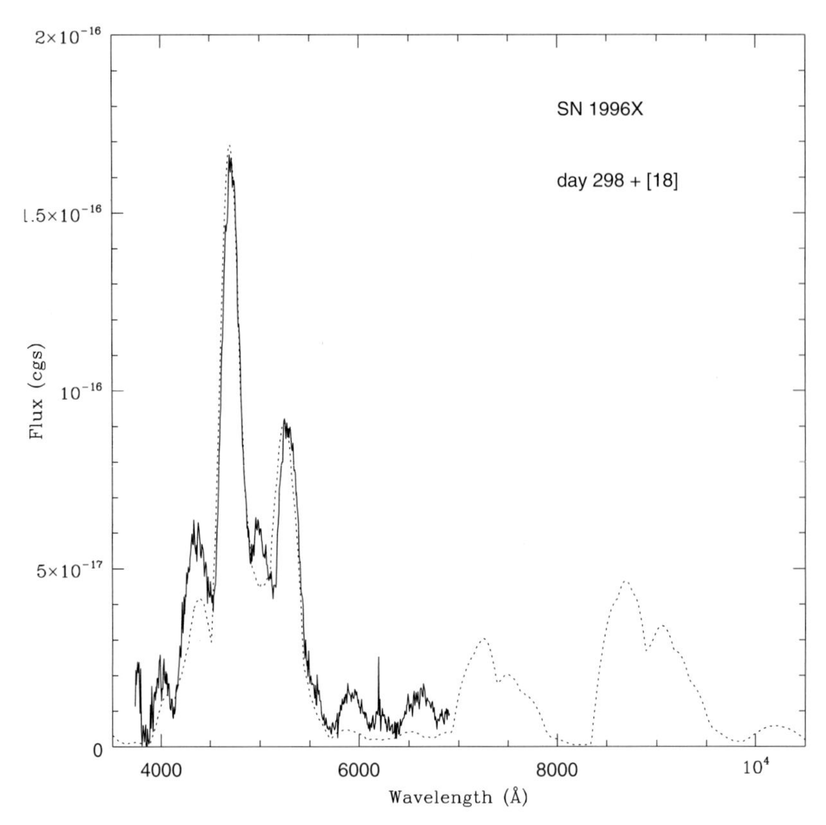

Fig. 4. The spectrum of the type Ia SN1996X 298 days after maximum brightness (*solid line*) is compared with a synthetic spectrum calculated with the *RL NEBULAR* code (*dotted line*) [112].

Deflagrations are expected to produce macroscopic mixing and clumping in the ejecta. The first application of the *CLUMPYSYN* code has shown that the uniformity of the observed depth of the 6100 Å Si II absorption in SNIa near maximum brightness provides limits on the sizes and numbers of clumps [123].

Although the polarization of most SNIa is low, the significant polarization of the SN1991bg-like event SN1999by raises the question of whether the production of weak SNIa may involve rapid white dwarf rotation or merging [61] (Plate 1, top panel).

4 Type Ib Supernovae

SNIb are thought to result from core-collapse in massive stars that have lost all or almost all of their hydrogen envelopes. The photospheric spectra contain the usual low excitation SN features such as Ca II and Fe II together with strong He I lines that are nonthermally excited by the decay products of ^{56}Ni and ^{56}Co [48,84,120]. Recent work on photospheric spectrum calculations includes a

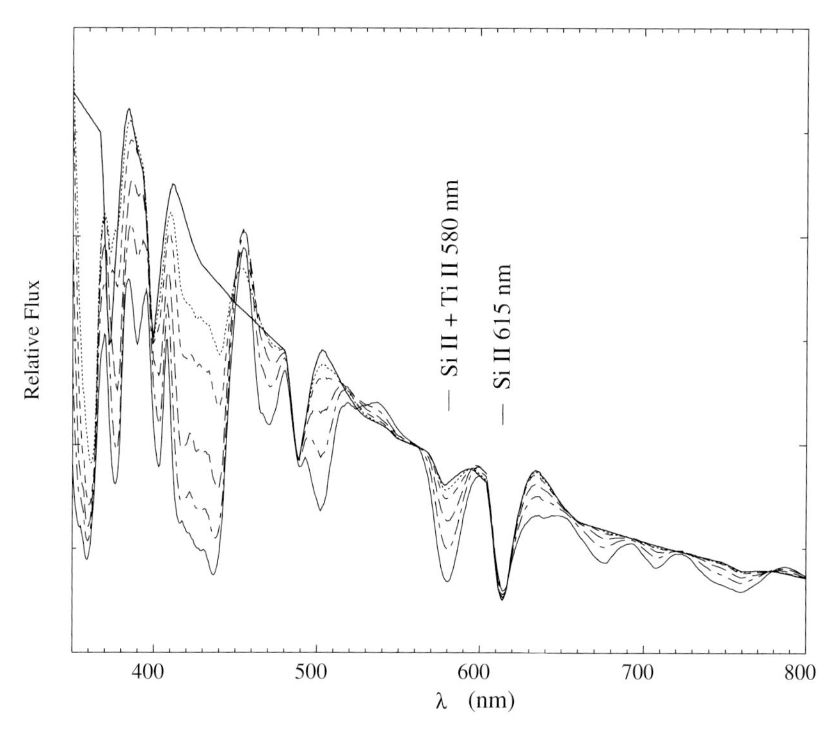

Fig. 5. *SYNOW* synthetic spectra with only Si II lines (*heavy line*) and also with Ti II lines of various strengths (*thin lines*). The 580 nm absorption is strongly affected by Ti II lines [44].

comparison of one *NLTE* synthetic spectrum of a hydrodynamic model with a spectrum of SN1984L [140], a *SYNOW* study of line identifications in SN1999dn [30], and a comparative *SYNOW* study of the spectra of a dozen SNIb [13].

SYNOW spectra can match the observed spectra rather well (Fig. 7). Consistent with the observational homogeneity of SNIb found by [89], the sample studied by [13] obeys a surprisingly tight relation between the velocity at the photosphere as inferred from Fe II lines and the time relative to maximum brightness. The masses and kinetic energies of the events in the sample appear to be similar, and not much room is left for any influence of departures from spherical symmetry on the velocity at the photosphere. After maximum brightness the He I lines usually are detached, but the minimum velocity of the ejected helium is at least as low as 7,000 $\mathrm{km\ s^{-1}}$ (Fig. 8). The spectra of SN2000H, SN1999di, and SN1954A contain detached hydrogen absorption features forming at $\sim 12,000\ \mathrm{km\ s^{-1}}$ (Fig. 8), and hydrogen appears to be present in SNIb in general, although in most events it becomes too weak to identify soon after maximum brightness. The hydrogen line optical depths used to fit the spectra of SN2000H, SN1999di, and SN1954A are not high, so only a mild reduction would be required to make these events look like typical SNIb. Similarly, the He I line

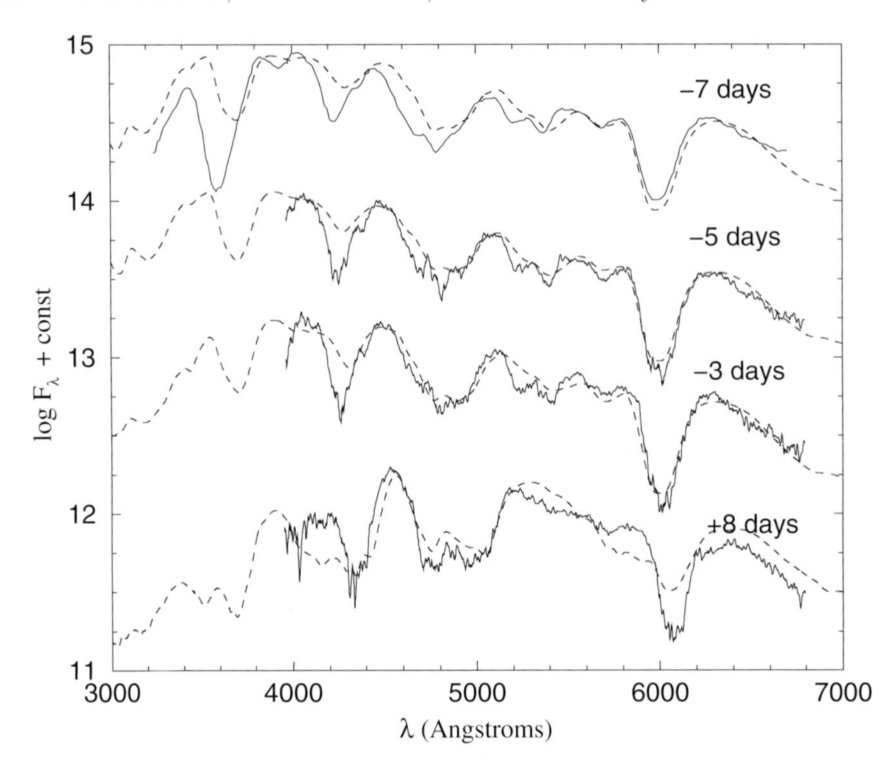

Fig. 6. *PHOENIX* synthetic spectra for the delayed detonation model CS15DD3 (*dashed lines*) and observed spectra of the high-velocity type Ia SN1984A (*solid lines*) [71].

optical depths in typical SNIb are not very high, so a moderate reduction would make them look like SNIc.

The number of SNIb for which good spectral coverage is available is still small. More events need to be observed to explore the extent of the spectral homogeneity and to determine whether there is a continuum of hydrogen line strengths. Also needed are detailed *NLTE* calculations for hydrodynamical SNIb models having radially stratified composition structures, to determine the hydrogen and helium masses and the distribution of the ^{56}Ni that excites the helium.

To our knowledge there have not been any synthetic spectrum calculations for the SNIb nebular phase since 1989 [43]. Most of the recent effort has been devoted to inferring the composition from emission line profiles and fluxes [89]. The oxygen mass, evidence for hydrogen, and evidence for asymmetry from the nebular spectra of SN1996N have been discussed by [118].

5 Type Ic Supernovae

SNIc are thought to result from core-collapse in massive stars that either have lost their helium layer or fail to nonthermally excite their helium. If the helium

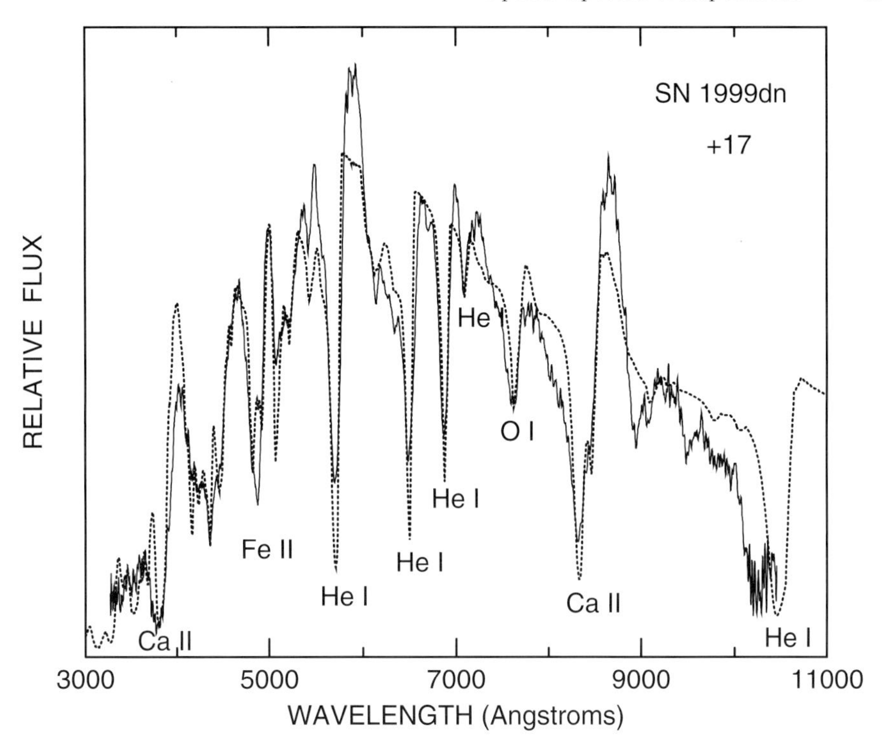

Fig. 7. The spectrum of the type Ib SN1999dn 17 days after maximum brightness (*solid line*) is compared with a *SYNOW* synthetic spectrum (*dotted line*) that has $v_{phot} = 6,000$ km s^{-1} [13].

layer is gone, the outer layers of the ejecta are expected to be mainly carbon and oxygen. SNIc are spectroscopically more diverse than SNIb [28,89]. The photospheric spectra are dominated by the low excitation SN features that appear in SNIb. The spectra of the best studied ordinary type Ic, SN1994I, can be reasonably well matched by *SYNOW* spectra [101] (Fig. 9). Discrepancies between the SN1994I spectra and *PHOENIX* spectra indicate that the appropriate hydrodynamical model for SN1994I has not yet been found [5]. The important issue of whether helium is detectable in SNIc spectra is not yet resolved, because an absorption near 10000 Å that sometimes is attributed to He I λ10,830 could be produced by one of several other ions (see Fig. 9 and [5,101]), and identifications of very weak optical lines [26] are difficult to confirm [89].

The peculiar, hyperenergetic type Ic SN1998bw that was associated with a γ-ray burst is discussed elsewhere in this volume. SN1997ef was a less extreme example of a hyperenergetic SNIc. The spectra of these events, although difficult to interpret because of the severe Doppler broadening and blending, can be matched fairly well by the *SYNOW* [12] and *ML MONTE CARLO* [94] codes (Fig. 10), and the resulting spectroscopic estimates of the kinetic energy (in the spherical approximation) are far in excess of the canonical 10^{51} erg [12,94,96,107].

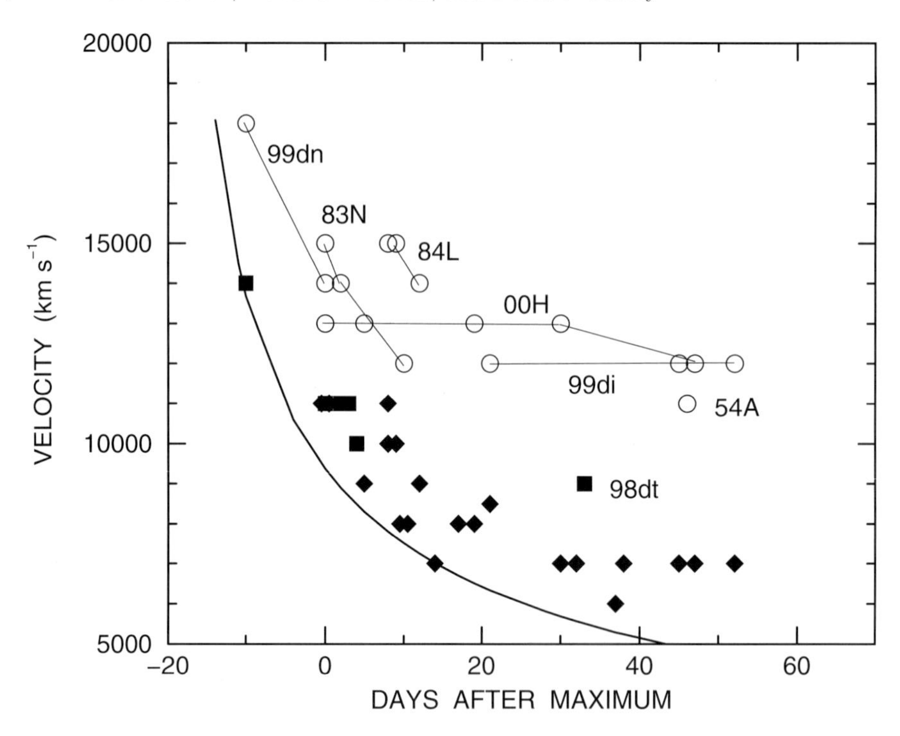

Fig. 8. For a sample of 12 SNIb, the minimum velocity of the He I lines (*filled squares* when undetached, *filled diamonds* when detached) and the minimum velocity of the hydrogen lines (*open circles*, always detached) are plotted against time after maximum brightness. The *heavy curve* is a power law fit to the velocity at the photosphere as determined by Fe II lines [13].

Detailed *NLTE* synthetic spectrum calculations have not yet been carried out for the photospheric spectra of hyperenergetic SNIc, and the only recent synthetic spectrum calculations for nebular phase SNIc have been for SN1998bw with the *RL NEBULAR* code [96].

The type Ic SN1999as, one of the most luminous reliably measured SNe to date, contained highly blueshifted ($\sim 11,000$ km s^{-1}) but narrow (~ 2000 km s^{-1}) absorption features [52] that cannot be produced by spherically symmetric SN ejecta. Whether these unusual features are produced by an ejecta clump in front of the photosphere or by circumstellar matter accelerated by the SN is not yet clear.

Detailed *NLTE* calculations for hydrodynamical models of typical and hyperenergetic SNIc are needed to constrain the helium mass and the ejected mass and kinetic energy.

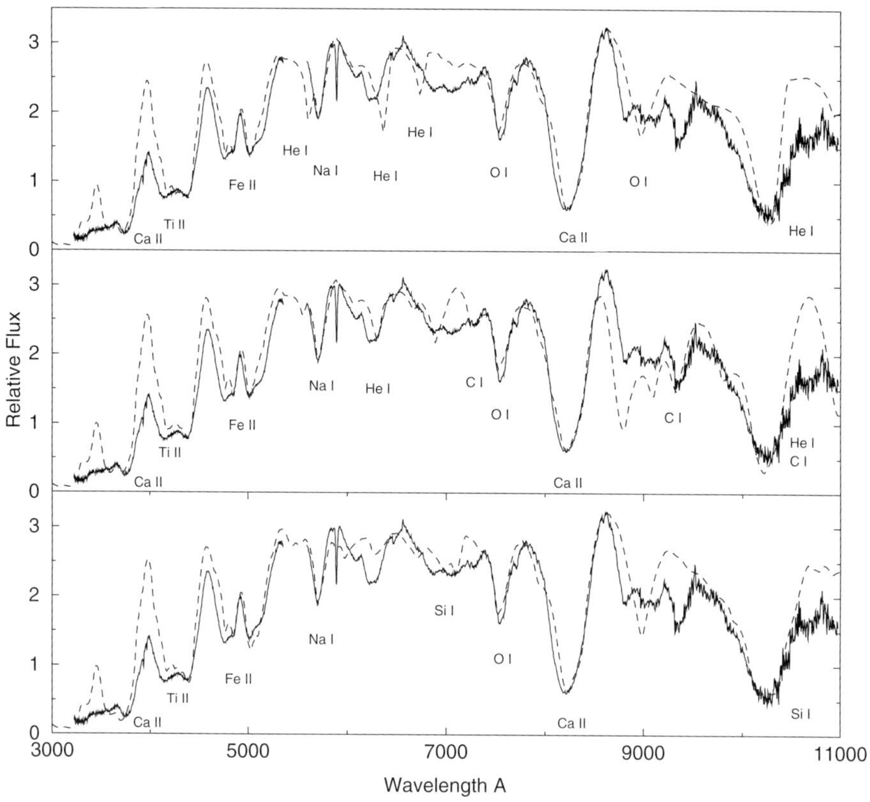

Fig. 9. The spectrum of the typical type Ic SN1994I 7 days after maximum brightness is compared with three *SYNOW* synthetic spectra that have $v_{phot} = 10,000$ km s^{-1}. To account for the observed infrared feature, the *synthetic spectra* include He I lines detached at 15,000 km s^{-1} (*top*); undetached C I lines and He I lines detached at 18,000 km s^{-1} (*middle*); and Si I lines detached at 14,000 km s^{-1} (*bottom*) [101].

6 Type II Supernovae

SNII are the most spectroscopically diverse of the SN types, partly because the optical effects of CSI range from negligible in some events to dominant in others. In the absence of CSI and the presence of a substantial hydrogen envelope (SNIIP and SNIIL), the spectrum evolves from an almost featureless continuum when the temperature is high to one that contains first hydrogen and helium lines and then also the usual low excitation lines as the temperature falls. The photospheric spectra of SN1987A were, of course, extensively studied [31,55,66,114]. Recently it has been shown that the observed profile of the He I λ10830 line in SN1987A is a sensitive probe of the degree of mixing of ^{56}Ni out to relatively high velocities, $\sim 3,000$ km s^{-1} [33] (see also [34] for similar work on the type IIP SN1995V). *PHOENIX* calculations [103] show that the hydrogen lines of SN1987A also are a sensitive probe of the nickel mixing and may require some ^{56}Ni as fast as

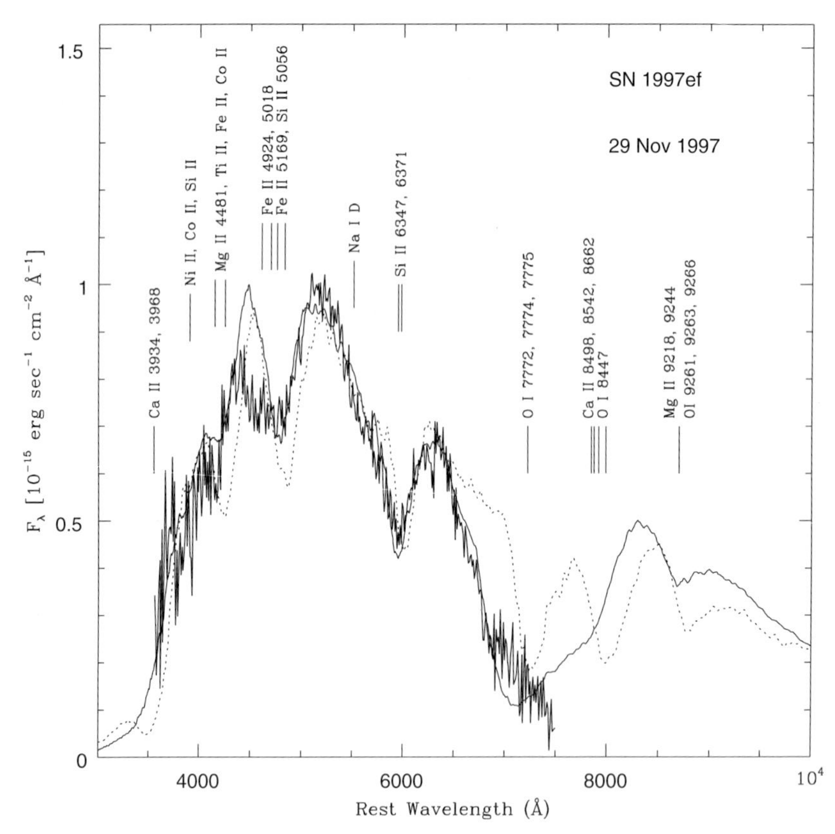

Fig. 10. The spectrum of the hyperenergetic type Ic SN1997ef about 11 days before maximum brightness (*thick line*) is compared with two synthetic spectra calculated with the *ML MONTE CARLO* code. The *thin solid line* and the *dotted line* are for models having kinetic energies of 8×10^{51} and 1.75×10^{52} erg, respectively [94].

5,000 km s^{-1}. As discussed in these three papers, quantitative conclusions are sensitive to the details of the macroscopic mixing of the hydrogen, helium, and nickel.

SN1999em is a recent SNIIP that has been extremely well observed and that is not complicated by CSI or large departures from spherical symmetry [76]. *Hubble Space Telescope (HST)* UV spectra [4] of this event are discussed elsewhere in this volume. Two extensive independent sets of optical spectra and photometry have been used to apply the expanding photosphere method for determining the distance [47,75]. *SYNOW* and *PHOENIX* synthetic spectra have been compared to the photospheric spectra [4]. Complicated observed P Cygni profiles of hydrogen and He I lines are well reproduced by the *PHOENIX* calculations when the density gradient is sufficiently shallow (Fig. 11). The total extinction can be inferred by comparing the observed and *synthetic spectra* at early times when the temperature is high [4].

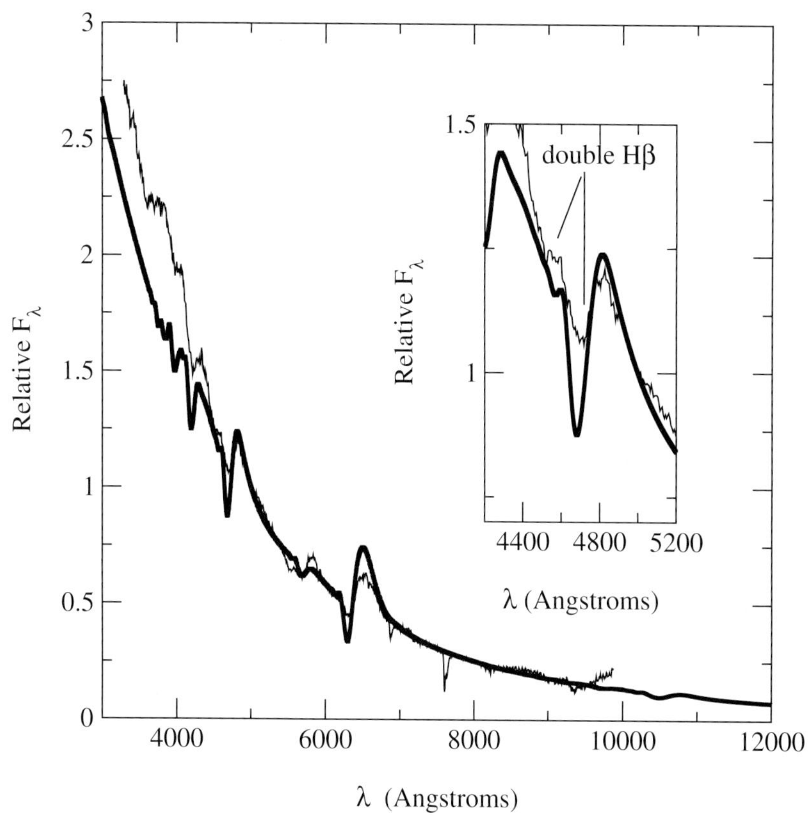

Fig. 11. The spectrum of the SNII SN1999em four days before maximum brightness (*thin line*) is compared to a *PHOENIX* synthetic spectrum (*thick line*) with $Z = Z_\odot/100$. All lines in the synthetic spectrum are produced by hydrogen and helium. The inset shows a shoulder in both the observed and calculated $H\beta$ profiles [4].

The unusually weak SNIIP SN1997D has attracted considerable attention in connection with the possibility of black hole formation and fallback. The *ML MONTE CARLO* code has been used to fit the spectrum with an extremely low velocity at the photosphere of only 970 km s^{-1} [124]. It has been pointed out that, because of the unusually low ionization, Rayleigh scattering may have significant effects on the photospheric spectrum [25]. The low expansion velocity offered a unique opportunity to identify lines in the nebular spectra [7]. Nebular *synthetic spectra* have been calculated [25] and it has been pointed out that the *Sobolev approximation* may break down in the early nebular phase because $H\alpha$ damping wings can have significant effects [20].

SNIIb contain conspicuous hydrogen lines at early times but not at late times, indicating that these events lost nearly all of their hydrogen before exploding. In the well studied SN1993J, He I lines became conspicuous after maximum brightness, as the hydrogen lines faded. The photospheric and nebular spectra were modeled in detail ([2,58,60,127] and references therein). More recent work on SN1993J has concentrated on inferring properties of the clumps and the nature of the CSI directly from line profiles and fluxes [88]. A *SYNOW* line-identification study of extensive observations of the type IIb SN1996cb [109] has been carried out [29].

SNIIn have narrow lines that form in low velocity CSM. Optical and infrared spectra of SN1998S have been intensively studied [1,23,32,70,77]. *HST* ultraviolet spectra [70] are discussed in elsewhere in this volume. *PHOENIX* synthetic spectra calculated for a simple model of the circumstellar shell accounted reasonably well for the early circumstellar-dominated spectra (Plate 1, bottom panel). The inferences from spectroscopy are that the progenitor of SN1998S underwent several distinct mass loss episodes during the decades and centuries before explosion, and that the SN ejecta or the CSM (or both) were significantly asymmetric. A detailed study of the nebular spectra of the type IIn SN1995N has revealed the existence of three distinct kinematic components (see Fig. 12 for nebular phase *synthetic spectra* calculated for the $\sim 5,000$ km s^{-1} intermediate velocity component that corresponds to unshocked SN ejecta) and supported the suggestion that SNIIn are produced by red supergiants that collapse during their superwind phases [42] (see also [19]).

Synthetic spectra have not yet been compared to spectra of hyperenergetic SNIIn such as SN1997cy [46,125] and SN1999E [37], both of which may have been associated with γ-ray bursts. These events may prove to be related to SN1988Z, which has been modeled in terms of CSI involving either dense clumps or a dense equatorial wind (see, e.g., [24,139] and the chapter by Sramek and Weiler in this volume).

7 Prospects

Observationally, in addition to a large increase in the sheer number of optical spectra, we can look forward to more and better infrared and nebular spectra, spectra obtained shortly after explosion, and spectropolarimetry.

Fast spherically symmetric (1D) photospheric phase codes such as *SYNOW* and *ML MONTE CARLO* will continue to be valuable for making rapid interpretations of new flux spectra and for certain comparative studies. Monte Carlo 1D *NLTE* codes may be constructed [86]. As computational resources permit, detailed 1D codes such as *PHOENIX* will be used to generate large grids of models and spectra to expedite the comparison with observed spectra. *PHOENIX* will be modified to take into account the radiative coupling between CSM and SN ejecta, and as forbidden lines are added to the line list *PHOENIX* also will become a powerful 1D nebular-phase code.

The major challenge for the future is to develop codes for calculating spectra for arbitrary geometries (3D) of both the SN ejecta and the CSM (and learning how to use such codes effectively). For flux spectra, the already existing *CLUMPYSYN* code will be followed in the near future by 3D Monte Carlo codes and, eventually, by 3D *PHOENIX*-level codes. Spectropolarization calculations

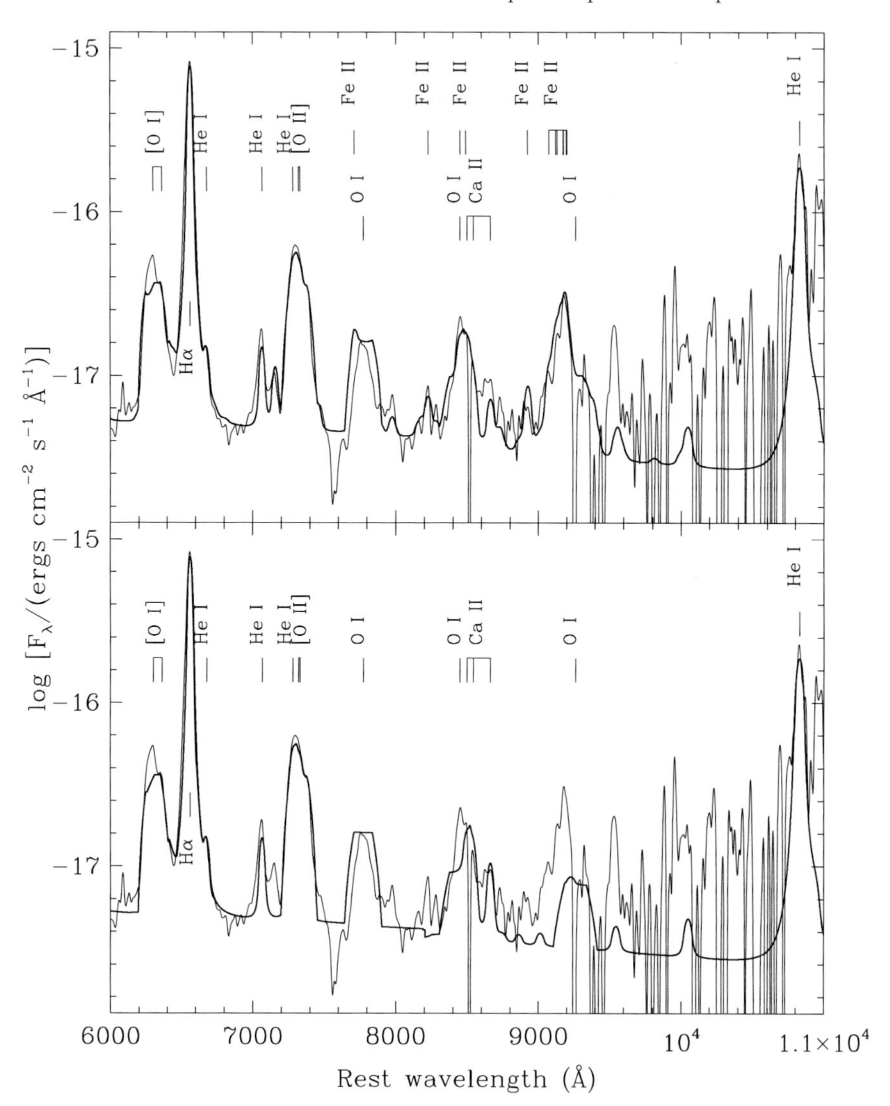

Fig. 12. The spectrum of the type IIn SN1995N nearly five years after maximum brightness is compared to *synthetic spectra* calculated with (*upper*) and without (*lower*) the influence of Lyα fluorescence on the Fe II lines taken into account [42].

for 3D configurations are certain to emerge as a major focus of SN spectroscopy. Monte Carlo codes for calculating 3D polarization spectra also will become available in the near future.

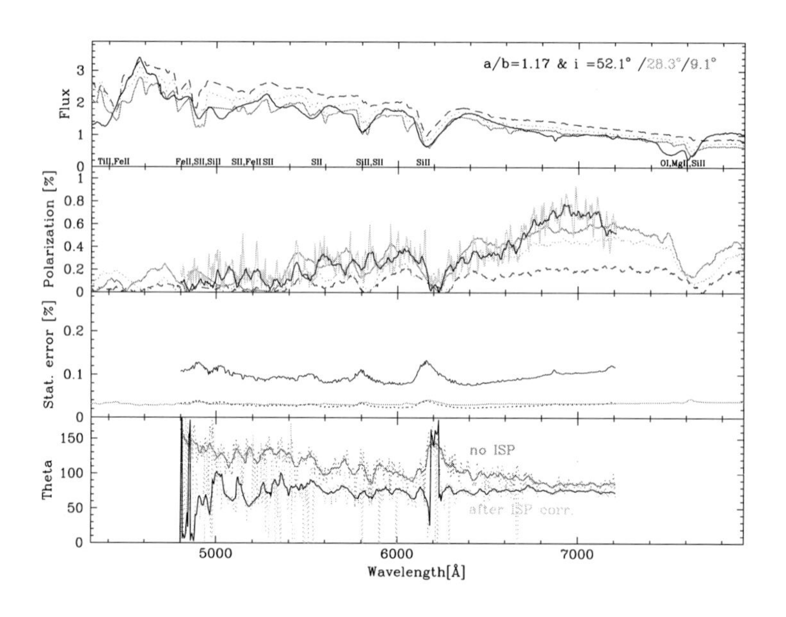

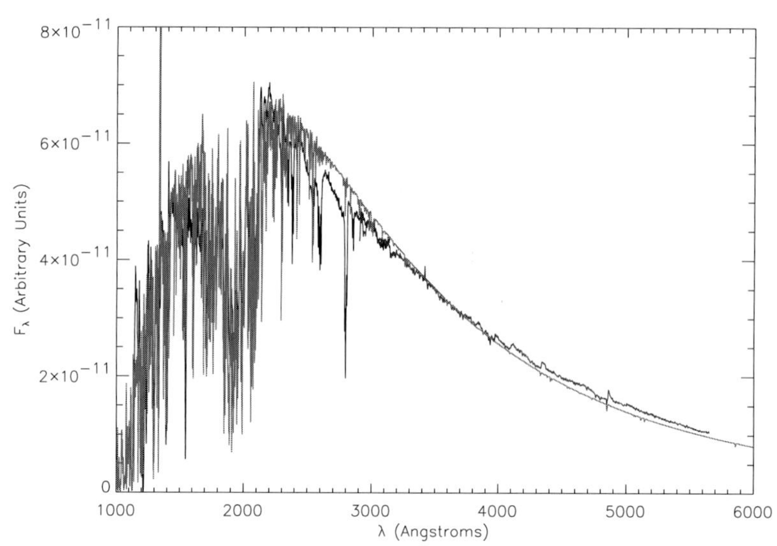

Plate 1 (top). Flux spectra (*top panel, top segment, black line*) and raw and smoothed polarization spectra (*top panel, second segment, light blue and black lines*) of the peculiar, weak type Ia SN1999by near maximum brightness, are compared with synthetic spectra computed for an oblate ellipsoidal model with axis ratio 1.17 and three inclinations of the symmetry axis from the line of sight (from [61], where further explanation of this figure can be found).

Plate 2 (bottom). The UV (*bottom panel, black line*) and optical (*bottom panel, blue line*) spectra of the type IIn SN1998S about four days before maximum brightness are compared with a *PHOENIX* synthetic spectrum (*bottom panel, red line*) calculated for a circumstellar shell having an r^{-2} density distribution and a constant expansion velocity of 1,000 km s^{-1} [70].

Acknowledgments

We are grateful to Doug Leonard for providing an unpublished figure, and to him and Lifan Wang for helpful comments. This work was supported by NASA, NSF, and the Physics Department of New Mexico Tech.

References

1. G.C. Anupama, T. Sivarani, G. Pandey: Astron. Astrophys. **367**, 506 (2001)
2. E. Baron et al.: Astrophys. J. **441**, 170 (1995)
3. E. Baron, P.H. Hauschildt, A. Mezzacappa: Mon. Not. R. Astron. Soc. **278**, 763 (1996)
4. E. Baron et al.: Astrophys. J. **545**, 444 (2000)
5. E. Baron, D. Branch, P.H. Hauschildt, A.V. Filippenko, R.P. Kirshner: Astrophys. J. **527**, 739 (1999)
6. E. Baron, P.H. Hauschildt, P. Nugent, D. Branch: Mon. Not. R. Astron. Soc. **283**, 297 (1996)
7. S. Benetti et al.: Mon. Not. R. Astron. Soc. **322**, 361 (2001)
8. S. Benetti, E. Cappellaro, I.J. Danziger, M. Turatto, F. Patat, M. Della Valle: Mon. Not. R. Astron. Soc. **294**, 448 (1998)
9. S. Benetti, M. Turatto, E. Cappellaro, I.J. Danziger, P.A. Mazzali: Mon. Not. R. Astron. Soc. **305**, 811 (1999)
10. E.J.C. Bowers et al.: Mon. Not. R. Astron. Soc. **290**, 663 (1997)
11. D. Branch: Pub. Astron. Soc. Pacific **113**, 169 (2001)
12. D. Branch: 'Direct Analysis of Spectra of Type Ic Supernovae'. In: *Supernovae and Gamma Ray Bursts, Space Telescope Science Institute Symposium on Supernovae and Gamma Ray Bursts at Baltimore, MD, USA, May 3–6, 1999* ed. by M. Livio, N. Panagia, K. Sahu (Cambridge University, Cambridge 2001) p. 96
13. D. Branch et al.: Astrophys. J. **566**, 1005 (2002)
14. D. Branch, D.J. Jeffery, M. Blaylock, K. Hatano: Pub. Astron. Soc. Pacific **112**, 217 (2000)
15. J.P. Cassinelli, D.A. Hummer: Mon. Not. R. Astron. Soc. **154**, 9 (1971)
16. S. Chandrasekhar: *R*adiative Transfer (Dover Publications, New York 1960)
17. R.A. Chevalier, C. Fransson: Astrophys. J. **420**, 268 (1994)
18. N.N. Chugai: Sov. Astron. Lett. **18(3)**, 168 (1992)
19. N.N. Chugai: Astron. Rep. **41**, 672 (1997)
20. N.N. Chugai: Astrophys. J. **531**, 411 (2000)
21. N.N. Chugai: Astrophys. J. **326**, 1448 (2001)
22. N.N. Chugai, A.A. Andronova, V.P. Utrobin: Astron. Zhurnal **22**, 672 (1996)
23. N.N. Chugai, S.I. Blinnikov, A. Fassia, P. Lundqvist, W.P.S. Meikle, E.I. Sorokina: Mon. Not. R. Astron. Soc. **330**, 473 (2002)
24. N.N. Chugai, I.J. Danziger: Mon. Not. R. Astron. Soc. **268**, 173 (1994)
25. N.N. Chugai, V.P. Utrobin: Astron. Astrophys. **354**, 557 (2000)
26. A. Clocchiatti et al: Astrophys. J. **483**, 675 (1997)
27. A. Clocchiatti et al: Astrophys. J. **529**, 661 (2000)
28. A. Clocchiatti et al: Astrophys. J. **553**, 886 (2001)
29. J.S. Deng, Y.L. Qiu, J.Y. Hu: Astrophys. J. submitted (2003)
30. J.S. Deng, Y.L. Qiu, J.Y. Hu, K. Hatano, D. Branch: Astrophys. J. **540**, 452 (2000)

31. R. Eastman, R.P. Kirshner: Astrophys. J. **347**, 771 (1989)
32. A. Fassia et al.: Mon. Not. R. Astron. Soc. **325**, 907 (2001)
33. A. Fassia, W.P.S. Meikle: Mon. Not. R. Astron. Soc. **302**, 314 (1999)
34. A. Fassia, W.P.S. Meikle, T.R. Geballe, N.A. Walton, D.L. Pollacco, R.G.M. Rutten, C. Tinney: Mon. Not. R. Astron. Soc. **299**, 150 (1998)
35. R. Fesen et al.: Astron. J. **117**, 725 (1999)
36. A.V. Filippenko: Ann. Rev. Astron. Astrophys. **35**, 309 (1997)
37. A.V. Filippenko: 'Optical Observations of Type II Supernovae'. In: *Cosmic Explosions, 10th Annual October Astrophysics Conference at College Park, MD, USA, 11–13 October 1999*, ed. by S. Holt, W.W. Zhang (AIP, New York 2000) p. 40
38. A.K. Fisher: Direct Analysis of Type Ia Supernova Spectra. PhD Thesis, University of Oklahoma, Norman (2000)
39. A.K. Fisher, D. Branch, K. Hatano, E. Baron: Mon. Not. R. Astron. Soc. **304**, 67 (1999)
40. C. Fransson: Astron. Astrophys. **132**, 115 (1984)
41. C. Fransson: 'The Late Emission from Supernovae'. In: *Supernovae, NATO Advanced Study Institute on Supernovae at Les Houches, France, 31 July–01 September 1990*, ed. by S.A. Bludman, R. Mochkovitch, J. Zinn-Justin (Elsevier, Amsterdam 1994) p. 677
42. C. Fransson et al.: Astrophys. J. **572**, 350 (2002)
43. C. Fransson, R.A. Chevalier: Astrophys. J. **343**, 323 (1989)
44. P.M. Garnavich et al.: Astrophys. J. submitted (2003)
45. C.L. Gerardy, R.A. Fesen, P. Höflich, J.C. Wheeler: Astron. J. **119**, 2968 (2000)
46. L.M. Germany, D.J. Riess, E.M. Sadler, B.P. Schmidt, C.W. Stubbs: Astrophys. J. **533**, 320, (2000)
47. M. Hamuy et al.: Astrophys. J. **558**, 615 (2001)
48. R.P. Harkness et al.: Astrophys. J. **317**, 355 (1987)
49. K. Hatano, D. Branch, A. Fisher, E. Baron, A. V. Filippenko: Astrophys. J. **525**, 881 (1999)
50. K. Hatano, D. Branch, A. Fisher, J. Millard, E. Baron: Astrophys. J. Suppl. **121**, 233 (1999)
51. K. Hatano, D. Branch, E.J. Lentz, E. Baron, A.V. Filippenko, P.M. Garnavich: Astrophys. J. **543**, L49 (2000)
52. K. Hatano, D. Branch, K. Nomoto, J.S. Deng, K. Maeda, P. Nugent, G. Aldering: Bull. Am. Astron. Soc. **198**, 3902 (2001)
53. P.H. Hauschildt, E. Baron: J.: J. Comp. Appl. Math. **109**, 41 (1999)
54. M. Hernandez et al.: Mon. Not. R. Astron. Soc. **319**, 223 (2000)
55. P. Höflich : Proc. Astron. Soc. Australia **7**, 434 (1988)
56. P. Höflich: Astron. Astrophys. **246**, 481 (1991)
57. P. Höflich, A. Khokhlov, L. Wang: 'Aspherical Supernova Explosions: Hydrodynamics, Radiation Transport, and Observational Consequences'. In: *Relativistic Astrophysics, 20th Texas Symposium on Relativistic Astrophysics at Austin, TX, USA, 10–15 December 2000*, ed. by J.C. Wheeler, H. Martel (New York, AIP 2001) p. 459
58. P. Höflich, J.C. Wheeler, D.C. Hines, S.R. Tramell: Astrophys. J. **459**, 307 (1996)
59. P. Höflich, J.C. Wheeler, F.K. Thielemann: Astrophys. J. **495**, 617 (1998)
60. J.C. Houck, C. Fransson: Astrophys. J. **456**, 811 (1996)
61. D.A. Howell, P. Höflich, L. Wang, J.C. Wheeler: Astrophys. J. **556**, 302 (2001)
62. D.J. Jeffery: Astrophys. J. Suppl. **71**, 951 (1989)

63. D.J. Jeffery: Astrophys. J. Suppl. **77**, 405 (1991)
64. D.J. Jeffery: Astrophys. J. **415**, 734 (1993)
65. D.J. Jeffery: in preparation (2003)
66. D.J. Jeffery, D. Branch: 'Analysis of Supernova Spectra'. In: *Supernovae, 6th Jerusalem Winter School for Theoretical Physics at Jerusalem, Israel, 28 December 1988–05 January 1989*, ed. by J.C. Wheeler, T. Piran, S. Weinberg (World Scientific, Singapore 1990) p. 149
67. S. Jha et al.: Astrophys. J. Suppl. **125**, 73 (1999)
68. A.H. Karp, G. Lasher, K.L. Chan, E.E. Salpeter: Astrophys. J. **214**, 161 (1977)
69. D. Kasen, D. Branch, E. Baron, D.J. Jeffery: Astrophys. J. **565**, 380 (2002)
70. E.J. Lentz et al.: Astrophys. J. **547**, 406 (2001)
71. E.J. Lentz, E. Baron, D. Branch, P. Hauschildt: Astrophys. J. **547**, 402 (2001)
72. E.J. Lentz, E. Baron, D. Branch, P. Hauschildt: Astrophys. J. **557**, 266 (2001)
73. E.J. Lentz, E. Baron, D. Branch, P. Hauschildt, P.E. Nugent: Astrophys. J. **530**, 966 (2000)
74. D.C. Leonard et al.: Pub. Astron. Soc. Pacific **114**, 35 (2002)
75. D.C. Leonard, A.V. Filippenko: Pub. Astron. Soc. Pacific **113**, 920 (2001)
76. D.C. Leonard, A.V. Filippenko, D.R. Ardila, M.S. Brotherton: Astrophys. J. **553**, 861 (2001)
77. D.C. Leonard, A.V. Filippenko, A.J. Barth, T. Matheson: Astrophys. J. **536**, 239 (2000)
78. D.C. Leonard, A.V. Filippenko, T. Matheson: 'Probing the Geometry of Supernovae with Spectropolarimetry'. In: *Cosmic Explosions, 10th Annual October Astrophysics Conference at College Park, MD, USA, 11-13 October 1999*, ed. by S. Holt, W.W. Zhang (AIP, New York 2000) p. 165
79. H. Li, R. McCray, R.A. Sunyaev: Astrophys. J. **419**, 824 (1993)
80. W.D. Li et al.: Astron. J. **117**, 2709 (1999)
81. W.D. Li et al.: Pub. Astron. Soc. Pacific **113**, 1178 (2001)
82. W.D. Li, A.V. Filippenko, R.R. Treffers, A.G. Riess, J. Hu, Y. Qiu: Astrophys. J. **546**, 734 (2000)
83. W. Liu, D.J. Jeffery, D.R.Schultz: Astrophys. J. **494**, 812 (1998)
84. L.B. Lucy: Astrophys. J. **383**, 308 (1991)
85. L.B. Lucy: Astron. Astrophys. **345**, 211 (1999)
86. L.B. Lucy: Mon. Not. R. Astron. Soc. **326**, 95 (2001)
87. T. Matheson, A.V. Filippenko, R. Chornock, D.C. Leonard, W. Li: Astron. J. **119**, 2303 (2000)
88. T. Matheson, A.V. Filippenko, Ho, L.C., Barth, A.J., D.C. Leonard: Astron. J. **120**, 1499 (2000)
89. T. Matheson, A.V. Filippenko, W. Li, D.C. Leonard, J.C. Shields: Astron. J. **121**, 1648 (2001)
90. P.A. Mazzali: Astrophys. J. **499**, L49 (1998)
91. P.A. Mazzali: Astron. Astrophys. **363**, 705 (2000)
92. P.A. Mazzali: Mon. Not. R. Astron. Soc. **321**, 341 (2001)
93. P.A. Mazzali, N. Chugai, M. Turatto, L.B. Lucy, I.J. Danziger, E. Cappellaro, M. Della Valle, S. Benetti: Mon. Not. R. Astron. Soc. **284**, 151 (1997)
94. P.A. Mazzali, K. Iwamoto, K. Nomoto: Astrophys. J. **545**, 407 (2000)
95. P.A. Mazzali, L.B. Lucy: Mon. Not. R. Astron. Soc. **295**, 428 (1998)
96. P.A. Mazzali, K. Nomoto, F. Patat, K. Maeda: Astrophys. J. **559,**, 1047 (2001)
97. M.L. McCall: Mon. Not. R. Astron. Soc. **210**, 829 (1984)
98. R. McCray: Ann. Rev. Astron. Astrophys. **31**, 175 (1993)

99. W.P.S. Meikle et al.: Mon. Not. R. Astron. Soc. **281**, 263 (1996))
100. D. Mihalas: *S*tellar Atmospheres (W. H. Freeman, San Francisco 1978)
101. J. Millard et al.: Astrophys. J. **527**, 746 (1999)
102. P.A. Milne, L.-S. The, M.D. Leising: Astrophys. J. **559**, 1019 (2001)
103. R. Mitchell, E. Baron, D. Branch, P. Lundqvist, S. Blinnikov, P.H. Hauschildt, C.S.J. Pun: Astrophys. J. **556**, 979 (2001)
104. M. Modjaz, W. Li, A.V. Filippenko, J.Y. King, D.C. Leonard, T. Matheson, R.R. Treffers: Pub. Astron. Soc. Pacific **113**, 308 (2001)
105. K. Nomoto, F.-K. Thielemann, Y. Yokoi: Astrophys. J. **286**, 644 (1984)
106. P. Nugent, M.M. Phillips, E. Baron, D. Branch, P. Hauschildt: Astrophys. J. **455**, L147 (1995)
107. F. Patat et al.: Astrophys. J. **555**, 900 (2001)
108. P. Pinto, R.G. Eastman:: Astrophys. J. **530**, 757 (2000)
109. Y. Qiu, W. Li, Q. Qiao, J. Hu: Astron. J. **117**, 736 (1999)
110. P. Ruiz-Lapuente: Astrophys. J. **439**, 60 (1995)
111. P. Ruiz-Lapuente, L.B. Lucy: Astrophys. J. **400**, 127 (1992)
112. M.E. Salvo, E. Cappellaro, P.A. Mazzali, S. Benetti, I. J. Danziger, M. Turatto: Mon. Not. R. Astron. Soc. **321**, 254 (2001)
113. I. Salamanca, R. Cid-Fernandes, G. Tenorio-Tagle, E. Telles, R.J. Terlevich, C. Munoz-Tunon: Mon. Not. R. Astron. Soc. **300**, 17 (1998)
114. W. Schmutz, D.C. Abbott, R.S. Russell, W.-R. Hamann, U. Wessolowski: Astrophys. J. **355**, 255 (1990)
115. K. Serkowski: In *Interstellar Dust and Related Topics, IAU Symposium 52 at Albany, NY, 29 May–02 June 1972* ed. by J.M. Greenberg, H.C. van de Hulst (Reidel, Dordrecht 1973) p. 145
116. P.R. Shapiro, P.G. Sutherland: Astrophys. J. **263**, 902 (1982)
117. J. Sollerman, R.J. Cumming, P. Lundqvist: Astron. Astrophys. **493**, 933 (1998)
118. J. Sollerman, B. Leibundgut, J. Spyromilio: Astron. Astrophys. **337**, 207 (1998)
119. J. Spyromilio, B. Leibundgut, R. Gilmozzi: Astron. Astrophys. **376**, 188 (2001)
120. D.A. Swartz, A.V. Filippenko, K. Nomoto, J.C. Wheeler: Astrophys. J. **411**, 313 (1993)
121. D.A. Swartz, P.G. Sutherland, R.P. Harkness J.C. Wheeler: Astrophys. J. **446**, 766 (1995)
122. F.-K. Thielemann, K. Nomoto, K. Yokoi: Astron. Astrophys. **158**, 17 (1986)
123. R.C. Thomas, D. Kasen, D. Branch, E. Baron: Astrophys. J. **567**, 1037 (2002)
124. M. Turatto et al.: Astrophys. J. **498**, L129 (1998)
125. M. Turatto et al.: Astrophys. J. **534**, L57 (2000)
126. M. Turatto, A. Piemonta, S. Benetti, E. Cappellaro, P.A. Mazzali, I.J. Danziger, F. Patat: Astron. J. **116**, 2431 (1998)
127. V.P. Utrobin: Astron. Astrophys. **306**, 231 (1996)
128. S.D. Van Dyk, R.A. Sramek, K.W. Weiler, N. Panagia: Astrophys. J. Lett. **419**, L69 (1993)
129. J. Vinkó, L.L. Kiss, B. Csák, G. Fürész, R. Sxabó, J.R. Thomson, S.W. Mochnacki: Astron. J. **121**, 3127 (2001)
130. L. Wang, J.C. Wheeler: Astrophys. J. **462**, L27 (1996)
131. L. Wang, J.C. Wheeler, P. Höflich: Astrophys. J. **476**, L27 (1997)
132. L. Wang, D.A. Howell, P. Höflich, J.C. Wheeler: Astrophys. J. **550**, 1030 (2001)
133. L. Wang et al.: Astrophys. J. **579**, 671 (2002)
134. K.W. Weiler, N. Panagia, R.A. Sramek: Astrophys. J. **364**, 611 (1990)
135. K.W. Weiler, N. Panagia, M.J. Montes, R.A. Sramek: Ann. Rev. Astron. Astrophys. **40**, 387 (2002)

136. J.C. Wheeler, S. Benetti: 'Supernovae'. In: *Allen's Astrophysical Quantities, 4th edition* ed. by A.N. Cox (Springer, New York 2000) p. 451
137. J.C. Wheeler, P. Höflich, R.P. Harkness, J. Spyromilio: Astrophys. J. **496**, 908 (1998)
138. J.C. Wheeler, D.A. Swartz, R.P. Harkness: Phys. Rep. **227**, 113 (1993)
139. C.L. Williams, N. Panagia, C.K. Lacey, K.W. Weiler, R.A. Sramek, S.D. Van Dyk: Astrophys. J. **(581)**, 396 (2002)
140. S.E. Woosley, R.G. Eastman: 'Type Ib and Ic Supernovae: Models and Spectra'. In: *Thermonuclear Supernovae, NATO Advanced Study Institute on Thermonuclear Supernovae, at Aiguablava, Spain, 20–30 June 1995*, ed. by P. Ruiz-Lapuente, R. Canal, J. Isern (Kluwer, Dordrecht 1997) p. 821

Optical Light Curves of Supernovae

Bruno Leibundgut[1] and Nicholas B. Suntzeff[2]

[1] European Southern Observatory, Karl-Schwarzschild-Strasse 2, D-85748 Garching, Germany
[2] Cerro Tololo Inter-American Observatory, Casilla 603, La Serena, Chile

Abstract. Photometry is the most easily acquired information about supernovae. The light curves constructed from regular imaging provide signatures not only for the energy input, the radiation escape, the local environment and the progenitor stars, but also for the intervening dust. They are the main tool for the use of supernovae as distance indicators through the determination of the luminosity.

The light curve of SN1987A still is the richest and longest observed example for a core-collapse supernova. Despite the peculiar nature of this object, as explosion of a blue supergiant, it displayed all the characteristics of type II supernovae. The light curves of type Ib/c supernovae are more homogeneous, but still display the signatures of explosions in massive stars, among them early interaction with their circumstellar material.

Wrinkles in the near-uniform appearance of thermonuclear (type Ia) supernovae have emerged during the past decade. Subtle differences have been observed especially at near infrared (NIR) wavelengths. Interestingly, the light curve shapes appear to correlate with a variety of other characteristics of these supernovae.

The construction of bolometric light curves provides the most direct link to theoretical predictions and can yield sorely needed constraints for the models. First steps in this direction have been already made.

1 Physics of Supernova Light Curves

The temporal evolution of the energy release by supernovae (SNe) is one of the major sources of information about the nature of these events. The brightness information is relatively easy to obtain and, hence, light curves have been one of the main stays of supernova research. It is not only the light curves themselves, but also the absolute luminosity and the color evolution that have provided major insights into the supernova phenomenon. Through light curves it has been possible to distinguish between progenitor models, infer some aspects of the progenitor evolution, measure the power sources, detail the explosion models, and probe the local environment of the supernova explosions.

The observational data have been substantially increased over the last decade (for a status before 1990 see [51]). In particular, there are now large databases with light curve data for type Ia supernovae (SNIa). The data on type II supernovae has been extensively expanded as well, but there is still a clear lack of light curves for peculiar objects. Over the last decade the supernova family has further acquired new members and subclassifications (see the chapter by Turatto in this volume).

The energetic display of a supernova can have several different input sources. The most important power comes in almost all cases from the radioactive decay of material newly synthesized in the explosion. The major contributor is ^{56}Ni, the main product of burning to nuclear statistical equilibrium at the temperatures and densities encountered in supernovae. This nucleus is unstable and decays with a half-life of 6.1 days due to electron capture to ^{56}Co emitting γ-photons with energies of 750 keV, 812 keV, and 158 keV. The cobalt isotope is also unstable and decays with a half-life of 77.1 days through electron capture (81%) and β-decay (19%) to ^{56}Fe. In this process γ-photons with energies of 1.238 MeV and 847 keV are emitted. The kinetic energy of the electrons is about 600 keV (for more details on the radioactive decays see [3,26,78]). The γ-rays are down-scattered or thermalized in the ejecta until they emerge as optical or NIR photons [2,48,85].

The light curves depend on the size and mass of the progenitor star and the strength of the explosion. Additional energy input, which results in modulations of the emerging radiation, comes from shock cooling, recombination of the ionized ejecta, collision of the shock with circumstellar medium (CSM) and possible accretion onto a compact remnant. Light curves are further shaped by the time-variable escape fraction of γ-rays, dust formation and absorption in the interstellar medium. In some cases, forward scattered light can change the light curves, e.g., through light echos and fluorescence of nearby gas ionized by the X-ray/UV shock breakout of the explosion.

With this panoply of different energy contributors and modulators, light curve displays are very rich indeed. Despite the plethora of possibilities, light curves of different supernova types are rather distinct, although not sufficiently so for a solid classification. They are, however, important tools to learn about the physics of supernovae.

Light curves are discussed in [37,51,56,80,81]. Reviews concentrating more specifically on SNIa light curves can be found in [57,58,73,107], while the light curves of core-collapse supernovae have mostly been summarized in relation to SN1987A [4,72]. Additional well-sampled data sets are available for SN1993J, SN1998S and SN1999em (see references below).

The following sections give a brief overview of observational data sources (Sect. 2), describe the light curves of the main supernova types (core-collapse supernovae in Sect. 3 and thermonuclear supernovae in Sect. 4) and the physics behind the light curve shapes. We will discuss bolometric light curves in Sect. 5 before we summarize in Sect. 6.

2 Observations

With modern area detectors, light curves of supernovae have become much easier to assemble. While early light curves have been compiled from photographic plates [60,82,84] observations are now recorded with CCDs. The increased sensitivity has allowed astronomers to successfully move to smaller-size telescopes and to improve the temporal sampling. In parallel, the move to more robotic telescopes has increased the number of supernova discoveries tremendously (see, e.g., the chapter by Cappellaro in this volume). There are many observational programs for supernovae currently in progress (for a description of some see [58]), which contribute photometry for many supernovae. Most successful are efforts with semi-automated telescopes. The robotic telescopes of the *Lick Observatory and Tenagra Observatory Supernova Search (LOTOSS)* have discovered and followed many supernovae in the last few years [65,90]. At the *Cerro Tololo Inter-American Observatory (CTIO)* the *Yale-AURA-Lisbon-Ohio (YALO)* telescope has regularly provided light curves for nearby supernovae (see, e.g., [54,112]). light curves have been further contributed by the Padova group (see, e.g., [9,97,116,117] and the results of the *Harvard-Smithsonian Center for Astrophysics (CfA)* team [50,92]). There are additional contributions on SN1993J [8,11,27,64,74,90,120] and more recently SN1998S [32] and SN1999em [45,63]. Many amateur groups are also collecting supernova light curves. These data are mostly maintained on Web sites (see [58] for a small collection of sites).

Infrared observations are still rare. A complete compilation of the available photometry for SNIa before 2000 has been provided by Meikle [73]. More data are being added (see, e.g., [46,54]). For core-collapse supernovae complete light curves are available only for SN1987A [108,110], SN1993J [120], SN1998S [32], and SN1999em [45]. There are several programs starting up that will concentrate on NIR light curves with robotic telescopes.

3 Core Collapse Supernovae

Once the shock, which results from the reversal of the core collapse, breaks out at the surface of the progenitor star the fireworks begin. The rapid evolution of the core burning just before the collapse is hidden from the surface due to the long time scales in the atmosphere. The brightness of the shock break out is mostly determined by the temperature in the shock and the size of the progenitor star [20,31,53]. This early peak lasts typically from a few hours to a couple of days and has been observed only for SN1987A [4], SN1993J [90], and SN1999em [45,63]. After a rapid, initial cooling the supernova enters a phase when its temperature and luminosity remain fairly constant [29,45,63]. For supernovae with large progenitors the resulting light curve shows a , while the evolution of supernovae from smaller stars first exhibits a decline before the supernova brightens again to reach the plateau. Examples for the former are SN1990E [98] and SN1999em [45,63], while the type II SN1987A (see, e.g., Fig. 1) together with the SNIb/c belong to the latter (see, e.g., [56]). The plateau originates from a balance between the receding photosphere in the expanding ejecta [29]. During this phase the supernova is powered by the recombination of hydrogen previously ionized in the supernova shock. The length of the plateau phase is determined by the depth of the envelope (i.e., the envelope mass and the explosion energy), which is reflected in the expansion velocity of the ejecta [19,89]. For some objects this plateau phase is conspicuously absent [80]. Most prominent among these are SN1979C [15] and SN1980K [7].

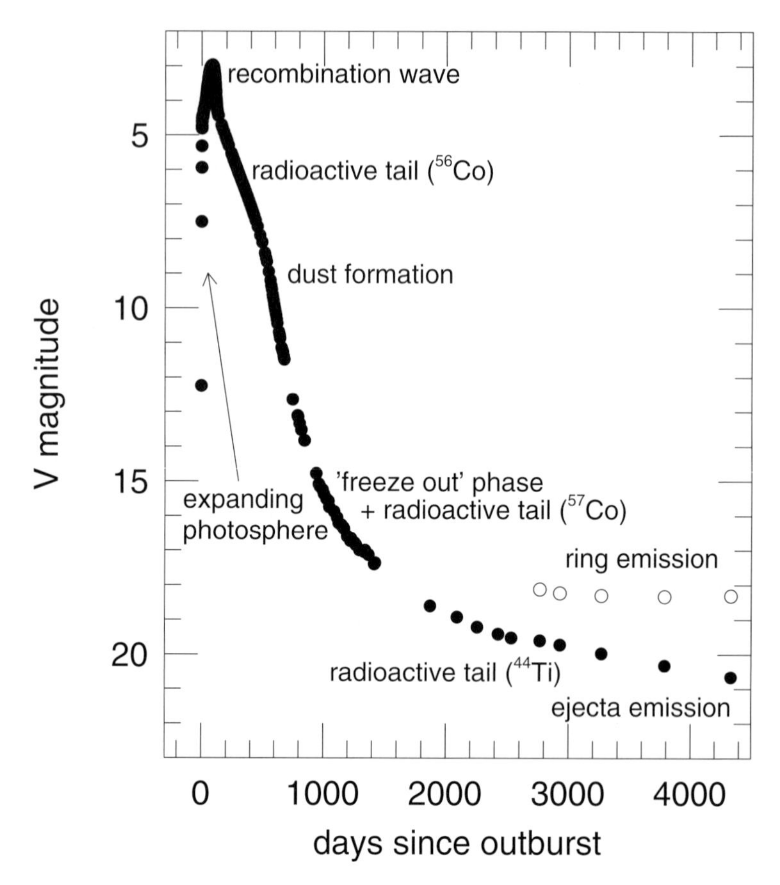

Fig. 1. V light curve of SN1987A. The various phases are labeled.

Once the photosphere has receded deep enough, additional heating from the radioactive decay of ^{56}Ni and ^{56}Co extends the plateau for a brief time [123]. Afterwards the light curve is powered solely by the radioactive decay in the remaining nebula. The γ-rays are captured in the ejecta and converted into optical photons, which can escape freely. At this moment the supernova light curve drops onto the "radioactive tail." This happens typically after about 100 days (see Fig. 1).

For a complete trapping of the γ-rays the luminosity of the late decline gives an indication of the amount of ^{56}Ni and ^{56}Co decays powering the light curve [12,99]. This can be checked with the decline rate of the bolometric light curve, which should reflect the ^{56}Co decay time.

From such measurements a rather large range of nickel masses has been derived [99,103,116]. These phases are especially interesting as they may show signatures of significant fallback of the inner explosion material onto the forming compact object, neutron star or black hole, in the explosion [6,10,122].

Especially for SNIb/c the decline rates are steeper [21,90,104,105] at these times, which is an indication that some of the γ-rays escape from the ejecta without any energy deposition.

Very few objects have been followed beyond about 200 days and the situation has not changed very much since about 10 years ago [80,113]. The photometry of such objects becomes very difficult as they fade into the glare of the underlying galaxy. The remarkable exception is, of course, SN1987A on which all very late phase information is based. This supernova suffered from dust forming within the ejecta, which resulted in an increase of the decline rate in the optical as light was shifted to the infrared [111]. This occured after about 450 days and could also be observed as a shift of emission lines towards the blue as the redder parts of the lines were absorbed. After about 800 days the light curves started to flatten again [111] due to energy release of ionized matter [40]. This so-called "freeze-out" stems from tenuous material which was ionized during the original explosion but recombines on time scales longer than the expansion time. At later times, the flattening is caused by the energy input from long-lived ^{57}Co (half-life of 270 days) and ^{44}Ti which has a half-life of about 60 years [26].

As is apparent from Fig. 1, the very late times are, in fact, dominated by the emission from the circumstellar inner ring, which was ionized by the shock breakout [39]. Around 1500 days after the explosion the ring emission is stronger than that from the supernova ejecta itself.

The closeness of SN1987A has permitted us to resolve the ring emission and also light echos from interstellar dust from the supernova ejecta [109,124]. For any other supernova these contributions can not be separated and would influence the light curve shape.

Some supernovae do not follow the path described above. They often have a much slower evolution and also display narrow lines7 in their spectra (see, e.g., [37,41]). These objects remain bright for a long time and must be powered by a different energy source. The mostly likely explanation for these objects is interaction of the shock with a dense CSM. In this process, kinetic energy is converted to light and hence an additional energy source can be tapped for the light curve. Since many of these objects are dominated by line emission and the line strengths critically depend on density and composition of both the supernova ejecta and the CSM, filter light curves can vary significantly from object to object. A few classical cases are known so far: SN1978K [96], SN1986J [61,95], SN1988Z [114] and SN1995N [41].

Of similar nature are probably also objects which can be observed for decades after their explosions although at a much lower luminosity. The prime examples are SN1979C [35,119], SN1980K [34,36,61], SN1970G [33], and SN1957D [68,69]. A summary of the observational characteristics of these objects can be found in Leibundgut [55] and the detailed physics of the SN-CSM interaction is given in the chapter by Chevalier and Fransson in this volume. It is noteworthy that all these objects also emit radio waves (see the chapter by Sramek and Weiler in this volume).

On no occasion has it been possible, so far, to observe the emergence of a pulsar within a supernova. Even for SN1987A the data do not require any input from a pulsar. Some objects which have been observed for decades, like SN1957D, SN1970G, SN1979C, SN1980K, still do not require the energy input corresponding to the expectations from a pulsar powered plerion (see, e.g., [55]). It is more likely that these objects are powered by interaction of the shock with the CSM. In all of these cases the spectrum clearly shows that we are still observing supernova light and not an underlying stellar association. This is further supported by changes in the light curves as observed for SN1957D [69] and SN1980K [36].

4 Thermonuclear Supernovae

The observational situation for type Ia SNe is quite different. There have been several focused searches for SNIa, which have produced large sets of well-sampled light curves (see, e.g., [42,44,46,50,66,73,88,91,92,97,107,112]. These data samples have produced a detailed view of SNIa. A recent summary of SNIa light curves can be found in Leibundgut [58] and Meikle [73].

The incineration of a white dwarf, which is the most favored model today (see, e.g., [47,123]), does not predict an observable shock breaking out at the surface. The rise in brightness is due to the increase in size of the ejecta and lasts for almost three weeks [1,23,42,93] with the color and temperature rather constant. The earliest observed supernovae are SN1990N and SN1998bu, which were observed about 17 days before maximum [58,93]. The light curves are shaped by the progressing diffusion of photons out of the ejecta (see, e.g., [85]).

The maximum is reached first at NIR wavelengths [24,73] and is followed a few days later in the optical. While the blue light curves display a monotonic decrease of the brightness for the first month after maximum, the NIR bands I, J, H, and K display a prominent second maximum after about 20 days for most SNIa [30,73], which is often also observed as "shoulders" in the V and R filter light curves. This second maximum is conspicuously missing for objects of the type Ia faint subclass with SN1991bg and SN1999de as the most prominent examples [38,62,77,115].

SNIa show a strong color evolution towards the red through the maximum phase. Despite the difficulty that this poses for an exact measurement, the intrinsic color of SNIa appears to be very uniform [88] and this is often used to determine the amount of reddening towards the supernova.

After about 40 days the light curves settle onto an exponential (in luminosity) decline for several months. This has been interpreted as the optically thin phase when the ejecta nebula captures fewer and fewer γ-rays and the optical and NIR light curves decline faster than the ^{56}Co decay rate (see, e.g., [49,59,85]). After about 150 days the light curves change slope once more (see, e.g., [58]) when the importance of the positron channel in the ^{56}Co decay sets in (see, e.g., Sect. 1, [5,75]. The decline at these phases should tell us about the magnetic fields in the explosion as they determine whether the positrons are captured or escape the ejecta [22,75,94]. Currently, the best estimate of these late light curves predicts a positron escape similar to that of the photons [76].

Light echos can start to dominate SNIa light curves several hundred days after the explosion. There are now at least two objects with clear signatures of light echos, SN1991T [100,106] and SN1998bu [18]. Their light curves have flattened almost completely as the peak light is scattered off nearby dust clouds and, since we observe time and intensity integrated light, the brightness does not change until the edge of the dust layer is reached or the scattering angle increases to the point where the scattering efficiency decreases. High spatial resolution imaging shows rings around these supernovae ([106] and Garnavich, private communication) very similar to the ones observed around SN1987A.

Despite the highly complicated hydrodynamics and the radiative transport in SNIa ejecta the bolometric light curves can provide important insights. Most of the emission from SNIa is emitted in the optical and NIR region [23,107]. By sampling the emission from the atmospheric cutoff near 3600 Å to 1 μm we are capturing about 80% of the energy emitted by these objects outside the γ-ray region. Although the color changes significantly through the observable life of a SNIa, not much light is emitted in the near-UV (see, e.g., [52,57,107]) or the infrared [23]. In fact, the V light curve is a rather good surrogate of the bolometric light curve after maximum [24]. In particular, during the late declines the V light curves have been used to calculate the Ni mass synthesized in the explosions [17] and also to estimate the positron escape [76].

Figure 2 illustrates the connection between observed bolometric light curves (dotted line – constructed from observations of SN1992bc) and theoretical models. The expected decay lines for ^{56}Ni and ^{56}Co are indicated by the thin, gray lines. At early times all the energy is trapped in the ejecta and as the surface increases the brightness increases as well. Around maximum the energy released is almost identical to the decay energy from ^{56}Ni and ^{56}Co combined. This has been pointed out long ago by Arnett [2] and has been confirmed by Pinto and Eastman [85]. A compilation of typical values of early models has been given in Branch [14]. This is an important feature of SNIa and can be used to measure the total amount of radioactive material synthesized in the explosion (see, e.g., [24]). For a brief time after maximum the energy output from the supernova exceeds the prediction from a complete trapping of the radioactive energy while "old" photons still leak out at the surface. At later phases more and more γ-rays are lost and the decline is faster than the line indicating a full trapping (dash-dotted). The dashed line indicates a light curve in which no energy from the original γ decays is converted to optical light. Only when 3% of the β decay sets in can there be some leveling of the light curve. The observed light curve evolves between these extreme cases.

Correlations. The SNIa light curve shape has been recognized as correlating with the peak luminosity [87]. This has become the linchpin for distance determinations using SNIa (see, e.g., the chapter by Perlmutter and Schmidt in this volume). The normalization of the peak luminosity allows the determination

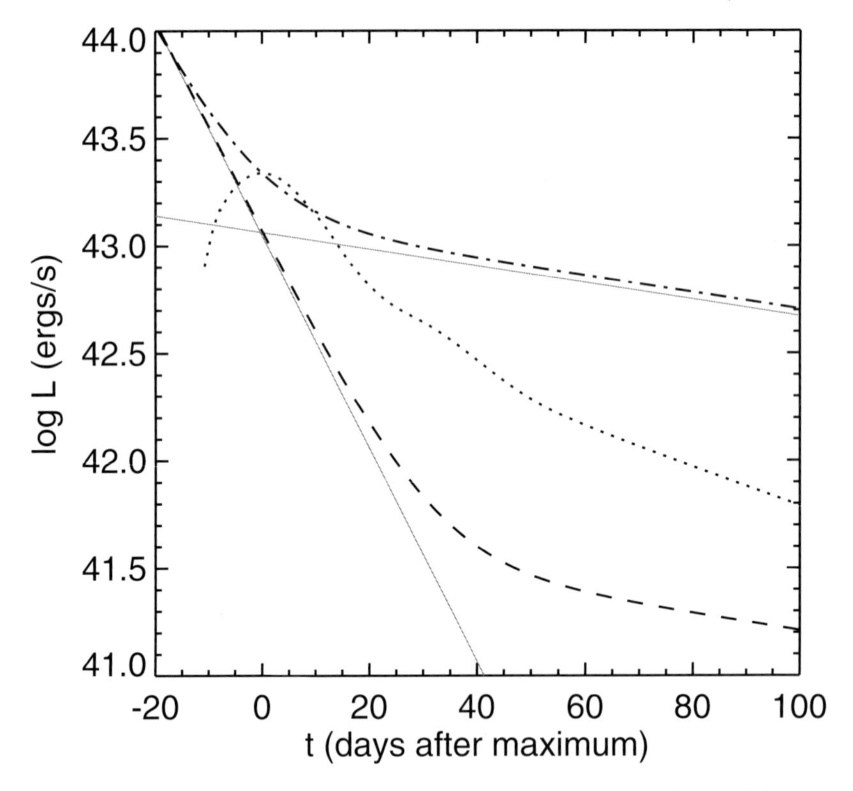

Fig. 2. Bolometric light curve of a SNIa. The dotted line is the observed bolometric light curve. The thin lines represent the ^{56}Ni and ^{56}Co decay lines, while the dashed line would be the expected curve if all γ-rays from the radioactive decay escape the ejecta and only the positrons are converted into optical emission. The dash-dotted line indicates the expectation of full trapping of all decay energy [23].

of cosmological parameters. However, the correlation is not as clear-cut as one would wish. There are three implementations [83,88,91], which are currently not consistent with each other [28,58].

There are other parameters which correlate with the peak luminosity of SNIa. They are the rise time to maximum [93], color near maximum light [88], line strengths of the primary Ca and Si absorption lines [79], the velocities as measured in Fe lines at late phases [70], the host galaxy morphology [101], and host galaxy colors [16,43]. There may be indications that the secondary peak in the I light curves and the shoulder in the bolometric light curves correlate with the absolute luminosity [24,44,91].

With SN2000cx we now also have a clear example of an object which does not follow these simple rules. For most objects the rise and decline rates correlate fairly well [24,93], but SN2000cx violates this rule [66]. While it had a rapid rise, its light curve did decline more slowly than many other SNIa. It is, at the moment, not clear what fraction of SNIa show a similar behavior as the rise phase is often not observed and the correlations have been based on the decline rate. Nevertheless, the fact that for some SNIa the rise and decline rates do correlate [24,93] hints at intrinsic differences within this class.

The physical reason for these correlations are not yet clear. Possibilities are differences in the amount ^{56}Ni synthesized and the distribution of the radioactive material, and hence the heating of the ejecta [49,71,86].

5 Bolometric Light Curves

The connection of the observations to the explosion and radiation physics has to come through bolometric light curves as they represent the total energy output of a supernova. This integrated quantity is easily constructed from the available multi-filter photometry and is also rather easily extracted from model calculations. Detailed calculations of the emerging spectrum have proven difficult for all supernova types. While the physics of the radiation escape for core-collapse supernovae during the plateau phase, with their large envelopes and a rather well-defined photosphere (see, e.g., the chapter by Branch et al. in this volume), is fairly simple to calculate, possible chemical abundance mixing and deviations from spherical geometry can spoil direct comparison with the observations. For thermonuclear supernovae the spectrum formation is due to multiple scattering of the γ-rays in a non-thermal environment and hence the photons rather "leak" out at the surface than originate from a thermalized photosphere. Hence, the calculation of a filter light curve requires a detailed spectrum formation calculation which, in the case of SNIa, is currently impossible at the needed detail.

The temporal evolution of the integrated light gives a rather straight forward comparison between theory and observation. Important explosion parameters can be calculated from these integrated quantities. Foremost, the amount of energy available from the nucleosynthesis can be determined from bolometric light curves. The nickel masses derived for all supernovae are an essential input of the explosion models. So far, this observational input has been missing. It is foreseeable that in the near future this will change and observational constraints on the models will become available.

The bolometric light curve of SN1987A has been instrumental in decoding the various phases of the energy release (see Fig. 1 and [72,111]). The luminosity of the plateau phase indicates the size of the progenitor and the explosion energy, while the length of the plateau phase is mostly dominated by the mass of the exploded star [89]. For the core-collapse supernovae the luminosity on the radioactive tail, after the light curve leaves the plateau due to the recombination, together with the time since explosion gives the ^{56}Ni mass directly. This method has been employed for several supernovae and it could be shown that there are rather large differences among these objects. This is particularly interesting as it might provide insight into fall-back onto the forming neutron star.

It should be possible to combine some of these measurements into a picture of the explosion. One can investigate how the progenitor mass correlates with the explosion energy and the amount of fall-back onto the forming compact remnant.

In extreme cases, it might be possible to show that a black hole formed in the explosion [6]. It is essential to show that the event has all the signatures of a massive stellar explosion to avoid confusion between different explosion types. Comparing the explosion parameters with direct information of the progenitor star, which are becoming available more often now (see, e.g., [102]), will be an important link between the post-event deductions and the models.

The bolometric light curves of thermonuclear supernovae can be used to extract physical parameters of the explosions (see, e.g., [118]). According to Arnett's rule [2,85], the peak luminosity of SNIa reflects the amount of radioactive ^{56}Ni produced in the explosion. This fact has been used by Contardo et al. [24] (see also [23]) to derive the ^{56}Ni masses for several supernovae. They claim that there is a significant range of ^{56}Ni produced in the explosions (up to a factor of ten when extreme objects are included). A similar result was derived by Cappellaro et al. [17] based on the late light curves and the assumption that the V light curve can serve as a surrogate for the bolometric luminosity. Independent confirmation of this result comes from the observations of infrared spectra of SNIa [13].

The late decline can be further used to determine the escape fraction of the γ-rays. A steeper decline indicates a faster decrease of the column density. Three possible explanations for this effect could be: 1) different explosion energies, 2) a varying distribution of nickel in the explosion, or 3) differences in ejecta mass. All possibilities indicate fundamental variations in the explosions. Interestingly, the expansion velocity of the iron in the ejecta correlates with peak luminosity of SNIa [70], however not in the way expected from simple models. The brighter supernovae have larger expansion velocities indicating, for a fixed ejecta mass, a higher explosion energy. These are also the objects which produce more ^{56}Ni in the explosions. In this case, the distribution of the iron-peak elements must be different to explain the slower decline at late times [23]. An alternative explanation could be that the ejecta mass for SNIa is not the same in all events, quite a radical suggestion in view of the currently favored models of Chandrasekhar-mass white dwarf/indexstar!white dwarf progenitors (see, e.g., [67]). Bolometric light curves of more objects will be needed to confirm such a result.

6 Summary

SN light curves are essential for our understanding of supernova physics. The acquisition of light curves has become relatively easy and the use of robotic telescopes and pipeline reductions will further advance. The success of recent supernova searches is increasing the available data set dramatically and we will soon be able to investigate detailed statistics on light curve parameters.

The variety of light curve shapes of core-collapse supernovae provides evidence for the many physical effects which can influence the outcome of these events. The rapid, and sometimes violent, evolution of their massive progenitor stars and their death within their cradle provides for vastly different environments and displays. We can still learn a great deal from detailed observations.

The physics background has developed considerably over the past decade and the nearby example of SN1987A, offering extremely detailed observations, has led to major new insights into this phenomenon.

The short lifetime of core-collapse supernovae can be used to directly measure the star formation rates as a function of look back time (see, e.g., [25] and the chapter by Cappellaro in this volume). However, their light and color curves are required for a secure identification.

The more uniform appearance of thermonuclear type Ia explosions has been challenged by the observations in redder filter bands and additional examples. Although these objects still are rather uniform, their light curves have allowed us to show that significant variations in these explosions must exist. The correlation of the light curve shape with the luminosity has provided a convenient way to normalize these objects and make them the best cosmological distance indicator available at the moment (see the chapter by Perlmutter and Schmidt in this volume). Light curves are at the heart of the cosmological applications of SNIa. It remains a major task to establish the physical understanding of these events in the coming years.

References

1. G. Aldering, R. Knop, P. Nugent: Astron. J. **119**, 2110 (2000)
2. W.D. Arnett: Astrophys. J. **253**, 785 (1982)
3. W.D. Arnett: *Supernovae and Nucleosynthesis* (Princeton University Press, Princeton 1996)
4. W.D. Arnett, J.N. Bahcall, R.P. Kirshner, S.E. Woosley: Ann. Rev. Astron. Astrophys. **27**, 629 (1989)
5. T.S. Axelrod: In: *Type I Supernovae*, ed. by J.C. Wheeler (University of Texas, Austin 1980) p. 80
6. S. Balberg, L. Zampieri, S.L. Shapiro: Astrophys. J. **541**, 860 (2000)
7. R. Barbon, F. Ciatti, L. Rosino: Astron. Astrophys. **116**, 35 (1982)
8. R. Barbon, S. Benetti, E. Cappellaro, F. Patat, M. Turatto, T. Iijima: Astron. Astrophys. Suppl. Ser. **110**, 513 (1995)
9. S. Benetti, M. Turatto, E. Cappellaro, I.J. Danziger, P.A. Mazzali: Mon. Not. R. Astron. Soc. **305**, 811 (1999)
10. S. Benetti et al.: Mon. Not. R. Astron. Soc. **322**, 361 (2001)
11. P.J. Benson et al.: Astron. J. **107**, 1453 (1994)
12. P. Bouchet, M.M. Phillips, N.B. Suntzeff, C. Gouiffes. R. Hanuschik, D.H. Wooden: Astron. Astrophys. **245**, 490 (1991)
13. E.J.V. Bowers, W.P.S. Meikle, T.R. Geballe, N.A. Walton, P.A. Pinto, V.S. Dhillon, S.B. Howell, M.K. Harrop-Allin: Mon. Not. R. Astron. Soc. **290**, 663 (1997)
14. D. Branch: Astrophys. J. **392**, 35 (1992)
15. D. Branch, S.W. Falk, M.L. McCall, P. Rybski, A. Uomoto, B.J. Wills: Astrophys. J. **244**, 780 (1981)
16. D. Branch, W. Romanishin, E. Baron: Astrophys. J. Lett. **465**, L73 (1996)
17. E. Cappellaro, P.A. Mazzali, S. Benetti, I.J. Danziger, M. Turatto, M. Della Valle, F. Patat: Astron. Astrophys. **328**, 203 (1998)
18. E. Cappellaro et al.: Astrophys. J. Lett. **549**, L215 (2001)

19. N.N. Chugai: Sov. Astron. Lett. **16**, 457 (1991)
20. A. Clocchiatti, J.C. Wheeler, E.S. Barker, A.V. Filippenko, T. Matheson, J.W. Liebert: Astrophys. J. **446**, 167 (1995)
21. A. Clocchiatti et al.: Astrophys. J. **553**, 886 (2001)
22. S.A. Colgate, A.G. Petschek, J.T. Kriese: Astrophys. J. Lett. **237**, L81 (1980)
23. G. Contardo: Bolometric light curves of Type Ia Supernovae. PhD Thesis, Technical University, Munich (2001)
24. G. Contardo, B. Leibundgut, W.D. Vacca: Astron. Astrophys. **359**, 876 (2000)
25. T. Dahlén, C. Fransson: Astron. Astrophys. **350**, 349 (1999)
26. R. Diehl, F.X. Timmes: Pub. Astron. Soc. Pacific **110**, 637 (1998)
27. V.T. Doroshenko, Yu.S. Efimov, N.M. Shakhovskoi: Astron. Lett. **21**, 580 (1995)
28. P.S. Drell, T.J. Loredo, I. Wasserman: Astrophys. J. **530**, 593 (2000)
29. R.G. Eastman, B.P. Schmidt, R.P. Kirshner: Astrophys. J. **466,**, 911 (1996)
30. J.H. Elias, J.A. Frogel, J.A. Hackwell, S.E. Persson: Astrophys. J. Lett. **251**, L13 (1981)
31. S.W. Falk, W.D. Arnett: Astrophys. J. Suppl. **33**, 515 (1977)
32. A. Fassia et al.: Mon. Not. R. Astron. Soc. **318**, 1093 (2000)
33. R.A. Fesen: Astrophys. J. Lett. **413**, L109 (1993)
34. R.A. Fesen, R.H. Becker: Astrophys. J. **351**, 437 (1990)
35. R.A. Fesen, D.M. Matonick: Astrophys. J. **407**, 110 (1993)
36. R.A. Fesen et al.: Astron. J. **117**, 725 (1999)
37. A.V. Filippenko: Ann. Rev. Astron. Astrophys. **35**, 309 (1997)
38. A.V. Filippenko et al.: Astron. J. **104**, 1534 (1992)
39. C. Fransson, A. Cassatella, R. Gilmozzi, R.P. Kirshner, N. Panagia, G. Sonneborn, W. Wamsteker: Astrophys. J. **336**, 429 (1989)
40. C. Fransson, C. Kozma: Astrophys. J. Lett. **408**, L25 (1993)
41. C. Fransson et al.: Astrophys. J. **572**, 350 (2002)
42. G. Goldhaber et al.: Astrophys. J. **558**, 359 (2001)
43. M. Hamuy, M.M. Phillips, J. Maza, N.B. Suntzeff, R.A. Schommer, R. Avilés: Astron. J. **109**, 1 (1995)
44. M. Hamuy et al.: Astron. J. **112**, 2408 (1996)
45. M. Hamuy et al.: Astrophys. J. **558**, 615 (2001)
46. M. Hernandez et al.: Mon. Not. R. Astron. Soc. **319**, 223 (2000)
47. W. Hillebrandt, J.C. Niemeyer: Ann. Rev. Astron. Astrophys. **38**, 191 (2000)
48. P. Höflich: Astrophys. J. **443**, 89 (1995)
49. P. Höflich, A.M. Khokhlov, J.C. Wheeler, M.M. Phillips, N.B. Suntzeff, M. Hamuy: Astrophys. J. Lett. **472**, L81 (1996)
50. S. Jha et al.: Astrophys. J. Suppl. **125**, 73 (1999)
51. R.P. Kirshner: In: *Supernovae*, ed. by A.G. Petschek (Springer, New York 1990) p. 59
52. R.P. Kirshner et al.: Astrophys. J. **415**, 589 (1993)
53. R.I. Klein, R.A. Chevalier: Astrophys. J. Lett. **223**, L109 (1978)
54. K. Krisciunas et al.: Astron. J. **122**, 1616 (2001)
55. B. Leibundgut: In: *Circumstellar Media in Late Stages of Stellar Evolution*, ed. by R. Clegg, I. Stevens, P. Meikle (Cambridge University Press, Cambridge 1994), p. 100
56. B. Leibundgut: In: *The Lives of Neutron Stars*, ed. by A. Alpar, Ü. Kiziloglu, J. van Paradijs (Kluwer, Dordrecht 1995) p. 3
57. B. Leibundgut: In: *IAU Colloquium 145: Supernovae and Supernova Remnants*, ed. by R. McCray, (Cambridge University Press, Cambridge 1996) p. 11

58. B. Leibundgut: Astron. Astrophys. Rev. **10**, 179 (2000)
59. B. Leibundgut, P.A. Pinto: Astrophys. J. **401**, 49 (1992)
60. B. Leibundgut, G.A. Tammann, R. Cadonau, D. Cerrito: Astron. Astrophys. Suppl. Ser. **89**, 537 (1991a)
61. B. Leibundgut, R.P. Kirshner, P.A. Pinto, M.P. Rupen, R.C. Smith, J.E. Gunn, D.P. Schneider: Astrophys. J. **372**, 531 (1991b)
62. B. Leibundgut et al.: Astron. J. **105**, 301 (1993)
63. D.C. Leonard et al.: Pub. Astron. Soc. Pacific **114**, 35 (2002)
64. J.R. Lewis et al.: Mon. Not. R. Astron. Soc. **266**, L27 (1994)
65. W.D. Li, A.V. Filippenko, R.R. Treffers, A.G. Riess, J. Hu, Y. Qiu: Astrophys. J. **546**, 734 (2001a)
66. W.D. Li et al.: Pub. Astron. Soc. Pacific **113**, 1178 (2001b)
67. M. Livio: In: *Type Ia Supernovae: Theory and Cosmology*, ed. by J.C. Niemeyer, J.W. Truran, (Cambridge University Press, Cambridge 1999) p. 33
68. K.S. Long, W.P. Blair, W. Krzeminski: Astrophys. J. Lett. **340**, L25 (1989)
69. K.S. Long, P.F. Winkler, W.P. Blair: Astrophys. J. **395**, 632 (1992)
70. P.A. Mazzali, E. Cappellaro, I.J. Danziger, M. Turatto, S. Benetti: Astrophys. J. Lett. **499**, L49 (1998)
71. P.A. Mazzali, K. Nomoto, E. Cappellaro, T. Nakamura, H. Umeda, K. Iwamoto: Astrophys. J. **547**, 988 (2001)
72. R. McCray: Ann. Rev. Astron. Astrophys. **31**, 175 (1993)
73. W.P.S. Meikle: Mon. Not. R. Astron. Soc. **314**, 782 (2000)
74. N.V. Melova, D.Yu. Tsvetkov, S.Yu. Shugarov, V.F. Esipov, N.N. Pavlyuk: Astron. Lett. **21**, 670 (1995)
75. P.A. Milne, L.-S. The, M. Leising: Astrophys. J. Suppl. **124**, 503 (1999)
76. P.A. Milne, L.-S. The, M. Leising: Astrophys. J. **559**, 1019 (2001)
77. M. Modjaz et al.Pub. Astron. Soc. Pacific **113**, 308 (2001)
78. D.K. Nadyozhin: In: *Supernovae and Cosmology*, ed. by L. Labhardt, B. Binggeli, R. Buser (University of Basel, Basel 1998) p. 125
79. P. Nugent, M.M. Phillips, E. Baron, D. Branch, P. Hauschildt: Astrophys. J. Lett. **455**, L147 (1995)
80. F. Patat, R. Barbon, R. Cappellaro, M. Turatto: Astron. Astrophys. Suppl. Ser. **98**, 443 (1993)
81. F. Patat, R. Barbon, R. Cappellaro, M. Turatto: Astron. Astrophys. **282**, 731 (1994)
82. F. Patat, R. Barbon, E. Cappellaro, M. Turatto: 1997, Astron. Astrophys. **317**, 423 (1997)
83. S. Perlmutter et al.: Astrophys. J. **483**, 565 (1997)
84. M.J. Pierce, G.H. Jacoby: Astron. J. **110**, 2885 (1995)
85. P.A. Pinto, R.G. Eastman: Astrophys. J. **530**, 757 (2000)
86. P.A. Pinto, R.G. Eastman: New Astron. **6**, 307 (2001)
87. M.M. Phillips: Astrophys. J. Lett. **413**, L105 (1993)
88. M.M. Phillips, P. Lira, N.B. Suntzeff, R.A. Schommer, M. Hamuy, J. Maza: Astron. J. **118**, 1766 (1999)
89. D.V. Popov: Astrophys. J. **414**, 712 (1993)
90. M.W. Richmond, R.R. Treffers, A.W. Filippenko, Y. Paik: Astron. J. **112**, 732 (1996)
91. A.G. Riess et al.: Astron. J. **116**, 1009 (1998a)
92. A.G. Riess et al.: Astron. J. **117**, 707 (1999a)
93. A.G. Riess et al.: Astron. J. **118**, 2675 (1999b)

94. P. Ruiz-Lapuente, H.C. Spruit: Astrophys. J. **500**, 360 (1998)
95. M.P. Rupen, J.H. van Gorkom, G.R. Knapp, J.E, Gunn, D.P. Schneider: Astron. J. **94**, 61 (1987)
96. S. Ryder, L. Staveley-Smith, M.A. Dopita, R. Petre, E. Colbert, D. Malin, E. Schlegel: Astrophys. J. **416**, 167 (1993)
97. M.E. Salvo, E. Cappellaro, P.A. Mazzali, S. Benetti, I.J. Danziger, F. Patat, M. Turatto: Mon. Not. R. Astron. Soc. **321**, 254 (2001)
98. B.P. Schmidt et al.: Astron. J. **105**, 2236 (1993)
99. B.P. Schmidt et al.: Astron. J. **107**, 1444 (1994a)
100. B.P. Schmidt, R.P. Kirshner, B. Leibundgut, L.A. Wells, A.C. Porter, P. Ruiz-Lapuente, P. Challis, A.V. Filippenko: Astrophys. J. Lett. **434**, L19 (1994b)
101. B.P. Schmidt et al.: Astrophys. J. **507**, 46 (1998)
102. S.J. Smartt, G.F. Gilmore, N. Trentham, C.A. Tout, C.M. Frayn: Astrophys. J. Lett. **556**, L29 (2001)
103. J. Sollerman, R.J. Cumming, P. Lundqvist: Astrophys. J. **493**, 933 (1998a)
104. J. Sollerman, B. Leibundgut, J. Spyromilio: Astron. Astrophys. **337**, 207 (1998b)
105. J. Sollerman, C. Kozma, C. Fransson, B. Leibundgut, P. Lundqvist, F. Ryde, P. Woudt: Astrophys. J. Lett. **537**, L127 (2000)
106. W.B. Sparks, F.D. Maccetto, N. Panagia, F.R. Boffi, D. Branch, M.L. Hazen, M. Della Valle: Astrophys. J. **523**, 585 (1999)
107. N.B. Suntzeff: In: *IAU Colloquium 145: Supernovae and Supernova Remnants*, ed. by R. McCray (Cambridge University Press, Cambridge 1996) p. 41
108. N.B. Suntzeff, P. Bouchet: Astron. J. **99**, 650 (1990)
109. N.B. Suntzeff, S. Heathcote, W.G. Weller, N. Caldwell, J.P. Huchra: Nature **334**, 135 (1988)
110. N.B. Suntzeff, M.M. Phillips, D.L. DePoy, J.H. Elias, A.R. Walker: Astron. J. **102**, 1118 (1991)
111. N.B. Suntzeff, M.M. Phillips, D.L. DePoy, J.H. Elias, A.R. Walker: Astrophys. J. Lett. **384**, L33 (1992)
112. N.B. Suntzeff et al.: Astron. J. **117**, 1175 (1999)
113. M. Turatto, E. Cappellaro, R. Barbon, M. Della Valle, S. Ortolani, L. Rosino: Astron. J. **100**, 771 (1990)
114. M. Turatto, E. Cappellaro, I.J. Danziger, S. Benetti, C. Gouiffes, M. Della Valle: Mon. Not. R. Astron. Soc. **262**, 128 (1993)
115. M. Turatto, S. Benetti, E. Cappellaro, I.J. Danziger, M. Della Valle, C. Gouiffes, P.A. Mazzali, F. Patat: Mon. Not. R. Astron. Soc. **283**, 1 (1996)
116. M. Turatto et al.: Astrophys. J. Lett. **498**, L129 (1998)
117. M. Turatto, A. Piemonte, S. Benetti, E. Cappellaro, P.A. Mazzali, I.J. Danziger, F. Patat: Astron. J. **116**, 2431 (1998)
118. W.D. Vacca, B. Leibundgut: Astrophys. J. Lett. **471**, L37 (1996)
119. S.D. Van Dyk et al.: Pub. Astron. Soc. Pacific **111**, 313 (1999)
120. T. Wada, M. Ueno: Astron. J. **113**, 231 (1997)
121. J.C. Wheeler, S. Benetti: In: *Allen's Astrophysical Quantities, 4th edition*, ed. by A.N. Cox (AIP Press, New York 2000) p. 451
122. S.W. Woosley, F.X. Timmes: 1996, Nucl. Phys. A **606**, 137 (1996)
123. S.E. Woosley, T.A. Weaver: Ann. Rev. Astron. Astrophys. **24**, 205 (1986)
124. J. Xu, A.P.S. Crotts, W.E. Kunkel: Astrophys. J. **451**, 806 (1995)

X-Ray Supernovae

Stefan Immler[1] and Walter H.G. Lewin[2]

[1] Department of Astronomy and Astrophysics, Pennsylvania State University, University Park, PA 16803, USA
[2] Center for Space Research and Department of Physics, Massachusetts Institute of Technology, Cambridge, MA 02139-4307, USA

Abstract. We present a review of X-ray observations of supernovae. By observing the ($\sim 0.1 - 100$ keV) X-ray emission from young supernovae, physical key parameters such as the circumstellar matter density, mass-loss rate of the progenitor, and temperature of the outgoing and reverse shocks can be derived as a function of time. Despite intensive search over the last $\sim$ 25 years, only 15 supernovae have been detected in X-rays. We review the individual X-ray observations of these supernovae and discuss their implications for our understanding of the physical processes giving rise to the X-ray emission.

1 Introduction

To date, several thousand supernovae (SNe) have been discovered in the optical, whereas only ~30 SNe have been detected in the radio and 15 in the X-ray band ($\sim 0.1 - 100$ keV). Neutrinos have been recorded from one nearby supernova only. The reasons for this large diversity lies in the simple facts that the flux levels of the various constituents, as well as the sensitivity of the available detectors and telescopes, are vastly different and that systematic or automatic SN searches for SNe are only made in the optical. Prior to the beginning of automatic SN searches in the late 1980s [13], about two dozen SNe were reported per year in the optical. This number has now grown to several hundred per year (e.g., 248 confirmed detections in the year 2001[1]). The detection of two dozen neutrinos from SN1987A [4,31] was only possible because of its close proximity at a distance of $\sim$ 50 kpc in the *Large Magellanic Cloud (LMC)*. SNe have been detected up to a redshift of $z \sim 1.7$ (SN1997ff [56]) in the optical and IR, and up to $z \sim 0.022$ (SN1988Z; ~90 Mpc) in both the radio [72] and X-rays [16].

Based on the presence or absence of hydrogen lines in their spectra, SNe are classified as type II (SNII) and type I (SNI), respectively, (see the chapter by Turatto in this volume and the references therein). No type Ia (SNIa) has been detected to date in either the radio or in X-rays. SNIa are believed to be nuclear detonations of carbon+oxygen (C+O) white dwarfs which exceed the Chandrasekhar limit through accretion ([48,83] and references therein). Due to recurrent pulsations of the progenitor, fast shells (several $\times 100$ km s^{-1}) are ejected from the progenitor. During periods of quiescence, no stellar wind is

[1] From the Asiago Supernova Catalogue at *http://merlino.pd.astro.it/~supern*

blown into the circumstellar medium (CSM) due to the low mass of the progenitor. This leads to the formation of a complicated, yet low density CSM structure with interacting shells of different expansion speeds. The relatively slowly expanding ejecta ($\lesssim 5,000$ km s^{-1}) can hence not form a shock region that could give rise to thermal X-ray or radio emission (see below). Due to this lack of even a single detection of X-ray emission from a type Ia SN and since from models X-ray emission is not expected at a detectable flux level, we will not discussed SNIa further.

This still leaves us with a number of different SN types: SNIb, SNIc, SNIIP, SNIIL, SNIIb, and SNIIn, all of which are believed to be the result of the core-collapse of a massive star ($\gtrsim 10$ $M_{\odot}$ ZAMS; see, e.g., [6,29] and references therein). The differences between these types is not always clear, and there are numerous examples where an original SN type designation was later changed as the spectra and the light curves evolved in time. (SN1998bw, for example, was classified as both type Ib and type Ic at different times during its evolution.) In general, the more detailed the information available on a particular SN, the more difficult it seems to become to place it in a category of known types. This has led to the introduction of the "peculiar" categories of SNe, abbreviated "pec," and it is not surprising that two of the best studied SNe, SN1987A and SN1993J, have both been designated as "pec" by some authors (see, e.g., [19,65]) at the same point in their evolution.

The X-ray luminosities of all detected SNe are in the range $10^{37} - 10^{41}$ erg s^{-1}. A list of all 15 X-ray SNe is given in Table 1[2]. The X-ray emission begins to dominate the total luminosity of the SN starting at an age of about one year. The X-ray emission, as well as the radio emission, is largely the result of interaction of the ejecta with the CSM established by the wind of the SN progenitor [10,69]. Progenitors of type II SNe have high mass-loss rates ($\dot{M} \sim 10^{-4} - 10^{-6}$ $M_{\odot}$ yr^{-1}) and low wind velocities of typically $w_{\rm wind} \sim 10$ km s^{-1} while type Ib/c SNe likely originate from more compact stars with lower mass-loss rates ($\dot{M} \sim 10^{-5} - 10^{-7}$ $M_{\odot}$ yr^{-1}) and significantly higher wind velocities of $w_{\rm wind} \sim 1,000$ km s^{-1}.

Shocks are formed as the ejecta plow into the CSM: the "circumstellar" or "forward shock" (also called the "blastwave") and the so-called "reverse shock." In the early phases after the explosion, the speed of the ejecta (i.e., of the circumstellar shock) is of the order of $\sim 10^4$ km s^{-1}. The speed of the reverse shock is $\sim 10^3$ km s^{-1} lower. Depending on the density profile of the ejecta as they emerge from the star, the density behind the reverse shock (i.e., at larger radii than that of the reverse shock) can be $5 - 10$ times higher than the density behind the circumstellar shock (i.e., at smaller radii than the forward shock front; see, e.g., Fig. 1 in the chapter by Chevalier and Fransson in this volume). The temperature behind the forward shock can be as high as $10^9 - 10^{10}$ K, whereas the temperature behind the reverse shock is significantly lower ($10^7 - 10^8$ K). The soft X-ray ($\lesssim 5$ keV) emission is therefore generally explained in terms

[2] A frequently updated list of X-ray SNe and references is available at: *http://www.astro.psu.edu/~immler/supernovae_list.html*

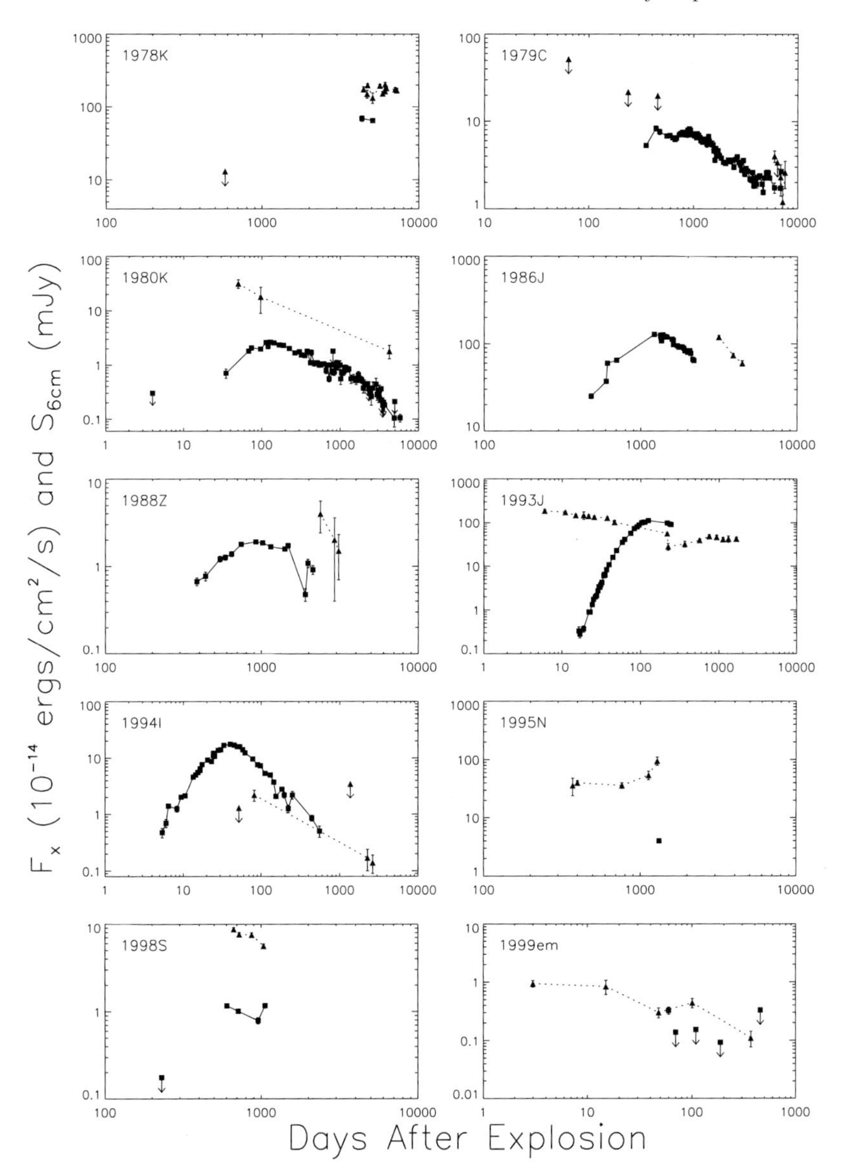

Fig. 1. Light curves for the 6 cm radio (*VLA* and *ATCA*; filled squares, solid lines) and soft X-ray (0.2 – 2.4 keV; filled triangles and dashed lines) bands of nine type II and one type Ic SN.

of thermal radiation from the reverse shock, whereas the higher energy X-rays ($\gtrsim$10 keV) are more likely to arise in the forward shock region. Because of the much higher density in the reverse shock region, the X-ray emission from this region dominates over that of the forward shock by the time the expanding shell has become optically thin enough to allow the X-rays from the reverse shock region to escape.

The radio emission is believed to come from the region behind the forward shock (see the chapter by Chevalier and Fransson in this volume and the references therein), where the temperature is very high and the density low. In Fig. 1 we plot the 6 cm radio and soft X-ray band ($0.1 - 2.4$ keV) light curves for several SNe. Note that the radio emission is absorbed at early times (see below), but later the radio and X-ray emission are roughly proportional and show similar rates of decline. (Sramek and Weiler discuss the radio emission from SNe elsewhere in this volume.)

The signatures of circumstellar interaction in the radio, optical, and X-ray regimes have been found for a number of type II SNe, such as the type IIL SN1979C [18,33,77,79] and SN1980K [9,17,42,77,80]. In type IIn SNe, Hα typically exhibits a very narrow component (FWHM $\lesssim 200$ km s^{-1}) superimposed on a base of intermediate width (FWHM $\sim 1,000 - 2,000$ km s^{-1}). Sometimes a very broad component (FWHM $\sim 5,000 - 10,000$ km s^{-1}) is also present [20]. The narrow optical lines are clear evidence for the presence of slowly moving dense CSM, probably photo-ionized by the intense flash of UV radiation. Examples of type IIn SNe are SN1986J [5,32,42,58,78], SN1995N [2,44,74], and SN1998S [21,27,45,54,75]. Each of these SNe are discussed in more detail below.

Several nearby type Ib/c SNe have been detected in the radio: SN1990B [72], SN1994I [73], SN1997X [76], SN1998bw [26,39] and SN2002ap [3]. SN1994I [34,36], SN1998bw [50,51] and SN2002ap [57,68] have also been detected in X-rays. These SNe may be special cases – in particular, the strong evidence for association between SN1998bw and GRB980425 makes it unique (see the chapters by Galama, by Frontera, and by Weiler et al. in this volume).

2 X-Ray Production Mechanism

As discussed above, the most important process for the generation of a substantial amount of X-ray emission is the interaction of the ejecta with the CSM. However, depending on the age of the SN, there can be other mechanisms for the production of X-ray emission:

- Radio activity of the ejecta.
- An initial burst of X-rays during the break-out of the shock from the surface of the star, producing short time-scale ($\sim 1,000$ s) hard X-rays (~ 100 keV) and a very soft (~ 0.02 keV) blackbody continuum.
- Inverse Compton scattering of relativistic electrons with UV photons produced by the outburst.
- Pulsar driven X-ray emission from the plerionic supernova remnant (SNR) in the case of a hot, fast spinning neutron star (if the latter is formed in the process of core-collapse); this last model is applicable to older SNRs, such as the Crab Nebula.

The second through fourth cases will not be discussed further since they are not relevant to the X-ray SNe reviewed in this chapter. We will restrict ourselves to a brief overview of the CSM interaction and radioactive decay mechanisms.

2.1 The Circumstellar Interaction

The thermal X-ray luminosity, L_x, produced by the shock heated CSM is the product of the emission measure, EM, and the cooling function, $\Lambda(T, Z, \Delta E)$, where T is the CSM plasma temperature, Z represents the elemental abundance distribution, and ΔE is the X-ray energy bandwidth. For spherically symmetric conditions

$$L_x = \Lambda(T, Z, \Delta E)\mathrm{d}V n^2, \tag{1}$$

where $\mathrm{d}V$ is the volume, $n = \rho_{\mathrm{CSM}}/m$ is the number density of the shocked CSM, and m is the mean mass per particle (2.1×10^{-27} kg for a H+He plasma). Assuming a constant supernova shock speed v_s,

$$r = v_s t, \tag{2}$$

where r is the radius of the shock and t represents the time elapsed since the explosion. If the CSM density ρ_{CSM} is dominated by a wind from by the progenitor star, the continuity equation requires

$$\dot{M} = 4\pi r^2 \rho_w(r) w_{\mathrm{wind}}(r) \tag{3}$$

through a sphere of radius r, where $\rho_w(r)$ and $w_{\mathrm{wind}}(r)$ are the wind density and wind velocity, respectively, at radius r. After the SN shock plows through the CSM, its density is

$$\rho_{\mathrm{CSM}} = 4\rho_w \tag{4}$$

from [25]. We thus obtain

$$L_x = 4/(\pi m^2)\Lambda(T) \times (\dot{M}/w_{\mathrm{wind}})^2 \times (v_s t)^{-1} \tag{5}$$

and can use the observed X-ray luminosity at time t after the outburst to measure the ratio $\dot{M}/w_{\mathrm{wind}}$.

Each X-ray measurement at time t is thus related to a corresponding distance r from the site of the explosion. This site had been reached by the wind at a time depending on w_{wind}, or the age of the wind $t_w = t v_s / w_{\mathrm{wind}}$.

Usually, $v_s \gg w_{\mathrm{wind}}$ so that with t of only a few years we can look back a large time span in the evolution of the progenitor's wind and can use our measurements as a "time machine" to probe the progenitor's mass-loss history. Assuming that w_{wind} did not change over t_w, we can measure the mass-loss rate back in time. Integration of the mass-loss rate along the path of the expanding shell thus gives the mean density inside a sphere of radius r. For a constant wind velocity w_{wind} and mass-loss rate $\dot{M}$, a $\rho_{\mathrm{CSM}} = \rho_0 (r/r_0)^{-s}$ profile with $s = 2$ is expected.

After the expanding shell has become optically thin, it is expected that emission from the SN ejecta itself, heated by the reverse shock, will dominate the X-ray output of the interaction region due to its higher emission measure and higher density. For uniformly expanding ejecta the density structure is a function of its expansion velocity, v, and the time after the explosion, t or

$$\rho_{\rm ejecta} = \rho_0 (t/t_0)^{-3} (v/v_0)^{-n} \tag{6}$$

with ρ_0 the ejecta density at time t_0 and velocity v_0 [25]. For a red supergiant progenitor, the velocity gradient power-law is rather steep with index $n \sim 20$ [25,69]. For constant n, the radius $r_{\rm c}$ of the discontinuity surface between the forward and the reverse shocks evolves in time t with $r_{\rm c} \propto t^m$, where $m = (n-3)/(n-s)$ is the deceleration parameter.

2.2 X-Ray Emission from Radioactive Decay

The SN ejecta are radioactive. Gamma-rays due to the decay of ^{56}Ni (half-life 6.1 days) $\rightarrow$ ^{56}Co (half-life 77 days) $\rightarrow$ ^{56}Fe and ^{57}Ni (half-life 36 hours) $\rightarrow$ ^{57}Co (half-life 270 days) $\rightarrow$ ^{57}Fe could, in principle, be detectable during the first few years after the explosion. The dominant lines produced in the decay of ^{56}Co are 511 keV (40%), 847 keV (100%), 1.04 MeV (15%), 1.24 MeV (66%), 1.76 MeV (15%), 2.02 MeV (11%), 2.60 MeV (17%), and 3.25 MeV (13%). The dominant lines produced in the decay of ^{57}Co are 14 keV (9%), 122 keV (87%), and 136 keV (11%). The numbers in brackets represent the probabilities that, for each decaying nucleus, the specific energy γ-rays are produced.

Several of these lines have been detected from SN1987A (see the chapter by McCray in this volume and the references therein) but not from any other SN. This, of course, is not surprising given the very modest flux levels of the γ-rays and the large distances to all modern SNe except SN1987A. Hard ($\gtrsim$10 keV) X-rays were observed from SN1987A as early as 168 – 191 days after the outburst [15,67] with at least part of these X-rays believed to be the result of Compton scattering of γ-rays [40]. The light curves and spectral evolution of the X-rays and γ-rays are reasonably well understood (up to $\sim$ 300 days after the outburst) if one assumes mixing of ^{56}Co into the hydrogen rich envelope. The X-ray light curve at later times may be more problematic [41]. Whether a neutron star or a black hole was formed during the outburst of SN1987A is still a matter of debate.

In the case of a type Ia SN occurring within a distance of $\sim$50 kpc, one would also expect to see γ-rays from the radioactive by-products of the detonation (subsonic speed of the shock through the progenitor) or deflagration (supersonic speed) of the C+O white dwarf. This, however, has not been observed since no such nearby example has yet occurred.

Table 1. List of X-Ray Supernovae

Supernova	Type Galaxy	d (Mpc)	$t-t_0^{\mathrm{a}}$ (days)	F_{x} (10^{-14} erg cm^{-2} s^{-1})	L_{x} (10^{38} erg s^{-1}) Band	Instruments	Ref.
SN1978K	II NGC1313	4.5	4,500	204	48 (0.1–2.4 keV)	*ROSAT*,*ASCA* *XMM*	[49,60–62]
SN1979C	IIL NGC4321	17.1	5,900	3.9	14 (0.1–2.4 keV)	*ROSAT*,*CHANDRA* *XMM*	[34,38,55]
SN1980K	IIL NGC6946	5.1	35	1.5	0.5 (0.1–2.4 keV)	*Einstein*,*ROSAT*	[9,59,60]
SN1986J	IIn NGC891	9.6	3,300	119	140 (0.1–2.4 keV)	*ROSAT*,*ASCA*	[5,11,32,60]
SN1987A[b]	IIP LMC	0.05	–	–	–	*ROSAT*,*CHANDRA* *XMM*	[8,14,28,30]
SN1988Z	IIn MCG+03-28-022	89	2,370	0.9	110 (0.1–2.4 keV)	*ROSAT*	[1,16]
SN1993J	IIb NGC3031	3.6	6	130	20 (0.5–2 keV)	*ROSAT*,*ASCA*	[35,69,84]
SN1994I	Ic NGC5194	7.7	79	2.3	1.6 (0.3–2 keV)	*ROSAT*,*CHANDRA*	[34,36]
SN1994W	IIP NGC4041	25	1,180	11	85 (0.1–2.4 keV)	*ROSAT*	[63]
SN1995N	IIn MCG-2-38-017	24	440	40	175 (0.1–2.4 keV)	*ROSAT*,*ASCA*	[23]
SN1998bw	Ic ESO184-G82	38	0.4	23 – 40	40 – 70 (2–10 keV)	*BeppoSAX*	[50,51]
SN1998S	IIn NGC3877	17	64	27	92 (2–10 keV)	*CHANDRA* *XMM*	[54]
SN1999em	IIP NGC1637	7.8	4	1.0	0.7 (2–10 keV)	*CHANDRA*	[54]
SN1999gi	IIP NGC3184	8.7	29	0.1	0.1 (0.5–10 keV)	*CHANDRA*	[64]
SN2002ap	Ic NGC628	10	4	0.2	0.2 (0.1–10 keV)	*XMM*	[57,68]

[a]Day after outburst corresponding to the maximum observed X-ray flux.
[b]X-ray flux is still rising.

3 Overview of X-Ray Supernovae

In the following sections we give a brief overview of X-ray observations of all X-ray SNe except SN1987A, which is discussed in the chapter by McCray in this volume.

3.1 SN1978K in NGC1313

Due to its discovery in the early phase of the *Röntgensatellit (ROSAT)* mission, and the relatively high flux level, SN1978K is one of the best studied SNe in the X-ray regime. While an *Einstein Observatory – HEAO-2 (Einstein)* observation ~ 1.5 yr after the outburst only provided an upper limit to the soft ($0.5-2$ keV) band X-ray luminosity ($L_x \lesssim 2 \times 10^{38}$ erg s^{-1} [61]), SN1978K was successfully monitored with *ROSAT* on 13 different dates ranging from $\sim 12-20$ yr after the outburst [61,62]. During this period, the SN showed no apparent evolution in X-ray flux ($L_x \sim 4 \times 10^{39}$ erg s^{-1}.

The *Advanced Satellite for Cosmology and Astrophysics (ASCA) Gas Imaging Spectrometer (GIS)* and *Solidstate Imaging Spectrometer (SIS)* and the *ROSAT Position Sensitive Proportional Counter (PSPC)* observations provided the first broad-band ($0.1-10$ keV), medium resolution X-ray spectra of a young SN [49]. The spectra showed no emission lines and could be equally well described by either a thermal bremsstrahlung spectrum or thermal plasma emission with a temperature of ~ 3 keV and sub-solar abundances ($Z \sim 0.2 Z_\odot$). Alternatively, the spectrum could be described by a power-law with photon index ~ 2.2 and a hydrogen absorbing column of $N_H \sim 1 \times 10^{21}$ cm^{-2}. From the X-ray light curve the index n for the ejecta power-law distribution, $\rho_{ejecta} \propto r^{-n}$, could be constrained to be in the range $4-12$ for an assumed CSM density gradient $\rho_{CSM} \propto r^{-s}$ with index $s = 1.5-2$ [62]. Estimates for the mass-loss rate from the presupernova star are $\dot{M} \sim 1 \times 10^{-4}$ M$_\odot$ yr^{-1}, as is expected for a massive progenitor.

A recent *XMM-Newton (XMM)* observation showed that SN1978K is still at the same flux level as observed during the last ~ 22 yr with no apparent evolution ($L_x = 4 \times 10^{39}$ erg s^{-1} [37]). The superior photon collecting area and spectral resolution of the *XMM* instruments with a total of $\sim 20,000$ net counts from SN1978K in the combined *EPIC-PN* and *EPIC-MOS* spectra demonstrates the dramatic technological progress that has been made in the development of X-ray instruments over the last two decades. The best-fit model to the spectrum gives a two-temperature thermal emission component with $kT_{low} \sim 0.8$ keV and $kT_{high} \sim 3$ keV. The *XMM* data clearly show, for the first time, emission from the forward (high temperature component) and reverse shock (low temperature component) regions.

3.2 SN1979C in NGC4321 (M100)

An X-ray source was detected in a *ROSAT High Resolution Imager (HRI)* observation of M100 at the position of SN1979C ~ 16 yr after the outburst (see Fig. 2), with a ($0.1-2.4$ keV) luminosity of $L_x = 1.3 \times 10^{39}$ erg s^{-1} [33], and in a *ROSAT HRI* follow-up observation ~ 19 yr after the outburst. For three earlier *Einstein* observations, taken on days 64, 239 and 454 after the outburst, only 3σ upper limits could be established ($< 1.8 \times 10^{40}$ erg s^{-1}, $< 7.6 \times 10^{39}$ erg s^{-1}, and $< 6.9 \times 10^{39}$ erg s^{-1}, respectively). The *ROSAT* data imply a mass-loss rate of $\dot{M} \sim 1 \times 10^{-4}$ M$_\odot$ yr^{-1}, similar to mass-loss rates found for

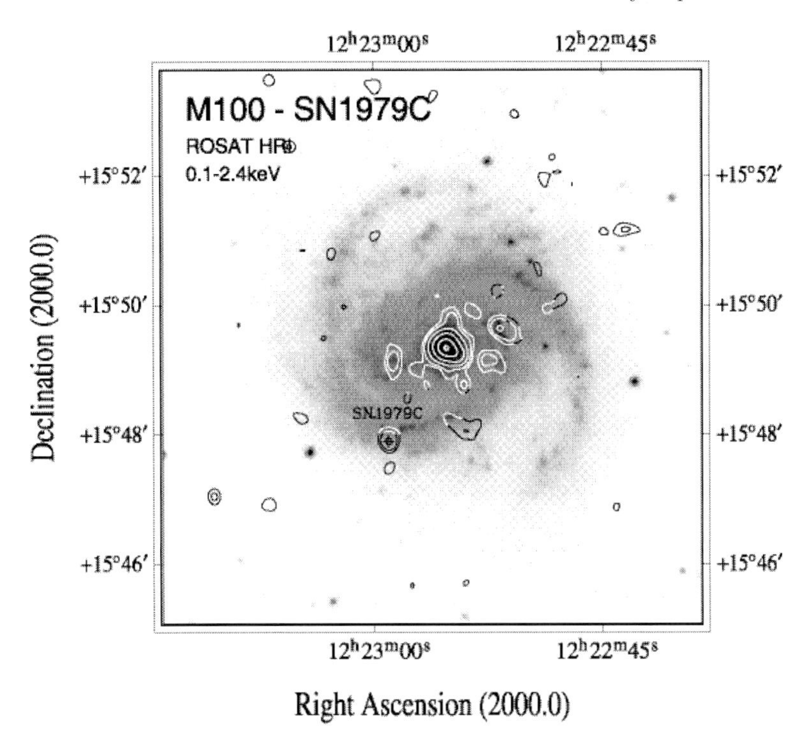

Fig. 2. *ROSAT HRI* soft $(0.1-2.4\ \text{keV})$ band X-ray contours of SN1979C in M100, overlayed onto an optical image.

other massive SN progenitors (e.g., SN1978K, SN1986J, SN1988Z and SN1998S) and in agreement with the mass-loss rate inferred from *Very Large Array (VLA)*[3] radio observations [79]. Since neither a *ROSAT PSPC* "last light" observation obtained by the author, nor *ASCA* and *Chandra X-ray Observatory – AXAF (CHANDRA)* pointings [38,55] gave enough photon statistics to constrain the spectrum, no information is available about the spectral characteristics of the X-ray emission.

Both the *ROSAT PSPC* and *CHANDRA* observations, however, showed that the emission is rather soft $(< 2\ \text{keV})$ and likely originates from shock heated material in the reverse shock. The combined *ROSAT*, *ASCA* and *CHANDRA* data indicate a rather slow X-ray rate of decline (decrease from $L_{\text{x}} = 1.3 \times 10^{39}$ erg s^{-1} to 0.9×10^{39} erg s^{-1} over 4.3 years in the $0.1-2.4$ keV band [38]) consistent with the rate of decline $(t^{-0.7})$ observed in the radio regime over the first ~ 10 years after the outburst [79]. Since both the X-ray and radio emission processes are considered to be linked to the interaction of the SN shock with

[3] The *VLA* telescope of the National Radio Astronomy Observatory is operated by Associated Universities, Inc. under a cooperative agreement with the National Science Foundation.

the high-density envelope of matter (see Sect. 1), one would expect the X-ray evolution to follow the observed radio evolution.

3.3 SN1980K in NGC6946

Historically, the SNIIL SN1980K in NGC6946 was the first SN detected in X-rays [9]. SN1980K was measured by an *Einstein Imaging Proportional Counter (IPC)* observation with a $(0.2-4\ \mathrm{keV})$ luminosity of $L_{\mathrm{X}} = 4.8 \times 10^{38}\ \mathrm{erg\ s^{-1}}$ 35 days after the outburst [9,60]. A follow-up observation on day 82 showed that the X-ray source had faded by a factor of ~ 5. Despite the rather large *IPC* error box of $\sim 1'$, the positional coincidence of the X-ray source with the optical position of the SN was later confirmed with the *ROSAT PSPC* [59]. Spectral analysis of the *Einstein* and *ROSAT* data showed that the emission was rather soft (*Einstein*, $kT \sim 0.5\ \mathrm{keV}$ [9]; *ROSAT*: $kT \sim 0.4\ \mathrm{keV}$ [60]) and in relatively good agreement with theoretical predictions ($kT \sim 0.7\ \mathrm{keV}$ [10]). Given the low photon statistics, which physical processes were producing the X-rays could not be well constrained. The data, however, could be used to give estimates as to the mass-loss rate of the SN progenitor ($\dot{M} \sim 0.5-5 \times 10^{-6}\ \mathrm{M_{\odot}\ yr^{-1}}$) and to demonstrate that SNe do not significantly contribute to the overall diffuse X-ray background.

3.4 SN1986J in NGC891

SN1986J was observed with the *ROSAT HRI* and with *ASCA* over a period covering $\sim 9-13$ yr after the outburst [5,11,32]. Contrary to many other X-ray SNe (e.g., SN1978K, SN1993J, SN1994I and SN1998S), SN1986J showed a rather fast X-ray rate of decline ($L_{\mathrm{X}} \propto t^{-2}$ [32]). Two *ASCA* spectra indicated that the emission was rather hard ($kT = 5-7\ \mathrm{keV}$) compared to other X-ray SNe (see, e.g., Sects. 3.3 and 3.2). The *ASCA* spectra also clearly showed a Fe-K emission line at 6.7 keV with a width of $< 20,000\ \mathrm{km\ s^{-1}}$ (FWHM). The spectral properties and rate of decline were used to test two different models. In the CSM interaction model the X-ray emission is thought to originate at the reverse shock which is passing through the outer layers of the SN ejecta [10]. While the predicted X-ray luminosity ($\sim 10^{40}\ \mathrm{erg\ s^{-1}}$ for a mass-loss rate of $\sim 10^{-4}\ \mathrm{M_{\odot}\ yr^{-1}}$ with an assumed wind velocity of $10\ \mathrm{km\ s^{-1}}$) and line width ($\sim 10,000\ \mathrm{km\ s^{-1}}$ FWHM) for the shocked plasma are in agreement with the model prediction, the predicted temperature is significantly lower ($\sim 1\ \mathrm{keV}$) than observed. In an alternative model proposed by Chugai [11], the emission arises from shocked interstellar clumps after the forward shock plows through the CSM. This model predicts a higher temperature of the shocked regions, in agreement with the data, and a similar mass-loss rate for the progenitor. Potential problems, however, arise in this model from the fact that the shocked clouds should have narrow emission lines (a few $\times 100\ \mathrm{km\ s^{-1}}$). Such narrow emission lines have, in fact, been observed in optical spectra of SN1986J (e.g., Hα emission lines with a FWHM of $530\ \mathrm{km\ s^{-1}}$ [42,58]), and could support the Chugai [11] model if low speed clumps are overtaken by a fast shock with a velocity of $\sim 2,000$ km s^{-1}, which is required for the observed temperature [32]. Weiler et al. [78] also find a clumpy CSM necessary to explain their radio data.

3.5 SN1988Z in MCG+03-28-011

At a distance of ~ 90 Mpc (assuming $H_0 = 75$ km s^{-1} Mpc^{-1}), SN1998Z is the most distant and most luminous X-ray ($L_{\rm x} \sim 10^{41}$ erg s^{-1}, 6.5 yr after the outburst) and radiosupernova!radio!luminosity ($L_{\rm 6cm} \sim 9 \times 10^{32}$ erg s^{-1} Hz^{-1} on day 1,253) SN discovered to date [16,72]. Given the observed X-ray luminosity, a CSM density of $\sim 10^6$ cm^{-3} was inferred at a radius of $r = 5 \times 10^{16}$ cm from the site of the explosion [16]. The radio data further implied that the dense cocoon resulted from a high mass-loss rate by the progenitor of $\dot{M} \sim 10^{-4}$ M$_\odot$ yr^{-1} in the late stages of a massive ($20-30$ M$_\odot$) star [72]. Two *ROSAT HRI* follow-up observations [1] showed that the SN rapidly faded by a factor of ~ 5 within 2 yr (see Fig. 1).

3.6 SN1993J in NGC3031 (M81)

ROSAT observations of SN1993J gave strong observational support for models that attribute the X-ray emission to the interaction of outgoing and reverse shocks with the CSM. Whereas the early X-ray spectrum on day 6 was hard ($T \sim 10^{8.5}$ K) and absorbed by a Galactic foreground column of $N_{\rm H} \sim 4 \times 10^{20}$ cm^{-2}, a softer component ($T \sim 10^7$ K) dominated by day $\sim$200 [25,84]. *ASCA* data from day 8 in the broader $1-10$ keV X-ray band were also best characterized by hard ($\sim 10^8$ K) thermal emission [70]. Very hard X-ray photons were recorded during days $9-15$ and $23-36$ using the *Oriented Scintillation Spectrometer Experiment (OSSE)* onboard the *CGRO*. During these dates, SN1993J reached ($50-150$ keV band) luminosities of $L_{\rm x} = 5.5 \times 10^{40}$ erg s^{-1} and 3.0×10^{40} erg s^{-1}, respectively, before falling below the *OSSE* detection threshold during a longer observation on days $93-121$ [43]. The harder component is probably due to the circumstellar shock region, whereas the softer component must have come from the region behind the reverse shock. The emergence of this radiation could be attributed to the decreased absorption by a cool shell [25]. The early (< 200 days) *ROSAT* and *ASCA* data were successfully used to estimate the mass-loss rate of the SN1993J progenitor ($\dot{M} \sim 4 \times 10^{-5}$ M$_\odot$ yr^{-1}) and indicate that the CSM density profile might be flatter ($\rho_{\rm CSM} \propto r^{-s}$ with $1.5 \lesssim s \lesssim 1.7$) than expected for constant wind velocity, constant mass-loss rate conditions ($s = 2$) [25].

Intriguing new results came from the analysis of the entire *ROSAT* data set, covering a period from 6 days to 7 years after the outburst of SN1993J. The combined *ROSAT PSPC* and *HRI* light curve is best fitted by a $L_{\rm x} \propto t^{-0.27}$ X-ray rate of decline (see, e.g., Fig. 1). Since each X-ray observation at time t is related to the corresponding distance r from the site of the explosion,

$$r = v_{\rm s} t \tag{7}$$

with shock front velocity v_s, and to the age of the stellar wind,

$$t_w = tv_s / w_{wind}, \tag{8}$$

as in the radio (see the chapter by Sramek and Weiler in this volume), the *ROSAT* measurements could be used as a "time machine" to look back into the history of the stellar wind. During the observed period, the SN shell had reached a radius of 3×10^{17} cm from the site of the explosion, corresponding to $\sim 10^4$ years in the progenitor's stellar wind history. Contrary to an expected CSM density profile of $\rho_{CSM} = \rho_0 (r/r_0)^{-s}$ profile with index $s = 2$ for a constant wind velocity w_{wind} and constant mass-loss rate $\dot{M}$, the data revealed a significantly flatter profile of $\rho_{CSM} \propto r^{-1.63}$. After ruling out alternative scenarios that might explain the data (e.g., variations in velocity and/or temperature of the shocked CSM or a non-spherical geometry caused by a binary companion), it was concluded that the mass-loss rate of the progenitor has decreased constantly from $\dot{M} = 4 \times 10^{-4}$ $M_\odot$ yr^{-1} to 4×10^{-5} $M_\odot$ yr^{-1} ($w_{wind}/10$ km s^{-1}. A similar result was found by Van Dyk et al. [73] from radio observations. The observed evolution (see Fig. 3) clearly reflects either a decrease in the mass-loss rate, an increase in the wind speed, or a combination of both, indicating that the progenitor likely was making a transition from the red to the blue supergiant phase during the late stage of its evolution. (See, e.g., the chapter by Sramek and Weiler in this volume and [82].) This scenario for the evolution of the SN1993J progenitor has interesting similarities with that of SN1987A, whose progenitor Sk-69 202 apparently reached the blue supergiant (BSG) phase, after significant mass-transfer to a companion, $\sim 10^4$ yr prior to the explosion (see [24,53] and Fig. 5 in the chapter by Sramek and Weiler in this volume).

3.7 SN1994I in NGC5194 (M51)

SN1994I represents a unique case since it is the first SNIc to be detected in soft X-rays. Based on early *ROSAT HRI* observations, evidence for soft (0.1 – 2.4 keV) X-ray emission was found on day 83 after the outburst [34]. However, due to the spatial resolution of the *HRI* (5″ on-axis) and the close location to the X-ray bright nucleus of M51 (18″ offset), as well as the high level of diffuse X-ray emission in the bulge of M51, the results were not entirely conclusive. Two follow-up observations 6 – 7 yr after the outburst, using the superior resolution and sensitivity of *CHANDRA*, revealed an X-ray source consistent with the radio position of the SN [36]. The *CHANDRA* observation yielded a $\sim 6\sigma$ detection with a luminosity of $L_x \sim 1 \times 10^{37}$ erg s^{-1} in the 0.3 – 2 keV band. Combined with the earlier *ROSAT* data, the X-ray light curve was parameterized as $L_x \propto (t - t_0)^\beta e^{-\tau}$ with $\tau \propto (t - t_0)^\delta$, where the external absorption (optical depth) at early times is represented by the $e^{-\tau}$ term with decline rate δ and the luminosity decline rate is exponential with index β. A best fit X-ray rate of decline with indices $\beta \sim -1$ and $\delta \sim -1$ was found. The index β is in good agreement with the radio rate of decline of SN1994I ($\beta = -1.6$). Assuming that the X-ray emission is due to the shocked CSM, a mass-loss rate

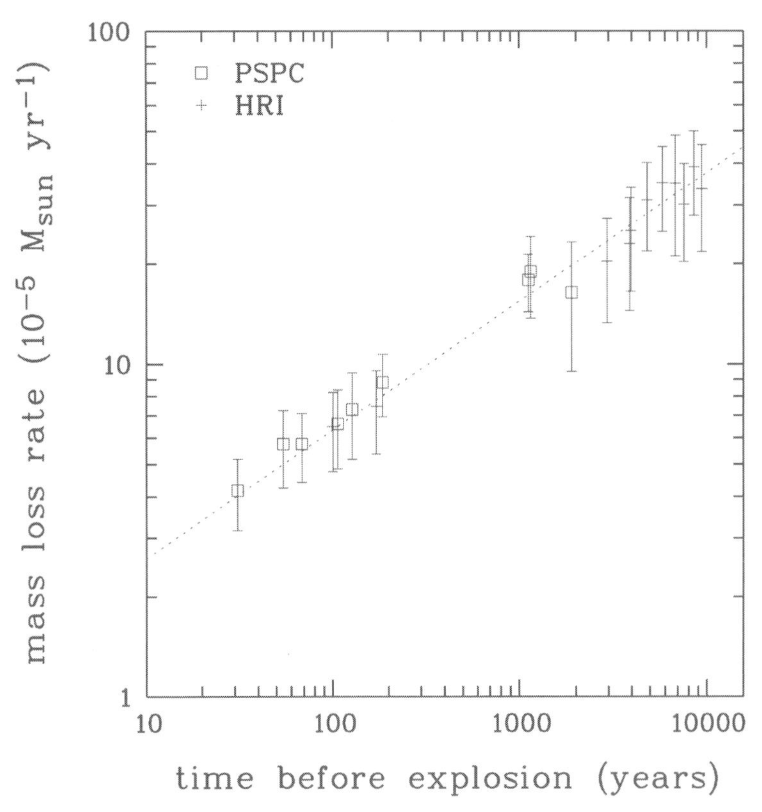

Fig. 3. Mass-loss rate history of the SN1993J progenitor. *ROSAT PSPC* data are marked by boxes, *HRI* data are indicated by crosses [35].

of $\dot{M} \sim 1 \times 10^{-5}(w_{\rm wind}/10\ {\rm km\ s^{-1}})\ {\rm M_\odot\ yr^{-1}}$ is consistent with all *ROSAT* and *CHANDRA* detections on days 82, 2,271 and 2,639 and a *ROSAT* upper limit on day 1,368 after the outburst. Also, the combined *ROSAT* and *CHANDRA* data gave a best fit CSM profile of $\rho_{\rm CSM} \propto r^{-(1.9\pm0.1)}$ for $r = 0.1-3.8\times10^{17}$ cm, consistent with what is expected for a constant stellar wind velocity and constant mass-loss rate from the progenitor (r^{-2}). A comparison of the CSM profiles of SN1994I and SN1993J, the only SNe for which the CSM profile was ever constructed from X-ray observations [35], is illustrated in Fig. 4.

3.8 SN1994W in NGC4041

A *ROSAT HRI* observation of SN1994W was performed 1,180 days after the outburst [63]. An X-ray source, with a luminosity of $L_{\rm x} \sim 8 \times 10^{39}$ erg s^{-1}, was found to coincide with the position of SN1994W. (Note that, given the positional coincidence, the low probability of $\sim 3 \times 10^{-3}$ of chance coincidence with a background or foreground object, and the indications of CSM interaction based on narrow absorption lines in the optical spectra, a detection of SN1994W

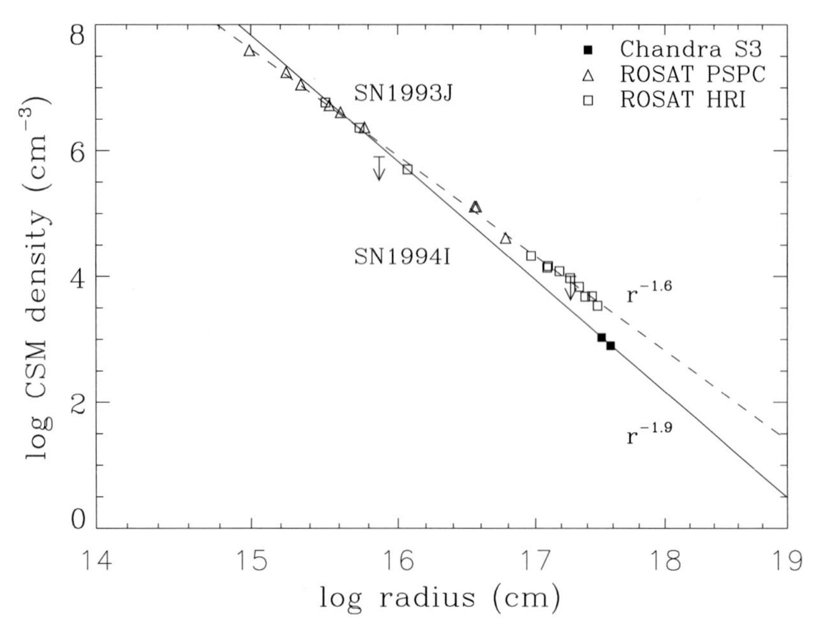

Fig. 4. CSM density profiles of SN1994I and SN1993J as a function of shell expansion radius [36].

in X-rays appears likely. However, since there are no other X-ray observations of the host galaxy NGC4041, confirmation of the detection is still pending.) Unfortunately, since SN1994W is the only X-ray SN not yet detected in the radio, no additional information about the CSM interaction is available. Interestingly, a CSM density of $\rho_{\rm CSM} \gtrsim 10^8 \ {\rm cm}^{-3}$ can be inferred from the optically thick Fe II lines [66] at radii of 8.5×10^{14} cm and 1.1×10^{15} cm (corresponding to days 31 and 57 after the outburst). The results are in remarkable agreement with the CSM density of $\sim 10^{7.5} - 10^8 \ {\rm cm}^{-3}$ for SN1994I and SN1993J at $r \sim 10^{15}$ cm (see, e.g., Fig. 4) and appears to be a common CSM density for massive progenitors ($\gtrsim 10 \ {\rm M}_\odot$ ZAMS) at this radius from the site of the explosion.

3.9 SN1995N in MCG-2-38-017

SN1995N was observed with the *ROSAT HRI* ~ 1 and 2 years after the outburst, followed by an *ASCA* observation some months later [23]. The *ASCA* spectrum could be well described by a $kT \sim 10$ keV thermal bremsstrahlung spectrum or a power-law with photon index ~ 1.7, both absorbed by a Galactic foreground column of $N_{\rm H} \sim 10^{21} \ {\rm cm}^{-2}$. Assuming that the spectrum has not changed, SN1995N faded by ~ 30% between the two *ROSAT* observations and subsequently brightened by a factor of two by the time of the *ASCA* observations. Together with the high observed luminosity ($\sim 10^{41} \ {\rm erg \ s}^{-1}$), this might be indicative that the SN shock has plowed through a dense and structured CSM.

3.10 SN1998bw in ESO184-G82

Variable, hard ($2-10$ keV) X-ray emission was found with the Italian/Dutch satellite *BeppoSAX* at the position of the unusual SN1998bw, which might be associated with a γ-ray burst event GRB980425 [50,51]. The variability suggests that the detected X-ray flux is associated with SN1998bw, although the limited angular resolution of the *BeppoSAX Narrow Field Instruments (NFI)* does not allow separation of the possible contribution from the SN host galaxy. Due to the likely association of SN1998bw with GRB980425, we refer to the Chapters by Galama, by Frontera, and by Weiler et al. in this volume for a more detailed review.

3.11 SN1998S in NGC3877

A wealth of X-ray data has been collected for the type IIn SN1998S with *CHANDRA* and *XMM* [37,54]. At an age of $\sim 2-3$ yr after the outburst, SN1998S is still bright in X-rays (*CHANDRA* data on day 668, $L_{\rm x} \sim 9 \times 10^{39}$ erg s^{-1} [54]; *XMM* data on day 1,202, $L_{\rm x} \sim 3 \times 10^{39}$ erg s^{-1} ($2-10$ keV) [37]) and increasing in cm radio flux density. The X-ray light curve based on five *CHANDRA* and one *XMM* observation is best described by an X-ray rate of decline of $L_{\rm x} \propto t^{-1.3}$. The inferred mass-loss rate is $\dot{M} \sim 2 \times 10^{-4}$ M$_\odot$ yr^{-1}. Spectral analysis of the *CHANDRA* data show a best fit temperature of $kT \sim 10$ keV and high over abundance of heavy elements such as Ne, Al, Si, S, Ar and Fe ($3-30$ times solar). The *CHANDRA* spectrum and best fit model are shown in Fig. 5. The high quality spectra, in combination with theoretical modeling, allowed, for the first time, use of emission line ratios of elements produced before the core-collapse and during the explosion to constrain the progenitor's mass. The observed O/Si ratio of $1-12$, Mg/Si of $0-0.7$ and Ne/Si of $0.6-14$ led to an estimate of the progenitor's mass of ~ 18 M$_\odot$ [54]. A more recent *XMM* observation revealed the emergence of a soft (~ 0.8 keV) component from the reverse shock in addition to the harder component (~ 9 keV) already observed with *CHANDRA*, and a prominent and broad Si emission line at 1.89 keV with an equivalent width of ~ 50 eV (FWHM) [37].

3.12 SN1999em in NGC1637

SN1999em is the only SNIIP detected both in the radio and X-rays, although at a very low level ($L_{\rm 6cm} = 2.2 \times 10^{25}$ erg s^{-1} Hz^{-1} on day 34; $L_{\rm x} = 7 \times 10^{37}$ erg s^{-1} on day 4 [54]). The *CHANDRA* X-ray data indicated a non-radiative interaction of the SN ejecta with the CSM, and a low mass-loss rate of $\dot{M} \sim 2 \times 10^{-6}$ M$_\odot$ yr^{-1}. Five *CHANDRA* observations, performed on days $4-368$ after the outburst, showed a temperature evolution during this period, with a softening of the emission from ~ 5 keV to ~ 1 keV for an assumed thermal bremsstrahlung spectrum. This evolution confirmed theoretical predictions of a change in temperature and indicate a rather flat density profile of the ejecta. Given an observed evolution of $L_{\rm x} \propto t^{-1}$, the SN was below the detection limit of *CHANDRA* for observations on days 495 and 633 after the outburst [54].

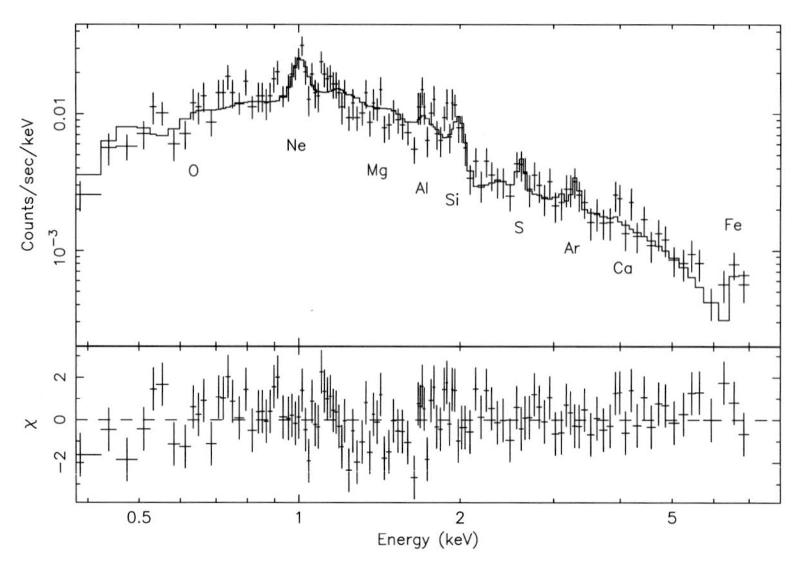

Fig. 5. *CHANDRA* ACIS-S3 spectrum of SN1998S [54]. The solid line gives the best fit model to the data. Labels indicate the location of emission lines.

3.13 SN1999gi in NGC3184

SN1999gi was detected in two *CHANDRA* observations on days 29 and 54 after the outburst [64]. Similar to SN1987A and SN1999em [54], SN1999gi only reached a relatively low luminosity ($\sim 10^{37}$ erg s^{-1}, $0.5-10$ keV band), implying a low mass-loss rate of $\dot{M} \sim 10^{-6}$ M$_\odot$ yr^{-1}. Like SN1999em, SN1999gi is a type IIP so that these two examples imply that type IIP SNe explode in a low density CSM environment [64]. By contrast, progenitors of SNIIL (e.g., SN1979C and SN1980K), SNIIb (SN1993J) and SNIIn (SN1986J, SN1988Z, SN1995N, and SN1998S) can establish a substantial CSM prior to the explosion. Chugai [12], assuming that the mass-loss rate is a function of the progenitor's zero age main sequence (ZAMS) mass, has proposed that the mass-loss rate rises to a peak near ~ 10 M$_\odot$ and falls with increasing mass, leading to a SNIIn-SNIIL-SNIIP sequence. Although the X-ray and radio data presently available have very limited statistics, they may lend support to this hypothesis. As more data is collected, it remains to be seen if this proposal can be sustained.

3.14 SN2002ap in NGC628 (M74)

At an age of 4 days, the SNIc [22] SN2002ap was detected both in the UV and in X-rays with the *XMM Optical Monitor (OM)* (wavelength range $245-320$ nm) and *EPIC-PN* [57,68]. Preliminary analysis of the *EPIC-PN* image showed a 3.5σ excess at a distance of $4''$ from the optical position of SN2002ap, with a positional error $< 10''$ at that stage of the data analysis. Located at a distance of 10 Mpc in the host galaxy NGC628, the recorded *EPIC-PN* count rate of $(9.0 \pm 2.5) \times 10^{-4}$ cts s^{-1} corresponds to a $(0.1-10$ keV band) luminosity of

$L_x \sim 2 \times 10^{37}$ erg s^{-1} for an assumed power-law spectrum with photon index 2 and a Galactic foreground column of $N_H = 5 \times 10^{20}$ cm^{-2} [57,68]. A similarly low X-ray band luminosity has been inferred for the type Ic SN1994I (see Sect. 3.7).

4 Summary

By observing electromagnetic radiation from SNe, we can, in principle, obtain information about:

- The nature and mass of the progenitor.
- The possible presence of a companion.
- The history of mass-loss and the speed of the progenitor's wind going back tens of thousands of years.
- The inhomogeneities in the CSM.
- The density gradient of the ejecta as they emerge in the early phase of the SNe.
- The geometry, physical conditions and composition of the ejected matter.
- The temperature and density in the region behind the reverse shock.
- The speed of the reverse shock as a function of time.
- The geometry, physical conditions and composition of the CSM.
- The speed of the circumstellar shock as a function of time.

A large part of the information comes from optical observations which historically have played a key role in our understanding and knowledge of SNe. The classification of the various SN types is exclusively linked to the optical (see the chapter by Turatto in this volume). A wealth of information also comes directly from the observed IR/optical/UV line profiles and fluxes (see the chapter by Branch et al. in this volume).

The radio emission, which is believed to arise largely from the region behind the forward shock, provides additional physical information (see the chapter by Sramek and Weiler in this volume). Radio results are, by and large, complementary to what we learn from the IR/optical/UV observations. In any wavelength range, there is not always a unique interpretation that explains the complicated spectral behavior. It is therefore perhaps not surprising that Filippenko's earlier review [20] is largely phenomenological – the physics is not always clear, although great progress is being made.

The X-ray observations are also valuable in that they preferentially probe the very hot regions ($T \sim 10^{6-10}$ K) and can thus provide information unavailable in any other wavelength range. Very roughly, we can expect three phases of X-ray emission:

- There should be a brief burst of high energy X-rays from type Ib/c and type II SNe.
- A blackbody continuum of ~ 0.02 keV should accompany the high temperature flash associated with the break out of the shock through the stellar surface.

- Weeks or months later, high energy X-rays may be detected when the expanding ejecta have become optically thin to X-rays, and/or when the ejecta plow into the CSM established by the stellar wind in previous phases of mass-loss of the progenitor.

In the case of type Ia, only weak X-ray emission is expected from the prompt thermal detonation (or deflagration) of the compact white dwarf. Such brief, prompt bursts have not yet been observed due to the lack of X-ray all sky monitors and the response time of orbiting X-ray observatories (days).

For the type Ib/c and type II SNe, the blastwave that is formed in the interaction with the CSM can have a very high temperature ($T \gtrsim 10^9$ K). radio emission is also believed to come from the region behind this shock. The reverse shock is also radiative and a dense, cool ($T \lesssim 10^4$ K) shell can form behind it [10]. If the density gradient of the outer part of a type Ib/c or type II SN is very large, the density at the reverse shock will be high and X-rays from the reverse shock will be heavily absorbed in this dense, cool shell. When this cool region has expanded sufficiently to become transparent to X-rays, X-ray emission from the reverse shock will start to dominate. Radio emission is not expected to come from this reverse shock region although it is remarkable how well the radio and the X-ray emission correlate (see Fig. 1).

If there are multiple density enhancements in the CSM due to spasmodic bursts of stellar wind from the progenitor, then the X-ray and radio emission will have complex evolution at later times (see, e.g., [46,47,81]. In the case of SN1987A the outgoing shock is just now beginning to plow into the equatorial ring of CSM seen earlier with the *Hubble Space Telescope (HST)* [7,8,52].

Immler et al. [35] have assembled a coherent picture of the various phases of X-ray emission for the case of SN1993J. They were even able to reconstruct the evolution of the progenitor of SN1993J, describing $\sim 10^4$ yr of mass-loss history. They found evidence that the progenitor was making a transition from the red to the blue supergiant phase during the last stages of evolution before explosion. Such a final evolution for the SN1993J progenitor is interesting to compare with that of SN1987A, whose progenitor completely entered the blue supergiant phase after significant mass transfer to a companion (see the chapter by McCray in this volume). The X-ray data for SN1993J extend over many years and demonstrate the potential of long term X-ray monitoring of SNe to probe the CSM interaction and the evolution of the progenitor.

In the case of type IIn SNe (e.g., SN1998S), it may not be easy to distinguish X-ray line emission from the region behind the reverse shock as opposed to line emission from a clumpy CSM [11]. The clumpy CSM may have been accelerated to $\sim 10^2 - 10^3$ km s^{-1}, in which case the X-ray lines should be Doppler broadened, but recent *CHANDRA* and *XMM* observations have not had sufficient photon statistics to address this question. The *CHANDRA* observations of SN1998S, however, did allow for the unprecedented determination of the abundances of various elements such as O, Ne, Mg, Al, Si, S, Ar, Ca, and Fe [54]. Comparing the observed abundance ratios with those of the models by Thielemann et al.[71] seems to indicate that the mass of the progenitor was ~ 18 $M_\odot$.

This is the first time that a mass determination has been made using X-ray data alone.

Acknowledgments

We wish to thank Dave Pooley for constructing Fig. 1 from archival, published, and propriety right data of the authors, and Roger Chevalier for helpful discussions.

References

1. I. Aretxaga et al.: Mon. Not. R. Astron. Soc. **309**, 343 (1999)
2. S. Benetti, P. Bouchet, H. Schwarz: IAUC 6170 (1995)
3. E. Berger, S.R. Kulkarni, D.A. Frail: GCN 1237 (2002)
4. R.M. Bionta et al.: Phys. Rev. Lett. **58**, 1494 (1997)
5. J.N. Bregman, R.A. Pildis: Astrophys. J. Lett. **398**, L107 (1992)
6. D.N. Burrows, J. Hayes, B.A. Fryxell: Astrophys. J. **450**, 830 (1995)
7. D.N. Burrows et al.: Astrophys. J. **452**, 680 (1995)
8. D.N. Burrows et al.: Astrophys. J. Lett. **543**, L149 (2000)
9. C.R. Canizares, G.A. Kriss, E.D. Feigelson: Astrophys. J. Lett. **253**, L17 (1982)
10. R.A. Chevalier, C. Fransson: Astrophys. J. **420**, 268 (1994)
11. N.N. Chugai: Astrophys. J. Lett. **414**, L101 (1993)
12. N.N. Chugai: Astron. Rep. **41**, 672 (1997)
13. S. Colgate: Sky & Tel. **74**, 229 (1987)
14. K. Dennerl et al.: Astron. Astrophys. **365**, L202 (2001)
15. T. Dotani, K. Hayashida, H. Inoue, M. Itoh, K. Koyama: Nature **330**, 230 (1987)
16. A.C. Fabian, R. Terlevich: Mon. Not. R. Astron. Soc. **280**, L5 (1996)
17. R.A. Fesen, R.H. Becker: Astrophys. J. **351**, 437 (1990)
18. R.A. Fesen, D.M. Matonick: Astrophys. J. **407**, 110 (1993)
19. A.V. Filippenko, T. Matheson, A.J. Barth: Astron. J. **108**, 2220 (1994)
20. A.V. Filippenko: Ann. Rev. Astron. Astrophys. **35**, 309 (1997)
21. A.V. Filippenko, E.C. Moran: IAUC 6830 (1998)
22. A.V. Filippenko, R. Chornock: IAUC 7825 (2002)
23. D.W. Fox et al.: Mon. Not. R. Astron. Soc. **319**, 1154 (2000)
24. C. Fransson, A. Cassatella, R. Gilmozzi, R.P. Kirshner, N. Panagia, G. Sonneborn, W. Wamsteker: Astrophys. J. **336**, 429 (1989)
25. C. Fransson, P. Lundqvist, R.A. Chevalier: Astrophys. J. **461**, 993 (1996)
26. T.J. Galama et al.: Nature **395**, 670 (1998)
27. P. Garnavich,S. Jha, R. Kirshner, J. Huchra: IAUC 6832 (1998)
28. P. Gorenstein, J.P. Hughes, W.H Tucker: Astrophys. J. Lett. **420**, L25 (1994)
29. M. Hashimoto, K. Iwamoto, K. Nomoto: Astrophys. J. Lett. **414**, L105 (1993)
30. G. Hasinger, B. Aschenbach, J. Trümper: Astron. Astrophys. **312**, L9 (1996)
31. K. Hirata, T. Kajita, M. Koshiba, M. Nakahata, Y. Oyama: Phys. Rev. Lett. **58**, 1490 (1987)
32. J.C. Houck, J.N. Bregman, R.A. Chevalier, K. Tomisaka: Astrophys. J. **493**, 431 (1998)
33. S. Immler, W. Pietsch, B. Aschenbach: Astron. Astrophys. **331**, 601 (1998)
34. S. Immler, W. Pietsch, B. Aschenbach: Astron. Astrophys. **336**, L1 (1998)

35. S. Immler, B. Aschenbach, Q.D. Wang: Astrophys. J. Lett. **561**, L107 (2001)
36. S. Immler, A.S. Wilson, Y. Terashima: Astrophys. J. **573**, 27 (2002)
37. S. Immler: 'X-Ray Emission from Young SNe', In: *High Energy Processes and Phenomena in Astrophysics, IAU Symp. 214 at Suzhou, China, 6-10 August 2002*, ed. by X-d Li, Z-r Wang, V. Trimble (Astron. Society of the Pacific 2003) in press
38. P. Kaaret: Astrophys. J. **560**, 715 (2001)
39. S.R. Kulkarni et al.: Nature **395**, 663 (1998)
40. S. Kumagai, T. Shigeyama, K. Nomoto, M. Itoh, J. Nishimura: Astron. Astrophys. **197**, L7 (1988)
41. S. Kumagai, T. Shigeyama, K. Nomoto, M. Itoh, J. Nishimura, S. Tsuruta: Astrophys. J. **345**, 412 (1989)
42. B. Leibundgut, R.P. Kirshner, P.A. Pinto, M.P. Rupen, R.C. Smith, J.E. Gunn, D.P. Schneider: Astrophys. J. **372**, 531 (1991)
43. M. Leising et al.: Astrophys. J. Lett. **431**, L95 (1994)
44. W.H.G. Lewin, H.-U. Zimmermann, B. Aschenbach: IAUC 6445 (1996)
45. W.-D. Li: IAUC 6829 (1998)
46. M.J. Montes, S.D. Van Dyk, K.W. Weiler, R.A. Sramek, N. Panagia: Astrophys. J. **506**, 874 (1998)
47. M.J. Montes, K.W. Weiler, S.D. Van Dyk, R.A. Sramek, N. Panagia, R. Park: Astrophys. J. **532**, 1124 (2000)
48. K. Nomoto, F.-K. Thielemann, K. Yokoi: Astrophys. J. **286**, 644 (1984)
49. R. Petre, K. Okada, T. Mihara, K. Makishima, E.J.M Colbert: Pub. Astron. Soc. Japan **46**, L115 (1994)
50. E. Pian et al.: Astron. Astrophys. Suppl. Ser. **138**, 463 (1999)
51. E. Pian et al.: Astrophys. J. **536**, 778 (2000)
52. P. Podsiadlowski: Nature **350**, 654 (1992)
53. P. Podsiadlowski, J.J.L. Hsu, P.C. Joss, R.R. Ross: Nature **364**, 509 (1992)
54. D. Pooley et al.: Astrophys. J. **572**, 932 (2002)
55. A. Ray, R. Petre, E.M. Schlegel: Astron. J. **122**, 966 (2001)
56. A.G. Riess et al.: Astrophys. J. **560**, 49 (2001)
57. P. Rodriguez-Pascual, R. Gonzalez-Riestra, B. Gonzalez-Garcia, M. Santos-Lleo, M. Guainazzi, N. Schartel: IAUC 7821 (2002)
58. M.P. Rupen, J.H. van Gorkom, G.R. Knapp, J.E. Gunn, D.P. Schneider: Astron. J. **94**, 61 (1987)
59. E.M. Schlegel: Astron. J. **108**, 1893 (1994)
60. E.M. Schlegel: Rep. Prog. Phys. **58**, 1375 (1995)
61. E.M. Schlegel:, R. Petre, E.J.M. Colbert: Astrophys. J. **456**, 187 (1996)
62. E.M. Schlegel, S. Ryder, L. Staveley-Smith, R. Petre, E. Colbert, M. Dopita, D. Campbell-Wilson: Astron. J. **118**, 2689 (1999)
63. E.M. Schlegel: Astrophys. J. Lett. **527**, L85 (1999)
64. E.M. Schlegel: Astrophys. J. Lett. **556**, L25 (2001)
65. L.A.L. da Silva: Astrophys. Space Sci. **165**, 255 (1990)
66. J. Sollerman, R.J. Cumming, P. Lundqvist: Astrophys. J. **493**, 933 (1998)
67. R. Sunyaev et al.: Nature **330**, 230 (1987)
68. F.K. Sutaria, A. Ray, P. Chandra: astro-ph 0210623
69. T. Suzuki, K. Nomoto: Astrophys. J. **455**, 658 (1995)
70. Y. Tanaka: IAUC 5753 (1993)
71. F.-K. Thielemann, K. Nomoto, M. Hashimoto: Astrophys. J. **460**, 468 (1996)
72. S.D. Van Dyk, K.W. Weiler, R.A. Sramek, N. Panagia: Astrophys. J. Lett. **419**, L69 (1993)

73. S.D. Van Dyk, K.W. Weiler, N. Panagia, M.P. Rupen, R.A. Sramek: IAUC 5979 (1994)
74. S.D. Van Dyk, R.A. Sramek, K.W. Weiler, M.J. Montes, N. Panagia: IAUC 6386 (1996)
75. S.D. Van Dyk, C.K. Lacey, R.A. Sramek, K.W. Weiler: IAUC 7322 (1999)
76. S.D. Van Dyk: private communication (2001)
77. K.W. Weiler, R.A. Sramek, N. Panagia, J.M. van der Hulst, M. Salvati: Astrophys. J. **301**, 790 (1986)
78. K.W. Weiler, N. Panagia, R.A. Sramek: Astrophys. J. **364**, 611 (1990)
79. K.W. Weiler, S.D. van Dyk, J.L. Discenna, N. Panagia, R.A. Sramek: Astrophys. J. **380**, 161 (1991)
80. K.W. Weiler, S.D. van Dyk, N. Panagia, R.A. Sramek: Astrophys. J. **398**, 248 (1992)
81. K.W. Weiler, N. Panagia, M.J. Montes: Astrophys. J. **562**, 670 (2001)
82. K.W. Weiler, N. Panagia, M.J. Montes, R.A. Sramek: Ann. Rev. Astron. Astrophys. **40**, 387 (2002)
83. S.E. Woosley, T.A. Weaver: Ann. Rev. Astron. Astrophys. **24**, 205 (1986)
84. H.-U. Zimmermann et al.: Nature **367**, 621 (1994)

Ultraviolet Supernovae

Nino Panagia

ESA/Space Telescope Science Institute, 3700 San Martin Drive, Baltimore, MD 21218, USA

Abstract. The observations of supernovae made with *IUE* and *HST* are reviewed and discussed. The various supernova types are characterized and discussed in the light of their ultraviolet properties. Special emphasis is placed on the results obtained for SN1987A in more than fifteen years of assiduous monitoring with both *IUE* and *HST*. It is shown that ultraviolet data have led to the clarification of a number of crucial issues such as the nature of the progenitor and its evolution and the recovery of the ultraviolet echo of the explosion. The observations made with *HST* show its great potential for exceptional results in the study of supernovae, both in the local Universe and at high redshifts.

1 Introduction

Supernovae (SNe) are the explosive death of massive stars as well as moderate mass stars in binary systems. They enrich the interstellar medium (ISM) of galaxies with heavy elements (only C and N can efficiently be produced and ejected into the ISM by red giant winds and by planetary nebulae, as well as by pre-SN massive star winds): type Ia SNe (SNIa) provide mostly Fe and iron peak elements and type II SNe (SNII) mostly O and alpha-elements (see below for type definitions). Therefore, SNe are the primary contributors to the chemical evolution of the Universe. Moreover, SN ejecta carry $\sim 10^{51}$ erg in the form of kinetic energy, which constitutes a large injection of energy into the ISM of a galaxy. In fact, the total mechanical energy of a Milky Way-class galaxy (a late-type spiral with $\sim 10^{11}$ $M_\odot$) is approximately given by the product

$$M_{gal} \times v_{rot}^2 \simeq 10^{43} \times 4 \times 10^{14} = 4 \times 10^{57} erg, \quad (1)$$

which is equivalent to a mere 4×10^6 SN events. At a rate of, say, four SNe per century, SN explosions could approximately double the energy content of a galaxy in about 100 Myrs (if there where no losses and dissipation). This energy input is thus very important for the evolution of the entire galaxy, both dynamically and for star formation through cloud compression/energetics.

SNe are bright events that can be detected and studied up to very large distances. Therefore,

- SNe can be used trace the evolution of the Universe, and they can serve as measuring sticks to determine cosmologically interesting distances, either as "standard candles" (SNIa, which at maximum are about 10 billion times

brighter than the Sun, have a luminosity dispersion of only $\sim 10\%$) or by employing a refined Baade-Wesselink method [2,87] (SNII, in which strong lines provide ideal conditions for the application of the method, yield a distance accuracy of $\sim 20\%$).
- The intense radiation from SNe can be used to study the ISM and intergalactic medium (IGM) properties through measurements of the absorption lines. Since most of the strong absorption lines are found in the UV, this is best done observing SNII at early phases when the UV continuum is still quite strong. Additional studies in the optical (mostly Ca II and Na I lines) are possible using all bright SNe, but only by combining optical and UV observations can one obtain the whole picture and, therefore, SNII are the preferred targets for these studies.
- Finally, the strong light pulse from SNe (the typical half-power width of a light curve in the optical is about one month for SNIa and about two to three months for SNII; in the UV the light curve evolution is much faster) can used to probe the intervening ISM in the SN parent galaxy by observing the brightness and the time evolution of associated light echoes.

Within this framework, UV spectroscopy is crucially important in order to:

- Study the metallicity of individual SNe.
- Study the metallicity of the intervening ISM/IGM.
- Study the kinematics of the fast moving (i.e., the outermost layers) of the ejecta through the analysis of strong UV lines with P Cygni profiles.
- Study the overall energetics of the SN explosion at early phases (from shock breakout to optical maximum for all types of SNe, but most importantly for SNII).
- Study the strong emission lines produced in the interaction of the ejecta with pre-SN circumstellar material (e.g., collisionally excited C IV 1550 Å, N IV] 1470 Å, O III] 1665 Å, N III] 1750 Å, C III] 1909 Å, etc.).

In the following sections we shall review the observed properties of supernovae in the ultraviolet domain (i.e., typically 1100 – 3300 Å) and discuss the implications about the nature and origins of SNe and related topics, without entering into detailed aspects of the physics of radiative transfer and model atmospheres, which are discussed in the chapter by Branch et al. in this volume. In particular, Sect. 2 summarizes the available UV spectroscopic data, Sects. 3, 4 and 5 present and discuss the observations and the properties of type Ia, type Ib/c, and type II supernovae, respectively. Section 6 is entirely devoted to SN1987A, i.e., both the various aspects that have been clarified thanks to UV observations started promptly after the discovery and conducted assiduously over more than 15 years, and the phenomena that are expected to occur in the next decades because of interaction of the ejecta with the surrounding medium. Finally, Sect. 7 briefly reviews some applications of SN UV observations to the study of the distant Universe.

2 Ultraviolet Observations

The launch of the *International Ultraviolet Explorer (IUE)* satellite in early 1978 marked the beginning of a new era for SN studies because of its capability of measuring the ultraviolet emission of objects as faint as $m_B = 15$. Moreover, just around that time other powerful astronomical instruments became available, such as the *Einstein Observatory – HEAO-2 (Einstein)* X-ray measurements, the *Very Large Array (VLA)*[1] for radio observations, and a number of telescopes either dedicated to infrared observations (e.g., the *United Kingdom Infra-Red Telescope (UKIRT)* and *NASA Infrared Telescope Facility (IRTF)* on Mauna Kea) or equipped with new and highly efficient IR instrumentation (e.g., the *Anglo-Australian Telescope (AAT)* and *European Southern Observatory (ESO)* telescopes).

As a result, starting in the late 1970s a wealth of new information became available that, thanks to the coordinated effort of astronomers operating at widely different wavelengths, provided us with fresh insights into the properties and the nature of supernovae of all types. Eventually, the successful launch of the *Hubble Space Telescope (HST)* on 24 April 1990 [76] opened new possibilities for the study of supernovae, allowing us to study SNe with an accuracy unthinkable before and out to the edge of the visible Universe.

Historically, the first supernova observed with *IUE* was SN1978G in IC5021, near the end of the first year of *IUE* operation. It was a type II SN discovered somewhat after maximum, so that the observations could not be carried out for a long time.

In 1979 a joint *ESA UK Science and Engineering Research Council (SERC)* Target of Opportunity (ToO) program was started for observing bright supernovae, defined as $B_{max} < 12$ and in April 1979 a bright supernova, SN1979C in NGC4321, was discovered, promptly observed, and followed with *IUE* for more than three months [54]. During its operational lifetime, *IUE* observed all bright supernovae plus a number of fainter ones for a total of 23 objects, of which 7 are type II, 12 type Ia, and 4 type Ib/c. However, only seven, SN1979C, SN1980K, SN1981B, SN1983N, SN1987A, SN1990N, and SN1992A, were bright enough to permit the measurement of fair quality UV spectra and/or the following of their time evolution in some detail.

The advent of the *HST* has added another dimension to the UV investigation of SNe by permitting the quantitative study of the high energy part of their spectral energy distributions.

A summary of all UV spectroscopic observations is presented in Table 1. For each supernova, columns 1–4 give the SN designation, its type, B-band magnitude (or V-band when noted) at maximum light, and the parent galaxy (information extracted from the *Asiago Supernova Catalog* or the *NASA*'s *Infrared Processing and Analysis Center (IPAC)* Extragalactic Database). Column

[1] The *VLA* telescope of the National Radio Astronomy Observatory is operated by Associated Universities, Inc. under a cooperative agreement with the National Science Foundation.

Table 1. Summary of Ultraviolet Spectroscopic Observations of Supernovae

SN	Type	B_{max}	Galaxy	Telescope	Observing Period	No. SW	No. LW
1978G	II	< 12.9 V	IC5201	*IUE*	30 Nov.–11 Dec. 1978	–	2
1979C	IIL	11.6	NGC4321	*IUE*	21 Apr.–04 Aug. 1979	12	19
				HST	30 Jan. 1997	–	3
1980K	IIL	11.6	NGC6946	*IUE*	30 Oct.–09 Dec. 1980	12	34[a]
				HST	95/12/03–96/09/10	–	4
1980N	Ia	∼12.5	NGC1316	*IUE*	11 Dec. 1980–16 Jan. 1981	1	8
1981B	Ia	12.0	NGC4536	*IUE*	09 Mar.–05 Apr. 1981	2	5
1982B	Ia	13.7	NGC2268	*IUE*	18 Feb. 1982	1	2
1983G	Ia	12.9	NGC4753	*IUE*	08 Apr.–25 Apr. 1983	1	7
1983N	Ib	11.6	NGC5236	*IUE*	04 Jul. 1983–31 Jul. 1984	11	16
1984J	II	13.2 V	NGC1559	*IUE*	13 Aug. 1984	1	–
1985F	Ib	12.1	NGC4618	*IUE*	18 May 1985	–	1
1985L	II	13.0	NGC5033	*IUE*	28 Jun.–17 Jul. 1985	1	2
1986G	Ia	12.5	NGC5128	*IUE*	06 May–29 May 1986	–	6
1987A	II pec	4.5	*LMC*	*IUE*	24 Feb. 1987–20 Feb. 1994	282[a]	619[a]
				HST	01 Apr. 1992–present[b]	42	42
1988Z	IIn	<16.4	MCG+03-28-022	*HST*	02 Apr. 1996	–	5
1989B	Ia	11.8 V	NGC3627	*IUE*	31 Jan.–01 Feb. 1989	1	4
1989M	Ia	12.0 V	NGC4579	*IUE*	06 Jul.–20 Jul. 1989	–	7
1990B	Ib	16.1 V	NGC4568	*IUE*	25 Jan.–04 Feb. 1990	–	3
1990M	Ia	12.4 V	NGC5493	*IUE*	18 Jun. 1990	1	1
1990N	Ia	13.2 V	NGC4639	*IUE*	26 Jun.–14 Jul. 1990	–	9
1991T	Ia pec	11.7	NGC4527	*IUE*	27 Apr.–20 Jun. 1991	–	5
1991bg	Ia pec	14.0 V	NGC4374	*IUE*	26 Jun.–14 Jul. 1991	–	1
1992A	Ia	12.56	NGC1380	*IUE*	14 Jan.–10 Feb. 1992	–	10
				HST	24 Jan.–05 Nov. 1992	6	7
1993J	IIb	10.8	NGC3031	*IUE*	30 Mar.–18 May 1993	20[a]	32[a]
				HST	08 Apr. 1993–18 Apr. 2000	17	16[a]
1994I	Ic	12.9 V	NGC5194	*IUE*	03 Apr.–20 Apr. 1994	3	7
				HST	19 Apr.–14 May 1994	–	9
1994Y	IIn	14.2	NGC5371	*HST*	29 Aug. 1995	3	3
1995N	II	<17.5 V	—galMCG-02-38-017	*HST*	10 Sep. 1996–06 Feb. 1997	5	3
1998S	IIn	12.2 V	NGC3877	*HST*	16 Mar. 1998–03 Jul. 1999	15	15[a]
1999em	IIP	<13.5	NGC1637	*HST*	05 Nov. 1999–24 Jan. 2000	12[a]	9[a]
2001ay	Ia	<16.7 V	IC4423	*HST*	02 May–09 May 2001	–	4
2001ba	Ia	<15.9	MCG-05-28-01	*HST*	08 May–16 May 2001	–	3
2001eh	Ia	<16.5	UGC1162	*HST*	25 Sep.–04 Oct. 2001	1	9
2001el	Ia	<13.7 V	NGC1448	*HST*	29 Oct.–04 Dec. 2001	–	12
2001ep	Ia	<17.1	NGC1699	*HST*	27 Oct.–03 Nov. 2001	5	5
2001ex	Ia	<17.2	UGC3595	*HST*	29 Oct.–03 Dec. 2001	–	12
2001ig	IIb, Ib/c	<14.5	NGC7424	*HST*	14 Dec.–22 Dec. 2001	5	5
2002ap	Ic pec	14.5 V	NGC628	*HST*	01 Feb.–13 Feb. 2002	5	5

[a]High resolution spectroscopy obtained as well.
[b]As of June 2002.

5 specifies the UV telescope used, and column 6 lists the time interval over which observations were made. The last two columns, 7 and 8, give the total number of spectra obtained in either the short wavelength (SW, typically 1150 – 1950 Å for *IUE*) or the long wavelength (LW, about 1950 – 3250 Å for *IUE*) range. For some of the SNe observed with the *HST*, near-UV spectra, extending to 2900 Å

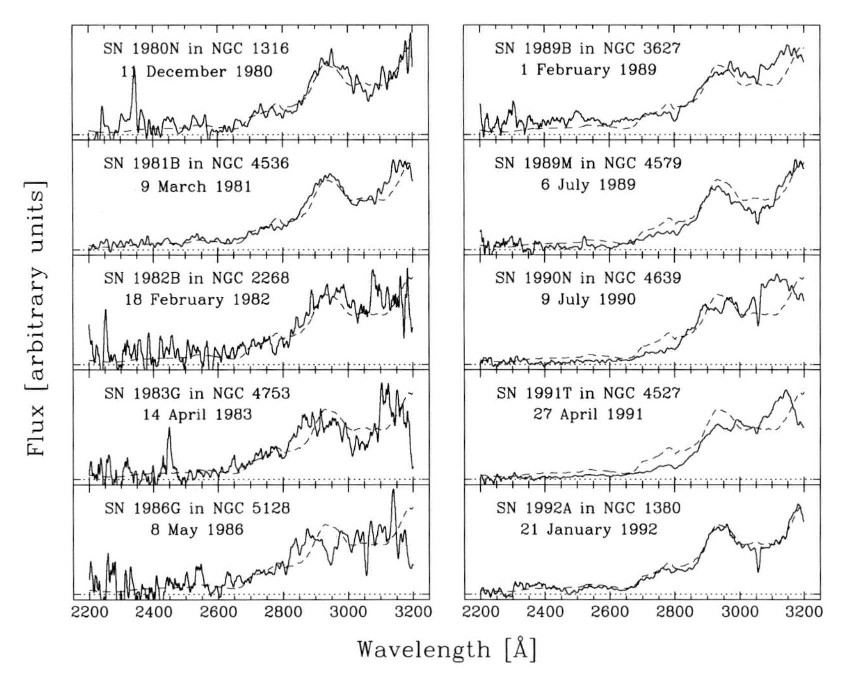

Fig. 1. Ultraviolet spectra of a sample of type Ia supernovae observed with *IUE* around maximum light.

on the short wavelength side, were obtained and they are included in the list as LW spectra.

3 Type Ia Supernovae

Type Ia supernovae (SNIa) are characterized by a lack of hydrogen in their spectra at all epochs and by a number of typically broad, deep absorption bands, most notably the Si II 6150 Å (actually the blue-shifted absorption of the 6347, 6371 Å Si II doublet (often referred to as the Si II 6355 Å feature; see e.g., [20]), which dominate their spectral distributions at early epochs. SNIa are found in all types of galaxies, from giant ellipticals to dwarf irregulars. However, the SNIa explosion rate, normalized relative to the galaxy H or K band luminosity and, therefore, relative to the galaxy mass, is much higher (up to a factor of 16 when comparing the extreme cases of irregulars and ellipticals [53]) in late type galaxies than in early type galaxies (see also the chapter by Cappellaro in this volume). This suggests that, contrary to common belief, SNIa belong to a relatively young (age less that 1 Gyr), moderately massive ($\sim 3.5\ M_\odot <$ M(SNIaprogenitor) $<$ $8\ M_\odot$) stellar population, and that in present day ellipticals SNIa are mostly the result of capture of dwarf galaxies by massive ellipticals [53].

Although 12 type Ia SNe were observed with *IUE*, only two events, namely SN1990N and SN1992A, had extensive time coverage, whereas all others were ob-

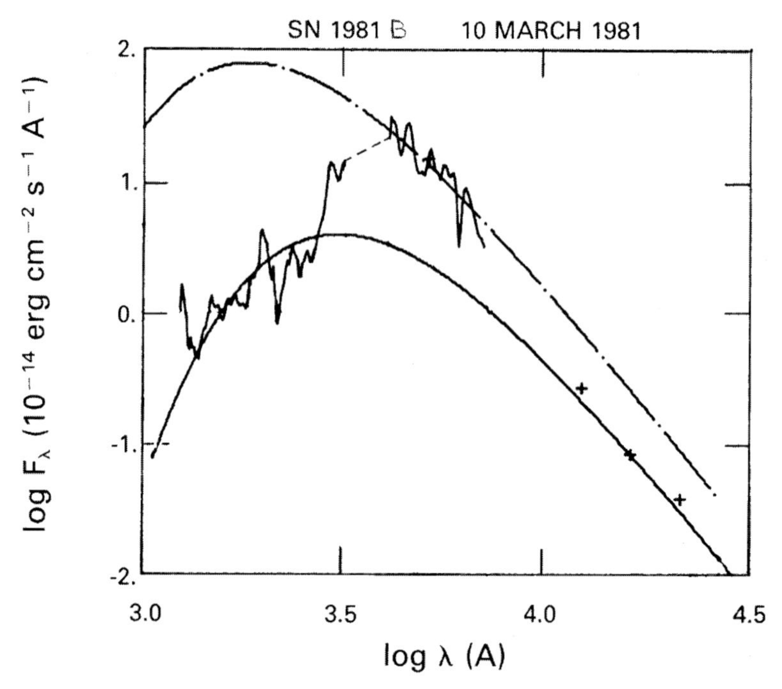

Fig. 2. Ultraviolet, optical, and near-infrared spectrum of SN1981B near maximum light. The two curves correspond to black-bodies at 15,800 K (dash-dotted line) and 9,400 K (solid line), which fit the optical spectrum and connect the UV and NIR spectra, respectively [51].

served only around maximum light either because of their intrinsic UV faintness or because of satellite pointing constraints. Even so, one can reach important conclusions of general validity, which are confirmed by the detailed data obtained for a few SNIa.

The UV spectra of type Ia SNe are found to decline rapidly with frequency, making it hard to detect any signal at short wavelengths. This aspect is illustrated in Fig. 1, which displays the LW spectra of the a sample of 10 type Ia SNe observed with *IUE*. In all cases the observing epoch is within three days of the optical maximum. It appears that the spectra do not have a smooth continuum but rather consist of a number of "bands" which are observed with somewhat different strengths. The most prominent feature is the emission that peaks at ~ 2950 Å with a half-power width of ~ 100 Å, i.e., $\Delta v \simeq 10^4$ km s^{-1}. Actually, this band is likely to be the result of strong absorptions occurring on both sides of the apparent emission (i.e., Mg II centered at ~ 2800 Å and Fe II ~ 3060 Å) and having half-power widths ~ 100 Å or, correspondingly, expansion velocities $\sim 10^4$ km s^{-1}. A similarly prominent emission band is seen at $\lambda \sim 1890$ Å in the only spectrum obtained at short wavelengths for SN1981B. Several other absorption features can be recognized, which are present at all epochs of observation. Although some of them may be identified with multiplets of Fe I,

Fe II and Mg II, no detailed study has yet been made for the majority of the absorptions. Nevertheless, the very fact that the spectrum is so similar for the first three SNe, and at all epochs, is already an important result. This supports the idea of an overall homogeneity in the properties of type Ia SNe.

On the other hand, some clear deviations from "normal" can be recognized for some SNIa. While the UV spectra of most SNIa shown in Fig. 1 are quite similar, and virtually indistinguishable from the spectrum of SN1992A near maximum light, one notices that both SN1983G and SN1986G display excess flux around 2850 Å, and a deficient flux around 2950 Å. This suggests that the Mg II resonance line is much weaker, which may indicate a lower abundance of Mg in these fast-decline, under-luminous SNIa. On the other hand, SN1990N, SN1991T, and, possibly, SN1989M show excess flux around ~2750 Å and ~2950 Å and a clear deficit around ~3100 Å, which may be ascribed to enhanced Mg II and Fe II features in these slow-decline, over-luminous SNIa.

In any case, their UV spectra are all much weaker than a blackbody extrapolation of their optical spectra would predict (see, e.g., Fig. 2 for the case of SN1981B [51]). This is due to strong line blanketing, mostly by Fe II and Co II lines, which both in the UV and in the near infrared (NIR) expose only atmospheric layers at temperatures about 2/3 the effective temperature of the supernova [9].

The best studied SNIa event so far is the "normal" type Ia supernova SN1992A in the S0 galaxy NGC1380 that was observed as a ToO by both *IUE* and *HST* [35]. The *HST Faint Object Spectrograph (FOS)* spectra, from 5 to 45 days past maximum light, are the best UV spectra available for a SNIa (Fig. 3) and reveal, with good signal to noise ratio, the spectral region blueward of ~2650 Å [35]. The UV photometry taken with the *HST Faint Object Camera (FOC)* in the F175W, F275W, and F342W bands shows light curves that resemble the SNIa template U-band light curve [44]. Using data from SN1992A and SN1990N, Kirshner et al. [35] constructed a SNIa template light curve for the flux region near 2750 Å (Fig. 3) that is quite detailed from 14 days before maximum light to 22 days after maximum light and that extends with less detail to 77 days after maximum light. This light curve also resembles the template U-band light curve although with a somewhat faster decline.

A local thermodynamic equilibrium (LTE) analysis of the SN1992A spectra, using a modified version of Woosley's delayed detonation model DD4 [90], shows that the features in the region shortward of ~2650 Å are P Cygni absorptions due to blends of iron peak element multiplets and the Mg II resonance multiplet. Newly synthesized Mg, S, and Si probably extend to velocities at least as high as ~19,000 km s^{-1}. Newly synthesized Ni and Co may dominate the iron peak elements out to ~13,000 km s^{-1} in the ejecta of SN1992A as in the DD4 model. On the other hand, an analysis of the O I λ7773 line in SN1992A and other SNIa implies that the oxygen rich layer in typical SNIa extends over a velocity range of at least $\sim 11,000-19,000$ km s^{-1}, but none of the "canonical" models, including model DD4, has an O-rich layer that completely covers this range.

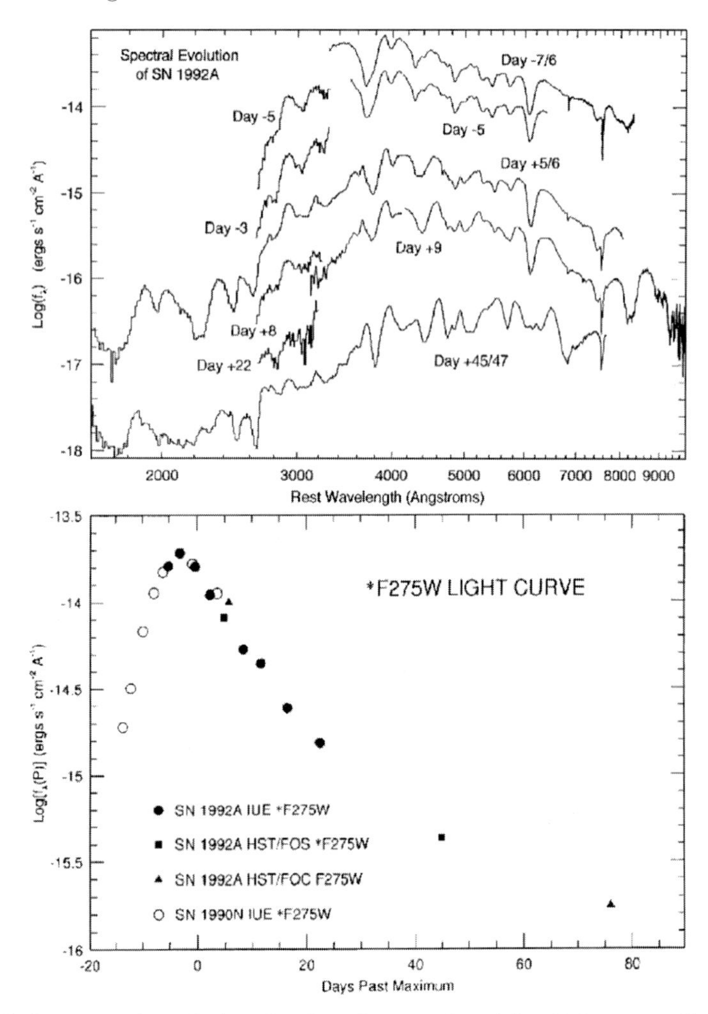

Fig. 3. Early spectral evolution (top) and near ultraviolet light curve (bottom) of SN1992A [35].

Even higher velocities were inferred by Jeffery et al. [42] for the over-luminous, slow-decline SNIa SN1990N and SN1991T through an LTE analysis of their photospheric epoch optical and UV spectra. In particular, SN1990N was found to have matter moving as fast as $40,000\ \text{km s}^{-1}$, while SN1991T had matter moving as fast as $20,000\ \text{km s}^{-1}$. The exponential density profile with e-folding velocity of $3,160\ \text{km s}^{-1}$ suggested by deflagration and delayed/late-detonation explosion models is adequate for calculating fits to the observed spectra. Findings on the iron peak elements suggest that in both SNe some nuclear burning continued into the outer ejecta or that the newly synthesized elements were mixed into the outer ejecta. The observed spectra of SN1990N are consistent with a composition of the inner envelope like that of the deflagration model W7

[50]. Intermediate-mass elements with abundances higher than solar are probably in the outer ejecta of both SNe, while Si, S, and Ca are under abundant in SN1991T relative to SN1990N by factors of order 3, 3, and 120, respectively [42].

It thus appears that type Ia supernovae are consistently weak UV emitters, and even at maximum light their UV spectra fall well below a blackbody extrapolation of their optical spectra. Broad features due to P Cygni absorption of Mg II and Fe II are present in all SNIa spectra, with remarkable constancy of properties for normal SNIa and systematic deviations for slow-decline, over-luminous SNIa (enhanced Mg II and Fe II absorptions) and fast-decline, under-luminous SNIa (weaker Mg II lines).

4 Type Ib/c Supernovae

Type Ib/c supernovae (SNIb/c) are similar to SNIa in not displaying any hydrogen lines in their spectra and are dominated by broad P Cygni-like metal absorptions, but they lack the characteristic 6150 Å trough of SNIa. The finer distinction into SNIb and SNIc was introduced by Wheeler and Harkness [88] and is based on the strength of He I absorption lines, most importantly He I (5876), so that the spectra of SNIb display strong He I absorptions and those of SNIc do not. Because of limited statistics, we will consider SNIb and SNIc together. SNIb/c have been found only in spiral galaxies, often (but not always) associated with spiral arms and/or H II regions. They are generally believed to be the result of the evolution of massive stars in close binary systems (see below).

Although the properties of some peculiarly red and under-luminous SNI (SN1962L and SN1964L) were already noticed by Bertola and collaborators in the mid-1960s [5,6], the first widely recognized member and prototype of the SNIb class was SN1983N in NGC5236 (M83). Because of its bright magnitude (B $\sim$ 11.6 mag at maximum light), SN1983N is one of the best studied SNe with *IUE*. Preliminary results can be found in [51], while a complete account of the UV, optical and IR observations obtained in the first two months after discovery will hopefully be published in the near future (N. Panagia, in preparation).

The UV spectrum of SN1983N closely resembles that of type Ia SNe at comparable epochs and, as such, only a minor fraction of the SN energy is radiated in the UV (see, e.g., Fig. 4). In particular, only $\sim$ 13% of the total luminosity was emitted by SN1983N shortward of 3400 Å at the time of the UV maximum (see, e.g., Fig. 5). Moreover, there is no indication of any stronger emission in the UV at very early epochs; this implies that the initial radius of the SN, i.e., the radius the stellar progenitor had when the shock front reached the photosphere, was definitely $\ll 10^{13}$ cm and probably $< 10^{12}$ cm, ruling out a red supergiant (RSG) progenitor. From the bolometric light curve (Fig. 5) Panagia [51] estimated that $\sim 0.15\ M_\odot$ of ^{56}Ni was synthesized in the explosion.

SN1994I was discovered on 02 April 1994 in the grand design spiral galaxy M51 and was promptly observed both with *IUE* (as early as 03 April) and

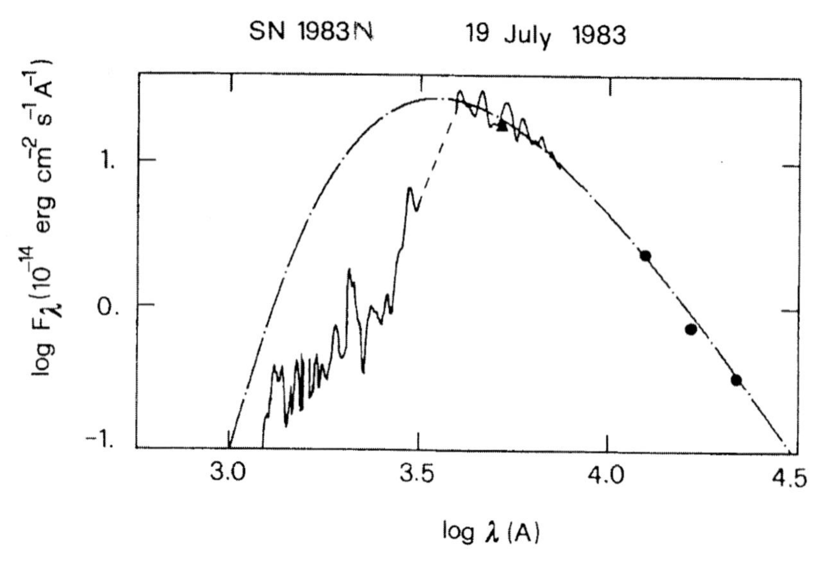

Fig. 4. The spectrum of SN1983N around maximum light (19 July 1983), dereddened with $E(B-V) = 0.16$. Both the UV and optical spectra have been smoothed with a 100 Å bandwidth. The triangle is the *IUE Fine Error Sensor (FES)* photometric point, and the dots represent the J, H, and K-band data. The dash-dotted curve is a blackbody curve at $T = 8300$ K [51].

with *HST FOS* (19 April). The UV spectra were remarkably similar to those obtained for SN1983N and, although they were taken only at two epochs well past maximum light (10 days and 35 days), they were of high quality. Millard et al. [49] analyzed the photospheric phase spectra and compared them with synthetic spectra generated by the parameterized supernova synthetic spectrum code *SYNOW* [8,21]. The observed optical spectra were well matched by synthetic spectra based on the assumption of spherical symmetry. The infrared absorption feature observed near 10250 Å, which previously had been attributed to He I λ10830 and regarded as strong evidence that SN1994I ejected some helium, could not be accounted for with He I alone and could be a blend of He I and C I lines or, alternatively, it can be fitted by Si I lines without compromising the fit in the optical region. From synthetic spectra matching the observed spectra from 4 days before to 26 days after the time of maximum brightness, the inferred velocity at the photosphere decreased from 17,500 to 7,000 km s^{-1}. Simple estimates of the kinetic energy carried by the ejected mass gave values that were near the canonical supernova energy of 10^{51} erg. Such velocities and kinetic energies for SN1994I are "normal" for SNe and are much lower than those found for the peculiar type Ic SN1997ef and SN1998bw (see, e.g., [10]) which appear to have been hyper-energetic.

Thus, as type Ia, type Ib/c supernovae are weak UV emitters with their UV spectra much fainter than a blackbody extrapolation of both optical and near

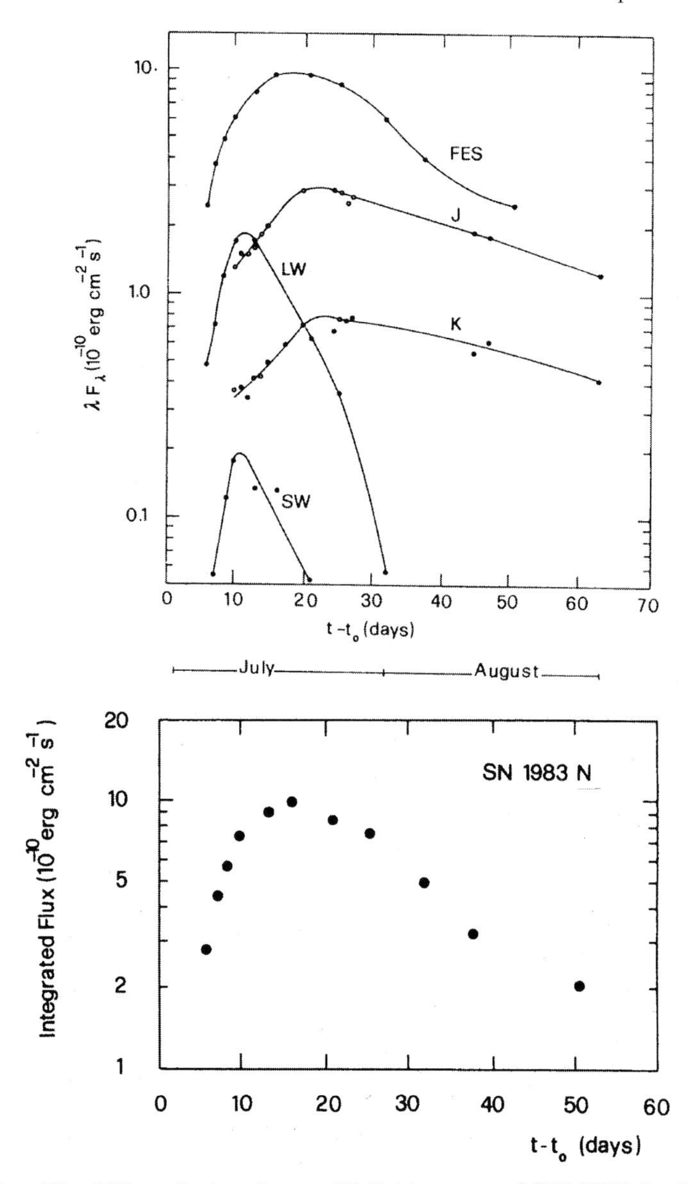

Fig. 5. Top: The UV, optical and near IR light curves of SN1983N in the form of λF_λ as a function of time elapsed since explosion [51]. The curves labeled as SW and LW represent the average fluxes measured in the short ($\lambda\lambda 1150 - 1950$ Å) and long ($\lambda\lambda 1950$–3250 Å) wavelength *IUE* cameras, respectively. Bottom: The bolometric light curve of SN1983N obtained by direct integration of the spectra dereddened with $E(B-V) = 0.16$ [51].

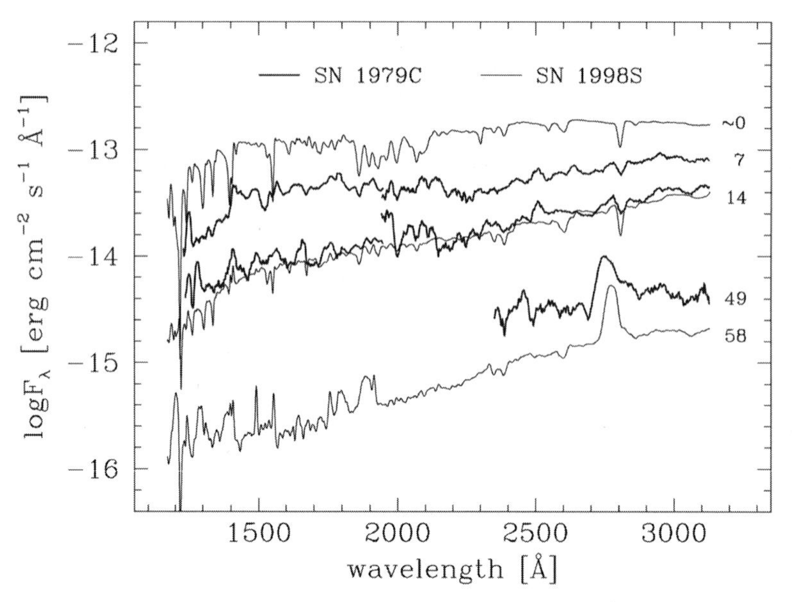

Fig. 6. A sample of UV spectra of SN1979C (adapted from [54]) and SN1998S (from the *Supernova INtensive Study (SINS)* project, unpublished) from near maximum light up to about two months after maximum. The labels denote the epochs, relative to the maximum, in days.

infrared (NIR) spectra. The mass of ^{56}Ni synthesized in a typical SNIb/c is, therefore, $\sim 0.15\ M_\odot$.

5 Type II Supernovae

Type II supernovae (SNII) display prominent hydrogen lines in their spectra (Balmer series in the optical) and their spectral energy distributions are mostly a continuum with relatively few broad P Cygni-like lines superimposed, rather than being dominated by discrete features as is the case of all type I supernovae. SNII are believed to be the result of a core collapse of massive stars, exploding at the end of their RSG phase. SN1987A was both a confirmation and an exception to this model. It was clearly the product of the collapse of a massive star, but it exploded when it was a blue supergiant (BSG) (see, e.g., [1] and Sect. 6), not an RSG.

Among the five SNII which were observed with *IUE*, only two, SN1979C and SN1980K, were bright enough to allow a detailed study of their properties in the UV. They were both of the so-called "linear" type (SNIIL), which is characterized by an almost straight-line decay of the B and V-band light curves, rather than of the more common "plateau" type (SNIIP) which display a flattening in their light curves starting a few weeks after maximum light.

For both SN1979C and SN1980K, the UV emission was strong for a considerable time past maximum light [54] with the UV flux higher than the extrapolation of the optical spectrum with a black body curve and with a clear excess shortward of $\lambda = 2000$ Å. Such an excess is found at all epochs, although it is less pronounced at early times. Fransson [22] has shown that the UV excess may be photospheric radiation which has been Compton-scattered by energetic, thermal electrons ($T \sim 10^9$ K) at the shock front where the ejecta interact with pre-existing circumstellar material. This model can explain both the extra radiation observed at short wavelengths and the high ionization implied by some emission lines (e.g., N IV] 1486 Å, C IV 1550 Å, etc.) observed in the spectrum of SN1979C [23,54].

The UV spectrum declines at a much faster rate than the optical emission and the UV spectral distribution becomes softer with time. However, at late times there is a hint of a slowing UV decline rate while the optical emission keeps declining steadily for more than a year (see Figs. 6 and 7). This behavior is also found in SN1998S and SN1987A, and is possibly due to the much higher UV opacity of the ejecta which become optically thin in the optical at early epochs and permits the optical emission to decline quite quickly. The UV appears to remain optically thick much longer and, therefore, the decline in UV brightness is much more gradual.

From a comparison of the line profiles observed in the UV and in the visual for SN1979C with theoretical calculations, Fransson et al. [23] concluded that the UV emission lines of highly ionized species are produced in the upper atmosphere, as well as Hα or Mg II λ 2800 Å, although just in the outermost layers where the density is lower and the ionizing radiation flux is higher. The line profiles imply an expansion velocity of $8,400$ km s^{-1}, which is only marginally lower than that measured in the optical (i.e., $9,200$ km s^{-1} [54]). From an analysis of the N V 1240 Å, N IV] 1486 Å, C IV 1550 Å, N III] 1750 Å, and C III] 1909 Å line intensities, the abundance ratio of N to C have been estimated to be N/C ~ 8, i.e., ~ 30 times higher than the cosmic value. This strong enhancement of nitrogen relative to carbon suggests that the pre-supernova star was a massive supergiant which had undergone a long period of mass-loss, thereby exposing CNO processed material.

An overabundance of N is common in RSGs (see, e.g., [37]), has been found in the wind of the SN1987A progenitor [24], is present in its circumstellar rings [58], and is possibly observed in SN1999em (see below). These results suggest that a considerable N overabundance may be a signature of type II supernovae.

Evidence for circumstellar interaction was found by Fesen et al. [19] in the optical and UV spectra of both SN1979C and 1980K observed at much later epochs. In particular, low-dispersion *W.M. Keck Telescope I (KECK I)* spectra of SN1980K taken in August 1995, 14.8 yr after explosion, and November 1997, 17.0 yr after explosion at the *MDM Observatory* showed broad $5,500$ km s^{-1} emission lines of Hα, [O I] 6300, 6364 Å, and [O II] 7319, 7330 Å. Weaker but similarly broad lines included [Fe II] 7155 Å, [S II] 4068, 4072 Å, and a blend of [Fe II] lines at $5050-5400$ Å. The presence of strong [S II] 4068, 4072 Å emission with a lack of [S II] 6716, 6731 Å emission indicated electron densities of 10^5-10^6 cm^{-2}. From the 1997 spectrum of SN1980K, an Hα flux of $(1.3 \pm 0.2) \times 10^{-15}$

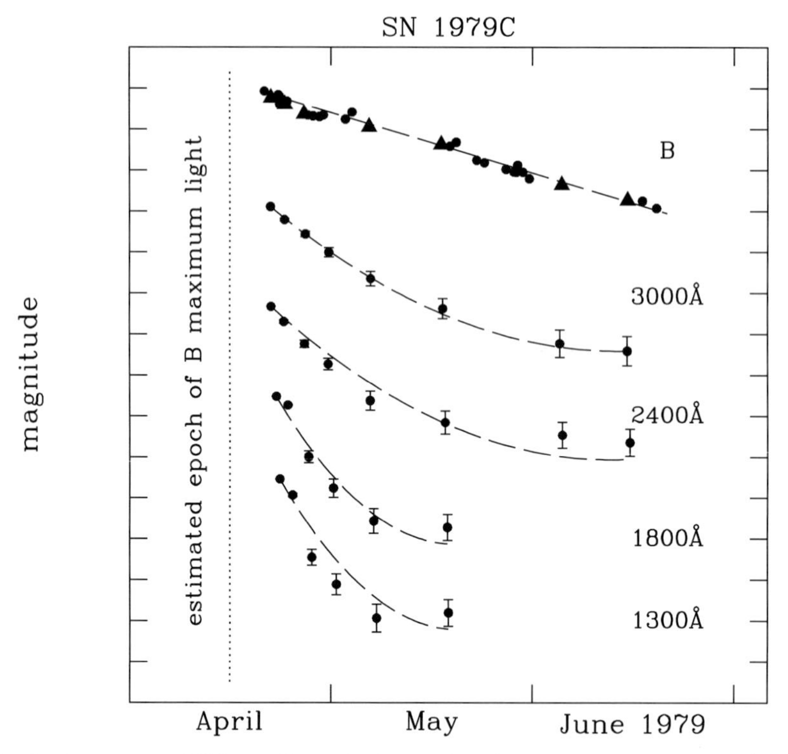

Fig. 7. Light curves of SN1979C in the B-band and in selected UV-bands (adapted from [54])

erg cm^{-2} s^{-1} was estimated, indicating a 25% decline during the period 1994 to 1997 from the 1987–1992 levels, possibly related to a reported decrease in its nonthermal radio emission.

On the other hand, a May 1993 *Multiple Mirror Telescope (MMT)* spectrum of SN1979C at age 14.0 yr showed a somewhat different spectrum from that of SN1980K. Broad, 6,000 km s^{-1} emission lines were seen, but with weaker Hα, stronger [O III] 4959, 5007 Å, more highly clumped [O I] and [O II] line profiles, no detectable [Fe II] 7155 Å emission, and a faint but very broad emission feature near 5750 Å. A 1997 *HST FOS*, near-UV spectrum (2200 – 4500 Å) of SN1979C showed strong lines of C II] 2324, 2325 Å, [O II] 2470 Å, and Mg II 2796, 2803 Å, along with weak [Ne III] 3969 Å, [S II] 4068, 4072 Å, and [O III] 4363 Å emission. The UV emission lines displayed a double-peak profile with the blueward peak substantially stronger than the red, suggesting appreciable dust extinction within the expanding ejecta ($E(B-V) = 0.11 - 0.16$ mag). The lack of detectable [O II] 3726, 3729 Å emission, together with [O III] I$(4959 + 5007)/I(4363) = 4$, implies electron densities of $10^6 - 10^7$ cm^{-3}.

It is clear that these type IIL supernovae like SN1979C and SN1980K show general spectral agreement with the lines expected in a circumstellar interaction model, even if specific models presently available show a number of differences with the observations. High electron densities ($10^5 - 10^7$ cm^{-3}) result in stronger collisional de-excitation than assumed in the models, thereby explaining the absence of several moderate to strong predicted lines such as [O II] 3726, 3729 Å, [N II] 6548, 6583 Å, and [S II] 6716, 6731 Å.

Independent evidence of interaction with a dense circumstellar medium (CSM) created by a pre-SN stellar wind is provided by the extensive series of studies of strong radio emission from many SNII by Weiler and collaborators (see, e.g., [77,82–85] and the chapter by Sramek and Weiler in this volume).

The SNII studied best in the UV so far is possibly SN1998S in NGC3877, a type II with relatively narrow emission lines (SNIIn). SN1998S was discovered several days before maximum. Its first UV spectrum, obtained on 16 March 1998, near maximum light, was very blue and displayed lines with extended blue wings, which indicate expansion velocities up to 18,000 km s^{-1} (see Fig. 8). The UV spectral evolution of SN1998S (Fig. 8) showed the spectrum to gradually steepen in the UV, from near maximum light on 16 March 1998 to about two weeks past maximum on 30 March, and the blue absorptions to weaken or disappear completely. About two months after maximum (13 May 1998) the continuum was much weaker, although its UV slope had not changed appreciably, and it had developed broad emission lines, the most noticeable being the Mg II doublet at about 2800 Å. This type of evolution is quite similar to that of SN1979C (see, e.g., Fig. 6) and suggests that the two sub-types are somewhat related to each other, especially in their circumstellar interaction properties.

A detailed analysis of early observations of SN1998S done by Lentz et al. [36] using the non-LTE atmosphere code PHOENIX [3,8], and modeling both the underlying supernova spectrum and the overlying circumstellar interaction region, produced spectra in good agreement with observations. The early spectra are well fitted by lines produced primarily in the circumstellar region itself, and later spectra are due primarily to the supernova ejecta. Intermediate spectra are affected by both regions. A mass-loss rate of order $\dot{M} \sim 0.0001 \times (v/100\text{km s}^{-1})$ $M_\odot$ yr^{-1} was inferred from these calculations but with a fairly large uncertainty.

Despite the fact that type II plateau (SNIIP) supernovae account for a large fraction of all SNII, so far SN1999em in NGC1637 is the only SNIIP that has been studied in some detail in the ultraviolet (see Table 1). Although caught at an early stage, SN1999em was already past maximum light (see, e.g., [32]), which may account for the fact that its spectrum on 05 November 1999 (see Fig. 9) was not as blue as that of SN1998S on 16 March 1998, when the latter was near maximum light. A preliminary analysis of the optical and UV spectra made by Baron et al. [4] using the *SYNOW* synthetic spectrum code indicates that, spectroscopically, this supernova appears to be a normal type II and the models are in excellent agreement with the observations. Also, the analysis suggests the presence of enhanced nitrogen as found in other SNII. More refined analysis of these spectra done with the full non-LTE general model atmosphere code *PHOENIX* [3] is not able to confirm or disprove a nitrogen enhancement but conclusively requires enhanced helium. Another important result is that very early spectra, such as those obtained for SN1999em, combined with sophisticated

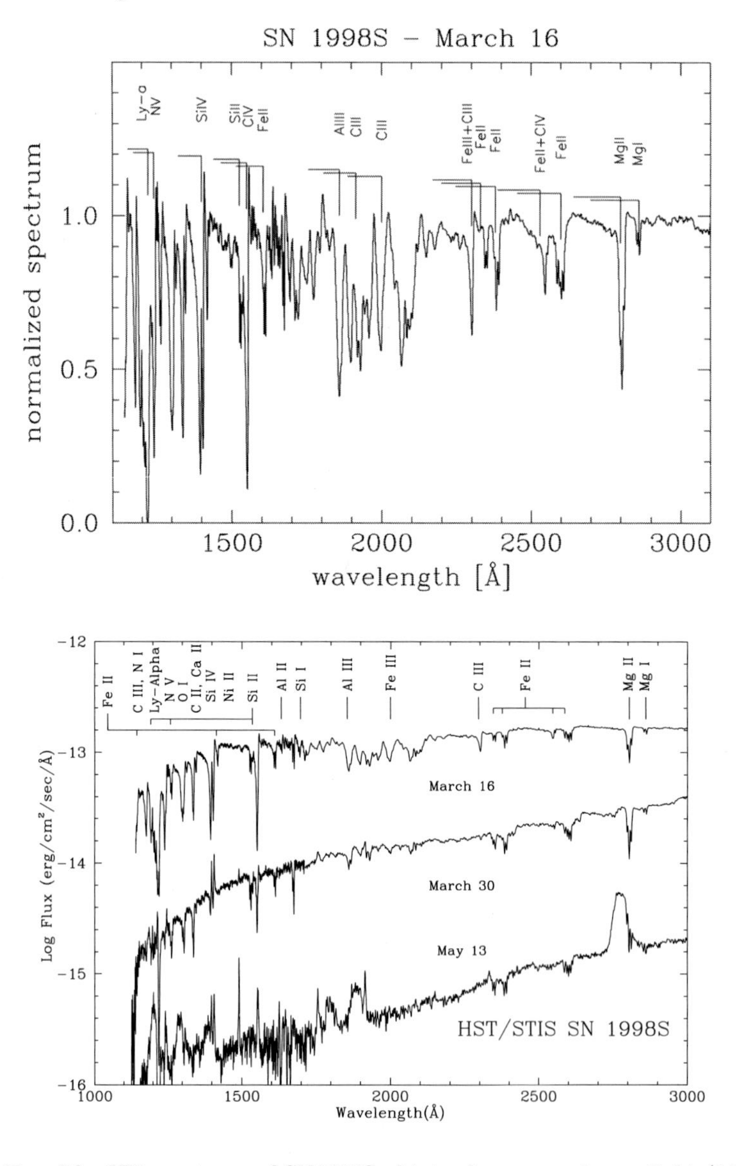

Fig. 8. Top: The UV spectrum of SN1998S obtained near maximum light (16 March 1998) and some line identifications. The broad blue wings of many lines indicate expansion velocities (as measured by the horizontal bars) up to 18,000 km s^{-1} or more. Bottom: UV spectral evolution of SN1998S (from the *SINS* project, unpublished). Shown are spectra obtained near maximum light (16 March 1998), about two weeks past maximum (30 March 1998), and about two months after maximum (13 May 1998).

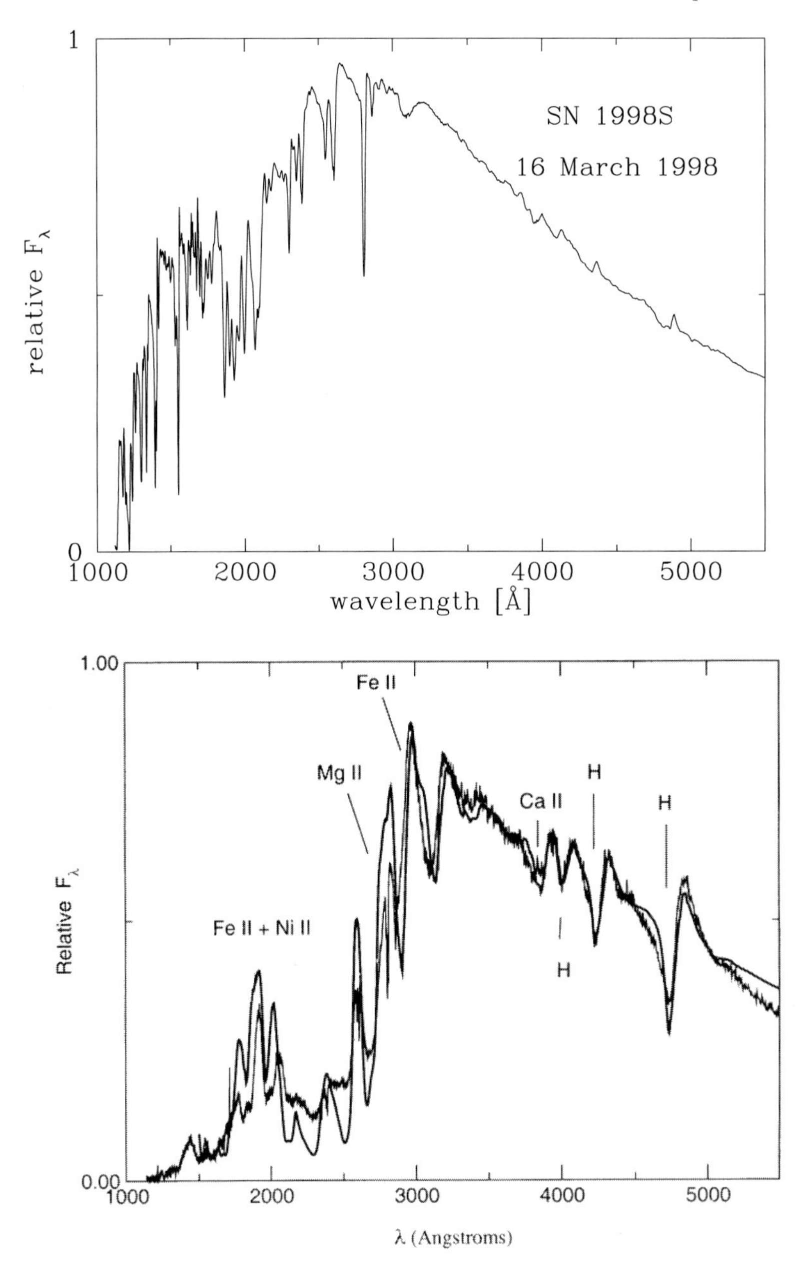

Fig. 9. Top: Observed spectra of SN1998S from the *HST STIS* and the *Fred L. Whipple Observatory (FLWO)* near maximum light (16 March 1998, adapted from [36]). Bottom: *SYNOW* fit to SN1999em UV+optical *HST* spectrum obtained on 05 November 1998, plus the optical spectrum obtained at the *FLWO* on 04 November 1998[4].

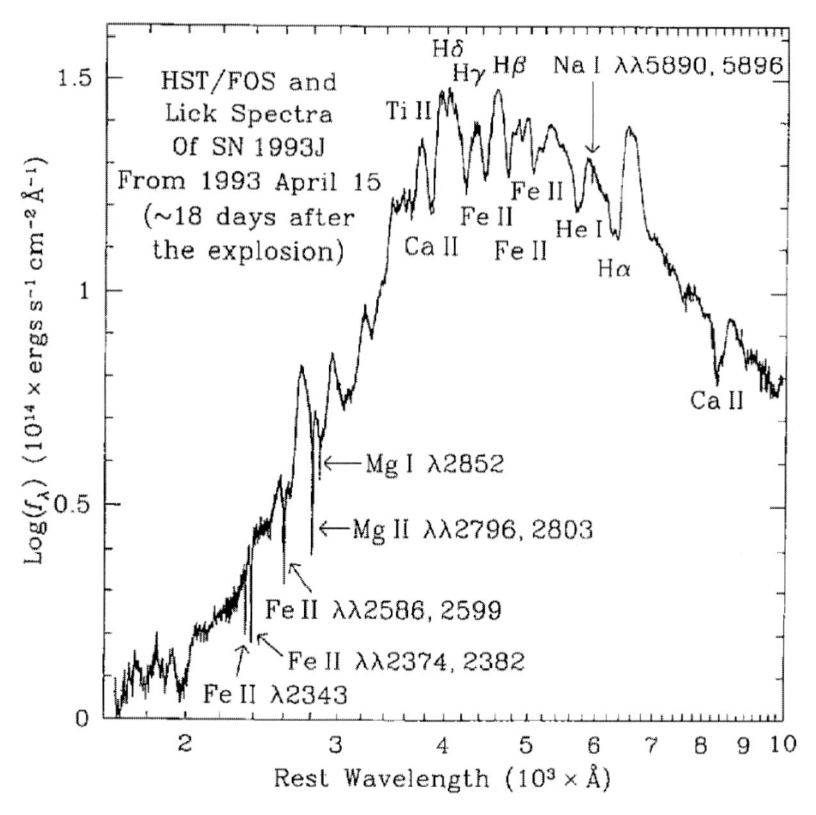

Fig. 10. *HST FOS* UV and *Lick Observatory* optical spectrum of SN1993J about 18 days after the explosion (15 April 1993 [43])

spectral modeling can lead to an independent estimate of the total reddening of the supernova, since, when the spectrum is very blue, dereddening leads to changes in the blue flux that cannot be reproduced by altering the "temperature" of the emitted radiation. Thus, detailed modeling of early spectra can shed light on both the abundances and total extinction of SNII, the latter improving their utility and reliability as distance indicators.

Another sub-type of the SNII family is the so-called type IIb SNe, dubbed so because at early phases their spectra display strong Balmer lines, typical of type II SNe, but at more advanced phases the Balmer lines weaken significantly or disappear altogether (see, e.g., [20]) and become more similar to the spectra of type Ib SNe. A prototypical member of this class is SN1993J that was discovered in early April 1993 in the nearby galaxy M81. A *HST* UV spectrum of SN1993J was obtained on 15 April 1993, about 18 days after explosion, and rather close to maximum light. The study of this spectrum [43] shows that the approximately 1650 – 2900 Å region is smoother than observed for SN1987A and SN1992A and lacks strong P Cygni absorptions caused by iron peak element lines

(Fig. 10). It is of interest to note that the UV spectrum of SN1993J is appreciably fainter than observed in most SNII, thus revealing its "hybrid" nature and some resemblance to a SNIb. Synthetic spectra calculated using a parameterized LTE procedure and a simple model atmosphere do not fit the UV observations. Radio observations suggest that SN1993J is embedded in a thick circumstellar envelope. The UV spectra of other supernovae that are believed to have thick circumstellar envelopes also have approximately 1650 – 2900 Å regions lacking in strong P Cygni absorptions. Interaction of supernova ejecta with circumstellar matter may be the origin of the smooth UV spectrum so that UV observations of supernovae could provide insight into the circumstellar environment of the supernova progenitors.

Thus, despite their different characteristics in the detailed optical and UV spectra, all type II supernovae of the various sub-types appear to provide clear evidence for the presence of a dense CSM and, in many cases, enhanced nitrogen abundance. Their UV spectra at early phases are very blue, possibly with strong UV excess relative to a blackbody extrapolation of their optical spectra.

6 UV Emission from SN1987A

Supernova 1987A was discovered on 24 February 1987 in the nearby, irregular galaxy, the *Large Magellanic Cloud (LMC)*, which is located in the southern sky. SN1987A is the first supernova to reach naked eye visibility since Kepler's SN in AD1604 (see the chapter by Green in this volume) and it is undoubtedly the best studied supernova event ever. Reviews of both early and more recent observations and their implications can be found in the chapter by McCray in this volume (see also [48]), in Arnett et al. [1], and in Gilmozzi and Panagia [28].

6.1 The Early Story

SN1987A brightened much faster than any known supernova, increasing in about one day from $\sim$ 12 to $\sim$ 5 mag at optical wavelengths, corresponding to an increase of about a factor of a thousand in luminosity before leveling off. Similarly in the ultraviolet, the flux initially was very high, even higher than in the optical, but by the first *IUE* observation, less than 14 hours after the discovery, the UV flux was already declining very quickly. It continued to drop by almost a factor of ten per day for several days. However, enough useful observations were obtained that SN1987A has become the most valuable test theories about SN explosions.

6.2 SN1987A Progenitor Star

From both the presence of hydrogen in the ejected matter and the detectable flux of neutrinos, it was clear that the star which exploded was quite massive, $\sim 20\ M_\odot$, and that the peculiarities in the development of the SN were due to the fact that the star which exploded was a BSG instead of the expected RSG star. In fact, the progenitor star can confidently be identified with Sk $-69°$ 202 [27,33].

On the other hand, the strong liklihood that Sk $-69°$ 202 was, in fact, in an RSG phase less than hundred thousand years before the explosion is shown by *IUE* observations indicating the presence of emission lines of nitrogen, oxygen, carbon and helium in the ultraviolet spectrum. The lines kept increasing in intensity with time and proved to be quite narrow, indicating that the emitting matter was moving at much lower speeds (less than a factor of hundred slower) than the supernova ejecta and provided a clear sign that matter was ejected by a presupernova, RSG wind phase.

6.3 IUE Observations of SN1987A

IUE observations of SN1987A were started a few hours after the announcement of its discovery, at both the *Goddard Space Flight Center (GSFC)* [33] and *Villafranca Satellite Tracking Station (VILSPA)* [79] *IUE* observatories. The two teams have since worked in close collaboration to provide a thorough coverage of the ultraviolet evolution of this SN at both high and low resolution. Also the *HST* has observed SN1987A since the time of its launch in 1990; the combined efforts of the *FOC Investigation Definition Team (IDT)*, *Wide-Field Planetary Camera 2 (WFPC2) IDT*, and *STIS IDT* teams, and the methodic and assiduous monitoring by the *SINS* project have provided an extensive data set for SN1987A.

The results obtained from studies of *IUE* observations can be summarized as follows:

A Compact Progenitor: Initially the UV flux was rather high, indicating high photospheric temperatures ($T > 14,000$ K). The fast UV flux decline (orders of magnitude in a few days; see, e.g., [55]) implies a small initial radius ($\sim$ few $\times$ 10 $R_{\odot}$) and, therefore, excludes a RSG progenitor at the time of explosion.

Progenitor Identification: The B3 I star Sk -69°202 was first suggested as the SN progenitor by Panagia et al. [55] and subsequently confirmed by the more detailed analyzes of Gilmozzi et al. [27] and Sonneborn et al. [74].

Milky Way and ISM Studies: The strong UV flux of the early epochs offered a unique opportunity to obtain high dispersion spectra of the ISM lines in our Galaxy and in the *LMC*. A large number of components for both high and low ionized species have been detected [16,17,86] indicating a rather complex structure for the intervening matter.

Evolution of the Progenitor: The presence of narrow emission lines of highly ionized species detected in short wavelength spectra since late May 1987 [34,56,80] has provided evidence for the progenitor having been a RSG before exploding in a BSG phase [24].

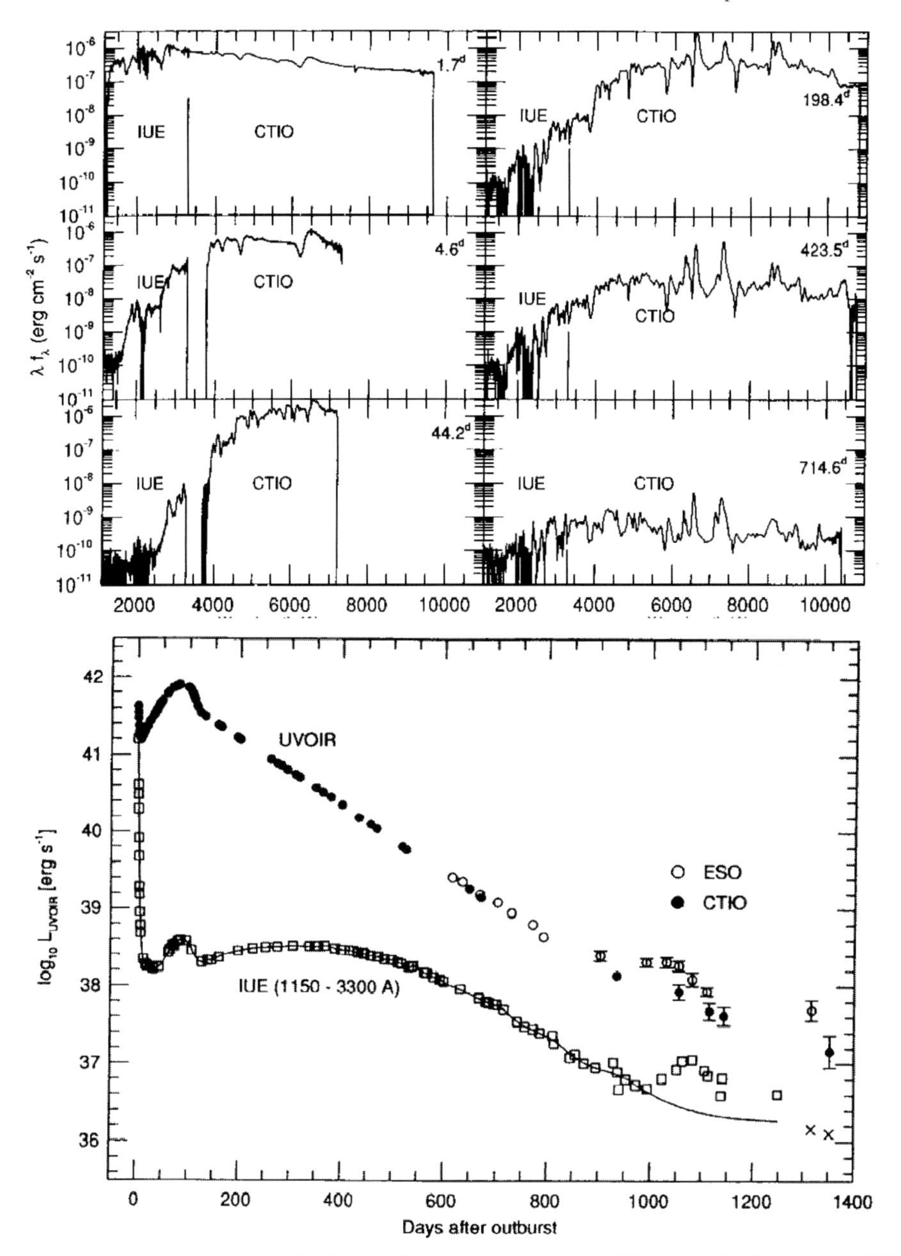

Fig. 11. Top: Evolution of the UV and optical spectrum of SN1987A [64]. Bottom: UV and bolometric light curves of SN1987A [64].

Energetics of the Radiation A complete catalog of SN1987A ultraviolet spectra obtained with *IUE* (751 spectra over the period 24 February 1987 [day 1.6] through 09 June 1992 [day 1567] has been presented and discussed by Pun et al.

[64]. They show that the UV flux plummeted during the earliest days of observations (Fig. 11) because of a drop in photospheric temperature and an increase in opacity. However, after reaching a minimum of 0.04% of the UV/optical/IR bolometric luminosity on day 44, the UV flux increased by 175 times to 7% of the bolometric luminosity by day 800 (Fig. 11). The UV colors reveal that the supernova started to get bluer in UV around the time when dust started to form in the ejecta, suggesting that the dust may be metal-rich.

UV Light Echo: Bright transient events such as novae and supernovae can give rise to a light echo. This is produced when light from the explosion illuminates nearby interstellar dust and is reflected in the direction of the observer. In the case of SN1987A echoes were predicted by Chevalier [12,13] and Schaefer [72], and discovered by Crotts [14] and Rosa [69] as two concentric rings around the supernova. A third inner ring was later discovered by Bond et al. [7]. Since discovery, the echoes have expanded and have been used to determine the three-dimensional cloud distribution and foreground geometry in the vicinity of SN1987A [91,92].

The geometry of light echos (first detected around Nova Per in 1901) is known; an echo seen at a projected angular separation θ is caused by a dust layer at a distance $r = ct[1 + (D\theta/ct)^2]/2$ in front of the supernova. For $ct << r$, $\theta \propto \sqrt{t}$ and the echo expands (superluminally) with time. The echo "thickness" is $\Delta\theta \propto \sqrt{\Delta r \Delta t}$ and is determined by whichever is dominant, the cloud thickness or the pulse duration.

The UV light curve was already plummeting by the time of the first *IUE* observation on 25 February 1987 and faded by a factor of $\sim$ 1000 in the next few days. Therefore, the expected UV echo is a reflection of the light emitted at the instant of the shock breakout (i.e., before discovery of the supernova!), while the optical echo is a reflection of the broad maximum reached three months later. Ultraviolet light emitted by SN1987A at the moment of the shock breakout, although not directly observed because the SN was discovered about 24 hours after the explosion, was studied by making *IUE* observations at a location a few arcseconds from a bright portion of an optical echo where there is a dust cloud a few hundred parsecs in front of the supernova (see Fig. 12 [28]). The spectrum of the UV echo shows a hot continuum and a wide P Cygni-like feature centered around 1500 Å (Fig. 13) which, if interpreted as C IV 1550 Å, implies an expansion velocity at the time of the shock breakout of $\sim 40,000$ km s^{-1} (in agreement with the first "direct" *IUE* spectrum taken 24 hours after the explosion which showed a Mg II line with a terminal velocity of $\sim 35,000$ km s^{-1}).

6.4 HST Observations

The *HST* was not in operation when the supernova exploded but, shortly after launch, first images taken with the *European Space Agency (ESA) FOC* on 23 and 24 August 1990, revealed the inner circumstellar ring (see, e.g., [38]).

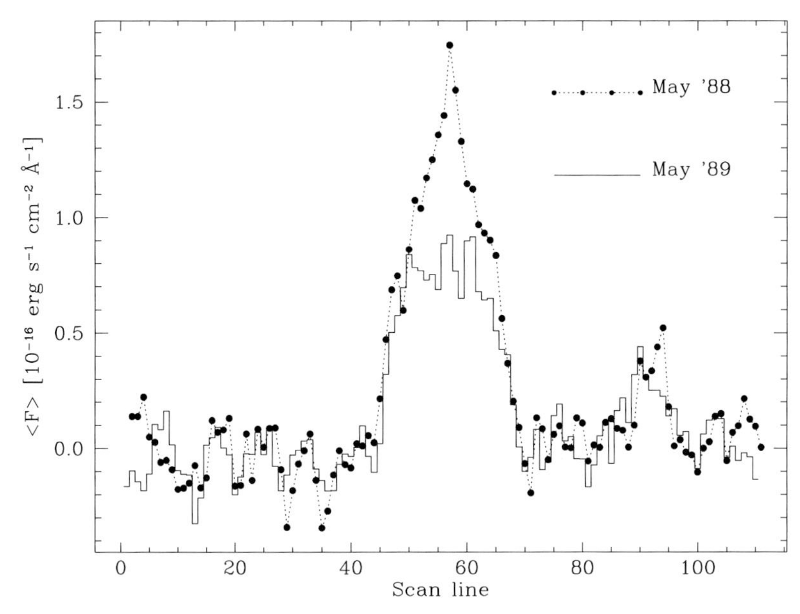

Fig. 12. Spatial flux distribution for SN1987A across the *IUE* slot aperture, integrated between 1250 and 1900 Å. The detection of the UV echo in 1988 is confirmed by its disappearance in 1989 due to its having expanded out of the aperture.

Since then, the *HST* observations of SN1987A, both imaging and spectrographic, have been carried out at least once a year, revealing a many important details.

Structure and Expansion History of the Ejecta: Supernova 1987A was first imaged on 23 and 24 August 1990, with the *ESA FOC* in the F/96 mode [38]. Later observations made with the *FOC* allowed Jakobsen et al. [39,40] to measure the angular expansion of the supernova ejecta. The results confirmed the validity of the expansion models proposed on the basis of spectroscopy. Additional observations made with the *WFPC2* on the re-refurbished *HST* confirmed the validity of the early expansion rate and revealed the presence of additional structures (Pun 1998, private communication, and [41]).

Spectroscopic Properties of the Ejecta at Late Times: *FOS* observations of SN1987A, made over the wavelength range 2000 – 8000 Å at 1862 and 2210 days after the supernova outburst, indicated that, at late times, the spectrum was formed in a cold gas that is excited and ionized by energetic electrons from the radioactive debris of the supernova explosion [81]. The profiles are all asymmetric, showing redshifted, extended tails with velocities up to 10,000 km s^{-1} in some strong lines. The blueshift of the line peaks is attributed to dust condensed from the SN1987A ejecta that is still distributed in dense opaque clumps.

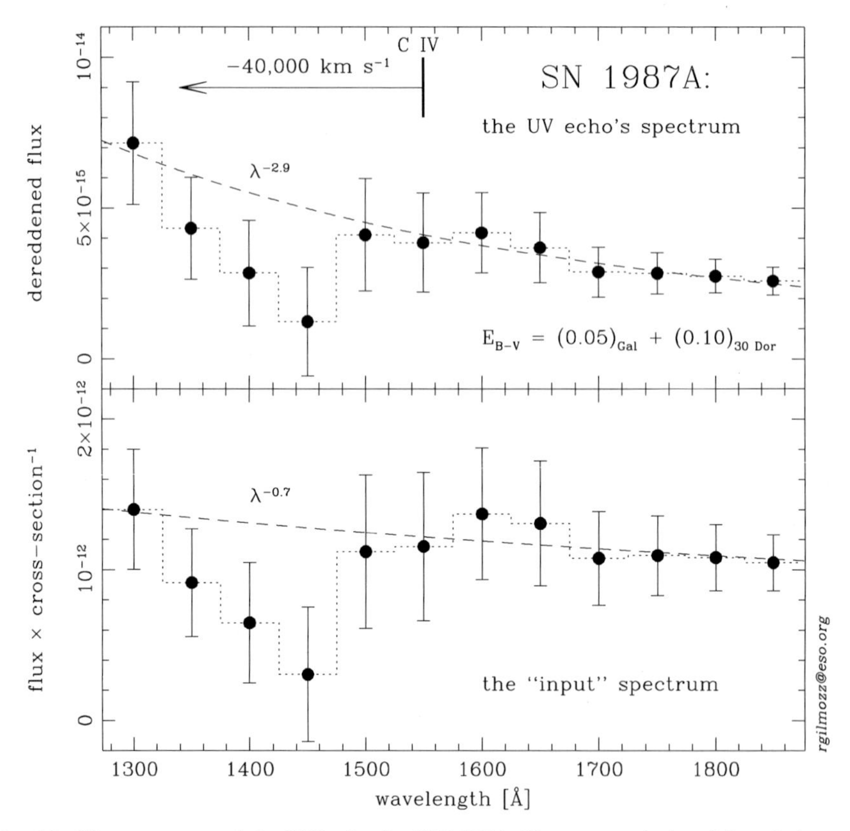

Fig. 13. The spectrum of the UV echo for SN1987A. Upper panel: dereddened observed data. Lower panel: The effect of forward scattering in the dust has been removed to assess the shape of the actual spectrum of the shock breakout.

Reddening and Extinction Law Towards SN1987A: The analysis of *HST FOS* observations of star 2 has permitted Scuderi et al. [73] not only to derive the detailed characteristics of this "companion" star and its coevality with the SN1987A progenitor, but also to determine the properties of the intervening interstellar material, in particular the extinction curve toward SN1987A. This information is crucially important for a correct interpretation of all observations of the supernova itself, as well as for the study of the stellar population around SN1987A [59,67,68].

Properties and Nature of the Circumstellar Rings: The study of the circumstellar rings, i.e., an equatorial ring (the "inner ring") about 0.86 $''$ in radius and inclined by about 45 degrees to the line of sight, plus two additional "outer rings" which are approximately, but not exactly, symmetrically placed relative to the equatorial plane and co-axial with the inner ring, with sizes 2 – 2.5 times larger than the inner ring is revealing. The presence of the inner ring was originally demonstrated by the *IUE* detection of narrow emis-

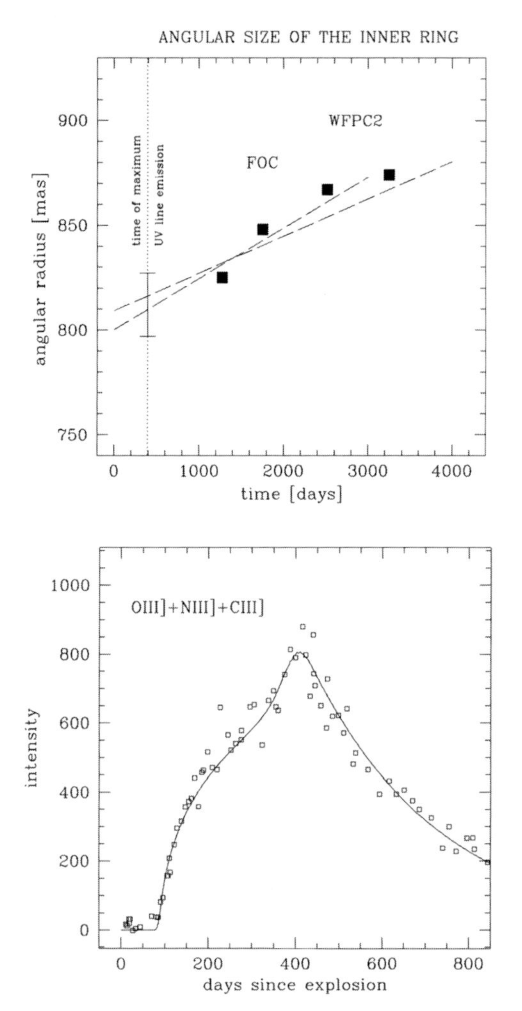

Fig. 14. Top: The surface brightness averaged radius of the SN1987A inner ring as measured in *FOC* images taken with the [O III] filter F501N, and the recent measurements made with the *STIS* in the light of the [O III] 5007 Å and [OI] 6300 Å lines. The two straight lines are the best fits to the *FOC* points only and to the *FOC* + *STIS* [O III] sizes, respectively. The error bar represents the range of possible values of the radius at the time of the UV maximum. Bottom: The sum of the observed intensities (in units of 10^{-15} erg cm^{-2} s^{-1}) of the O III] 1666 Å, N III] 1750 Å, and C III] 1909 Å lines is compared to its best-fit model light curve.

sion lines. Strenuous efforts done with ground based telescopes [15,78] provided early measurements of the shapes of the circumstellar rings and the rings were later imaged in more detail by *HST* using both the *FOC* and *WFPC2* (see, e.g., [11,38,61]). Detailed studies have suggested that the rings are characterized by strong nitrogen overabundances [24,58,75] and that they were ejected in two main episodes of paroxysmal mass-loss which occurred approximately 10,000 yr (the inner ring) and 20,000 yr (the outer rings) before the supernova explosion, respectively (see [47,58]).

Distance to SN1987A and the LMC: A valuable result of *IUE* and the *HST* observations of SN1987A is a geometric and, therefore, "direct" determination of the distance to the *LMC*. This result is critical in that it permits an accurate calibration of the Cepheid distance scale. Because Cepheids are possibly the most reliable, and certainly the most widely used, secondary distance standards for measuring distances up to several tens of Mpc, they play a crucial role in the determination of the cosmological distance scale. The Cepheid distance scale is based on study of their absolute luminosity, requiring accurate knowledge of their distance. Thus, determining the distance to the *LMC* is a fundamental step in determining distances to all cosmological objects.

Various methods have been employed to measure the distance to the *LMC*, with various degrees of success and/or accuracy (see, e.g., [46]). All methods, however, are indirect and depend on the calibration of other distance indicators. Moreover, different distance indicators appear to give discordant results, thus making the distance issue rather uncertain.

The presence of the inner ring around SN1987A has provided a unique opportunity to determine the distance to the *LMC* "directly" by using a purely geometric method – one measures the angular size of the ring from high resolution images and compares it to the absolute size estimated from the evolution of emission lines produced by the ionized gas in the ring (see, e.g., [18,29,30,52,57]).

The most recent analysis and discussion of the data has been done by Panagia [52] who has reanalyzed the results from the *IUE* emission line light curves and an extensive set of *HST* images. From these, the angular size of the [O III] ring appears to expand with time, growing from about 822 milliarcsec (mas) in August 1990 to about 872 mas in early 1996 (see, e.g., Fig. 14, top). This, however, is not a real expansion but is due to the fact that the inner layers of the ring are denser and, therefore, recombine and cool more quickly than the outer layers. The actual the size of ring at the time of the maximum UV line emission was likely smaller than the 1990 measurement and is estimated to have been $R = 808 \pm 17$ mas. The absolute size of the ring has been re-evaluated using a model in which the ring has a finite width ($\Delta R/R = 0.14$) and the intrinsic emissivity of each volume element decays exponentially. A best fit to the combined O III]+N III]+C III] light curve (Fig. 14, bottom) gives an absolute size of the ring of $R_{\rm abs} = (6.23 \pm 0.08) \times 10^{17}$ cm. A comparison of the absolute size to the angular size gives a distance to the supernova $d_{\rm SN1987A} = 51.4 \pm 1.2$ kpc and a distance modulus $m - M_{\rm SN1987A} = 18.55 \pm 0.05$. Allowing for a displacement of the SN1987A position relative to the *LMC* center, the distance to the barycenter of the *LMC* is estimated to be $d_{\rm LMC} = 52.0 \pm 1.3$ kpc, which corresponds to a distance modulus of $m - M_{\rm LMC} = 18.58 \pm 0.05$.

SN1987A, an Ongoing Experiment: It is clear that SN1987A constitutes an ideal laboratory for the study of supernovae and of cosmic explosions in general. While many observations have been taken and much is understood about SN1987A, its progenitor, and its environment, there are still important points that need further study and more observations. For example, the compact stellar remnant presumed to have been left behind by the explosion has not been detected and its nature remains a mystery. Also, the detailed path that led Sk -69° 202 to explode as a BSG has not been explained and/or understood in any satisfactory manner. However, the ever increasing interaction of the SN ejecta with the inner ring promises to provide us with ever increasing fluxes in all wavelength bands and with a unique laboratory for SN research for many decades to come.

7 UV Properties of Supernovae and Cosmology

SNIa are generally accepted to be virtually ideal standard candles (see, e.g., [31,60]) and can be used to measure distances up to redshifts of ≥ 1 (for a review, see, e.g., [45]). In particular, *HST* observations of Cepheids in the parent galaxies of SNIa have led to accurate determinations of their distances, and thus to the absolute magnitudes of normal SNIa at maximum light (see, e.g., [70,71]). With calibrated SNIa peak luminosities, it is possible to determine the distances to much more distant SNIa and obtain a direct comparison with the Hubble diagram (i.e., a plot of the observed magnitudes of SNIa versus their cosmological redshift velocities) for distant SNIa ($30,000\ \mathrm{km\ s^{-1}} > v > 3,000\ \mathrm{km\ s^{-1}}$). Such a comparison yields a Hubble constant (i.e., the expansion rate of the local Universe) of $H_0 = 59 \pm 6\ \mathrm{km\ s^{-1}\ Mpc^{-1}}$ [70]. Note that independent *HST* calibration of SNIa absolute magnitudes at maximum light has led to an estimated Hubble Constant of $H_0 = 71 \pm 8\ \mathrm{km\ s^{-1}\ Mpc^{-1}}$[25], so that there clearly remain some differences to be resolved. A "compromise" value of $H_0 = 65 \pm 7\ \mathrm{km\ s^{-1}\ Mpc^{-1}}$ is, therefore, often used.

Additionally, studying more distant SNIa (i.e., $z > 0.1$) has made it possible to extend our knowledge of other cosmological parameters. The preliminary results of two competing teams [62,63,65] agree in indicating a non-empty inflationary Universe with parameters lying along the line $0.8\Omega_M - 0.6\Omega_\Lambda = -0.2 \pm 0.1$. For a flat Universe this result corresponds to $\Omega_M \simeq 0.3$ and $\Omega_\Lambda \simeq 0.7$ which implies an age for the Universe within the interval 12.3–15.3 Gyr with a 99.7% confidence level [63].

Recently, Riess et al. [66] discussed the properties of SN1997ff, discovered by Gilliland et al. [26] in the *HST* Deep Field [89], and showed it was probably a type Ia supernova at a redshift of $z \simeq 1.7$. Fits to observations of SN1997ff provided constraints on the redshift-distance relation of SNIa and a powerful confirmation of the hypothesis that our Universe is currently expanding at an accelerated rate. However, the measurement errors were quite large and it is important to identify more high-z SNIa, even though the technical difficulties are challenging. One must make observations in the near IR (because of the redshift) of increasingly faint objects (because of distance) and be able to identify that they are indeed SNIa – traditionally done though spectroscopy. An additional

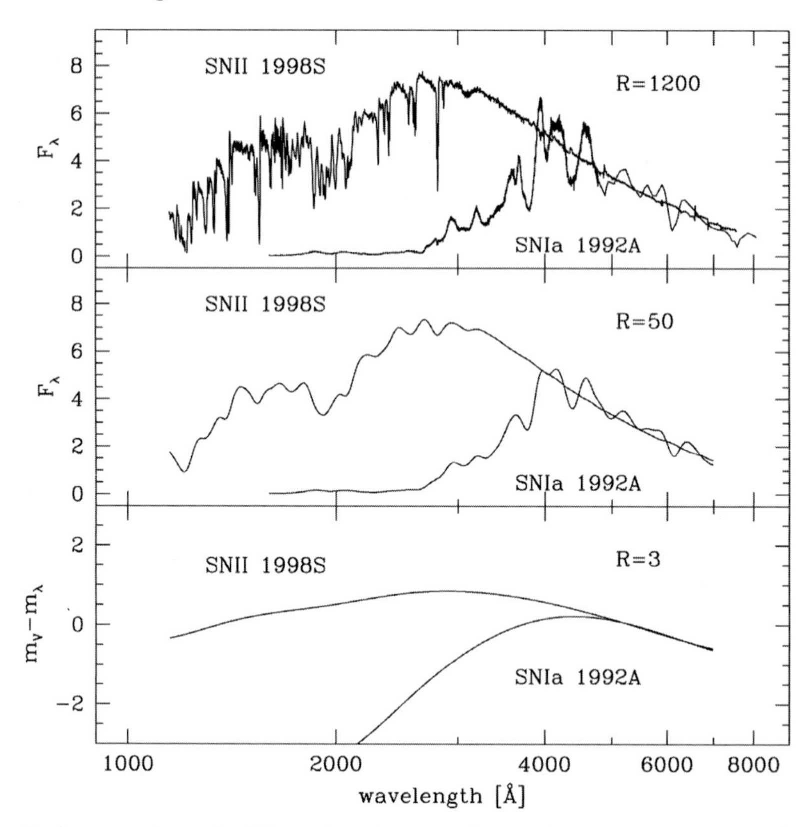

Fig. 15. Spectra of type Ia SN1992A and type II SN1998S near maximum light [35,36], normalized to have the same average flux in the rest-frame V-band. Upper panel: Linear flux scale, original spectral resolution of ~1500. Middle panel: Resolution degraded to $\lambda/\Delta\lambda = 50$ (low resolution spectroscopy). Lower panel: Magnitude scale ($m_\lambda = -2.5 \times log(F_\lambda) + const$), resolution $\lambda/\Delta\lambda = 4$ (broad-band photometry).

complication, of course, is to show that these very distant SNe share the same properties as their local Universe relatives.

Fortunately, one can discern SNI from SNII based on their UV spectral distributions (N. Panagia, in preparation), because SNII are strong UV emitters while all SNI, irrespective of whether they are Ia or Ib/c, have steeply declining spectra at short wavelengths as is illustrated in Fig. 15 for the cases of type Ia SN1992A and type II SN1998S. Figure 15 also shows the same spectra smoothed to a resolution of $\lambda/\Delta\lambda = 50$ (i.e., low resolution spectroscopy, $< F_\lambda >$; center panel), and smoothed to $\lambda/\Delta\lambda = 4$ (i.e., broad-band photometry expressed in magnitude difference relative to the V-band; bottom panel). We see that while characteristic features can still be recognized in the $\lambda/\Delta\lambda = 50$ spectra, the only property in the $\lambda/\Delta\lambda = 4$ spectra that distinguishes a SNIa from a SNII is the UV slope.

Table 2. Ultraviolet Spectral Parameters of SNe Near Maximum Light

SN	Type	λ_{max} (Å)	$\lambda_{1/2}$ (Å)	$\Delta m_{0.2}$
1998S	IIn	2805	1425	-0.52
1992A	Ia	4310	3255	-2.01

More quantitatively, the wavelength of the peak spectral emission, $\lambda_{\rm max}$, derived from the $\lambda/\Delta\lambda = 4$ spectral distributions of Fig. 15, is widely different for the two supernovae, 2805 Å for SN1998S and 4310 Å for SN1992A. Also, if we define $\lambda_{1/2}$ to be the wavelength at which the flux declines to 50% of the peak, we find values of $\lambda_{1/2} = 1425$ Å and $\lambda_{1/2} = 3255$ Å for SN1998S and SN1992A, respectively. This means that the UV spectrum has a steep decline for SNIa (and SNIb/c as well), and a shallower UV gradient for SNII. In particular, defining $\Delta m_{0.2}$ as the drop measured between $\lambda_{\rm max}$ and $\lambda_{\rm max}/1.58$ (a factor of $1.58 = 10^{0.2}$) approximately corresponds to the wavelength ratio of the I to the V-band, or the J to the I-band) values of $\Delta m_{0.2} = -0.52$ and $\Delta m_{0.2} = -2.01$ mag are found for SN1998S and SN1992A, respectively (here we define $m_\lambda = -2.5 \times log(< F_\lambda >) + const$).

Thus, it appears that even simple photometry may be adequate to distinguish SNI from SNII. Furthermore, once the SN type is identified from the UV slope, the redshift can be estimated to $\sim 20\%$ from the ratio of the $\lambda_{1/2}$ wavelength measured in the actual spectrum to the value appropriate for the SN (see Table 2). This method should indeed be tested with more local SNe such as SN1992A and SN1998S and further extended and "calibrated" with high-z SNe for which accurate spectroscopy is available. However, it appears that the use of photometry as a sort of very low resolution spectroscopy could allow us to extend the study of SNe to even higher redshifts without compromising the reliability of the results.

References

1. W.D. Arnett, J.N. Bahcall, R.P. Kirshner, S.E. Woosley: Ann. Rev. Astron. Astrophys. **27**, 629 (1989)
2. W. Baade: Astron. Nach. **228**, 359 (1926)
3. E. Baron, P.H. Hauschildt, P. Nugent, D. Branch: Mon. Not. R. Astron. Soc. **283**, 297 (1996)
4. E. Baron et al.: Astrophys. J. **545**, 444 (2000)
5. F. Bertola: Ann. Rev. Astron. Astrophys. **27**, 319 (1964)
6. F. Bertola, A. Mammano, M. Perinotto: Asiago Contr. **174**, 51 (1965)
7. H.E. Bond, N. Panagia, R. Gilmozzi, M. Meakes: Astrophys. J. Lett. **354**, L49 (1990)
8. D. Branch et al.: Astrophys. J. **566**, 1005 (2002)
9. D. Branch, K.L. Venkatakrishna: Astrophys. J. Lett. **306**, L21 (1986)

10. D. Branch: In: *The Largest Explosions Since the Big Bang: Supernovae and Gamma-Ray Bursts*, ed. by M. Livio, N. Panagia, K. Sahu (Cambridge University Press, Cambridge 2000) p. 96
11. C.J. Burrows et al.: Astrophys. J. **452**, 680 (1995)
12. R.A. Chevalier: Astrophys. J. Lett. **308**, L225 (1982)
13. R.A. Chevalier: In: *ESO Workshop on SN1987A*, ed. by J. Danziger (ESO, Garching 1987) p. 481
14. A. Crotts: Astrophys. J. Lett. **333**, L51 (1988)
15. A. Crotts, W.E. Kunkel, P.J. McCarthy: Astrophys. J. Lett. **347**, L61 (1989)
16. K. de Boer, M. Grewing, T. Richtler, W. Wamsteker, C. Gry, N. Panagia: Astron. Astrophys. **177**, L37 (1987)
17. A.K. Dupree, R.P. Kirshner, G.E. Nassiopoulos, J.C. Raymond, G. Sonneborn: Astrophys. J. **320**, 597 (1987)
18. E. Dwek, J.E. Felten: Astrophys. J. **387**, 551 (1992)
19. R.A. Fesen et al.: Astron. J. **117**, 725 (1999)
20. A. Filippenko: Ann. Rev. Astron. Astrophys. **35**, 309 (1997)
21. A.K. Fischer: Ph.D. Thesis, University of Oklahoma, Norman (2000)
22. C. Fransson: Astron. Astrophys. **133**, 264 (1984)
23. C. Fransson et al.: Astron. Astrophys. **132**, 1 (1984)
24. C. Fransson, A., Cassatella, R., Gilmozzi, R.P., Kirshner, N. Panagia, G., Sonneborn, W. Wamsteker: 1989, Astrophys. J. **336**, 429 (1989)
25. W.L. Freedman, et al.: Astrophys. J. **553**, 47 (2001)
26. R.L. Gilliland, P.E. Nugent, M.M. Phillips: Astrophys. J. **521**, 30 (1999)
27. R. Gilmozzi et al.: Nature **328**, 318 (1987)
28. R. Gilmozzi, N. Panagia: Mem. Soc. Astron. It. **70**, 583 (1999)
29. A. Gould: Astrophys. J. **452**, 189 (1995)
30. A. Gould, O. Uza: Astrophys. J. **494**, 118 (1998)
31. M. Hamuy, M.M. Phillips, N.B. Suntzeff, R.A. Schommer, J. Maza, R. Aviles: Astron. J. **112**, 2391 (1996)
32. M. Hamuy et al.: Astrophys. J. **558**, 615 (2001)
33. R.P. Kirshner, G. Sonneborn, D.M. Crenshaw, G.E. Nassiopoulos: Astrophys. J. **320**, 602 (1987)
34. R.P. Kirshner, G. Sonneborn, A. Cassatella, R. Gilmozzi, W. Wamsteker, N. Panagia: IAUC 4435 (1987)
35. R.P. Kirshner et al.: Astrophys. J. **415**, 589 (1993)
36. E.J. Lentz et al.: Astrophys. J. **547**, 406 (2001)
37. R.E. Luck, D.L. Lambert: Astrophys. J. **245**, 1018 (1981)
38. P. Jakobsen et al.: Astrophys. J. Lett. **369**, L63 (1991)
39. P. Jakobsen, F.D. Macchetto, N. Panagia: Astrophys. J. **403**, 736 (1993)
40. P. Jakobsen, R. Jedrzejewski, F.D. Macchetto, N. Panagia: Astrophys. J. Lett. **435**, L47 (1994)
41. R.A. Jansen, P. Jakobsen: Astron. Astrophys. **370**, 1056 (2001)
42. D.J. Jeffery, B. Leibundgut, R.P. Kirshner, S. Benetti, D. Branch, G. Sonneborn: Astrophys. J. **397**, 304 (1992)
43. D.J. Jeffery et al.: Astrophys. J. Lett. **421**, L27 (1994)
44. B. Leibundgut: Light Curves of Supernovae Type, I. PhD Thesis, Harvard University, Cambridge (1988)
45. F.D. Macchetto,N. Panagia: In: *Post-Hipparcos Cosmic Candles*, ed. by A. Heck, F. Caputo (Kluwer, Holland 1999) p. 225
46. B. Madore, W. Freedman, W.: Pub. Astron. Soc. Pacific **103**, 933 (1991)

47. S.P. Maran, G. Sonneborn, C.S.J. Pun, P. Lundqvist, R.C. Iping, T.R. Gull: Astrophys. J. **545**, 390 (2000)
48. R. McCray: Ann. Rev. Astron. Astrophys. **31**, 175 (1993)
49. J. Millard et al.: Astrophys. J. **527**, 746 (1999)
50. K. Nomoto, F.-K. Thielemann, K. Yokoi: Astrophys. J. **286**, 644 (1984)
51. N. Panagia: In: *Supernovae As Distance Indicators, Lect. Notes Phys. 224* (Springer, Berlin 1985) p. 226
52. N. Panagia: In: *IAU Symposium 190: New Views of the Magellanic Clouds*, ed. by Y.-H. Chu, N. Suntzeff, J. Hesser, D. Bohlender (ASP, San Francisco 1999) p. 549
53. N. Panagia: 'Supernovae'. In: *Experimental Physics of Gravitational Waves*, ed. by G. Calamai, M. Mazzoni, R. Stanga, F. Vetrano (World Scientific, Singapore 2000) p. 107
54. N. Panagia et al.: Mon. Not. R. Astron. Soc. **192**, 861 (1980)
55. N. Panagia, R. Gilmozzi, J. Clavel, M. Barylak, R. Gonzalez-Riestra, C. Lloyd, L. Sanz Fernandez de Cordoba, W. Wamsteker: Astron. Astrophys. **177**, L25 (1987)
56. N. Panagia, R. Gilmozzi, A. Cassatella, W. Wamsteker, R.P. Kirshner, G. Sonneborn: IAUC 4514 (1987)
57. N. Panagia, R. Gilmozzi, F. Macchetto, H.-M. Adorf, R.P. Kirshner: Astrophys. J. Lett. **380**, L23 (1991)
58. N. Panagia, S. Scuderi, R. Gilmozzi, P.M. Challis, P.M. Garnavich, R.P. Kirshner: Astrophys. J. Lett. **459**, L17 (1996)
59. N. Panagia, M. Romaniello, S. Scuderi, R.P. Kirshner: Astrophys. J. **539**, 197 (2000)
60. B.R. Parodi, A. Saha, A. Sandage, G.A. Tammann: Astrophys. J. **540**, 634 (2000)
61. P.C. Plait, P. Lundqvist, R.A. Chevalier, R.P. Kirshner: Astrophys. J. **439**, 730 (1995)
62. S. Perlmutter et al.: Nature **391**, 51 (1998)
63. S. Perlmutter et al.: Astrophys. J. **517**, 565 (1999)
64. C.S.J. Pun et al.: Astrophys. J. Suppl. **99**, 223 (1995)
65. A.G. Riess et al.: Astron. J. **116**, 1009 (1998)
66. A.G. Riess et al.: Astron. J. **560**, 49 (2001)
67. M. Romaniello: PhD Thesis, Scuola Normale Superiore, Pisa (1998)
68. M. Romaniello, N. Panagia, S. Scuderi, R.P. Kirshner: Astron. J. **123**, 915 (2002)
69. M. Rosa: IAUC 4564 (1988)
70. A. Saha, A. Sandage, G.A. Tammann, A.E. Dolphin, J. Christensen, N. Panagia, F.D. Macchetto: Astrophys. J. **562**, 314 (2001)
71. A. Sandage, A. Saha, G.A. Tammann, L. Labhardt, N. Panagia, F.D. Macchetto: Astrophys. J. Lett. **460**, L15 (1996)
72. B.E. Schaefer: Astrophys. J. Lett. **323**, L47 (1987)
73. S. Scuderi, N. Panagia, R. Gilmozzi, P.M. Challis, R.P. Kirshner: Astrophys. J. **465**, 956 (1996)
74. G. Sonneborn, B. Altner, R.P. Kirshner: Astrophys. J. Lett. **323**, L35 (1987)
75. G. Sonneborn, C. Fransson, P. Lundqvist, A. Cassatella, R. Gilmozzi, R.P. Kirshner, N. Panagia, W. Wamsteker: Astrophys. J. **477**, 848 (1997)
76. Space Telescope Science Institute: IAUC 5000 (1990)
77. S.D. Van Dyk, K.W. Weiler, R.A. Sramek, E.M. Schlegel, A.V. Filippenko, N. Panagia, B. Leibundgut: Astron. J. **111**, 1271 (1996)
78. J. E. Wampler, L. Wang, D. Baade, K. Banse, S. D'Odorico, C. Gouiffes, M. Tarenghi: Astrophys. J. Lett. **362**, L13 (1990)

79. W. Wamsteker et al.: Astron. Astrophys. **177**, L21 (1987)
80. W. Wamsteker, R. Gilmozzi, A. Cassatella, N. Panagia: IAUC 4410 (1987)
81. L. Wang et al.: Astrophys. J. **466**, 998 (1996)
82. K.W. Weiler, R.A. Sramek, N. Panagia, J.M. van der Hulst, M. Salvati: Astrophys. J. **301**, 790 (1986).
83. K.W. Weiler, R.A. Sramek: Ann. Rev. Astron. Astrophys. **26**, 295 (1988).
84. K.W. Weiler, N. Panagia, R.A. Sramek, J.M. van der Hulst, M.S. Roberts, L. Nguyen: Astrophys. J. **336**, 421 (1989)
85. K.W. Weiler, N. Panagia, M.J. Montes, R.A. Sramek: Ann. Rev. Astron. Astrophys. **40**, 387 (2002)
86. D.E. Welty, L.M. Hobbs, J.T. Lauroesch, D.C. Morton, L. Spitzer, D.G. York: Astrophys. J. Suppl. **124**, 465 (1999)
87. A.J. Wesselink: Bull. Astron. Inst. Neth. **10**, 91 (1946)
88. J.C. Wheeler, R.P. Harkness: : In: *Galaxy Distances and Deviations from Universal Expansion*, ed. by B.F. Madore, R.B. Tully (Reidel, Dordrecht 1986) p. 45
89. R.E. Williams et al.: Astron. J. **112**, 1335 (1996)
90. S. Woosley: In: *Gamma-Ray Line Astrophysics*, ed. by P. Durouchoux, N. Pantzos (AIP, New York 1991) p. 270
91. J. Xu, A.P.S. Crotts, W.E. Kunkel: Astrophys. J. **451**, 806 (1995)
92. J. Xu, A.P.S. Crotts: Astrophys. J. **511**, 262 (1999)

Radio Supernovae

Richard A. Sramek[1] and Kurt W. Weiler[2]

[1] National Radio Astronomy Observatory, P.O. Box 0, Socorro, NM 87801, USA
[2] Naval Research Laboratory, Code 7213, Washington, DC 20375-5320, USA

Abstract. Study of radio supernovae over the past 20 years includes two dozen detected objects and more than 100 upper limits. From this work it is possible to identify classes of radio properties, demonstrate conformance to and deviations from existing models, estimate the density and structure of the circumstellar material and, by inference, the evolution of the presupernova stellar wind, and reveal the last stages of stellar evolution before explosion. It is also possible to detect ionized hydrogen along the line of sight, to demonstrate binary properties of the stellar system, and to show clumpiness of the circumstellar material. More speculatively, it may be possible to provide distance estimates to radio supernovae.

1 Introduction to Radio Supernovae

A series of papers published over the past 20 years on radio supernovae (RSNe) has established the radio detection and, in a number of cases, radio evolution for approximately two dozen objects: 3 type Ib supernovae (SNe), 5 type Ic SNe (Because the differences between the SN optical classes are slight – type Ib show strong He I absorption while type Ic show weak He I absorption – and there are no obvious radio differences, we shall often refer to the classes as type Ib/c), and the rest type II SNe. A much larger list of more than 100 additional SNe have low radio upper limits (See Table 1 and *http://rsd-www.nrl.navy.mil/7214/weiler /kwdata/rsnhead.html*).

In this extensive study of the radio emission from SNe, several effects have been noted:

- type Ia SNe are not radio emitters to the detection limit of the *Very Large Array (VLA)*[1].
- type Ib/c SNe are radio luminous with steep spectral indices (generally $\alpha < -1$; $S \propto \nu^{+\alpha}$) and a fast turn-on/turn-off, usually peaking at 6 cm near or before optical maximum.
- type II SNe show a range of radio luminosities with flatter spectral indices (generally $\alpha > -1$) and a relatively slow turn-on/turn-off, usually peaking at 6 cm significantly after optical maximum.
- type Ib/c may be fairly homogeneous in some of their radio properties while type II, as in the optical, are quite diverse.

[1] The *VLA* telescope of the National Radio Astronomy Observatory is operated by Associated Universities, Inc. under a cooperative agreement with the National Science Foundation.

There are a large number of physical properties of SNe which can be determined from radio observations. *Very Long Baseline Interferometry (VLBI)* imaging shows the symmetry of the blastwave and the local circumstellar medium (CSM), estimates the speed and deceleration of the SN blastwave propagating outward from the explosion and, with assumptions of symmetry and optical line/radio-sphere velocities, allows independent distance estimates to be made (see, e.g., [3,24]).

Measurements of the multi-frequency radio light curves and their evolution with time show the density and structure of the CSM, evidence for possible binary companions, clumpiness or filamentation in the presupernova wind, mass-loss rates and changes therein for the presupernova stellar system and, through stellar evolution models, estimates of the zero age main sequence (ZAMS) presupernova stellar mass and the stages through which the star passed on its way to explosion. It has also been proposed [51] that the time from explosion to 6 cm radio maximum may be an indicator of the radio luminosity, and thus an independent distance indicator, for type II SNe and that type Ib/c SNe may be approximate radio standard candles at 6 cm radio peak flux density.

A summary of the radio information on SNe can be found at:
http://rsd-www.nrl.navy.mil/7214/weiler/sne-home.html.

2 Radio Supernova Models

All known RSNe appear to share common properties of:

- Nonthermal synchrotron emission with high brightness temperature.
- A decrease in absorption with time, resulting in a smooth, rapid turn-on first at shorter wavelengths and later at longer wavelengths.
- A power-law decline of the flux density with time at each wavelength after maximum flux density (optical depth ~ 1) is reached at that wavelength.
- A final, asymptotic approach of spectral index α ($S \propto \nu^{+\alpha}$) to an optically thin, nonthermal, constant negative value [45,47].

The characteristic RSN radio light curves such as those shown in Fig. 2 arise from the competing effects of slowly declining non-thermal radio emission and more rapidly declining thermal or non-thermal absorption yielding a rapid turn-on and slower turn-off of the radio emission at any single frequency. This characteristic light curve shape is also illustrated schematically in Fig. 1 of Weiler et al. [51]. Since absorption processes are greater at lower frequencies, transition from optically thick to optically thin (turn-on) occurs first at higher frequencies and later at lower frequencies. After the radiation is completely optically thin and showing the ongoing decline of the underlying emission process (turn-off), the non-thermal spectrum causes lower frequencies to have higher flux density. These two effects cause the displacement in time and flux density of the light curves at different frequencies also seen in Fig. 2.

Chevalier [7,8] has proposed that the relativistic electrons and enhanced magnetic field necessary for synchrotron emission arise from the SN blastwave interacting with a relatively high density circumstellar medium (CSM) which has

Table 1. Observed Supernovae (DT = Detection; LC = Light Curve Available)

SN	Type	Radio	SN	Type	Radio	SN	Type	Radio	SN	Type	Radio
1895B	I		1974G	Ia		1986I	IIP		1994W	IIn	
1901B	I		1975N	Ia		1986J	IIn	LC	1994Y	IIn	
1909A	II		1977B	?		1986O	Ia		1994ai	Ib/c	
1914A	?		1978B	II		1987A	IIpec	LC	1994aj	II	
1917A	I?		1978G	II		1987B	IIn		1994ak	IIn	
1921B	II		1978K	II	LC	1987D	Ia		1995G	IIn	
1921C	I		1979B	Ia		1987F	IIpec		1995N	IIn	DT
1923A	IIP	DT	1979C	IIL	LC	1987K	IIb		1995X	IIP	
1937C	Ia		1980D	IIP		1987M	Ib/c		1995ad	IIP	
1937F	II		1980I	Ia		1987N	Ia		1995ag	II	
1939C	II		1980K	IIL	LC	1988I	IIn		1995al	Ia	
1940A	IIL		1980L	?		1988Z	IIn	LC	1996L	IIL	
1945B	?		1980N	Ia		1989B	Ia		1996N	Ib/c	DT
1948B	II?		1980O	II		1989C	IIn		1996W	IIpec	
1950B	II?	DT	1981A	II		1989L	IIL		1996X	Ia	
1954A	I		1981B	Ia		1989M	Ia		1996ae	IIn	
1954J	II		1981K	II	LC	1989R	IIn		1996an	II	
1957D	II?	DT	1982E	Ia?		1990B	Ib/c	LC	1996aq	Ib/c	
1959D	II		1982R	Ib/c?		1990K	IIL		1996bu	IIn	
1959E	I		1982aa	?	LC	1990M	Ia		1996cb	IIb?	DT
1963J	I		1983G	I		1991T	Ia		1997X	Ib/c	DT
1966B	II		1983K	IIP/n		1991ae	IIn		1997ab	IIn	
1968D	II	DT	1983N	Ib/c	LC	1991ar	Ib/c		1997db	II	
1968L	IIP		1984A	I		1991av	IIn		1997dq	Ib/c	
1969L	IIP		1984E	IIL/n		1991bg	Ia		1997ei	Ib/c	
1969P	?		1984L	Ib/c	LC	1992A	Ia		1997eg	IIn	LC
1970A	II?		1984R	?		1992H	IIP		1998S	IIn	DT
1970G	IIL	LC	1985A	Ia		1992ad	IIP?	DT	1998bu	Ia	
1970L	I?		1985B	Ia		1992bd	II		1998bw	Ib/c	LC
1970O	?		1985F	Ib/c		1993G	II		1999D	II	
1971G	I?		1985G	IIP		1993J	IIb	LC	1999E	IIn	
1971I	Ia		1985H	II		1993N	IIn		1999em	IIP	DT
1971L	Ia		1985L	IIL	DT	1993X	II		2000P	IIn	
1972E	Ia		1986A	Ia		1994D	Ia				
1973R	IIP		1986E	IIL	LC	1994I	Ib/c	LC			
1974E	?		1986G	Ia		1994P	II				

been ionized and heated by the initial UV/X-ray flash. This CSM density (ρ), which decreases as an inverse power, s, of the radius, r, from the star, is presumed to have been established by a presupernova stellar wind with mass-loss rate, $\dot{M}$, and velocity, $w_{\rm wind}$, (i.e., $\rho \propto \frac{\dot{M}}{w_{\rm wind}\ r^s}$) from a massive stellar progenitor or companion. For a constant mass-loss rate and constant wind velocity $s = 2$. This ionized CSM is the source of some or all of the initial thermal gas absorption, although Chevalier [9] has proposed that synchrotron self-absorption (SSA) may play a role in some objects.

A rapid rise in the observed radio flux density results from a decrease in these absorption processes as the radio emitting region expands and the absorption processes, either internal or along the line-of-sight, decrease. Weiler et al. [47] have suggested that this CSM can be "clumpy" or "filamentary," leading to a slower radio turn-on, and Montes et al. [25] have proposed the possible presence of a distant ionized medium along the line-of-sight that is sufficiently distant from the explosion that it is unaffected by the blastwave and can cause a spectral turn-over at low radio frequencies. In addition to clumps or filaments, the CSM may be radially structured with significant density irregularities such as rings, disks, shells, or gradients.

3 RSN Parameterized Radio Light Curves

Weiler et al. [45,47] and Montes et al. [25] adopted a parameterized model which has been updated in Weiler et al. [53] to (Note that for consistency here we use the notation of F_ν(mJy) rather than the S(mJy) more commonly used by radio astronomers.):

$$F_\nu(\mathrm{mJy}) = K_1 \left(\frac{\nu}{5\mathrm{GHz}}\right)^{\alpha} \left(\frac{t-t_0}{1\mathrm{day}}\right)^{\beta} e^{-\tau_{\rm external}} \left(\frac{1-e^{-\tau_{\rm CSM_{clumps}}}}{\tau_{\rm CSM_{clumps}}}\right) \times \left(\frac{1-e^{-\tau_{\rm internal}}}{\tau_{\rm internal}}\right) \quad (1)$$

with

$$\tau_{\rm external} = \tau_{\rm CSM_{homog}} + \tau_{\rm distant}, \quad (2)$$

where

$$\tau_{\rm CSM_{homog}} = K_2 \left(\frac{\nu}{5\ \mathrm{GHz}}\right)^{-2.1} \left(\frac{t-t_0}{1\ \mathrm{day}}\right)^{\delta} \quad (3)$$

$$\tau_{\rm distant} = K_4 \left(\frac{\nu}{5\ \mathrm{GHz}}\right)^{-2.1} \quad (4)$$

and

$$\tau_{\rm CSM_{clumps}} = K_3 \left(\frac{\nu}{5\ \mathrm{GHz}}\right)^{-2.1} \left(\frac{t-t_0}{1\ \mathrm{day}}\right)^{\delta'} \quad (5)$$

with K_1, K_2, K_3, and K_4 determined from fits to the data and corresponding, formally, to the flux density (K_1), homogeneous (K_2, K_4), and clumpy or filamentary (K_3) absorption at 5 GHz one day after the explosion date t_0. The terms $\tau_{\rm CSM_{homog}}$ and $\tau_{\rm CSM_{clumps}}$ describe the attenuation of local, homogeneous CSM and clumpy CSM that are near enough to the SN progenitor that they are altered by the rapidly expanding SN blastwave. The $\tau_{\mathrm{CSM}_{\mathrm{homog}}}$ absorption is produced by an ionized medium that completely covers the emitting source ("homogeneous external absorption"), and the $(1-e^{-\tau_{\mathrm{CSM}_{\mathrm{clumps}}}})\tau_{\mathrm{CSM}_{\mathrm{clumps}}}^{-1}$ term describes the attenuation produced by an inhomogeneous medium ("clumpy absorption"; see [28] for a more detailed discussion of attenuation in inhomogeneous media). The τ_{distant} term describes the attenuation produced by a homogeneous medium which completely covers the source but is so far from the SN progenitor that it is not affected by the expanding SN blastwave and is constant in time. All external and clumpy absorbing media are assumed to be purely thermal, singly ionized gas which absorbs via free-free (f-f) transitions with frequency dependence $\nu^{-2.1}$ in the radio. The parameters δ and δ' describe the time dependence of the optical depths for the local homogeneous and clumpy or filamentary media, respectively.

The f-f optical depth outside the emitting region is proportional to the integral of the square of the CSM density over the radius. Since in the simple Chevalier model the CSM density (constant mass-loss rate, constant wind velocity) decreases as r^{-2}, the external optical depth will be proportional to r^{-3}, and since the blastwave radius increases as a power of time, $r \propto t^m$ with $m \leq 1$ (i.e., m $=1$ for undecelerated blastwave expansion), it follows that the deceleration parameter, m, is

$$m = -\delta/3. \tag{6}$$

The model by Chevalier [7,8] relates β and δ to the energy spectrum of the relativistic particles γ ($\gamma = 2\alpha - 1$) by $\delta = \alpha - \beta - 3$ so that, for cases where $K_2 = 0$ and δ is, therefore, indeterminate, one can use

$$m = -(\alpha - \beta - 3)/3. \tag{7}$$

Since it is physically realistic and may be needed in some RSNe where radio observations have been obtained at early times and high frequencies, Eq. (1) also includes the possibility for an internal absorption term[2]. This internal absorption (τ_{internal}) term may consist of two parts – synchrotron self-absorption (SSA; $\tau_{\mathrm{internal}_{\mathrm{SSA}}}$), and mixed, thermal f-f absorption/non-thermal emission ($\tau_{\mathrm{internal}_{\mathrm{ff}}}$).

$$\tau_{\mathrm{internal}} = \tau_{\mathrm{internal}_{\mathrm{SSA}}} + \tau_{\mathrm{internal}_{\mathrm{ff}}} \tag{8}$$

$$\tau_{\mathrm{internal}_{\mathrm{SSA}}} = K_5 \left(\frac{\nu}{5\ \mathrm{GHz}}\right)^{\alpha-2.5} \left(\frac{t-t_0}{1\ \mathrm{day}}\right)^{\delta''} \tag{9}$$

$$\tau_{\mathrm{internal}_{\mathrm{ff}}} = K_6 \left(\frac{\nu}{5\ \mathrm{GHz}}\right)^{-2.1} \left(\frac{t-t_0}{1\ \mathrm{day}}\right)^{\delta'''} \tag{10}$$

[2] Note that for simplicity an internal absorber attenuation of the form $\left(\frac{1-e^{-\tau_{\mathrm{CSM}_{\mathrm{internal}}}}}{\tau_{\mathrm{CSM}_{\mathrm{internal}}}}\right)$, which is appropriate for a plane-parallel geometry, is used instead of the more complicated expression (see, e.g., [30]) valid for the spherical case. The assumption does not affect the quality of the analysis because, to within 5% accuracy, the optical depth obtained with the spherical case formula is simply three-fourths of that obtained with the plane-parallel slab formula.

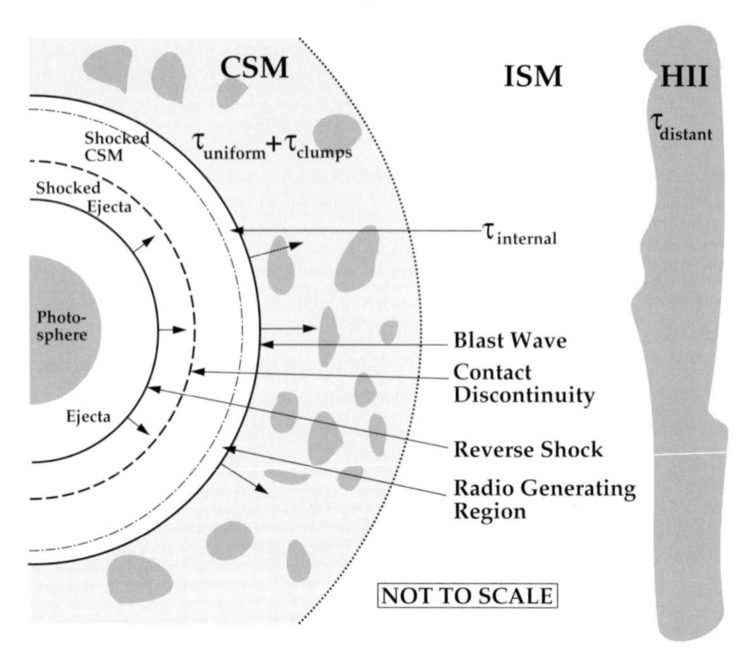

Fig. 1. Cartoon, not to scale, of the SN and its shocks, along with the stellar wind established circumstellar medium (CSM), the interstellar medium (ISM), and more distant ionized hydrogen (HII) absorbing gas. The radio emission is thought to arise near the blastwave front. The expected locations of the several absorbing terms in Eqs. (1)–(10) are illustrated.

with K_5 corresponding, formally, to the internal, non-thermal ($\nu^{\alpha-2.5}$) SSA and K_6 corresponding formally to the internal thermal ($\nu^{-2.1}$) free-free absorption mixed with nonthermal emission, at 5 GHz one day after the explosion date t_0. The parameters δ'' and δ''' describe the time dependence of the optical depths for the SSA and f-f internal absorption components, respectively.

A cartoon of the expected structure of an SN and its surrounding media is presented in Fig. 1 (see also [22]). The radio emission is expected to arise near the blastwave [10].

4 RSN Results

The success of the basic parameterization and modeling has been shown in the good correspondence between the model fits and the data for all subtypes of RSNe: *e.g.*, type Ib SN1983N (Fig. 2a [40]), type Ic SN1990B (Fig. 2b [42]), type II SN1979C (Fig. 3a [27,48,49]) and SN1980K (Fig. 3b [26,50]), and type IIn SN1988Z (Fig. 4 [43,55]). (Note that after day ~ 4000, the evolution of the radio emission from both SN1979C and SN1980K deviates from the expected model evolution, that SN1979C shows a sinusoidal modulation in its flux density prior

Table 2. Fitting Parameters for RSNe[a]

SN	Type	α	β	K_1	K_2	δ	K_3	δ'	Explosion Date
Type Ib/c									
SN1983N	Ib	-1.08	-1.55	3.30×10^3	3.01×10^2	-2.53			30-Jun-83
SN1984L	Ib	-1.15	-1.56	3.52×10^2	3.01×10^2	-2.59			13-Aug-84
SN1990B	Ic	-1.07	-1.24	1.77×10^2	1.24×10^4	-2.83			15-Dec-89
SN1994I[b]	Ic	-1.16	-1.57	8.70×10^3	3.43×10^1	-1.64	2.52×10^4	-2.70	30-Mar-94
SN1998bw[c]	Ib/c	-0.71	-1.38	2.37×10^3			1.73×10^3	-2.80	25-Apr-98
Type II									
SN1970G	IIL	-0.55	-1.87	1.77×10^6	1.80×10^7	-3.00			25-Jun-70
SN1978K[d]	II	-0.77	-1.41	1.14×10^7			3.34×10^8	-2.91	22-May-78
SN1979C[e]	IIL	-0.75	-0.80	1.72×10^3	3.38×10^7	-2.94			04-Apr-79
SN1980K	IIL	-0.60	-0.73	1.15×10^2	1.42×10^5	-2.69			02-Oct-80
SN1981K	II	-0.74	-0.70	7.61×10^1	1.00×10^4	-3.04			16-Jul-81
SN1982aa	II?	-0.73	-1.22	5.28×10^4	3.55×10^7	-2.96			28-Apr-79
SN1986J	IIn	-0.66	-1.65	1.19×10^7			3.06×10^9	-3.33	01-Jan-84
SN1988Z	IIn	-0.69	-1.25	1.47×10^4			5.39×10^8	-2.95	01-Dec-88
SN1993J	IIb	-1.07	-0.93	1.86×10^4	1.45×10^3	-2.02	6.31×10^4	-2.14	27-Mar-93

[a] In some cases the fitting parameters are determined with various input assumptions and/or from very limited data. Thus, all reference should be to the published literature.
[b] SN1994I fit includes $K_5 = 2.21 \times 10^4, \delta'' = -3.07, K_6 = 2.65 \times 10^2, \delta''' = -2.28$.
[c] SN1998bw fit includes $K_4 = 1.24 \times 10^{-2}$.
[d] SN1978K fit includes $K_4 = 1.18 \times 10^{-2}$.
[e] SN1979C is improved by inclusion of a sinusoidal component. See Weiler et al. [49] for discussion and parameter values.

to day ~4000, and that SN 1988Z changes its evolution characteristics after day ~ 1750.)

Thus, the radio emission from SNe appears to be relatively well understood in terms of blastwave interaction with a structured CSM as described by the Chevalier [7,8] model and its extensions by [25,45,47]. For instance, the fact that the homogeneous external absorption exponent δ is ~ -3, or somewhat less, for most RSNe is evidence that the absorbing medium is generally an $\rho \propto r^{-2}$ wind as expected from a massive stellar progenitor which explodes in the red supergiant (RSG) phase.

Additionally, in their study of the radio emission from SN1986J, Weiler et al. [47] found that the simple Chevalier [7,8] model could not describe the relatively slow turn-on. They therefore included terms described mathematically by $\tau_{\mathrm{CSM}_{\mathrm{clumps}}}$ in Eqs. (1) and (5). This extension greatly improved the quality of the fit and was interpreted by Weiler et al. [47] to represent the presence of filaments or clumps in the CSM. Such a clumpiness in the wind material was again required for modeling the radio data from SN1988Z [43,55] and SN1993J

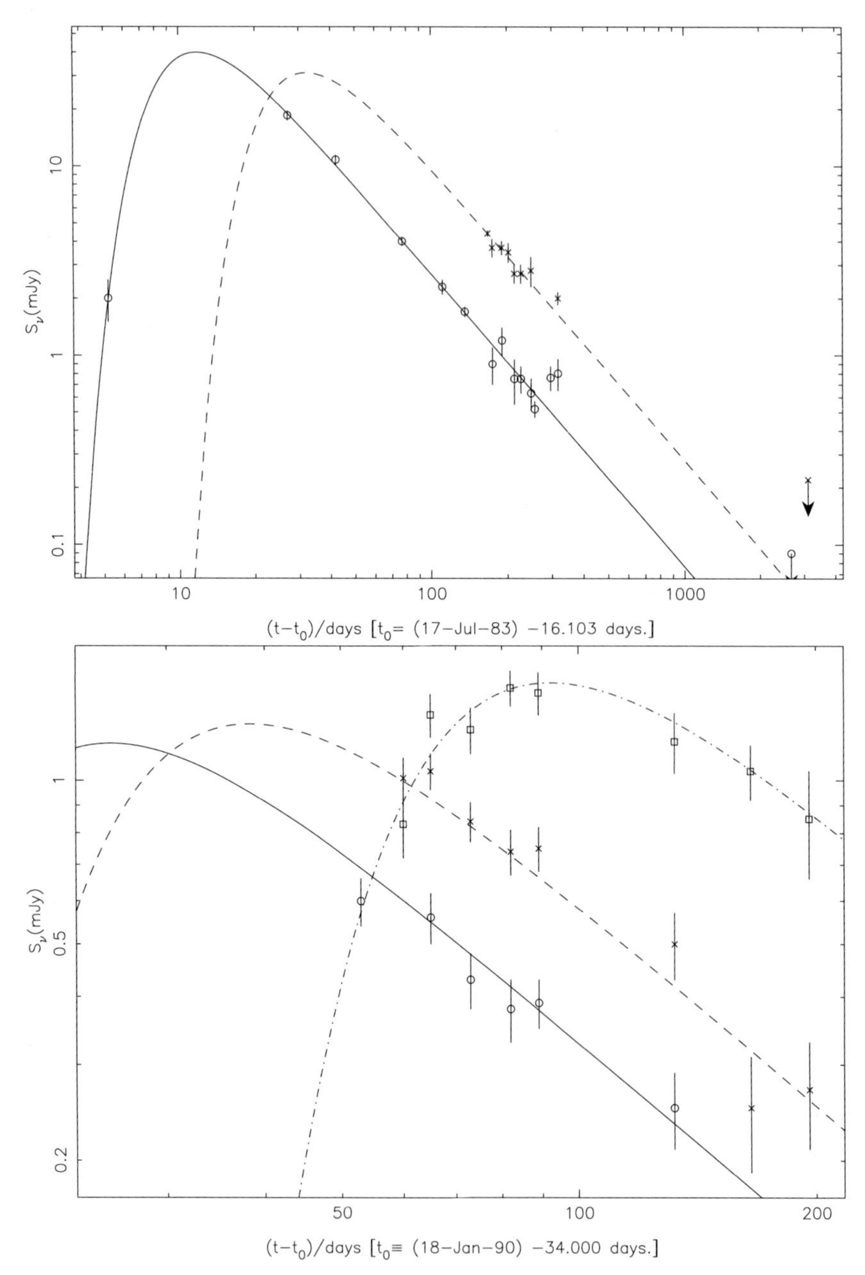

Fig. 2. Top (2a): type Ib SN1983N at 6 cm (4.9 GHz; *open circles, solid line*) and 20 cm (1.5 GHz; *stars, dashed line*). Bottom (2b): type Ic SN1990B. at 3.6 cm (8.4 GHz; *open circles, solid line*), 6 cm (4.9 GHz; *stars, dashed line*), and 20 cm (1.5 GHz; *open squares, dash-dot line*).

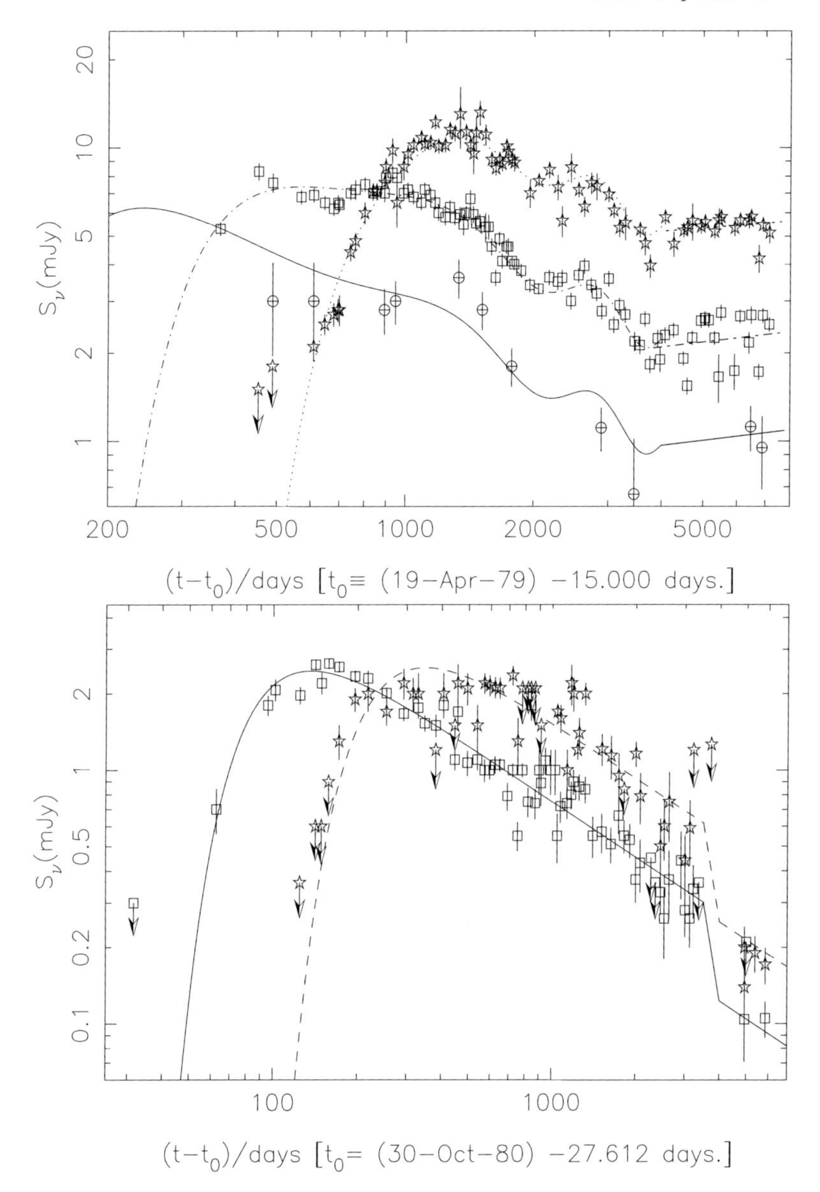

Fig. 3. Top (3a): type II SN1979C at 2 cm (14.9 GHz; *crossed circles, solid line*), 6 cm (4.9 GHz; *open squares, dash-dot line*), and 20 cm (1.5 GHz; *open stars, dotted line*) (Note that the radio flux density increases after day ~ 4000 and has a sinusoidal modulation before day ~ 4000 [27,48,49].) Bottom (3b): type II SN1980K at 6 cm (4.9 GHz; *open squares, solid line*), and 20 cm (1.5 GHz; *open stars, dashed line*). (Note a sharp drop in flux density after day ~ 4000 [26].)

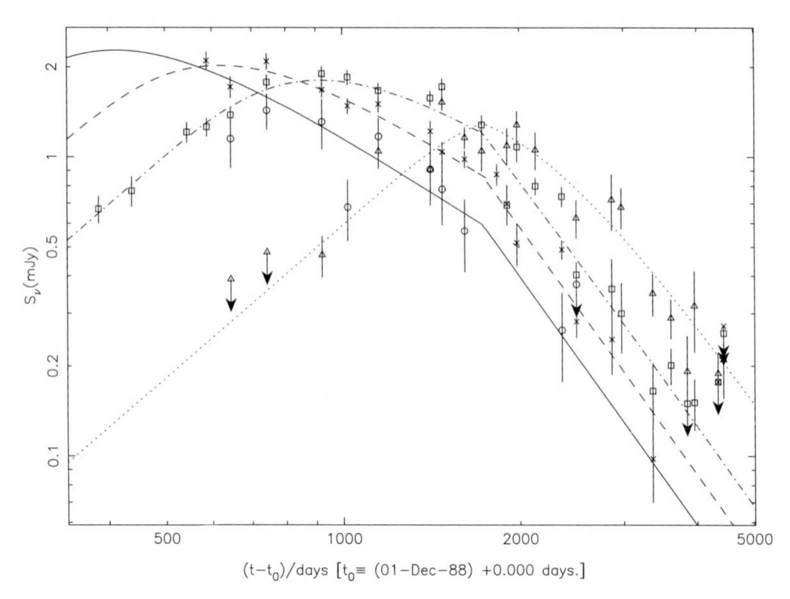

Fig. 4. Radio "light curves" for SN 1988Z in MCG +03-28-022. The four wavelengths, 2 cm (14.9 GHz, *open circles, solid curve*), 3.6 cm (8.4 GHz, *crosses, dashed curve*), 6 cm (4.9 GHz, *open squares, dot-dash curve*), and 20 cm (1.5 GHz, *open triangles, dotted curve*) are shown together with their best fit light curves. The age of the supernova is measured in days from the adopted [39] explosion date of 01 December 1988. A break in the model radio light curves can be seen around day 1750 [55].

[44]. Since that time, evidence for filamentation in the envelopes of SNe has also been found from optical and UV observations (see, e.g., [14,38]). The best fit parameters for a number of RSNe are listed in Table 2.

From this modeling there are several physical properties of SNe which can be determined from radio observations.

4.1 Mass-Loss Rate from Radio Absorption

From the Chevalier [7,8] model, the turn-on of the radio emission for RSNe provides a measure of the presupernova mass-loss rate to wind velocity ratio ($\dot{M}/w_{\rm wind}$). Weiler et al. [45] derived this ratio for the case of pure, external absorption by a homogeneous medium. However, Weiler et al. [52] propose several possible origins for absorption and generalize Eq. (16) of Weiler et al. [45] to

$$\frac{\dot{M}(\mathrm{M_\odot\ yr^{-1}})}{(w_{\rm wind}/10\ \mathrm{km\ s^{-1}})} = 3.0\times 10^{-6}\ <\tau_{\rm eff}^{0.5}>\ m^{-1.5}\left(\frac{v_{\rm i}}{10^4\ \mathrm{km\ s^{-1}}}\right)^{1.5}\times$$
$$\left(\frac{t_{\rm i}}{45\ \mathrm{days}}\right)^{1.5}\left(\frac{t}{t_{\rm i}}\right)^{1.5m}\left(\frac{T}{10^4\ \mathrm{K}}\right)^{0.68} \tag{11}$$

Since the appearance of optical lines for measuring SN ejecta velocities is often delayed a bit relative to the time of the explosion, they arbitrarily take $t_{\rm i} = 45$

days. Because observations have shown that, generally, $0.8 \leq m \leq 1.0$ and from Eq. (11) $\dot{M} \propto t_{\rm i}^{1.5(1-m)}$, the dependence of the calculated mass-loss rate on the date $t_{\rm i}$ of the initial ejecta velocity measurement is weak, $\dot{M} \propto t_{\rm i}^{<0.3}$, so that the best optical or VLBI velocity measurements available can be used without worrying about the deviation of the exact measurement epoch from the assumed 45 days after explosion. For convenience, and because many SN measurements indicate velocities of $\sim 10,000$ km s^{-1}, one usually assumes $v_{\rm i} = v_{\rm blastwave} = 10,000$ km s^{-1} and takes values of $T = 20,000$ K, $w_{\rm wind} = 10$ km s^{-1} (which is appropriate for a RSG wind), $t = (t_{\rm 6cm\ peak} - t_0)$ days from best fits to the radio data for each RSN, and m from Eq. (6) or (7), as appropriate.

The optical depth term $< \tau_{\rm eff}^{0.5} >$ used by Weiler et al. [45] is extended by Weiler et al. [52] and they identify at least three possible absorption regimes: 1) absorption by a homogeneous external medium, 2) absorption by a clumpy or filamentary external medium with a statistically large number of clumps, and 3) absorption by a clumpy or filamentary medium with a statistically small number of clumps. These three cases have different formulations for $< \tau_{\rm eff}^{0.5} >$.

Case 1: Absorption by a homogeneous External Medium is the simplest case and has been treated by Weiler et al. [45]. Their result is obtained by substituting

$$< \tau_{\rm eff}^{0.5} > = \tau_{\rm CSM_{homog}}^{0.5} \tag{12}$$

which is the homogeneous absorption described in Eq. (3).

Case 2: Absorption by a Statistically Large Number of Clumps or Filaments is applicable if the number density and the geometric cross section of clumps is large enough so that any line-of-sight from the emitting region intersects many clumps. Then one can use a statistical approach in a scenario that has numerous clumps immersed in a homogeneous medium. For the case of $\delta = \delta'$, it is clear that the fraction of clumpy material remains constant throughout the whole wind established CSM and, therefore, that the radio signal from the SN suffers an absorption $\tau_{\rm CSM_{homog}}$ from the homogeneous component of the CSM plus an additional absorption, with an even probability distribution between 0 and $\tau_{\rm CSM_{clumps}}$, from the clumpy or filamentary component of the CSM. In such a case the appropriate average over the possible extremes of the optical depth is taken as

$$< \tau_{\rm eff}^{0.5} > = 0.67\ [(\tau_{\rm CSM_{homog}} + \tau_{\rm CSM_{clumps}})^{1.5} - \tau_{\rm CSM_{homog}}^{1.5}]\tau_{\rm CSM_{clumps}}^{-1} \tag{13}$$

with $\tau_{\rm CSM_{homog}}$ and $\tau_{\rm CSM_{clumps}}$ described in Eqs. (3) and (5). Note that in the limit of $\tau_{\rm CSM_{clumps}} \longrightarrow 0$ then $< \tau_{\rm eff}^{0.5} > \longrightarrow \tau_{\rm CSM_{homog}}$ and in the limit of $\tau_{\rm CSM_{homog}} \longrightarrow 0$ then $< \tau_{\rm eff}^{0.5} > \longrightarrow 0.67\tau_{\rm CSM_{clumps}}^{0.5}$.

Case 3: Absorption by a Statistically Small Number of Clumps or Filaments is appropriate when the number density of clumps or filaments is small and the probability that the line of sight from a given clump intersects another clump is low. Then both the emission and the absorption will occur effectively within each clump. One still expects a situation with a range of optical depths from zero for clumps on the far side of the blastwave-CSM interaction region to a maximum corresponding to the optical depth through a clump for clumps on the near side of the blastwave-CSM interaction region. One also expects an attenuation of the form $(1 - e^{-\tau_{\mathrm{CSM_{clumps}}}})\tau^{-1}_{\mathrm{CSM_{clumps}}}$ but now $\tau_{\mathrm{CSM_{clumps}}}$ represents the optical depth along a clump diameter. Moreover, in this case the clumps occupy only a small fraction of the volume and have volume filling factor $\phi \ll 1$. Since the probability that the line of sight from a given clump intersects another clump is low, a condition between the size of a clump, the number density of clumps, and the radial coordinate can be written as

$$\eta\, \pi r^2\, R \;\approx\; N \;<\; 1 \tag{14}$$

where η is the volume number density of clumps, r is the radius of a clump, R is the distance from the center of the SN to the blastwave-CSM interaction region, and N is the average number of clumps along the line of sight with N appreciably lower than unity by definition. It is easy to verify that there is a relation between the volume filling factor ϕ, r, R and N, of the form

$$\phi \;=\; \frac{4}{3}\,\frac{r}{R}\,N. \tag{15}$$

One can then express the effective optical depth $< \tau^{0.5}_{\mathrm{eff}} >$ as

$$< \tau^{0.5}_{\mathrm{eff}} >= 0.47\; \tau^{0.5}_{\mathrm{CSM_{clumps}}}\, \phi^{0.5} N^{0.5} \tag{16}$$

where, for initial estimates, one can take $N \sim 0.5$ and a constant ratio $rR^{-1} \sim 0.33$ so that, from Eq. (15), $\phi \sim 0.22$.

While intermediate cases between these three will yield results with larger errors, it is felt, considering other uncertainties in the assumptions, that Eq. (11) with the relations for $< \tau^{0.5}_{\mathrm{eff}} >$ given in Eqs. (12), (13), and (16) yield reasonable estimates of the mass-loss rates of the presupernova star. Mass-loss rate estimates from radio absorption obtained in this manner tend to be $\sim 10^{-6}$ $\mathrm{M_\odot\ yr^{-1}}$ for type Ib/c SNe and $\sim 10^{-4} - 10^{-5}$ $\mathrm{M_\odot\ yr^{-1}}$ for type II SNe. Estimates for particular RSNe are listed in Table 3.

4.2 Mass-Loss Rate from Radio Emission

For comparison purposes, one can also try to estimate the presupernova mass-loss rate which established the CSM by considering the radio emission directly. From the Chevalier model and Weiler [46] one can write the mass-loss rate to wind velocity ratio for an RSN in the form:

$$\frac{\dot{M}(\mathrm{M_\odot\ yr^{-1}})}{(w_{\mathrm{wind}}/10\ \mathrm{km\ s^{-1}})} = 8.6{\times}10^{-9} \left(\frac{L_{\mathrm{6cm\ peak}}}{10^{26}\ \mathrm{ergs\ s^{-1}\ Hz^{-1}}}\right)^{0.71} \left(\frac{t_{\mathrm{6cm\ peak}} - t_0}{(\mathrm{days})}\right)^{1.14} \tag{17}$$

for type Ib/c SNe and

$$\frac{\dot{M}(\mathrm{M_\odot\ yr^{-1}})}{(w_{\mathrm{wind}}/10\ \mathrm{km\ s^{-1}})} = 1.0{\times}10^{-6} \left(\frac{L_{\mathrm{6cm\ peak}}}{10^{26}\ \mathrm{ergs\ s^{-1}\ Hz^{-1}}}\right)^{0.54} \left(\frac{t_{\mathrm{6cm\ peak}} - t_0}{(\mathrm{days})}\right)^{0.38} \tag{18}$$

Table 3. Estimated Mass-loss Rates for RSNe[a]

SN	Type	Explosion to 6 cm peak (days)	Flux density at 6 cm peak (mJy)	Peak 6 cm radio luminosity (erg s^{-1} Hz^{-1})	Absorption mass-loss rate ($M_\odot$ yr^{-1})
Type Ib/c					
SN1983N	Ib	11.6	40.10	1.41×10^{27}	8.74×10^{-7}
SN1984L	Ib	11.0	4.59	2.57×10^{27}	7.45×10^{-7}
SN1990B	Ic	37.5	1.26	5.64×10^{26}	2.69×10^{-6}
SN1994I	Ic	38.1	14.3	1.35×10^{27}	8.80×10^{-6}
SN1998bw	Ib/c	13.3	37.4	6.70×10^{28}	2.60×10^{-5}
Type II					
SN1970G	IIL	307.0	21.50	1.40×10^{27}	6.76×10^{-5}
SN1978K	II	802.0	518.0	1.25×10^{28}	1.52×10^{-4}
SN1979C	IIL	556.0	7.32	2.55×10^{27}	1.06×10^{-4}
SN1980K	IIL	134.0	2.45	1.18×10^{26}	1.28×10^{-5}
SN1981K	II	33.70	5.15	2.14×10^{26}	1.46×10^{-6}
SN1982aa[b]	II?	476.0	19.10	1.27×10^{29}	1.03×10^{-4}
SN1986J	IIn	1210.0	135.0	1.97×10^{28}	4.28×10^{-5}
SN1988Z	IIn	898.0	1.85	2.32×10^{28}	1.14×10^{-4}
SN1993J	IIb	180.0	95.20	1.50×10^{27}	2.41×10^{-5}

[a]See Table 2 for references.
[b]SN1982aa is not optically identified but behaves like an unusually radio luminous type II.

for type II SNe, assuming that the absorption $\tau_{6\mathrm{cm\ peak}} \sim 1$, from whatever origin, at the time of observed peak in the 6 cm flux density.

The coefficients in Eqs. (17) and (18) depend on the amount of kinetic energy that is transferred to accelerate relativistic electrons and on the details of the acceleration mechanism. Although Fermi acceleration is generally accepted as the relativistic electron acceleration process, and it is usually assumed that some fixed fraction of the explosion kinetic energy is transformed into relativistic synchrotron electrons (often assumed ~1%), the physics of these two aspects is not known in detail *a priori*. Therefore, Eqs. (17) and (18) can only be "calibrated" by using the values of well studied RSNe and the assumption that all RSNe of the same type will have similar characteristics. The constants in Eqs. (17) and (18) have thus been determined from averages within the two RSN subtypes (type Ib/c and type II) of those RSNe which have pure, homogeneous absorption (i.e., $K_3 = 0$), with the omission of SN1987A because of its blue supergiant (BSG) rather than RSG progenitor.

It should be kept in mind that, because the detailed mechanism of radio emission from SNe is not well understood and the estimates have to rely on *ad hoc* calibrations from the few RSNe which have well measured light curves, mass-loss rate estimated from Eqs. (17) and (18) can only complement, and perhaps support, the more accurate determinations done from radio absorption.

4.3 Changes in Mass-Loss Rate

A particularly interesting case of mass-loss from an RSN is SN1993J, where detailed radio observations are available starting only a few days after explosion (Fig. 5a). Van Dyk et al. [44] find evidence for a changing mass-loss rate (Fig. 5b) for the presupernova star which was as high as $\sim 10^{-4}$ $M_\odot$ yr^{-1} approximately 1000 years before explosion and decreased to $\sim 10^{-5}$ $M_\odot$ yr^{-1} just before explosion, resulting in a relatively flat density profile of $\rho \propto r^{-1.5}$.

Fransson & Björgsson [15] have suggested that the observed behavior of the f-f absorption for SN1993J could alternatively be explained in terms of a systematic decrease of the electron temperature in the circumstellar material as the SN expands. It is not clear, however, what the physical process is which determines why such a cooling might occur efficiently in SN1993J, but not in SNe such as SN1979C and SN1980K where no such behavior is required to explain the observed radio turn-on characteristics. Also, recent X-ray observations with the *Röntgensatellit (ROSAT)* of SN1993J indicate a non-r^{-2} CSM density surrounding the SN progenitor [20], with a density gradient of $\rho \propto r^{-1.6}$.

Moreover, changes in presupernova mass-loss rates are not unusual. Montes et al. [27] find that type II SN1979C had a slow increase in its radio light curve after day $\sim$ 4300 (see Fig. 3a) which implied by Day 7100 an excess in flux density by a factor of $\sim$1.7 with respect to the standard model, or a density enhancement of $\sim$ 30% over the expected density at that radius. This may be understood as a change of the average CSM density profile from r^{-2}, which was applicable until day $\sim$4300, to an appreciably flatter behavior of $\sim r^{-1.4}$ [27].

On the other hand, type II SN1980K showed a steep decline in flux density at all wavelengths (see Fig. 3b) by a factor of $\sim$ 2 occurring between day $\sim$3700 and day $\sim$4900 [26]. Such a sharp decline in flux density implies a decrease in ρ_{CSM} by a factor of $\sim$ 1.6 below that expected for a r^{-2} CSM density profile. If one assumes the radio emission arises from a $\sim 10^4$ km s^{-1} blastwave traveling through a CSM established by a ~ 10 km s^{-1} pre-explosion stellar wind, this implies a significant change in the stellar mass-loss rate, for a constant speed wind, at $\sim 12,000$ yr before explosion for both SNe.

4.4 Binary Systems

In the process of analyzing a full decade of radio measurements from SN1979C, Weiler et al. [48,49] found evidence for a significant, quasi-periodic, variation in the amplitude of the radio emission at all wavelengths of $\sim$ 15% with a period of 1575 days or $\sim$ 4.3 years (see Fig. 3a at age < 4000 days). They interpreted the variation as due to a minor ($\sim$ 8%) density modulation, with a period of $\sim$ 4000 years, on the larger, relatively constant presupernova stellar mass-loss rate. Since such a long period is inconsistent with most models for stellar pulsations, they concluded that the modulation may be produced by interaction of a binary companion in an eccentric orbit with the stellar wind from the presupernova star.

This concept was strengthened by more detailed calculations for a binary model from Schwarz and Pringle [37]. Since that time, the presence of binary companions has been suggested for the progenitors of SN1987A [33], SN1993J [34] and SN1994I [29], indicating that binaries may be common in presupernova systems.

4.5 Ionized Hydrogen Along the Line-of-Sight

A reanalysis of the radio data for SN1978K from Ryder et al. [35] clearly shows flux density evolution characteristic of normal type II SNe. Additionally, the data indicate the need for a time-independent, free-free absorption component. Montes et al. [25] interpret this constant absorption term as indicative of the presence of HII along the line-of-sight to SN1978K, perhaps as part of an HII region or a distant circumstellar shell associated with the SN progenitor. SN1978K had already been noted for its lack of optical emission lines broader than a few thousand $\mathrm{km\ s^{-1}}$ since its discovery in 1990 [35], indeed suggesting the presence of slowly moving circumstellar material.

To determine the nature of this absorbing region, a high-dispersion spectrum of SN1978K at the wavelength range 6530 – 6610 Å was obtained by Chu et al. [12]. The spectrum showed not only the moderately broad Hα emission of the SN ejecta, but also narrow nebular Hα and [N II] emission. The high [N II] 6583/Hα ratio of 0.8 – 1.3 region is a stellar ejecta nebula. The expansion velocity and emission measure of the nebula are consistent with those seen in ejecta nebulae of luminous blue variables. Previous low-dispersion spectra have detected a strong [N II] 5755Å line, indicating an electron density of $(3-12)\times 10^5\ \mathrm{cm^{-3}}$. These data suggest that the ejecta nebula detected towards SN1978K is probably part of a large, dense, structured circumstellar envelope of SN1978K.

4.6 Rapid Presupernova Stellar Evolution

SN radio emission that preserves its spectral index while deviating from the standard model is taken to be evidence for a change of the circumstellar density behavior from the canonical r^{-2} law expected for a presupernova wind with a constant mass-loss rate, $\dot{M}$, and a constant wind velocity, w_{wind}. Since the radio luminosity of a SN is proportional to $(\dot{M}/w_{\mathrm{wind}})^{(\gamma-7+12m)/4}$ [7] or, equivalently, to the same power of the circumstellar density (since $\rho \propto \frac{\dot{M}}{w_{\mathrm{wind}}\ r^2}$), a measure of the deviation from the standard model provides an indication of deviation of the circumstellar density from the r^{-2} law. Monitoring the radio light curves of RSNe also provides a rough estimate of the time scale of deviations in the presupernova stellar wind density. Since the SN blastwave travels through the CSM roughly 1,000 times faster than the stellar wind velocity which established the CSM ($v_{\mathrm{blastwave}} \sim 10{,}000\ \mathrm{km\ s^{-1}}$ versus $w_{\mathrm{wind}} \sim 10\ \mathrm{km\ s^{-1}}$), one year of radio light curve monitoring samples roughly 1000 years of stellar wind mass-loss history.

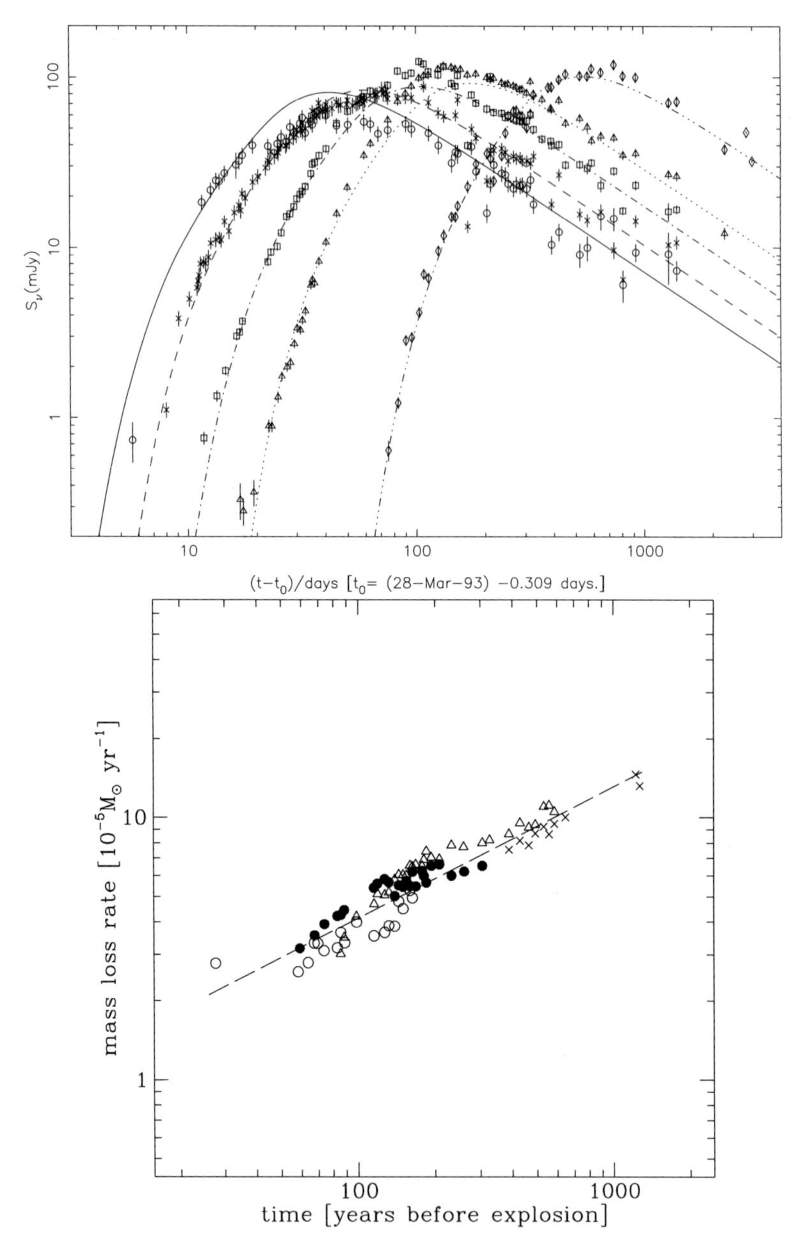

Fig. 5. Top (5a): type IIb SN1993J at 1.3 cm (22.5 GHz; *open circles, solid line*), 2 cm (14.9 GHz; *stars, dashed line*), 3.6 cm (8.4 GHz; *open squares, dash-dot line*), 6 cm (4.9 GHz; *open triangles, dotted line*), and 20 cm (1.5 GHz; *open diamonds, dash-triple dot line*). Bottom (5b): Changing mass-loss rate of the presumed red supergiant progenitor to SN1993J versus time before the explosion.

In addition to the changes in the radio light curves of SN1979C and SN1980K discussed in Sect. 4.3, the type IIn SN 1988Z can best be described by two evolution phases – an "early" phase which extends roughly from explosion through day 1479 and a "late" phase which extends roughly from day 2129 through the end of the data set. Fig. 4 shows a clear steepening of the light curves sometime between these two measurement epochs, but the actual "break" date is somewhat arbitrary due to uncertainties in the flux density measurements for this relatively faint source and the likely smoothness of any transition region. Williams et al. [55] have chosen to describe the flux density evolution separately for these two time intervals with the period from day $\sim$ 1500 to day $\sim$ 2000 as a transition. With the explosion date assumed to be 01 December 1998 from optical estimates [39], applying the fitting procedures separately to the early and late periods, with the data points between day 1500 and day 2000 included in both fits to provide a smooth transition, yields a spectral index (α) and clumpy absorption parameters (K_3 and δ') which are the same in the two time intervals within the fitting errors. However, the emission decay rate parameter β steepens significantly from ~ -1.3 for the early period to ~ -2.8 for the late period.

For a purely clumpy CSM ($K_2 = 0$), the sharp steepening of β around day 1750, while K_3 and δ' do not change, implies: 1) that the number density of clumps per unit volume (η) starts decreasing more rapidly with radius by approximately $R^{-1.5/m}$ (i.e., $\eta_{\mathrm{after\ day\ 1750}} = \eta_{\mathrm{before\ day\ 1750}} \left(\frac{R}{R_{\mathrm{day1750}}}\right)^{-1.5/m}$) with the average characteristics of the individual clumps remaining unchanged, and 2) that most of the absorption occurs within the emitting clumps themselves. In other words, the spatial distribution is so sparse that the average number of clumps along the line-of-sight is less than one ($N < 1$). This second condition is consistent with Case 3 from Weiler et al. [52]. It is interesting to note that Weiler et al. [52] also found that Case 3 applies to the radio light curves for the unusual SN1998bw/GRB980425.

The best observed example of rapid presupernova evolution is the type II SN1987A whose proximity makes it easily detectable even at very low radio luminosity. The progenitor to SN1987A was in a BSG phase at the time of explosion and had ended a RSG phase some ten thousand years earlier. After an initial, very rapidly evolving radio outburst [41] which reached a peak flux density at 6 cm $\sim$ 3 orders-of-magnitude fainter than other known type II RSNe (possibly due to sensitivity limited selection effects), the radio emission declined to a low radio brightness within a year. However, at an age of $\sim$ 3 years the radio emission started increasing again and continues to increase at the present time [1,2,16].

Although its extremely rapid development resulted in the early radio data at higher frequencies being very sparse, the evolution of the initial radio outburst is roughly consistent with the models described above in Eqs. (1)–(10) (i.e., a blastwave expanding into a circumstellar envelope). The density implied by such modeling is appropriate to a presupernova mass-loss rate of a few $\times\ 10^{-6}$ $M_\odot\ \mathrm{yr}^{-1}$ for a wind velocity of $w_{\mathrm{wind}} = 1,000\ \ \mathrm{km\ s}^{-1}$ (more appropriate to a

BSG progenitor), a blastwave velocity of $v_{\rm blastwave} = v_i = 35,000\ \ {\rm km\ s^{-1}}$, and a CSM temperature of $T = 20,000$ K.

Because the *Hubble Space Telescope (HST)* can actually image the denser regions of the CSM around SN1987A, we know that the current rise in radio flux density is caused by the interaction of the SN blastwave with the diffuse material at the inner edge of the well known inner circumstellar ring [16]. Since the density increases as the SN blastwave interaction region moves deeper into the main body of the optical ring, the flux density is expected to continue to increase steadily at all wavelengths. Recently, increases at optical and X-ray have also been reported [17,18]. Best estimates are that the blastwave-CSM interaction will reach a maximum by $\sim 2003 - 2005$.

4.7 Discussion of Circumstellar Medium Structure

For at least four type II SNe, SN1979C, SN1980K, SN1988Z, and SN1987A, there are sharp deviations from smooth modeling of the radio flux density occurring a few years after the explosion. (The smooth change of mass-loss rate for SN1993J is discussed above.) Since the SN blastwave is moving about 1000 times faster than the wind material of the RSG progenitor (i.e., $\sim 10,000\ \ {\rm km\ s^{-1}}$ versus $\sim 10\ \ {\rm km\ s^{-1}}$), it serves as a "time machine" to rapidly sample the mass-loss history of the presupernova star. Thus, the radio light curves imply a significant change in the presupernova stellar wind properties of these SNe several thousand years before the explosion. Such an interval is short compared to the lifetimes of typical RSN progenitors (say, 10–30 Myrs) but is a sizeable fraction of their RSG phase ($t_{\rm RSG} \sim 2 - 5 \times 10^5$ yrs), suggesting that a significant transition occurs in the evolution of presupernova stars just before the final explosive event.

SN1987A is an unusual case in that its BSG wind was almost certainly higher velocity ($\sim 10^3\ \ {\rm km\ s^{-1}}$) than the usually assumed RSG wind velocity ($\sim$10 ${\rm km\ s^{-1}}$). However, it also clearly underwent significant evolution in the last few thousand ($\sim 10^4$) years before explosion.

Additional evidence for altered mass-loss from type II SN progenitors over time intervals of several thousand years is provided by the detection of relatively narrow emission lines with typical widths of several $\times$ 100 ${\rm km\ s^{-1}}$ in some of the optical spectra (e.g., SN1978K [12,13,35], SN1997ab [36], SN1996L [4]), that indicate the presence of dense circumstellar shells ejected by the SN progenitors in addition to more diffuse, steady wind activity.

Since the radio emission is determined by the circumstellar density which is proportional to the mass-loss rate to stellar-wind velocity ratio ($\dot{M}/w$), one of these quantities, or both, is required to change by as much as a factor of 2 over the last few thousand years before the SN explosion. However, $\sim 10^4$ years is considerably shorter than the H and He burning phases, but much longer than any of the successive nuclear burning phases that a massive star goes through before core collapse (see, e.g., [11]) so that it is unlikely that the stellar luminosity (which determines the mass-loss rate, $\dot{M} \propto L^{1-1.5}$), can vary on a time scale needed to account for the observed changes.

On the other hand, the wind velocity, $w_{\rm wind}$, is roughly proportional to the square of the effective temperature ($w_{\rm wind} \propto T_{\rm eff}^2$, see, e.g., [31]) so that a change of a factor of ~ 2 in $w_{\rm wind}$ requires a change of a factor of only ~ 1.4 in $T_{\rm eff}$, (e.g., from $\sim 3,500$ K to $\sim 5,000$ K) or, correspondingly, a change from an early M to an early K supergiant spectrum. Such a transition would define a loop in the HR diagram reminiscent of the blue loops which are characteristic of the evolution of moderately massive stars (see, e.g., [6,21]) but classical blue loops are much slower and more extreme processes, occurring several $\times\, 10^5$ yr before the terminal stages of an RSG and involving temperature excursions from $\sim 3,500$ K to $> 10,000$ K.

The smaller temperature changes inferred from the radio data require a star to change only from a very red to a moderately red spectrum, and back, corresponding to a transition in the HR diagram which is more appropriately dubbed a "pink loop." A possibility for explaining the implied CSM density changes derives from a recent study by Panagia and Bono [32] who find that the pulsational instability of stars in the mass range $10-20$ $M_\odot$ appears, in some cases, to be of suitable period and magnitude to account for the observed radio light curve changes.

Another mechanism that may, in some cases, account for sudden changes of CSM density is the presence of a companion in a wide binary system. If the companion star has a sufficiently strong wind at the time of the primary star explosion, then the wind pressure could partially confine and compress the SN progenitor wind, increasing the average density at distances comparable to or larger than the binary system separation [5]. This mechanism has the potential of explaining the case of SN1979C, whose recent slow increase of radio flux density occurs roughly at the time when the SN blastwave is expected to have traveled a distance comparable to the binary system separation estimated from the sinusoidal modulation of the radio light curves. However, the sudden declines in the radio emission observed for SN1980K and SN1988Z cannot be explained by this mechanism.

4.8 Sphericity of an SN Explosion

It has often been suggested that SN explosions are non-spherical, and there is evidence in a number of stellar systems for jets, lobes, and other directed mass-loss phenomena. Also, the presence of polarization in the optical light from SNe (including SN1993J) has been interpreted for non-sphericity (see, e.g., [19] and the chapter by Branch et al. in this volume) and probably the most obvious evidence for non-spherical structure in an SN system is the very prominent inner ring around SN1987A. However, the most direct evidence for the structure of at least the blastwave from an SN explosion and the CSM with which it is interacting is from *VLBI* measurements on SN1993J. A series of images taken by Marcaide and co-workers (Fig. 6 [24]) over a period of two years from September 1994 through October 1996 show only a very regular ring shape indicative of a relatively spherical blastwave expanding into a relatively uniform CSM. It is

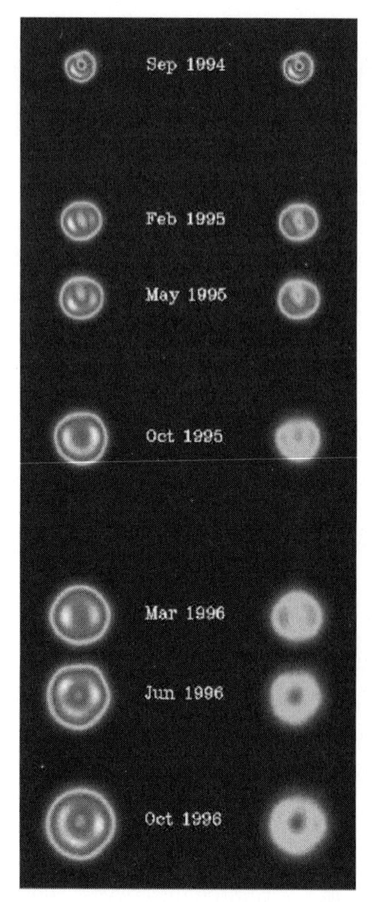

Fig. 6. VLBI radio images of SN1993J at 6 cm wavelength. Images on the left-hand side are normalized to the same peak brightness to emphasize structural changes. Images on the right-hand side are on a single brightness scale to illustrate the decrease in brightness with time [24].

possible, of course, that the early phases of the supernova explosion are asymmetric, but the interaction with a uniformly distributed CSM quickly transforms the shock interaction into a spherical blastwave which generates the observed radio emission.

4.9 Deceleration of Blastwave Expansion

Radio studies also offer the only possibility for measuring the deceleration of the blastwave from the SN explosion. So far, this has been directly possible for two objects, SN1987A and SN1993J, although in some cases the deceleration can be estimated from model fitting and Eqs. (6) or (7). Manchester et al. [23] have shown that the blastwave from the explosion of SN1987A traveled through the tenuous medium of the bubble created by the high speed wind of its BSG progenitor at an average speed of $\sim$ 10% of the speed of light ($\sim$ 35,000 km s^{-1}), but has decelerated dramatically to only $\sim$ 3,000 km s^{-1} since it has reached the inner edge of the prominent optical ring (Fig. 7a).

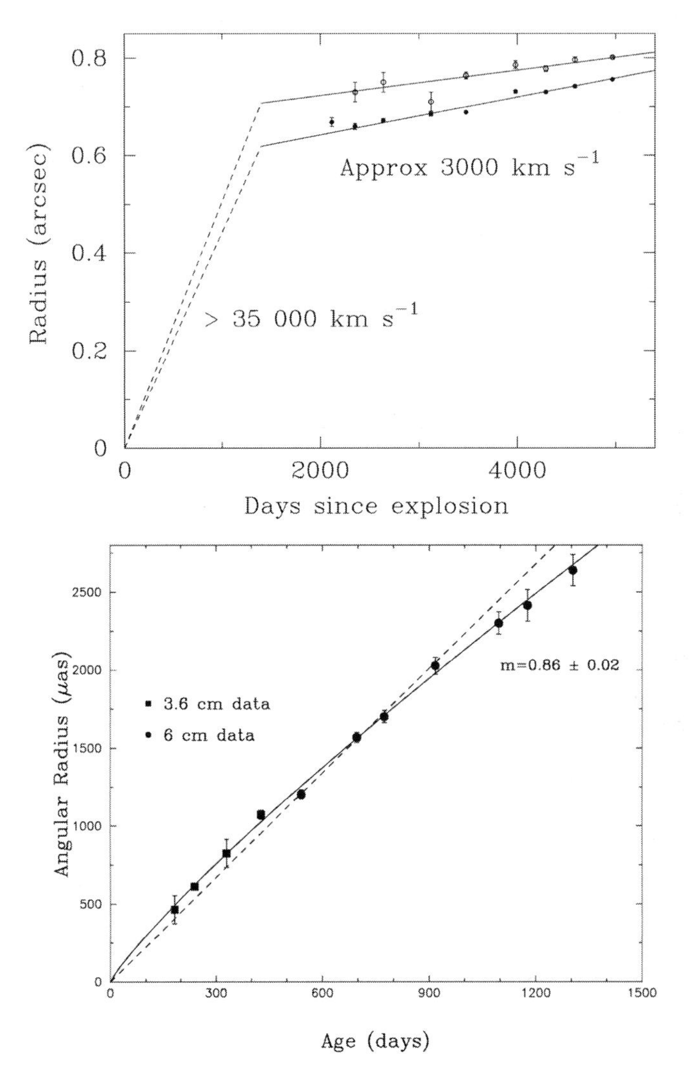

Fig. 7. Top (7a): The deceleration of the radio generating blastwaves from SN1987A; the average expansion speed of SN1987A is $\sim 0.1c$ until the blastwave reaches the inner edge of the optical ring when it slows abruptly to only $\sim$3,000 km s^{-1}. Bottom (7b): The deceleration of the radio generating blastwaves from SN1993J; the expansion speed of SN1993J is slower, but still quite high at $\sim$15,000 km s^{-1} and the deceleration is much more gradual.

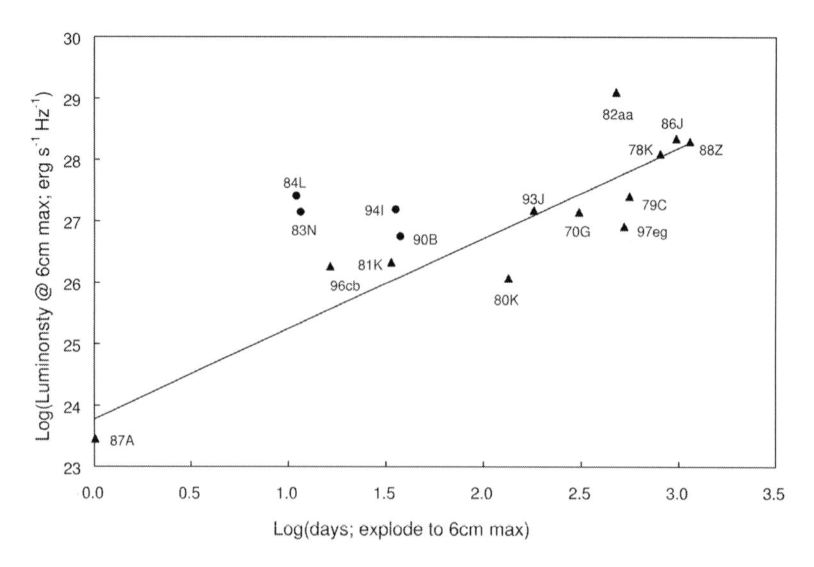

Fig. 8. Peak 6 cm luminosity ($L_{6\text{cm peak}}$) of RSNe versus time after explosion to reach that peak ($t_{6\text{cm peak}} - t_0$) for type Ib/c SNe (*filled circles*) and for type II SNe (*filled triangles*). The *solid line* (given by Eq. (17) in the text) is the unweighted, best fit to the 12 available type II RSNe. (See Weiler et al [51] for further discussion.)

Marcaide et al. [24], through the use of VLBI techniques, have been able to follow the expansion of the radio shell of SN1993J in detail (see Fig. 6) and also find that it is starting to decelerate, although the deceleration is not nearly as dramatic as that seen for SN1987A. While the expansion speed of SN1993J is quite high at $\sim$ 15,000 km s^{-1} [24], the deceleration is much more gradual (Fig. 7b) than that of SN1987A.

4.10 Peak Radio Luminosities and Distances

Weiler et al. [51] find from their long-term monitoring of the radio emission from SNe that the radio light curves evolve in a systematic fashion with a distinct peak flux density (and thus, in combination with a distance, a peak spectral luminosity) at each frequency and a well-defined time from explosion to that peak. Studying these two quantities at 6 cm wavelength, peak spectral luminosity ($L_{6\text{cm peak}}$) and time after explosion date (t_0) to reach that peak ($t_{6\text{cm peak}} - t_0$), they find that the two appear related (see Fig. 8). In particular, based on four objects, Weiler et al. [51] suggest that type Ib/c SNe may be approximate radio standard candles with a peak 6 cm spectral luminosity of

$$L_{6\text{cm peak}} \approx 1.3 \times 10^{27} \text{ erg s}^{-1} \text{ Hz}^{-1} \tag{19}$$

and, based on 12 objects, type II SNe appear to obey a relation

$$L_{6\text{cm peak}} \simeq 5.5 \times 10^{23} \, (t_{6\text{cm peak}} - t_0)^{1.4} \text{ erg s}^{-1} \text{ Hz}^{-1} \tag{20}$$

with time measured in days.

If these relations are supported by further observations, they may provide a means for determining distances to SNe, and thus to their host galaxies, from purely radio continuum observations.

5 Summary

We have assembled an overview of the radio emission from SNe and have shown that there are many similarities. In particular, the radio emission from SNe can be modeled in great detail and used to describe the physical properties of the CSM and the nature and final stages of evolution of the pre-supernova system.

Acknowledgments

KWW thanks the *Office of Naval Research (ONR)* for the 6.1 funding supporting this research. Additional information and data on radio emission from SNe can be found on *h*ttp://rsd-www.nrl.navy.mil/7214/weiler/ and linked pages.

References

1. L. Ball, D. Campbell-Wilson, D.F. Crawford, A.J. Turtle: Astrophys. J. **453**, 864 (1995)
2. L. Ball, D.F. Crawford, R.W. Hunstead, I. Kalmer, V.J. McIntyre: Astrophys. J. **549**, 599 (2001)
3. N. Bartel et al.: Nature **318**, 25 (1985)
4. S. Benetti, M. Turatto, E. Cappellaro, I.J. Danziger, P. Mazzali: Mon. Not. R. Astron. Soc. **305**, 811 (1999)
5. F.R. Boffi, N. Panagia: ASP Conf. Proc. **93**, 153 (1996)
6. E. Brocato, V. Castellani: Astrophys. J. **410**, 99 (1993)
7. R.A. Chevalier: Astrophys. J. **259**, 302 (1982a)
8. R.A. Chevalier: Astrophys. J. Lett. **259**, L85 (1982b)
9. R.A. Chevalier: Astrophys. J. **499**, 810 (1998)
10. R.A. Chevalier, C. Fransson: Astrophys. J. **420**, 268 (1994)
11. A. Chieffi, M. Limongi, O. Straniero: Astrophys. J. **502**, 737 (1998)
12. Y.-H. Chu, A. Caulet, M.J. Montes, N. Panagia, S.D. Van Dyk, K.W. Weiler: Astrophys. J. Lett. **512**, L51 (1999)
13. N.N. Chugai, I.J. Danziger, M. della Valle: Mon. Not. R. Astron. Soc. **276**, 530 (1995)
14. A. Filippenko, T. Matheson, A. Barth: Astron. J. **108**, 222 (1994)
15. C. Fransson, C.-I. Björgsson: Astrophys. J. **509**, 861 (1998)
16. B.M. Gaensler, R.N. Manchester, L. Staveley-Smith, A.K. Tzioumis, J.E. Reynolds, M.J. Kesteven: Astrophys. J. **479**, 845 (1997)
17. P. Garnavich, R. Kirshner, P. Challis: IAUC 6710 (1997)
18. G. Hasinger, B. Aschenbach, J. Truemper: Astron. Astrophys. **312**, 9 (1996)
19. P. Hoeflich, J.C. Wheeler, D.C. Hines, S.R. Tramaell: Astrophys. J. **459**, 307 (1996)
20. S. Immler, B. Aschenbach, Q.D. Wang: Astrophys. J. Lett. **561**, L107 (2001)
21. N. Langer, A. Maeder: Astron. Astrophys. **295**, 685 (1995)

22. T.A. Lozinskaya: In: *Supernovae and Stellar Wind in the Interstellar Medium* (American Institute of Physics, New York 1992) p. 190
23. R.N. Manchester, B.M. Gaensler, V.C. Wheaton, L. Staveley-Smith, A.K. Tzioumis, M.J. Kesteven, J.E. Reynolds, N.S. Bizunok: Pub. Astron. Soc. Australia **19**, 207 (2001)
24. J.M. Marcaide et al.: Astrophys. J. Lett. **486**, L31 (1997)
25. M.J. Montes, K.W. Weiler, N. Panagia: Astrophys. J. **488**, 792 (1997)
26. M.J. Montes, S.D. Van Dyk, K.W. Weiler, R.A. Sramek, N. Panagia: Astrophys. J. **506**, 874 (1998)
27. M.J. Montes, K.W. Weiler, S.D. Van Dyk, R.A. Sramek, N. Panagia, R. Park: Astrophys. J. **532**, 1124 (2000)
28. A. Natta N. Panagia: Astrophys. J. **287**, 228 (1984)
29. K. Nomoto, H. Yamaoka, O.R. Pols, E. van den Heuvel, K. Iwamoto, S. Kumagai, T. Shigeyama: Nature **371**, 227 (1994)
30. D.E. Osterbrock: In: *Astrophysics of Gaseous Nebulae* (Freeman, San Francisco 1974) p. 82
31. N. Panagia, F. Macchetto: Astron. Astrophys. **106**, 266 (1982)
32. N. Panagia, G. Bono: In: *The Largest Explosions since the Big Bang: Supernovae and Gamma Ray Bursts*, ed. by M. Livio, N. Panagia, K. Sahu (Cambridge University Press, Cambridge 2000) p. 184
33. Ph. Podsiadlowski: Pub. Astron. Soc. Pacific **104**, 717 (1992)
34. Ph. Podsiadlowski, J. Hsu, P. Joss, R. Ross: Nature **364**, 509 (1993)
35. S. Ryder, L. Staveley-Smith, M. Dopita, R. Petre, E. Colbert, D. Malin, E. Schlegel: Astrophys. J. **417**, 167 (1993)
36. I. Salamanca, R. Cid-Fernandes, G. Tenorio-Tagle, E. Telles, R.J. Terlevich, C. Munoz-Tunon: Mon. Not. R. Astron. Soc. **300**, L17 (1998)
37. D.H. Schwarz, J.E. Pringle: Mon. Not. R. Astron. Soc. **282**, 1018 (1996)
38. J. Spyromilio: Mon. Not. R. Astron. Soc. **266**, 61 (1994)
39. R.A. Stathakis, E.M. Sadler: Mon. Not. R. Astron. Soc. **250**, 786 (1991)
40. R.A. Sramek, N. Panagia, K.W. Weiler: Astrophys. J. Lett. **285**, L59 (1984)
41. A.J. Turtle et al.: Nature **327**, 38 (1987)
42. S.D. Van Dyk, R.A. Sramek, K.W. Weiler, N. Panagia: Astrophys. J. **409**, 162 (1993)
43. S.D. Van Dyk, R.A. Sramek, K.W. Weiler, N. Panagia: Astrophys. J. Lett. **419**, L69 (1993)
44. S.D. Van Dyk, K.W. Weiler, R. Sramek, M. Rupen, N. Panagia: Astrophys. J. Lett. **432**, L115 (1994)
45. K. Weiler, R. Sramek, N. Panagia, J. van der Hulst, M. Salvati: Astrophys. J. **301**, 790 (1986)
46. K.W. Weiler, N. Panagia, R.A. Sramek, J.M. van der Hulst, M.S. Roberts, L. Nguyen: Astrophys. J. **336**, 421 (1989)
47. K.W. Weiler, N. Panagia, R.A. Sramek: Astrophys. J. **364**, 611 (1990)
48. K. Weiler, S. Van Dyk, N. Panagia, R. Sramek, J. Discenna: Astrophys. J. **380**, 161 (1991)
49. K. Weiler, S. Van Dyk, J. Pringle, N. Panagia: Astrophys. J. **399**, 672 (1992a)
50. K. Weiler, S. Van Dyk, N. Panagia, R. Sramek: Astrophys. J. **398**, 248 (1992b)
51. K.W. Weiler, S.D. Van Dyk, M.J. Montes, N. Panagia, R.A. Sramek: Astrophys. J. **500**, 51 (1998)
52. K.W. Weiler, N. Panagia, R.A. Sramek, S.D. Van Dyk, M.J. Montes, C.K. Lacey: In: *The Largest Explosions since the Big Bang: Supernovae and Gamma Ray Bursts*, ed. by M. Livio, N. Panagia, K. Sahu (Cambridge University Press, Cambridge 2000) p. 198

53. K.W. Weiler, N. Panagia, M.J. Montes: Astrophys. J. **562**, 670 (2001)
54. K.W. Weiler, N. Panagia, M.J. Montes, R.A. Sramek: Ann. Rev. Astron. Astrophys. **40**, 387 (2002)
55. C.L. Williams, N. Panagia, C.K. Lacey, K.W. Weiler, R.A. Sramek, S.D. Van Dyk: Astrophys. J. **(581)**, 396 (2002)

Supernova Interaction with a Circumstellar Medium

Roger A. Chevalier[1] and Claes Fransson[2]

[1] Department of Astronomy, University of Virginia, P.O. Box 3818, Charlottesville, VA 22903, USA
[2] Stockholm Observatory, Department of Astronomy, SCFAB, SE-106 91 Stockholm, Sweden

Abstract. The explosion of a core-collapse supernova drives a powerful shock front into the wind from the progenitor star. A layer of shocked circumstellar gas and ejecta develops that is subject to hydrodynamic instabilities. The hot gas can be observed directly by its X-ray emission, some of which is absorbed and re-radiated at lower frequencies by the ejecta and the circumstellar gas. Synchrotron radiation from relativistic electrons accelerated at the shock fronts provides information on the mass loss density if free-free absorption dominates at early times or the size of the emitting region if synchrotron self-absorption dominates. Analysis of the interaction leads to information on the density and structure of the ejecta and the circumstellar medium, and the abundances in these media. The emphasis here is on the physical processes related to the interaction.

1 Introduction

The collision of supernova ejecta with dense surrounding gas can generate high pressure regions comparable to those found in the emission regions of active galactic nuclei (AGNs). The shock wave interactions are analogous to those occurring in γ-ray burst afterglows, although the supernova case involves nonrelativistic velocities. Multiwavelength observations, when combined with models for the emission, give detailed information on the outer supernova structure and the structure of the surrounding mass loss region. The stellar mass loss history leading up to the supernova explosion can be deduced from the surrounding structure.

The study of circumstellar interaction around supernovae has benefitted from several developments in the observational investigations of supernovae. One is the ability to observe supernovae in wavelength regions other than optical. Radio, X-ray, infrared, and ultraviolet observations have all been crucial for revealing various aspects of circumstellar interaction. Another is the ability to follow the optical emission from supernovae to late times with the new generation of large telescopes. The emission from circumstellar interaction typically has a lower decay rate than does the inner emission from gas heated by the initial explosion and by radioactivity, so that it eventually comes to dominate the emission. Finally, the interest in supernovae generated by SN1987A and other events has increased the rate of supernova discovery, including events which show signs of circumstellar interaction from a young age.

Many of the observational manifestations of circumstellar interaction are treated in other chapters of this volume. The emphasis here is on providing the theoretical basis to model the observed phenomena. The structure of the freely expanding supernova ejecta and of the surrounding medium provide the initial conditions for the interaction and are discussed in the first two sections. Sect. 4 deals with the hydrodynamics of the interaction and Sect. 5 treats the emission from hot gas. The effects of supernova radiation on the circumstellar gas are discussed in Sect. 6 and relativistic particles in Sect. 7. The discussion and conclusions are in Sect. 8.

2 Ejecta Structure

The density structure of a supernova is set up during the first days after the explosion. Over this timescale, the pressure forces resulting from the initial explosion, and some later power input from radioactivity, become too small to change the supernova density structure. The velocity profile tends toward that expected for free expansion, $v = r/t$, and the density of a gas element drops as t^{-3}. The pressure drops to a sufficiently low value that it is not a factor in considering shock waves propagating in the ejecta as a result of later interaction.

The density structure of the ejecta is a more complex problem. For core-collapse supernovae, where all the explosion energy is generated at the center of the star, the explosion physics leads to an outer, steep power-law density profile along with an inner region with a relatively flat profile [17,62]. The outer profile is produced by the acceleration of the supernova shock front through the outer stellar layers with a rapidly decreasing density. This part of the shock propagation does not depend on the behavior in the lower layers and the limiting structure is described by a self-similar solution. The structure depends on the initial structure of the star and thus depends on whether the progenitor star has a radiative or a convective envelope [62]. In the radiative case, which applies to Wolf-Rayet stars and to progenitors like that of SN1987A, the limiting profile is $\rho \propto r^{-10.2}$. In the convective case, which applies to red supergiant progenitors, the limiting profile is $\rho \propto r^{-11.7}$. These are the limiting profiles and the density profile over a considerable part of the supernova might be described by a somewhat flatter profile. Numerical calculations of the explosion of SN1987A have indicated $\rho \propto r^{-(8-9)}$ in the outer parts of the supernova [1].

The calculations of the density profile assume an adiabatic flow. As the shock wave approaches the surface of the star, radiative diffusion becomes a factor and the shock acceleration stops. In the case of SN1987A, theory as well as observations suggest that diffusion is not important until a velocity of $\sim 30,000$ km s^{-1} is reached [33]. A comparable limit applies to the relatively compact progenitors of the Wolf-Rayet star explosions [62]. In the case of red supergiant progenitors, which is the case applicable to most type II supernovae, a lower velocity limit is expected. At the point in the shock evolution where radiative diffusion becomes important, the loss of the radiative energy can lead to a large compression of the gas, to the point where the thermal pressure becomes important [7]. There is the possibility of a dense shell at the maximum velocity generated by the supernova. The shell may be broken up by Rayleigh-Taylor instabilities [35] and the final outcome of these processes is not known.

The overall result of these considerations is that the outer part of a core-collapse supernova can be approximated by a steep power law density profile, or $\rho_{\rm ej} \propto r^{-n}$ where n is a constant. After the first few days the outer parts of the ejecta expand with constant velocity, $V(m) \propto r$ for each mass element, m, so that $r(m) = V(m)t$ and $\rho(m) = \rho_o(m)(t_o/t)^3$. Therefore

$$\rho_{\rm ej} = \rho_o(t/t_o)^{-3}(V_o t/r)^n. \qquad (1)$$

This expression takes into account the free expansion of the gas.

Type Ia supernovae, which are believed to be the thermonuclear explosions of white dwarfs, are different because the supernova energy is not all centrally generated. A burning front spreads through the star and the energy released in this way gives rise to an explosion. The results of a number of explosion simulations show that the post-explosion density profile can be approximated by an exponential in velocity [27].

3 Stellar Mass Loss

Type II supernovae, with H lines in their spectra, are thought to be the explosions of massive stars that have reached the ends of their lives with their H envelopes still intact. In most cases, the stars explode as red supergiants, which are known to have slow, dense winds. Typical parameters are a mass loss rate of $\dot{M} = 10^{-6} - 10^{-4}$ $\mathrm{M}_\odot$ yr^{-1} and a wind velocity $w_{\rm wind} = 5 - 25$ km s^{-1}. If the mass loss parameters stay approximately constant leading up to the explosion, the circumstellar density is given by

$$\rho_{\rm cs} = \dot{M}/(4\pi w_{\rm wind} r^2). \qquad (2)$$

In the late stellar evolutionary phases, the evolution of the stellar core occurs on a rapid timescale, but the stellar envelope has a relatively long dynamical time, which can stabilize the mass loss properties.

The mechanisms by which mass is lost in the red supergiant phase are poorly understood, but some insight into the wind properties can be gained by considering observations of the winds. VY CMa, thought to have a zero age main sequence (ZAMS) mass of $30 - 40$ $\mathrm{M}_\odot$ [84], a mass loss rate $\sim 3 \times 10^{-4}$ $\mathrm{M}_\odot$ yr^{-1}, and $w_{\rm wind} \approx 39$ km s^{-1} [25,70], is an especially well-observed case. *Hubble Space Telescope (HST)* imaging shows that the density profile is approximately r^{-2} over the radius range $3 \times 10^{16} - 1.4 \times 10^{17}$ cm but that there is considerable structure superposed on this profile, including knots and filaments [70]. Observations of masers imply the presence of clumps with densities up to $\sim 5 \times 10^9$ cm^{-3} and suggest the presence of an expanding disk seen at an oblique angle [68]. VY CMa is an extreme mass loss object, but there is evidence for irregular mass loss in α Orionis and other red supergiants [76,77]. Among lower mass stars, the bipolar structure of planetary nebulae has been widely attributed to a high equatorial density in the red giant wind. In one case where we can observe the circumstellar surroundings of a supernova, SN1987A, an axisymmetric structure similar to that found in planetary nebulae has been observed [6].

Some massive stars, either through individual mass loss or through binary interaction, lose their H envelopes entirely and become Wolf-Rayet stars. Their ultimate explosions are thought to be observed as type Ib and type Ic supernovae. The progenitor stars are more compact in this case and have faster winds, with $w_{\rm wind} = 1,000 - 2,500$ km s^{-1} and $\dot{M} = 10^{-6} - 10^{-4}$ M$_{\odot}$ yr^{-1} [83]. The fast wind can create a bubble in the surrounding medium, which is typically the slow wind from a previous evolutionary phase, and the resulting shells have been observed around a number of Wolf-Rayet stars [45]. Their typical radii are a few pc.

The progenitors of type Ia supernovae are not known, so that observations of circumstellar interaction as well as other signatures of the companion star potentially contain important clues on the progenitor question [5,24,59]. In models where a white dwarf accretes mass from a companion star, the from the companion can provide a relatively dense circumstellar medium. In the case of a double degenerate progenitor, there may be a disk around the coalesced object, but the interaction is with the interstellar medium on larger scales. There have been no detections of circumstellar interaction around type Ia supernovae, so we do not discuss them further in this review, although the physical processes described here should apply if any surrounding wind is present.

4 Hydrodynamics

When the radiation dominated shock front in a supernova nears the stellar surface, a radiative precursor to the shock forms when the radiative diffusion time is comparable to the propagation time. There is radiative acceleration of the gas and the shock disappears when optical depth ~ 1 is reached [33]. The fact that the velocity decreases with radius implies that the shock will re-form as a viscous shock in the circumstellar wind. This occurs when the supernova has approximately doubled in radius.

The interaction of the ejecta, expanding with velocity $\gtrsim 10^4$ km s^{-1}, and the nearly stationary circumstellar medium results in a reverse shock wave propagating inwards (in mass), and an outgoing circumstellar shock. (Note that the "outgoing circumstellar shock" is also referred to by other authors as the "outgoing shock," the "forward shock," or the "blastwave.") The density in the circumstellar gas is given by Eq. (2). As discussed above, hydrodynamical calculations show that, to a good approximation, the ejecta density can be described by Eq. (1). A useful similarity solution for the interaction can then be found [8,9,65]. Here we sketch a simple derivation. More details can be found in these papers, as well as in the review [11].

Assume that the shocked gas can be treated as a thin shell with mass $M_{\rm s}$, velocity $V_{\rm s}$, and radius $R_{\rm s}$. Balancing the ram pressure from the circumstellar gas and the impacting ejecta, the momentum equation for the shocked shell of circumstellar gas and ejecta is

$$M_{\rm s}\frac{dV_{\rm s}}{dt} = 4\pi R_{\rm s}^2[\rho_{\rm ej}(V-V_{\rm s})^2 - \rho_{\rm cs}V_{\rm s}^2], \tag{3}$$

where $M_{\rm s}$ is the sum of the mass of the shocked ejecta and circumstellar gas. The swept up mass behind the circumstellar shock is $M_{\rm cs} = \dot{M}R_{\rm s}/w_{\rm wind}$, and that behind the reverse shock $M_{\rm rev} = 4\pi\ t_o^3 V_o^n (t/R_{\rm s})^{n-3}/(n-3)$, assuming that $R_{\rm s} >> R_p$, the radius of the progenitor. With $V = R_{\rm s}/t$ we obtain

$$\left[\frac{\dot{M}}{w_{\rm wind}}R_{\rm s} + \frac{4\pi\ \rho_o\ t_o^3\ V_o^n\ t^{n-3}}{(n-3)\ R_{\rm s}^{n-3}}\right]\frac{d^2R_{\rm s}}{dt^2} =$$
$$4\pi R_{\rm s}^2\left[\frac{\rho_o\ t_o^3\ V_o^n\ t^{n-3}}{R_{\rm s}^n}\left(\frac{R_{\rm s}}{t} - \frac{dR_{\rm s}}{dt}\right)^2 - \frac{\dot{M}}{4\pi\ w_{\rm wind}R_{\rm s}^2}\left(\frac{dR_{\rm s}}{dt}\right)^2\right]. \tag{4}$$

This equation has the power-law solution

$$R_{\rm s}(t) = \left[\frac{8\pi\rho_o\ t_o^3\ V_o^n\ w_{\rm wind}}{(n-4)(n-3)\ \dot{M}}\right]^{1/(n-2)} t^{(n-3)/(n-2)}. \tag{5}$$

The form of this similarity solution can be written down directly by dimensional analysis from the only two independent quantities available, $\rho_o\ t_o^3\ V_o^n$ and $\dot{M}/w_{\rm wind}$. The solution applies after a few expansion times, when the initial radius has been "forgotten." The requirement of a finite energy in the flow implies $n > 5$. More accurate similarity solutions, taking the structure within the shell into account, are given in [8,65]. In general, these solutions differ by less than $\sim 30\%$ from the thin shell approximation.

The maximum ejecta velocity close to the reverse shock depends on time as $V = R_{\rm s}/t \propto t^{-1/(n-2)}$. The velocity of the circumstellar shock, $dR_{\rm s}/dt$, in terms of V is $V_{\rm s} = V(n-3)/(n-2)$ and the reverse shock velocity, $V_{\rm rev} = V - V_{\rm s} = V/(n-2)$. Assuming cosmic abundances and equipartition between ions and electrons, the temperature of the shocked circumstellar gas is

$$T_{\rm cs} = 1.36\times 10^9\ \left(\frac{n-3}{n-2}\right)^2\left(\frac{V}{10^4\ {\rm km\ s^{-1}}}\right)^2\ {\rm K} \tag{6}$$

and at the reverse shock

$$T_{\rm rev} = \frac{T_{\rm cs}}{(n-3)^2}. \tag{7}$$

The time scale for equipartition between electrons and ions is

$$t_{\rm eq} \approx 2.5\times 10^7\ \left(\frac{T_{\rm e}}{10^9\ {\rm K}}\right)^{1.5}\ \left(\frac{n_{\rm e}}{10^7\ {\rm cm^{-3}}}\right)^{-1}\ {\rm s}. \tag{8}$$

One finds that the reverse shock is marginally in equipartition, unless the temperature is $\gtrsim 5\times 10^8$ K. The ion temperature behind the circumstellar shock is

$\gtrsim 6 \times 10^9$ K for $V_4 \gtrsim 1.5$, and the density a factor $\gtrsim 4$ lower than behind the reverse shock. Ion-electron collisions are therefore ineffective, and $T_e << T_{ion}$, unless efficient plasma instabilities heat the electrons collisionlessly (Fig. 1).

For typical parameters, the electron temperatures of the two shocks are very different, $\sim (1-3) \times 10^9$ K for the circumstellar shock and $10^7 - 5 \times 10^8$ K for the reverse shock, depending on n. The radiation from the reverse shock is mainly below ~ 20 keV, while that from the circumstellar shock is above ~ 50 keV.

The density behind the reverse shock is

$$\rho_{\rm rev2} = \frac{(n-4)(n-3)}{2} \rho_{\rm cs2}, \tag{9}$$

where the subscript 2 refers to the postshock gas, and is much higher than behind the circumstellar shock for $n \gtrsim 7$. There is a drop in density across the contact discontinuity, moving from the shocked ejecta to the circumstellar medium (see Fig. 1). The fact that low density gas is decelerating higher density gas leads to a Rayleigh-Taylor instability. Chevalier et al. [18] have calculated the structure using a two-dimensional PPM (piecewise parabolic method) hydrodynamic code. They indeed find that instabilities develop, with dense, shocked ejecta gas penetrating into the hotter, low density shocked circumstellar gas (Fig. 2). The instability mainly distorts the contact surface, and does not seriously affect the general dynamics. The calculation assumed that cooling is not important. If the gas at the reverse shock cools efficiently, the extent of the instability is similar, although the Rayleigh-Taylor fingers are narrower [13].

In view of the evidence for dense equatorial winds from red supergiant stars, Blondin et al. [4] simulated the interaction of a supernova with such a wind. They found that for relatively small values of the angular density gradient, the asymmetry in the interaction shell is greater than, but close to, that expected from purely radial motion. If there is an especially low density close to the pole, the flow qualitatively changes and a protrusion emerges along the axis, extending to $2-4$ times the radius of the main shell. Protrusions have been observed in the probable supernova remnant SNR$41.9+58$ in M82, although the nature of the explosion is not clear in this case [63].

In addition to asymmetric winds, there is evidence for supernova shock waves interacting with clumps of gas in the wind, as have been observed in some red supergiant winds (Sect. 3) and suggested from radio modeling (see the chapter by Sramek and Weiler in this volume). In some cases the clumps can be observed by their very narrow lines in supernova spectra, as in type IIn supernovae. The velocity of a shock wave driven into a clump, v_c, can be estimated by approximate pressure balance $v_c \approx v_s(\rho_s/\rho_c)^{1/2}$, where v_s is the shock velocity in the smooth wind with density ρ_s and ρ_c is the clump density. The lower shock velocity and higher density can lead to radiative cooling of the clump shock although the main shock is non-radiative. Optical line emission of intermediate velocity observed in type IIn (narrow line) supernovae like SN1978K, SN1988Z, and SN1995N can be explained in this way [21,22,44].

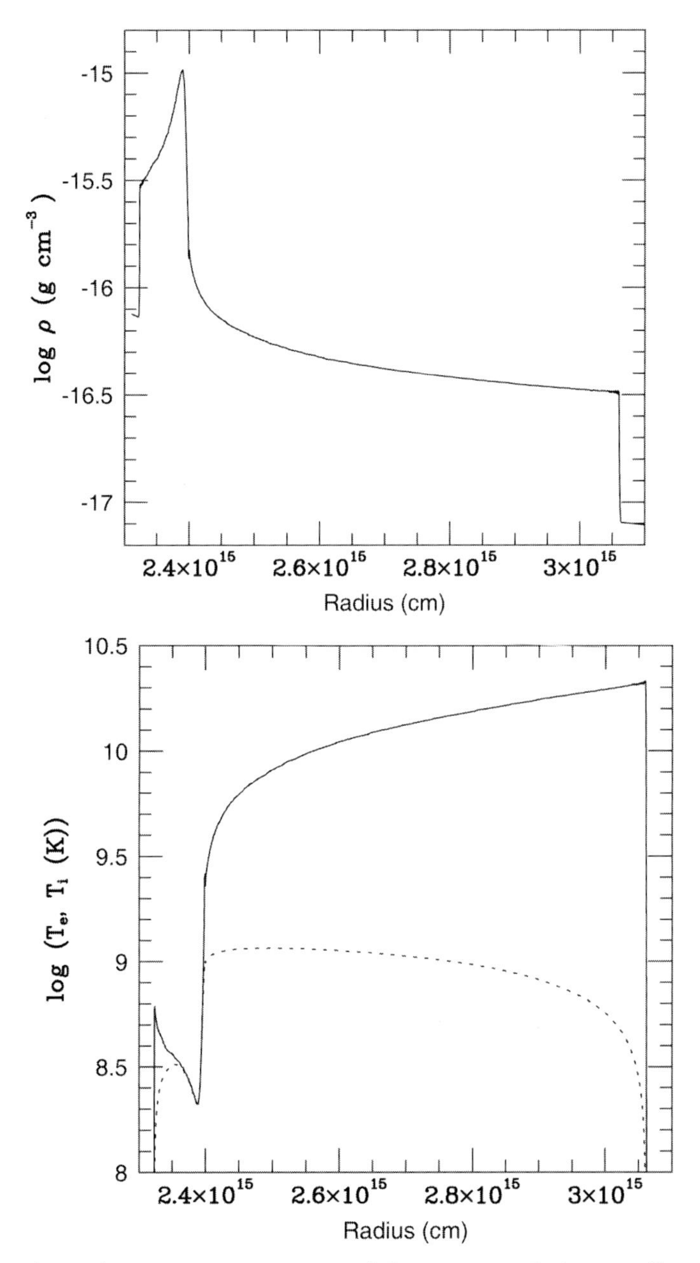

Fig. 1. Density and temperature structure of the reverse and circumstellar shocks for $n = 7$ and a velocity of 2.5×10^4 km s^{-1} at 10 days. Both shocks are assumed to be adiabatic. Because of the slow Coulomb equipartition the electron temperature (dotted line) is much lower than the ion temperature (solid line) behind the circumstellar shock.

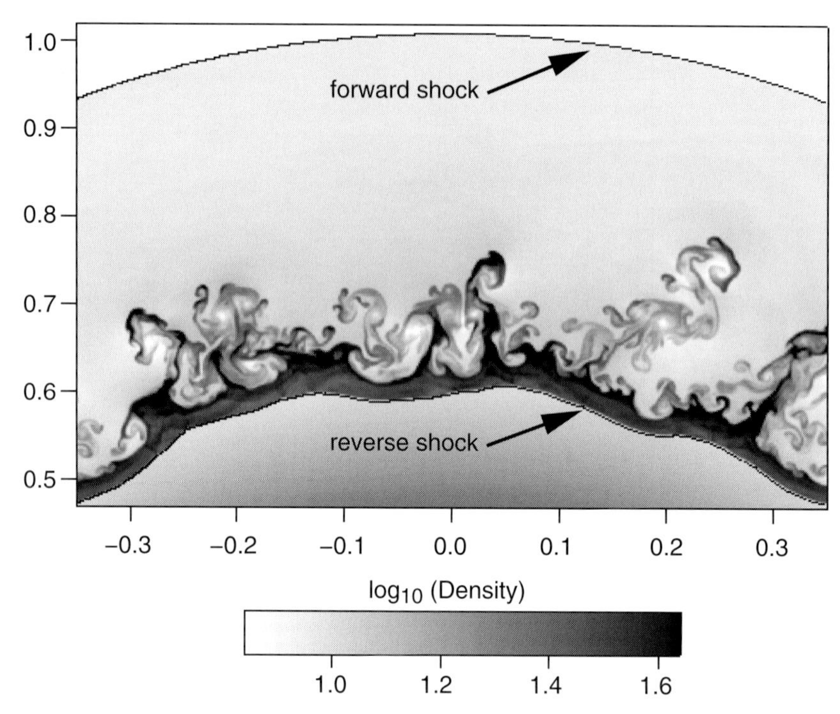

Fig. 2. Two-dimensional calculation of the shock structure for a supernova with $n = 6$ in a stellar wind (courtesy John Blondin).

The presence of many clumps can affect the of the interaction. Jun et al. [49] found that propagation in a region with clumps gives rise to widespread turbulence in the shocked region between the forward shock and the reverse shock, whereas the turbulence is confined to a region near the reverse shock for the non-clump case (Fig. 2). Their simulations are for interaction with a constant density medium, but the same probably holds true for interaction with a circumstellar wind.

5 Emission from the Hot Gas

During the first month, radiation from the supernova photosphere is strong enough for Compton scattering to be the main cooling process for the circumstellar shock. The photospheric photons have energies $\sim 3\ kT_{eff} \approx 1 - 10$ eV. The optical depth to electron scattering behind the circumstellar shock is

$$\tau_{\mathrm{e}} = 0.18\ \dot{M}_{-5}\ w_{\mathrm{wind1}}^{-1}\ V_4^{-1}\ t_{\mathrm{days}}^{-1}. \tag{10}$$

A fraction τ_{e}^N of the photospheric photons will scatter N times in the hot gas. In each scattering the photon increases its energy by a factor $\Delta\nu/\nu \approx 4\ kT_{\mathrm{e}}/m_{\mathrm{e}}c^2 \gtrsim 1$. The multiple scattering creates a power-law continuum that may reach as far up in energy as the X-ray regime. If relativistic effects can be ignored ($T_e \lesssim 10^9$ K), the spectral index is [37]

$$\alpha = -\left\{\frac{9}{4} - \frac{m_e c^2}{kT_e} \ln[\frac{\tau_e}{2}(0.9228 - \ln \tau_e)]\right\}^{1/2} + \frac{3}{2}. \qquad (11)$$

Typically, $-1 \gtrsim \alpha \gtrsim -3$. This type of emission may have been observed in the UV emission from SN1979C [37,66]. For $T_e \gtrsim 10^9$ K relativistic effects become important and considerably increase the cooling [54].

One can estimate the free-free luminosity from the circumstellar and reverse shocks from

$$L_i = 4\pi \int \Lambda_{ff}(T_e) n_e^2 r^2 dr \approx \Lambda_{ff}(T_i) \frac{M_i \rho_i}{(\mu_e m_H)^2}. \qquad (12)$$

where the index i refers to quantities connected either with the reverse shock or circumstellar shock. The density behind the circumstellar shock is $\rho_{cs2} = 4\,\rho_{cs} = \dot{M}/(\pi w_{wind} R_s^2)$. The swept up mass behind the circumstellar shock is $M_{cs} = \dot{M} R_s / w_{wind}$ and that behind the reverse shock $M_{rev} = (n-4) M_{cs}/2$. With $\Lambda_{ff} = 2.4 \times 10^{-27} \bar{g}_{ff}\, T_e^{0.5}$, we get

$$L_i \approx 3.0 \times 10^{39}\ \bar{g}_{ff}\ C_n \left(\frac{\dot{M}_{-5}}{w_{wind1}}\right)^2 \left(\frac{t}{10 \text{ days}}\right)^{-1} \text{erg s}^{-1}, \qquad (13)$$

where $\bar{g}_{ff}$ is the free-free Gaunt factor, including relativistic effects. For the reverse shock $C_n = (n-3)(n-4)^2/4(n-2)$, and for the circumstellar shock $C_n = 1$. This assumes electron-ion equipartition, which is highly questionable for the circumstellar shock (see Fig. 1). Because of occultation by the ejecta only half of the above luminosity escapes outward.

At $T_e \lesssim 2 \times 10^7$ K, line emission increases the cooling rate and $\Lambda \approx 3.4 \times 10^{-23}\ T_{e7}^{-0.67}$ erg s^{-1}cm^3. If the temperature at the reverse shock falls below $\sim 2 \times 10^7$ K, a thermal instability may occur and the gas cools to $\lesssim 10^4$ K, where photoelectric heating from the shocks balances the cooling. Using $t_{cool} = 3kT_e/\Lambda$, one obtains for the cooling time

$$t_{cool} = \frac{605}{(n-3)(n-4)(n-2)^{3.34}} \left(\frac{V_{ej}}{10^4 \text{ km s}^{-1}}\right)^{5.34} \left(\frac{\dot{M}_{-5}}{w_{wind1}}\right)^{-1} \times \left(\frac{t}{\text{days}}\right)^2 \text{ days}, \qquad (14)$$

assuming solar abundances [41]. From this expression it is clear that the cooling time is very sensitive to the density gradient, as well as the shock velocity and mass loss rate. SNe with high mass loss rates, like SN1993J, generally have radiative reverse shocks for $\gtrsim 100$ days, while SNe with lower mass loss rates, like the type IIP SN1999em, have adiabatic shocks from early times.

The most important effect of the cooling is that the cool gas may absorb most of the emission from the reverse shock. Therefore, in spite of the higher intrinsic luminosity of the reverse shock, little of this will be directly observable. The column density of the cool gas is given by $N_{\rm cool} = M_{\rm rev}/(4\pi R_{\rm s}^2 m_p)$, or

$$N_{\rm cool} \approx 1.0 \times 10^{21}(n-4) \left(\frac{\dot{M}_{-5}}{w_{\rm wind1}}\right) \left(\frac{V}{10^4 \ {\rm km \ s^{-1}}}\right)^{-1} \left(\frac{t}{100 \ {\rm days}}\right)^{-1} {\rm cm}^{-2}. \tag{15}$$

Because the threshold energy due to photoelectric absorption is related to $N_{\rm cool}$ by $E(\tau = 1) = 1.2(N_{\rm cool}/10^{22} \ {\rm cm}^{-2})^{3/8}$ keV, it is clear that the emission from the reverse shock is strongly affected by the cool shell, and a transition from optically thick to optically thin is expected during the first months, or year. As an illustration, we show in Fig. 3 the calculated X-ray spectrum at 10 days and at 200 days for SN1993J [41]. At early epochs the spectrum is dominated by the very hard spectrum from the circumstellar shock, which reaches out to $\gtrsim 100$ keV. At later epochs the soft spectrum from the reverse shock penetrates the cool shell, and the line dominated emission from the cooling gas dominates.

If cooling, the total energy emitted from the reverse shock is

$$\begin{aligned} L_{\rm rev} &= 4\pi R_{\rm s}^2 \ \frac{1}{2} \ \rho_{\rm ej} V_{\rm rev}^3 = \frac{(n-3)(n-4)}{4(n-2)^3} \ \frac{\dot{M} V^3}{w_{\rm wind}} \\ &= 1.6 \times 10^{41} \ \frac{(n-3)(n-4)}{(n-2)^3} \dot{M}_{-5} w_{\rm wind1}^{-1} V_4^3 \ \ {\rm erg \ s^{-1}}. \end{aligned} \tag{16}$$

For high $\dot{M}/w_{\rm wind}$ the luminosity from the reverse shock may contribute appreciably to, or even dominate, the bolometric luminosity.

Because $V \propto t^{-1/(n-2)}$, $L_{\rm rev} \propto t^{-3/(n-2)}$ in the cooling case. Although the total luminosity is likely to decrease in the cooling case, the increasing transparency of the cool shell, $\tau_{\rm cool} \propto t^{-1}$, can cause the observed flux in energy bands close to the low energy cutoff, $E(\tau = 1)$, to increase with time, as was seen, e.g., in SN1993J (Fig. 3).

Because of the low temperature, the spectrum of the reverse shock is dominated by line emission from metals (Fig. 3). An important point is that the observed spectrum is formed in gas with widely different temperatures, varying from the reverse shock temperature to $\sim 10^4$ K. A spectral analysis based on a one temperature model can be misleading.

Chugai [19] has proposed that the X-ray emission from the type IIn SN1986J is the result of the forward shock front moving into clumps, as opposed to the reverse shock emission. One way to distinguish these cases is by the width of emission lines; emission lines from the reverse shock wave are expected to be broad. It has not yet been possible to carry through with this test [46].

Another way of observing the hot gas is through the emission from collisionally heated dust grains in the gas [30]. Dust formation in the rapidly expanding ejecta is unlikely, so the forward shock front must be considered. Evaporation by the supernova radiation creates a dust-free zone around the supernova (see Sect. 6.3). The time for the supernova shock wave to reach the dust depends on the supernova luminosity and the shock velocity; it is probably at least several years. The infrared luminosity from dust can be up to ~ 100 times the X-ray

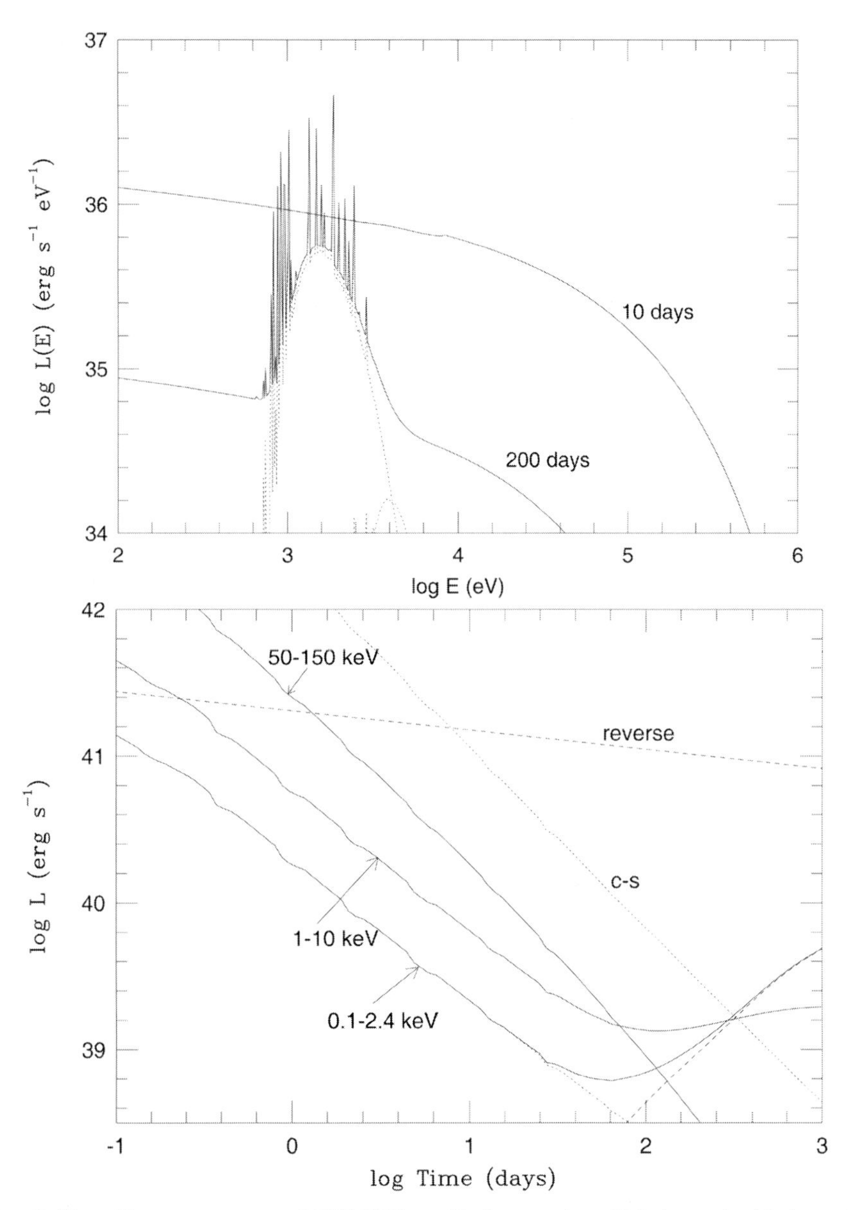

Fig. 3. Top: X-ray spectrum of SN1993J at 10 days and at 200 days. At 10 days the free-free emission from the outer shock dominates, while at 200 days the cool shell is transparent enough for the line dominated spectrum from the reverse shock to dominate instead. Bottom: The solid lines give the luminosity in the 0.1-2.4, 1-10, and 50-100 keV bands, corrected for absorption, as a function of time, while the dotted lines give the total emitted luminosity from the reverse and circumstellar shocks [41].

luminosity of the hot gas for typical parameters, and the dust temperature is a measure of the density of the gas [30]. If the X-ray emission from a supernova like SN1986J is from circumstellar clumps that are in the region where dust is present, there is the possibility of a large infrared luminosity.

6 Radiative Heating and Re-emission

6.1 Soft X-Ray Burst and Circumstellar Gas

The earliest form of circumstellar interaction occurs at shock breakout. As the shock approaches the surface, radiation leaks out on a time scale of less than an hour. The color temperature of the radiation is $\sim (1-5) \times 10^5$ K and the energy $\sim (1-10) \times 10^{48}$ erg [33,34,51,62]. This burst of EUV (extreme ultraviolet) radiation and soft X-rays ionizes and heats the circumstellar medium on a time scale of a few hours. In addition, the momentum of the radiation may accelerate the circumstellar gas to a high velocity. Most of the emission at energies $\gtrsim 100$ eV is emitted during the first few hours, and after 24 hours little ionizing energy remains.

The radiative effects of the soft X-ray burst were most clearly seen from the ring of SN1987A, where a number of narrow emission lines from highly ionized species, like N III–V, were first seen in the UV [43,73]. Later, a forest of lines came to dominate also the optical spectrum [79]. Imaging with *HST* (e.g., [6,47]) showed that the lines originated in the now famous inner circumstellar ring of SN1987A at a distance of ~ 200 light days from the SN. The presence of highly ionized gas implied that the gas must have been ionized and heated by the radiation at shock breakout. Because of the finite light travel time across the ring, the observed total emission from the ring is a convolution of the emission at different epochs from the various parts of the ring. Detailed modeling [55,56] shows that, while the ionization of the ring occurs on the time scale of the soft X-ray burst, the gas recombines and cools on a timescale of years, explaining the persistence of the emission decades after the explosion. The observed line emission provides sensitive diagnostics of both the properties of the soft X-ray burst and the density, temperature and abundances of the gas in the ring. In particular, the radiation temperature must have reached $\sim 10^6$ K, in good agreement with the most detailed recent modeling of the shock breakout [3]. Narrow emission lines are not unique to SN1987A, but have also been observed for several other SNe, in particular several type IIn SNe, such as SN1995N [44] and SN1998S [36].

The soft X-ray burst may also pre-accelerate the gas in front of the shock. In the conservative case that Thompson scattering dominates, the gas immediately in front of the shock will be accelerated to

$$V = 1.4 \times 10^3 \left(\frac{E}{10^{48}\ \mathrm{erg}}\right) \left(\frac{V_\mathrm{s}}{1 \times 10^4\ \mathrm{km\ s^{-1}}}\right)^{-2} \left(\frac{t}{\mathrm{days}}\right)^{-2}\ \mathrm{km\ s^{-1}}, \quad (17)$$

where E is the total radiative energy in the burst. Line absorption may further boost this [39]. If the gas is pre-accelerated, the line widths are expected to decrease with time. After about one expansion time ($\sim R_p/V$) the reverse and circumstellar shocks are fully developed, and the radiation from these will dominate the properties of the circumstellar gas. The fraction of this emission going inward is absorbed by the ejecta and there re-emitted as optical and UV radiation [37,38].

The X-ray emission from the shocks ionizes and heats both the circumstellar medium and the SN ejecta. Observationally, these components are distinguished easily by the different velocities. The circumstellar component is expected to have velocities typical of the progenitor winds, i.e., $\lesssim 1000 \ \mathrm{km\ s^{-1}}$, while the ejecta have considerably higher velocities. The density is likely to be of the order of the wind density $10^5 - 10^7 \ \mathrm{cm^{-3}}$, or higher if clumping is important. The ionizing X-ray flux depends strongly on how much of the flux from the reverse shock can penetrate the cool shell. The state of ionization in the circumstellar gas is characterized by the value of the ionization parameter [50],

$$\zeta = L_{\mathrm{cs}}/(r^2 n) = 10^2 (L_{\mathrm{cs}}/10^{40} \ \mathrm{erg\ s^{-1}})(r/10^{16} \ \mathrm{cm})^{-2}(n/10^6 \ \mathrm{cm^{-3}})^{-1}. \quad (18)$$

The comparatively high value of $\zeta \approx 10 - 10^3$ explains the presence of narrow coronal lines of [Fe V–XI] seen in objects like SN1995N [44].

6.2 SN Ejecta

The ingoing X-ray flux from the reverse shock ionizes the outer parts of the ejecta. The state of highest ionization, therefore, is close to the shock, with a gradually lower degree of ionization inwards. Unless clumping in the ejecta is important, the ejecta density is $\sim 10^6 - 10^8 \ \mathrm{cm^{-3}}$. In the left panel of Fig. 4 we show temperature and ionization structure of the ejecta, as well as the emissivity of the most important lines. The temperature close to the shock is $\sim 3 \times 10^4$ K. Calculations show that most of the emission here is emitted as UV lines of highly ionized ions, like Lyα, C III–IV, N III–V, and O III–VI. Inside the ionized shell there is an extended, partially ionized zone, similar to that present in the broad emission line regions of AGNs. Most of the emission here comes from Balmer lines.

As we have already discussed, the outgoing flux from the reverse shock is to a large extent absorbed by the cool shell between reverse shock and the contact discontinuity, if radiative cooling has been important. The whole region behind the reverse shock is in approximate pressure balance, and the density of this gas is therefore a factor $\sim 4T_{\mathrm{rev}}/T_{\mathrm{cool}} \approx 10^3 - 10^4$ higher than that of the ejecta. Because of the high density, the gas is only partially ionized and the temperature is only $(5 - 8) \times 10^3$ K. Most of the emission comes out as Balmer lines, Mg II and Fe II lines (Fig. 4, right panel). The thickness of the emitting region is also very small, $\sim 3 \times 10^{12}$ cm. In one dimensional models, the velocity is marginally smaller than the highest ejecta velocities. Instabilities in the shock are, however, likely to erase this difference.

An important diagnostic of the emission from the cool shell and the ejecta is the Hα line. This line arises as a result of recombination and collisional excitation. In [16], it is shown that $\sim 1\%$ of the reverse shock luminosity is emitted as

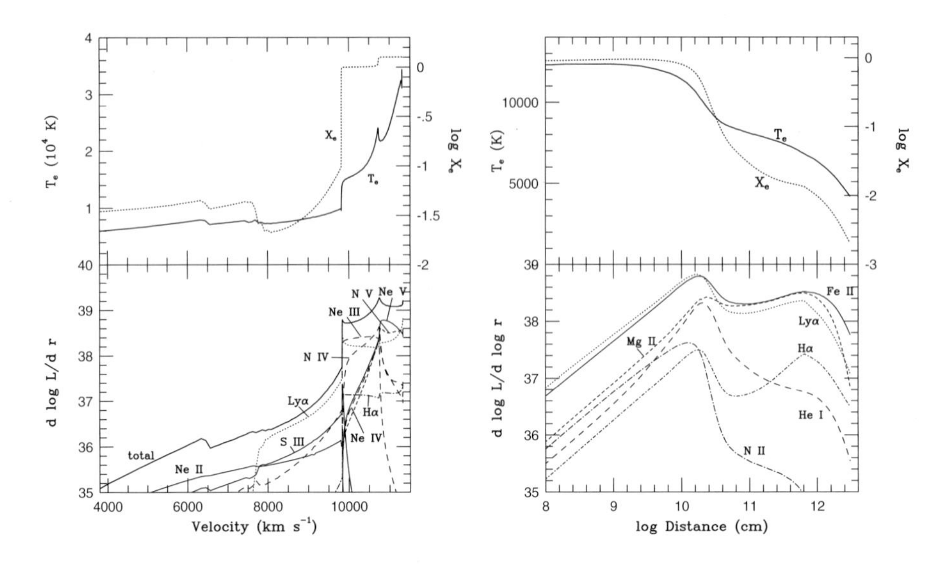

Fig. 4. Structure of the ejecta and cool shell ionized by the reverse shock at 500 days for parameters appropriate to SN1993J ($\dot{M} = 5 \times 10^{-5}$ $\mathrm{M}_\odot$ yr^{-1} for $w_{\mathrm{wind}} = 10$ km s^{-1}). Upper panels show the temperature and ionization of the ejecta (left panels) and the cool shell (right panels), while the lower panels show the corresponding luminosities per unit distance. Note the different length scales in the two panels. The ejecta region has low density, high temperature and ionization, while the cool shell has a high density, is extremely thin, has a low temperature, and is only partially ionized.

Hα, fairly independent of density and other parameters. Observations of this line permit us to follow the total luminosity from the reverse shock, complementary to the X-ray observations. In SN1993J, the Hα line had the box-like shape that is expected for shocked, cooled ejecta [60,61]. The top of the line showed structure that varied with time; this could be related to hydrodynamic instabilities of the reverse shocked gas.

6.3 Interaction with Dust

The winds from red supergiant stars are known to contain dust, so that infrared emission from radiatively heated dust and dust scattering of supernova light might be expected. The dust temperature is determined by a balance between radiative heating and emission from the grain surface. If the supernova radiates like a blackbody of temperature T_{sn} and if the dust absorption efficiency varies as λ^{-q}, the grain temperature is $T_{\mathrm{g}} = T_{\mathrm{sn}} W^{1/(4+q)}$, where W is the dilution factor for the supernova radiation [28]. For $q = 1$, we have

$$T_{\mathrm{g}} = 280 \left(\frac{T_{\mathrm{sn}}}{5000\ \mathrm{K}} \right)^{0.2} \left(\frac{L_{\mathrm{sn}}}{10^{42}\ \mathrm{erg\ s}^{-1}} \right)^{0.2} \left(\frac{r}{10^{18}\ \mathrm{cm}} \right)^{-0.4} \mathrm{K}, \tag{19}$$

where L_{sn} is the supernova luminosity.

Because typical evaporation temperatures for grain materials are in the range $1000 - 1500$ K, the dust near the supernova is evaporated. The radius out to which dust is evaporated, r_v, is probably determined by the luminosity at the time of shock breakout, when $L_{\rm sn}$ can be $\gtrsim 10^{44}$ erg s^{-1}. SN1979C and SN1980K were detected as infrared sources, and Dwek [28] estimates $r_v \approx 3 \times 10^{17}$ cm in these cases. The infrared light curve can be calculated from $T_{\rm g}$, taking into account light travel time effects. If the characteristic time that the supernova is bright is short compared to $2r_v/c$, there is a plateau phase until a time $2r_v/c$. The infrared flux then drops as t^{-2} for a highly extended wind and drops more rapidly if the cutoff in the wind, r_w, is close to r_v. For SN1979C and SN1980K, Dwek [28] found $r_w \lesssim 10^{18}$ cm. The ratio of the infrared emitted energy to the total emitted energy gives an estimate of the optical depth through the dusty wind. The optical depths for SN1979C and SN1980K are ~ 0.3 and 0.03, respectively, leading to minimum shell masses of $\sim 1 - 5$ $\rm M_\odot$ and $\sim 0.1 - 0.4$ $\rm M_\odot$ for the two supernovae [28]. The corresponding mass loss rates are consistent with those derived from radio observations (see [54,80,82] and the chapter by Sramek and Weiler in this volume). infrared emission that can be attributed to radiatively heated dust has recently been observed from SN1998S [36]. infrared dust echoes have the potential to give information on dust composition and distribution [29,32], but the observations do not yet exist to addres s these issues.

The dust grains that give an infrared echo can also give a scattered light echo [10]. The ratio between the total scattered light and the infrared light depends on the albedo of the dust. The light curve decreases with time before $2r_v/c$ in this case because of the strong forward scattering of typical grains. Chevalier [10] failed to find good evidence for scattered light echoes from SN1979C and SN1980K, although such echoes might have been observable. The implication may be that the dust grains have small albedos. The problem with detecting scattered light echoes is that any echo light may be dominated by light coming directly from the circumstellar interaction. Roscherr and Schaefer [69] examined the late emission from two type IIn supernovae and determined that it was due to shock interaction. Because of the ability to spatially resolve the emission in SN1987A, it has been possible to observe scattered light from the wind interaction nebula [23] and possibly a shell at the outer extent of the red supergiant wind [15].

7 Relativistic Particles

Unambiguous evidence for the presence of relativistic electrons comes from radio observations of SNe. A characteristic is the wavelength-dependent turn-on of the radio emission (see the chapter by Sramek and Weiler in this volume and [74,80,82]), first seen at short wavelengths, and later at longer wavelengths. This behavior is interpreted as a result of decreasing absorption due to the expanding emitting region [9].

Depending on the magnetic field and the density of the circumstellar medium, the absorption may be produced either by free-free absorption in the surrounding thermal gas, or by synchrotron self-absorption (SSA) by the same electrons that are responsible for the emission. The relativistic electrons are believed to be produced close to the interaction region, which provides an ideal environment for the acceleration of relativistic particles. The details of the acceleration and injection efficiency are still not well understood (see, e.g., [26] and references therein). Here we just parameterize the injection spectrum with the power-law index $p_{\rm i}$ and an efficiency, η, in terms of the postshock energy density. Without radiation or collisional losses the spectral index of the synchrotron emission will then be $\alpha = -(p-1)/2$, where $F_\nu \propto \nu^{+\alpha}$. Diffusive acceleration predicts that $p_{\rm i} = 2$ in the test particle limit. If the particle acceleration is very efficient and nonlinear effects are important, the electron spectrum can be steeper (see, e.g., [31]).

For free-free absorption, the optical depth $\tau_{\rm ff} = \int_{R_{\rm s}}^{\infty} \kappa_{\rm ff} n_{\rm e} n_{\rm i} dr$ from the radio emitting region close to the shock through the circumstellar medium decreases as the shock wave expands, explaining the radio turn-on. Assuming a fully ionized wind with constant mass loss rate and velocity, so that Eq. (2) applies, the free-free optical depth at wavelength λ is

$$\tau_{\rm ff}(\lambda) \approx 7.1 \times 10^2 \lambda^2 \left(\frac{\dot{M}_{-5}}{w_{\rm wind1}} \right)^2 T_5^{-3/2} V_4^{-3} t_{\rm days}^{-3} \tag{20}$$

here $\dot{M}_{-5}$ is the mass loss rate in units of 10^{-5} $\mathrm{M}_\odot$ yr^{-1}, $w_{\rm wind1}$ the wind velocity in units of 10 km s^{-1}, and T_5 the temperature of the circumstellar gas in units of 10^5 K. From the radio light curve, or spectrum, the epoch of $\tau_{\rm ff} = 1$ can be estimated for a given wavelength, and from the line widths in the optical spectrum the maximum expansion velocity, V, can be obtained. Because the effects of the radiation from the supernova have to be estimated from models of the circumstellar medium, the temperature in the gas is uncertain. Calculations show that initially the radiation heats the gas to $T_{\rm e} \approx 10^5$ K [54]. $T_{\rm e}$ then decreases with time, and after a year $T_{\rm e} \approx (1.5 - 3) \times 10^4$ K. In addition, the medium may recombine, which further decreases the free-free absorption. From $t[\tau(\lambda)_{\rm ff} = 1]$ the ratio $\dot{M}/w_{\rm wind}$ can be calculated. Because $\dot{M}/w_{\rm wind} \propto T_{\rm e}^{3/4} x_{\rm e}^{-1}$, errors in $T_{\rm e}$ and $x_{\rm e}$ may lead to large errors in $\dot{M}$. If the medium is clumpy, Eq. (20) may lead to an overestimate of $\dot{M}/w_{\rm wind}$. (See also the discussion of a clumpy CSM in the chapter by Sramek and Weiler in this volume.)

Under special circumstances (see below), SSA by the same relativistic electrons emitting the synchrotron radiation may be important [12,40,71]. The emissivity of the synchrotron plasma is given by $j(\lambda) = \mathrm{const}\ \lambda^\alpha B^{1-\alpha} N_{\rm rel}$ [53], while the optical depth to self-absorption is given by

$$\tau_{\rm s} = \mathrm{const} \times \lambda^{(5/2)-\alpha} B^{(3/2)-\alpha} N_{\rm rel} \ . \tag{21}$$

Here $N_{\rm rel}$ is the column density of relativistic electrons and B the magnetic field. The flux from a disk with radius $R_{\rm s}$ of relativistic electrons is, including the effect of SSA, given by

$$F_\nu(\lambda) \propto R_{\rm s}^2 S(\lambda)[1 - e^{-\tau_{\rm s}(\lambda)}], \tag{22}$$

where $S(\lambda) = j(\lambda)/\kappa(\lambda) = \text{const} \times \lambda^{-5/2}B^{-1/2}$ is the source function. In the optically thick limit we therefore have $F_\nu(\lambda) \approx \text{const} \times R_\text{s}^2\ \lambda^{-5/2}B^{-1/2}$, independent of N_rel. A fit of this part of the spectrum therefore gives the quantity $R_\text{s}^2\ B^{-1/2}$. The break of the spectrum determines the wavelength of optical depth unity, $\lambda(\tau_\text{s} = 1)$. Eq. (21) therefore gives a second condition on $B^{3/2-\alpha}\ N_\text{rel}$. If $R_\text{s}(t)$ is known in some independent way, one can determine both the magnetic field and the column density of relativistic electrons, independent of assumptions about equipartition, etc.

In some cases, most notably for SN1993J, the shock radius, R_s, can be determined directly from *Very Long Baseline Interferometry (VLBI)* observations. If this is not possible, an alternative is from observations of the maximum ejecta velocity seen in, e.g., the Hα line, which should reflect the velocity of the gas close to the shock. Because the SN expands homologously, $R_\text{s} = V_\text{max}t$. A fit of the spectrum at a given epoch can therefore yield both B and N_rel independently. From observations at several epochs the evolution of these quantities can then be determined.

Although the injected electron spectrum from the shock is likely to be a power-law with $p_\text{i} \approx 2$ ($\alpha \approx -0.5$), the integrated electron spectrum is affected by various loss processes. Most important, the synchrotron cooling time scale of an electron with Lorentz factor Γ is $t_\text{syn} \approx 9 \times 10^3 \Gamma^{-1} B^{-2}$ days. This especially affects the high energy electrons, steepening the index of the electron spectrum by one unit, $p = p_\text{i} + 1 \approx 3$ ($\alpha \approx -1$). Inverse Compton losses have a similar effect. At low energy, Coulomb losses may be important, causing the electron spectrum to flatten.

The best radio observations of any SN were obtained for SN1993J. This SN was observed from the very beginning until late epochs with the *Very Large Array (VLA)*[1] at wavelengths between 1.3 – 90 cm [78], producing a set of beautiful light curves (see also the chapter by Sramek and Weiler in this volume). In addition, the SN was observed with *VLBI* [2,57,58], resulting in an impressive sequence of images in which the radio emitting plasma could be directly observed (see, e.g., Fig. 6 in the chapter by Sramek and Weiler in this volume). These images show a remarkable degree of symmetry and clearly resolved the shell of emitting electrons. The evolution of the radius of the radio emitting shell could be well fitted by $R_\text{s} \propto t^{0.86}$, implying a deceleration of the shock front.

From a fit of the observed spectra for the different epochs, the magnetic field and number of relativistic electrons could be determined for each epoch, as described above [40]. In Fig. 5 we show the evolution of B and particle density n_rel, plotted as a function of the shock radius. The most remarkable thing is the smooth evolution of these quantities, showing that $B \approx 6.4(R_\text{s}/10^{16}\ \text{cm})^{-1}$ G, and $n_\text{rel} \propto \rho V^2 \propto t^{-2}$, the thermal energy density behind the shock. The magnetic field is close to equipartition, $B^2/8\pi \approx 0.14\ \rho V_\text{s}^2$, much higher than expected if the circumstellar magnetic field of the order of a few mG was just

[1] The *VLA* telescope of the National Radio Astronomy Observatory is operated by Associated Universities, Inc. under a cooperative agreement with the National Science Foundation.

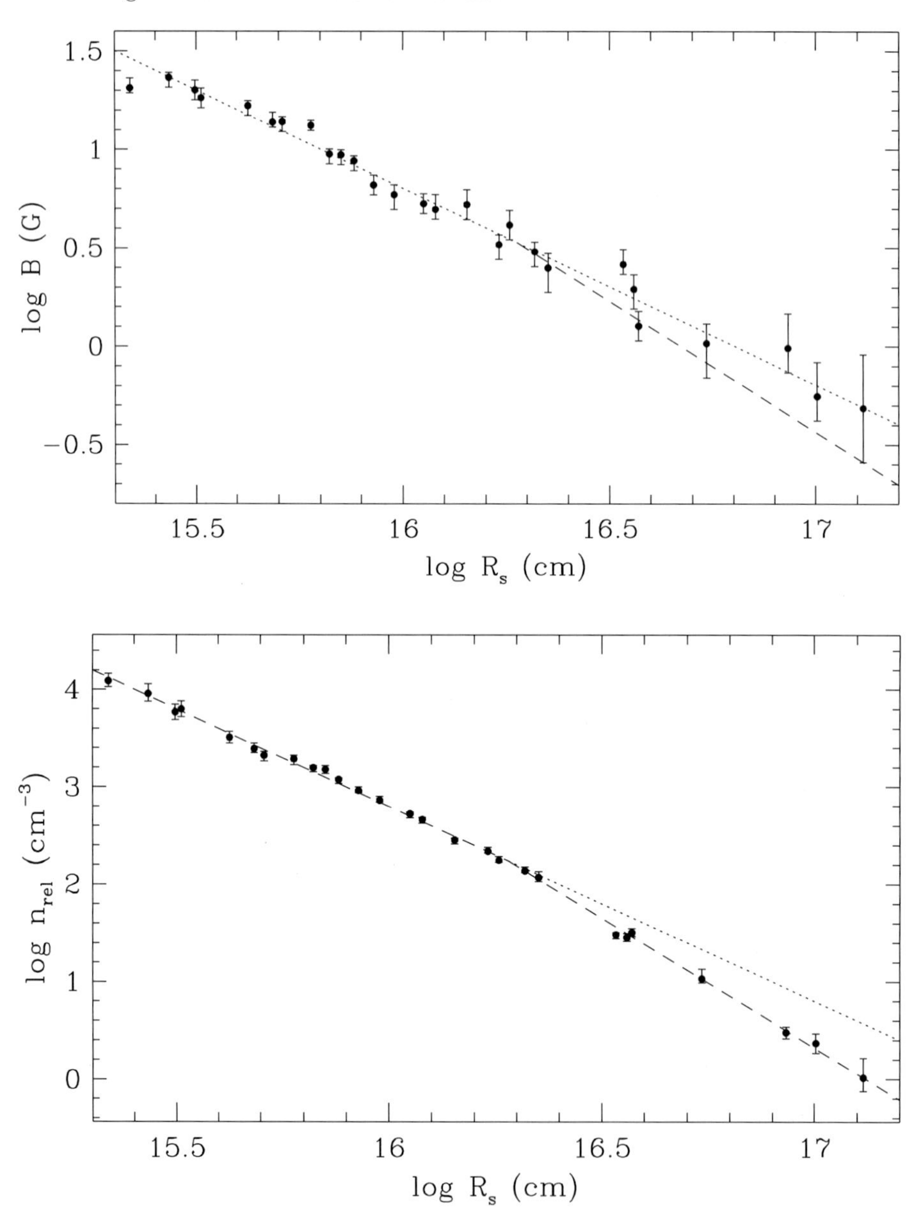

Fig. 5. Magnetic field (top) and density of relativistic electrons (bottom) as a function of the shock radius for SN1993J. The dashed lines show the expected evolution if the magnetic energy density and relativistic particle density scale with the thermal energy density, $B^2/8\pi \propto \rho V_s^2 \propto n_{\rm rel} \propto t^{-2}$, while the dotted lines show the case when $B \propto r^{-1}$ and $n_{\rm rel} \propto r^{-2}$[40].

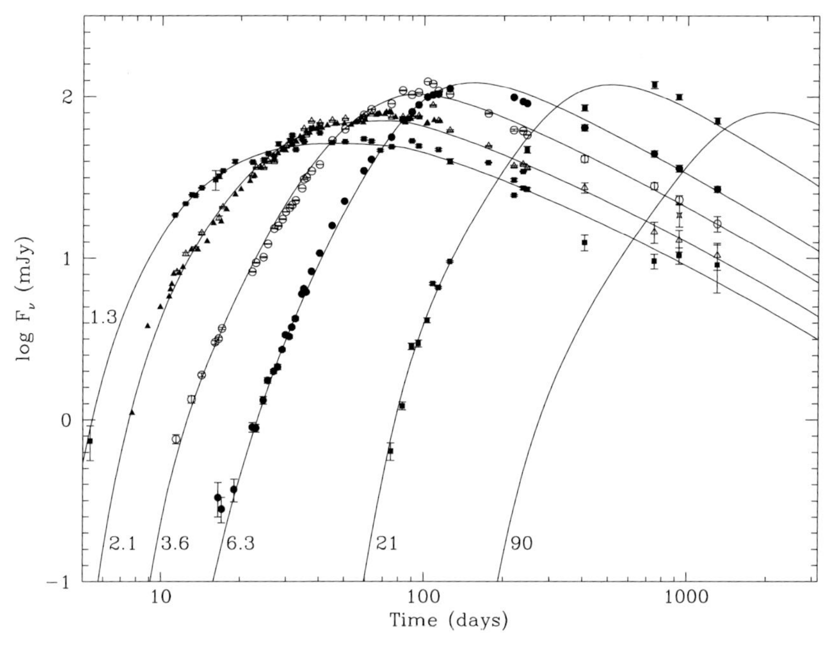

Fig. 6. Observed [78] and model radio light curves of SN1993J [40].

compressed, and strongly argues for field amplification, similar to what has been seen in simulations [48]. Contrary to earlier, simplified models for SN1993J based on free-free absorption only [41,78], the circumstellar density was consistent with $\rho \propto r^{-2}$.

In Fig. 6, we show the excellent fit of the resulting light curves. The high values of B implied that synchrotron cooling was important throughout most of the evolution for the electrons responsible for the cm emission, and also for the 21 cm emission before $\sim$ 100 days. At early epochs, Coulomb losses were important for the low energy electrons. The injected electron spectrum was best fitted with $p_{\rm i} = 2.1$.

The form of the light curves can be understood if, for simplicity, we assume equipartition, so that $B^2/8\pi = \eta\rho V_{\rm s}^2$. With $\rho \propto (\dot{M}/w_{\rm wind})\ R_{\rm s}^{-2}$ and $V_{\rm s} \propto R_{\rm s}/t$, we find that $B \propto (\dot{M}/w_{\rm wind})^{1/2}t^{-1}$. The optically thick part is therefore given by

$$F_\nu(\lambda) \propto\ R_{\rm s}^2\ \lambda^{-5/2}B^{-1/2} \propto (\dot{M}/w_{\rm wind})^{-1/4}\lambda^{-5/2}t^{(5n-14)/2(n-2)}, \qquad (23)$$

since $R_{\rm s} \propto t^{(n-3)/(n-2)}$. For large n, we get $F_\nu(\lambda) \propto t^{5/2}$. An additional curvature of the spectrum is produced by free-free absorption in the wind, although this only affects the spectrum at early epochs.

In the optically thin limit, $F_\nu(\lambda) \propto R_{\rm s}^2 j(\lambda) \propto R_{\rm s}^2\lambda^{-\alpha}B^{1-\alpha}N_{\rm rel}$. If losses are unimportant, $N_{\rm rel,tot} = 4\pi R_{\rm s}^2 N_{\rm rel}$, the total number of relativistic electrons, may either be assumed to be proportional to the total mass, if a fixed fraction of the shocked electrons are accelerated, or be proportional to the swept up thermal energy. In the first case, $N_{\rm rel,tot} \propto \dot{M} R_{\rm s}/w_{\rm wind}$, while in the second $N_{\rm rel,tot} \propto \dot{M} R_{\rm s} V_{\rm s}^2/w_{\rm wind}$, so that in general $N_{\rm rel,tot} \propto \dot{M} R_{\rm s} V_{\rm s}^{2\epsilon}/w_{\rm wind}$, where $\epsilon = 0$ or 1 in these two cases. Therefore, $F_\nu(\lambda) \propto \dot{M}/w_{\rm wind}\; R_{\rm s} V_{\rm s}^{2\epsilon} \lambda^{-\alpha} B^{1-\alpha}$. If the B-field is in equipartition, as above, and using $V = (n-3)/(n-2) R_{\rm s}/t \propto t^{-1/(n-2)}$ we find

$$F_\nu(\lambda) \propto (\dot{M}/w_{\rm wind})^{(3-\alpha)/2}\; \lambda^{-\alpha} t^{+\alpha-(1+2\epsilon)/(n-2)}. \tag{24}$$

If synchrotron cooling is important, a similar type of expression can be derived [40]. The main thing to note is, however, that the optically thin emission is expected to be proportional to the mass loss rate, $\dot{M}/w_{\rm wind}$, and that the decline rate depends on whether the number of relativistic particles scale with the number density or the thermal energy of the shocked gas, as well as spectral index. Observations of the decline rate can therefore test these possibilities.

Although a self-consistent model can be developed for SN1993J and other radio supernovae, the modeling of SN1986J and related objects has been unclear. Weiler et al. [81] proposed a model for SN1986J in which thermal absorbing material is mixed in with the nonthermal emission; one possibility for this is a very irregular shocked emission region with clumps or filaments. The absorption is described by a parameterization and is not derived from an *ab initio* physical model. Chugai and Belous [20] propose a model in which the absorption is by clumps. The narrow line optical emission also implies the presence of clumps, but they are different from those required for the radio absorption. The possible presence of clumps and irregularities introduces uncertainties into models for the radio emission, although rough estimates of the circumstellar density can still be obtained (see the chapter by Sramek and Weiler in this volume).

The relative importance of SSA and free-free absorption depends on a number of parameters and we refer to [12,40] for a more detailed discussion. The most important of these are the mass loss rate, $\dot{M}/w_{\rm wind}$, the shock velocity, $V_{\rm s}$, and the circumstellar temperature, $T_{\rm e}$. In general, a high shock velocity and a high circumstellar temperature favor SSA, while a high mass loss rate favors free-free absorption.

Assuming SSA, an interesting expression for the velocity of the shock can be derived which can be tested against the observations. If we assume equipartition, $N_{\rm rel} \propto \rho V_{\rm s}^2 R_{\rm s}$ and $B^2/8\pi \propto \rho V_{\rm s}^2$ we have $N_{\rm rel} \propto B^2 R_{\rm s}$. Using this in Eq. (21) we get $\tau_{\rm s} = {\rm const} \times \lambda^{5/2-\alpha} B^{7/2-\alpha}\; R_{\rm s}$. The peak in the light curve is given by $\tau_{\rm s} \approx 1$. If we approximate the flux at this point by the optically thick expression Eq. (23) and solve for B we get $B \propto F_\nu(\lambda)^{-2}\; R_{\rm s}^4\; \lambda^{-5}$. Inserting this expression in the condition $\tau_{\rm s} \approx 1$, we find $R_{\rm s}^{15-4\alpha} F_\nu(\lambda)^{-7+2\alpha}\; \lambda^{-15+4\alpha} \approx$ const. With $V_{\rm s} = (n-3)/(n-2) R_{\rm s}/t$ we finally have

$$V_{\rm s} \approx {\rm const}\; F_\nu(\lambda)^{(7-2\alpha)/(15-4\alpha)}\; \lambda\, t^{-1}, \tag{25}$$

where all parameters refer to their values at the peak of the light curve. Using this expression, we can plot lines of constant shock velocity into a diagram with peak radio luminosity versus time of peak flux, assuming that SSA dominates (Fig. 7). The positions of the lines depend only weakly on the equipartition assumption. Each SN can now be placed in this diagram to give a predicted

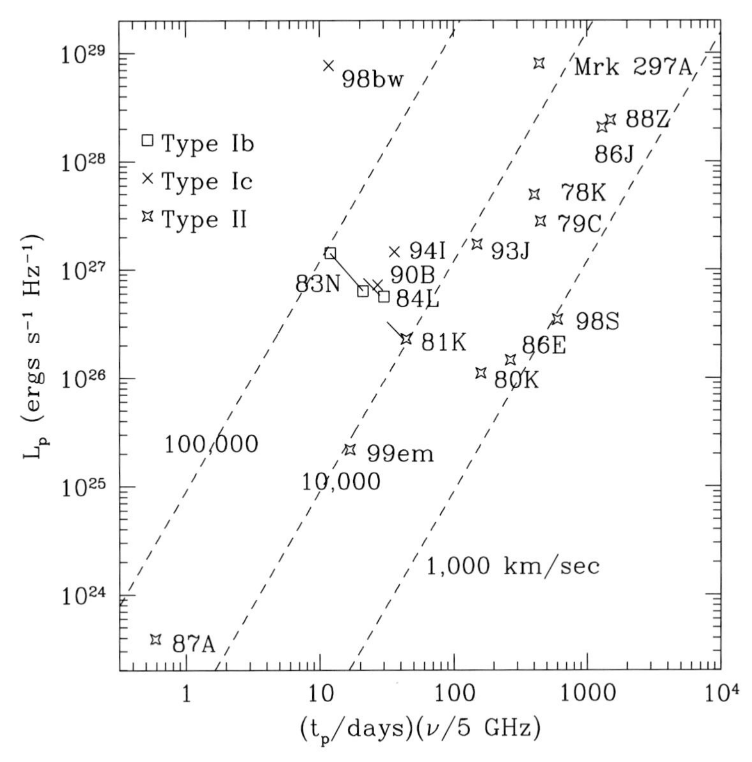

Fig. 7. Peak luminosity and corresponding epoch for the well-observed radio SNe. The dashed lines give curves of constant expansion velocity, assuming SSA (see [12]).

shock velocity. If this is lower than the observed value (as measured by *VLBI* or from line profiles) SSA gives a too low a flux and should, therefore, be relatively unimportant and free-free absorption should instead dominate. In Fig. 7 we show an updated version of the figure in [12]. The most interesting point is that most type Ib/c SNe like SN1983N, SN1994I, and SN1998bw fall into the high velocity category, while type IIL SNe like SN1979C and SN1980K, as well as the type IIn SNe like SN1978K, SN1988Z, and SN1998S, fall in the free-free group. SN1987A is clearly special with its low mass loss rate, but is most likely dominated by SSA [14,71,75]. Mrk 297A (SN1982aa) is an apparent radio supernova in the clumpy, irregular galaxy Mrk 297 [85] but its supernova type is not known.

8 Discussion and Conclusions

Circumstellar interaction of supernovae gives an important window on the nature of stars that explode and their evolution leading up to the explosion. Mass loss rates for the red supergiant progenitors of type II supernovae range from $\sim 2\times10^{-6}$ $M_\odot$ yr^{-1} for SN1999em [67] to $\gtrsim 2\times10^{-4}$ $M_\odot$ yr^{-1} for SN1979C and SN1986J [54,64,81]. Evidence for CNO processing has been found in a number of supernovae, including SN1979C [42], SN1987A [43], and SN1995N [44]. In some cases, the reverse shock appears to be moving into gas that is H poor and O rich, e.g., SN1995N [44]; this relates to the total amount of mass loss before the supernova. The complex circumstellar environment of SN1987A has become clear because of its proximity (see the chapter by McCray in this volume). For more distant supernovae, studies of polarization and spectral line profiles can reveal asymmetries, as in SN1998S [52].

The evidence on circumstellar interaction is especially useful when it can be combined with information from other aspects of the supernovae, such as their light curves and stellar environments. For example, from the pre-supernova stellar environment of SN1999em, Smartt et al. [72] deduced an initial mass of 12 ± 1 $M_{\odot}$. The supernova was of the plateau type IIP, implying that the hydrogen envelope was largely intact at the time of the explosion. This is consistent with the relatively low rate of mass loss deduced for the supernova progenitor [67].

In addition to information on the evolution of massive stars and their explosions, circumstellar interaction provides an excellent laboratory for the study of shock wave physics. Compared to older supernova remnants, the shock velocities are higher and the time evolution gives an additional dimension for study, although there is little spatial information in most cases. *VLBI* observations can, however, in this respect be extremely valuable, as demonstrated by SN1993J. An object where both the spatial and time dimensions are accessible is SN1987A, which has turned out to be an excellent source for the study of shock waves.

Acknowledgments

We are grateful to John Blondin for providing Fig. 2 and to Peter Lundqvist for comments on the manuscript. This work was supported in part by NASA grant NAG5-8232 and by the Swedish Research Council.

References

1. W.D. Arnett: Astrophys. J. **331**, 377 (1988)
2. N. Bartel et al.: Science **287**, 112 (2000)
3. S. Blinnikov, P. Lundqvist, O. Bartunov, K. Nomoto, K. Iwamoto: Astrophys. J. **532**, 1132 (2000)
4. J.M. Blondin, P. Lundqvist, R.A. Chevalier: Astrophys. J. **472**, 257 (1996)
5. D. Branch, M. Livio, L.R. Yungelson, F.R. Boffi, E. Baron: Pub. Astron. Soc. Pacific **107**, 1019 (1995)
6. C.J. Burrows et al.: Astrophys. J. **452**, 680 (1995)
7. R.A. Chevalier: Astrophys. J. **207**, 872 (1976)
8. R.A. Chevalier: Astrophys. J. **258**, 790 (1982)
9. R.A. Chevalier: Astrophys. J. **259**, 302 (1982)
10. R.A. Chevalier: Astrophys. J. **308**, 225 (1986)
11. R.A. Chevalier: In: *Supernovae*, ed. by A.G. Petschek, (Springer, Berlin 1990) p. 91
12. R.A. Chevalier: Astrophys. J. **499**, 810 (1998)

13. R.A. Chevalier, J.M. Blondin: Astrophys. J. **444**, 312 (1995)
14. R.A. Chevalier, V.V. Dwarkadas: Astrophys. J. Lett. **452**, L45 (1995)
15. R.A. Chevalier, R.T. Emmering: Astrophys. J. Lett. **342**, L75 (1989)
16. R.A. Chevalier, C. Fransson: Astrophys. J. **420**, 268 (1994)
17. R.A. Chevalier, N. Soker: Astrophys. J. **341**, 867 (1989)
18. R.A. Chevalier, J.M. Blondin, R.T. Emmering: Astrophys. J. **392**, 118 (1992)
19. N.N. Chugai: Astrophys. J. Lett. **414**, L101 (1993)
20. N.N. Chugai, M.L. Belous: Astron. Rep. **43**, 89 (1999)
21. N.N. Chugai, I.J. Danziger: Mon. Not. R. Astron. Soc. **268**, 173 (1994)
22. N.N. Chugai, I.J. Danziger, M. Della Valle: Mon. Not. R. Astron. Soc. **276**, 530 (1995)
23. A.P.S. Crotts, W.E. Kunkel, S.R. Heathcote: Astrophys. J. **438**, 724 (1995)
24. R.J. Cumming, P. Lundqvist, L.J. Smith, M. Pettini, D.L. King: Mon. Not. R. Astron. Soc. **283**, 1355 (1996)
25. W.C. Danchi et al.: Astron. J. **107**, 1469 (1994)
26. B.T. Draine, C.F. McKee: Ann. Rev. Astron. Astrophys. **31**, 373 (1993)
27. V.V. Dwarkadas, R.A. Chevalier: Astrophys. J. **497**, 807 (1998)
28. E. Dwek: Astrophys. J. **274**, 175 (1983)
29. E. Dwek: Astrophys. J. **297**, 719 (1985)
30. E. Dwek: Astrophys. J. **322**, 812 (1987)
31. D.C. Ellison, S.P. Reynolds: Astrophys. J. **382**, 242 (1991)
32. R.T. Emmering, R.A. Chevalier: Astron. J. **95**, 152 (1988)
33. L. Ensman, A. Burrows: Astrophys. J. **393**, 742 (1992)
34. S.W. Falk: Astrophys. J. Lett. **226**, L133 (1978)
35. S.W. Falk, W.D. Arnett: Astrophys. J. Lett. **180**, L65 (1973)
36. A. Fassia et al.: Mon. Not. R. Astron. Soc. **318**, 1093 (2000)
37. C. Fransson: Astron. Astrophys. **111**, 140 (1982)
38. C. Fransson: Astron. Astrophys. **133**, 264 (1984)
39. C. Fransson: Highlights of Astronomy **7**, 611 (1986)
40. C. Fransson, C.-I. Björnsson: Astrophys. J. **509**, 861 (1998)
41. C. Fransson, P. Lundqvist, R.A. Chevalier: Astrophys. J. **461**, 993 (1996)
42. C. Fransson, P. Benvenuti, W. Wamsteker, C. Gordon, K. Hempe, D. Reimers, G.G.C. Palumbo, N. Panagia: Astron. Astrophys. **132**, 1 (1984)
43. C. Fransson, A. Cassatella, R. Gilmozzi, R.P. Kirshner, N. Panagia, G. Sonneborn, W. Wamsteker: Astrophys. J. **336**, 429 (1989)
44. C. Fransson et al.: Astrophys. J. **572**, 350 (2002)
45. G. García-Segura, N. Langer, M.-M. MacLow: Astron. Astrophys. **316**, 133 (1996)
46. J.C. Houck, J.N. Bregman, R.A. Chevalier, K. Tomisaka: Astrophys. J. **493**, 431 (1998)
47. P. Jakobsen et al.: Astrophys. J. Lett. **369**, L63 (1991)
48. B. Jun, M.L. Norman: Astrophys. J. **472**, 245 (1996)
49. B. Jun, T.W. Jones, M.L. Norman: Astrophys. J. Lett. **468**, L59 (1996)
50. T.R. Kallman, R. McCray: Astrophys. J. Suppl. **50**, 263 (1982)
51. R.I. Klein, R.A. Chevalier: Astrophys. J. Lett. **223**, L109 (1978)
52. D.C. Leonard, A.V. Filippenko, A.J. Barth, T. Matheson: Astrophys. J. **536**, 239 (2000)
53. M.S. Longair: *High Energy Astrophysics* (Cambridge University Press, Cambridge 1992)
54. P. Lundqvist, C. Fransson: Astron. Astrophys. **192**, 221 (1988)
55. P. Lundqvist, C. Fransson: Astrophys. J. **380**, 575 (1991)

56. P. Lundqvist, C. Fransson: Astrophys. J. **464**, 924 (1996)
57. J.M. Marcaide et al.: Science **270**, 1475 (1995)
58. J.M. Marcaide et al.: Astrophys. J. Lett. **486**, L31 (1997)
59. E. Marietta, A. Burrows, B. Fryxell: Astrophys. J. Suppl. **128**, 615 (2000)
60. T. Matheson et al.: Astron. J. **120**, 1487 (2000)
61. T. Matheson, A.V. Filippenko, L.C. Ho, A.J. Barth, D.C. Leonard: Astron. J. **120**, 1499 (2000)
62. C.D. Matzner, C.F. McKee: Astrophys. J. **510**, 379 (1999)
63. A.R. McDonald, T.W.B. Muxlow, A. Pedlar, M.A. Garrett, K.A. Willis, S.T. Garrington, P. Diamond, P.N. Wilkinson: Mon. Not. R. Astron. Soc. **322**, 100 (2001)
64. M.J. Montes, K.W. Weiler, S.D. Van Dyk, N. Panagia, C.K. Lacey, R.A. Sramek, R. Park: Astrophys. J. **532**, 1124 (2000)
65. D.K. Nadyozhin: Astrophys. Space Sci. **112**, 225 (1985)
66. N. Panagia et al.: Mon. Not. R. Astron. Soc. **192**, 861 (1980)
67. D. Pooley et al.: Astrophys. J. **572**, 932 (2002)
68. A.M.S. Richards, J.A. Yates, R.J. Cohen: Mon. Not. R. Astron. Soc. **299**, 319 (1998)
69. B. Roscherr, B.E. Schaefer: Astrophys. J. **532**, 415 (2000)
70. N. Smith, R. M. Humphreys, K. Davidson, R.D. Gehrz, M.T. Schuster, J. Krautter: Astron. J. **121**, 1111 (2001)
71. V.I. Slysh: Sov. Astron. Lett. **16**, 339 (1990)
72. S.J. Smartt, G.F. Gilmore, C.A. Tout, S.T. Hodgkin: Astrophys. J. **565**, 1089 (2002)
73. G. Sonneborn, C. Fransson, P. Lundqvist, A. Cassatella, R. Gilmozzi, R.P. Kirshner, N. Panagia, W. Wamsteker: Astrophys. J. **477**, 848 (1997)
74. R.A. Sramek, K.W. Weiler: In: *Supernovae*, ed. by A.G. Petschek (Springer, Berlin 1990) p. 76
75. M. Storey, R.N. Manchester: Nature **329**, 421 (1987)
76. P.G. Tuthill, C.A. Haniff, J.E. Baldwin: Mon. Not. R. Astron. Soc. **285**, 529 (1997)
77. H. Uitenbroek, A.K. Dupree, R.L. Gilliland: Astron. J. **116**, 2501 (1998)
78. S.D. van Dyk, K.W. Weiler, R.A. Sramek, M.P. Rupen, N. Panagia: Astrophys. J. Lett. **432**, L115 (1994)
79. L. Wang: Astron. Astrophys. **246**, L69 (1995)
80. K.W. Weiler, R.A. Sramek: Ann. Rev. Astron. Astrophys. **26**, 295 (1988)
81. K.W. Weiler, N. Panagia, R.A. Sramek: Astrophys. J. **364**, 611 (1990)
82. K.W. Weiler, N. Panagia, M.J. Montes, R.A. Sramek: Ann. Rev. Astron. Astrophys. **40**, 387 (2002)
83. A.J. Willis: In: *Wolf-Rayet Stars and Interrelations with Other Massive Stars in Galaxies*, ed. by K.A. van der Hucht, B. Hidayat (Kluwer, Dordrecht 1991) p. 265
84. M. Wittkowski, N. Langer, G. Weigelt: Astron. Astrophys. **340**, L39 (1998)
85. Q.F. Yin: Astrophys. J. **420**, 152 (1994)

Measuring Cosmology with Supernovae

Saul Perlmutter[1] and Brian P. Schmidt[2]

[1] Physics Division, Lawrence Berkeley National Laboratory, University of California, Berkeley, CA 94720, USA
[2] Research School of Astronomy and Astrophysics, The Australian National University, via Cotter Rd, Weston Creek, ACT 2611, Australia

Abstract. Over the past decade, supernovae have emerged as some of the most powerful tools for measuring extragalactic distances. A well developed physical understanding of type II supernovae allow them to be used to measure distances independent of the extragalactic distance scale. Type Ia supernovae are empirical tools whose precision and intrinsic brightness make them sensitive probes of the cosmological expansion. Both types of supernovae are consistent with a Hubble Constant within ~10% of $H_0 = 70$ km s^{-1}Mpc^{-1}. Two teams have used type Ia supernovae to trace the expansion of the Universe to a look-back time more than 60% of the age of the Universe. These observations show an accelerating Universe which is currently best explained by a cosmological constant or other form of dark energy with an equation of state near $w = p/\rho = -1$. While there are many possible remaining systematic effects, none appears large enough to challenge these current results. Future experiments are planned to better character ize the equation of state of the dark energy leading to the observed acceleration by observing hundreds or even thousands of objects. These experiments will need to carefully control systematic errors to ensure future conclusions are not dominated by effects unrelated to cosmology.

1 Introduction

Understanding the global history of the Universe is a fundamental goal of cosmology. One of the conceptually simplest tests in the repertoire of the cosmologist is observing how a standard candle dims as a function of redshift. The nearby Universe provides the current rate of expansion, and with more distant objects it is possible to start seeing the varied effects of cosmic curvature and the Universe's expansion history (usually expressed as the rate of acceleration/deceleration). Over the past several decades a paradigm for understanding the global properties of the Universe has emerged based on General Relativity with the assumption of a homogeneous and isotropic Universe. The relevant constants in this model are the Hubble constant (or current rate of cosmic expansion), the relative fractions of species of matter that contribute to the energy density of the Universe, and these species' equation of state.

Early luminosity distance investigations used the brightest objects available for measuring distance – bright galaxies [3,39], but these efforts were hampered by the impreciseness of the distance indicators and the changing properties of the distance indicators as a function of look back time. Although many other methods for measuring the global curvature and cosmic deceleration exist (see, e.g., [66]), supernovae (SNe) have emerged as one of the preeminent distance methods due to their significant intrinsic brightness (which allows them to be observable in the distant Universe), ubiquity (they are visible in both the nearby and distant Universe), and their precision (type Ia SNe provide distances that have a precision of approximately 8%).

2 Supernovae as Distance Indicators

2.1 Type II Supernovae and the Expanding Photosphere Method

Massive stars come in a wide variety of luminosities and sizes and would seemingly not be useful objects for making distance measurements under the standard candle assumption. However, from a radiative transfer standpoint these objects are relatively simple and can be modeled with sufficient accuracy to measure distances to approximately 10%. The expanding photosphere method (EPM), was developed by Kirshner and Kwan [44], and implemented on a large number of objects by Schmidt et al. [86] after considerable improvement in the theoretical understanding of type II SN (SNII) atmospheres [15,16,99].

EPM assumes that SNII radiate as dilute blackbodies

$$\theta_{ph} = \frac{R_{ph}}{D} = \sqrt{\frac{F_\lambda}{\zeta^2 \pi B_\lambda(T)}}, \tag{1}$$

where θ_{ph} is the angular size of the photosphere of the SN, R_{ph} is the radius of the photosphere, D is the distance to the SN, F_λ is the observed flux density of the SN, and $B_\lambda(T)$ is the Planck function at a temperature T. Since SNII are not perfect blackbodies, we include a correction factor, ζ, which is calculated from radiate transfer models of SNII. SNe freely expand, and

$$R_{ph} = v_{ph}(t - t_0) + R_0, \tag{2}$$

where v_{ph} is the observed velocity of material at the position of the photosphere, and t is the time elapsed since the time of explosion, t_0. For most stars, the stellar radius ,R_0, at the time of explosion is negligible, and Eqs. (1–2) can be combined to yield

$$t = D\left(\frac{\theta_{ph}}{v_{ph}}\right) + t_0 \tag{3}$$

By observing a SNII at several epochs, measuring the flux density and temperature of the SN (via broad band photometry) and v_{ph} from the minima of the weakest lines in the SN spectrum, we can solve simultaneously for the time of explosion and distance to the SNII. The key to successfully measuring distances via EPM is an accurate calculation of $\zeta(T)$. Requisite calculations were performed by Eastman et al. [16] but, unfortunately, no other calculations of $\zeta(T)$ have yet been published for typical SNIIP progenitors.

Hamuy et al. [34] and Leonard et al. [52] have measured the distances to SN1999em, and they have investigated other aspects of EPM. Hamuy et al. [34]

challenged the prescription of measuring velocities from the minima of weak lines and developed a framework of cross correlating spectra with synthesized spectra to estimate the velocity of material at the photosphere. This different prescription does lead to small systematic differences in estimated velocity using weak lines but, provided the modeled spectra are good representations of real objects, this method should be more correct. At present, a revision of the EPM distance scale using this method of estimating v_{ph} has not been made.

Leonard et al. [51] have obtained spectropolarimetry of SN1999em at many epochs and see polarization intrinsic to the SN which is consistent with the SN have asymmetries of $10-20\%$. Asymmetries at this level are found in most SNII [101], and may ultimately limit the accuracy EPM can achieve on a single object ($\sigma \sim 10\%$). However, the mean of all SNII distances should remain unbiased.

Type II SNe have played an important role in measuring the Hubble constant independent of the rest of the extragalactic distance scale. In the next decade it is quite likely that surveys will begin to turn up significant numbers of these objects at $z \sim 0.5$ and, therefore, the possibility exists that SNII will be able to make a contribution to the measurement of cosmological parameters beyond the Hubble Constant. Since SNII do not have the precision of the SNIa (next section) and are significantly harder to measure, they will not replace the SNIa but will remain an independent class of objects which have the potential to confirm the interesting results that have emerged from the SNIa studies.

2.2 Type Ia Supernovae as Standardized Candles

SNIa have been used as extragalactic distance indicators since Kowal [42] first published his Hubble diagram ($\sigma = 0.6$ mag) for type I SNe. We now recognize that the old type I SNe spectroscopic class is comprised of two distinct physical entities: SNIb/c which are massive stars that undergo core collapse (or in some rare cases might undergo a thermonuclear detonation in their cores) after losing their hydrogen atmospheres, and SNIa which are most likely thermonuclear explosions of white dwarfs. In the mid-1980s it was recognized that studies of the type I SN sample had been confused by these similar appearing SNe, which were henceforth classified as type Ib [59,94,102] and type Ic [36]. By the late 1980s/early 1990s, a strong case was being made that the vast majority of the true type Ia SNe had strikingly similar light curve shapes [11,46–48], spectral time series [6,18,28,62], and absolute magnitudes [47,54]. There were a small minority of clearly peculiar type Ia SNe (e.g., SN1986G [63], SN1991bg [19,49], and SN1991T [19,78]), but these could be identified and removed by their unusual spectral features. A 1992 review by Branch and Tammann [7] of a variety of studies in the literature concluded that the intrinsic dispersion in B and V maximum for type Ia SNe must be < 0.25 mag, making them "the best standard candles known so far."

In fact, the Branch and Tammann review indicated that the magnitude dispersion was probably even smaller, but the measurement uncertainties in the available datasets were too large to tell. The *Calan/Tololo Supernova Search*

(CTSS), a program begun by Hamuy et al. [31] in 1990, took the field a dramatic step forward by obtaining a crucial set of high quality SN light curves and spectra. By targeting a magnitude range that would discover type Ia SNe in the redshift range $z = 0.01 - 0.1$, the *CTSS* was able to compare the peak magnitudes of SNe whose relative distance could be deduced from their Hubble velocities.

The *CTSS* observed some 25 fields (out of a total sample of 45 fields) twice a month for over three and one half years with photographic plates or film at the *Cerro Tololo Inter-American Observatory (CTIO)* Curtis Schmidt telescope, and then organized extensive follow-up photometry campaigns primarily on the *CTIO* 0.9 m telescope, and spectroscopic observation on either the *CTIO* 4 m or 1.5 m telescope. Toward the end of this search, Hamuy et al. [31] pointed out the difficulty of this comprehensive project: "Unfortunately, the appearance of a SN is not predictable. As a consequence of this we cannot schedule the followup observations *a priori*, and we generally have to rely on someone else's telescope time. This makes the execution of this project somewhat difficult." Despite these challenges, the search was a major success; with the cooperation of many visiting *CTIO* astronomers and *CTIO* staff, it contributed 30 new type Ia SN light curves to the pool [32] with an almost unprecedented control of measurement uncertainties.

As the *CTSS* data began to become available, several methods were presented that could select for the "most standard" subset of the type Ia standard candles, a subset which remained the dominant majority of the ever-growing sample [8]. For example, Vaughan et al. [97] presented a cut on the B-V color at maximum that would select what were later called the "Branch Normal" SNIa, with an observed dispersion of less than 0.25 mag.

Phillips [64] found a tight correlation between the rate at which a type Ia SN's luminosity declines and its absolute magnitude, a relation which apparently applied not only to the Branch Normal type Ia SNe, but also to the peculiar type Ia SNe. Phillips plotted the absolute magnitude of the existing set of nearby SNIa, which had dense photoelectric or CCD coverage, versus the parameter Δm_{15}(B), the amount the SN decreased in brightness in the B-band over the 15 days following maximum light. The sample showed a strong correlation which, if removed, dramatically improved the predictive power of SNIa. Hamuy et al. [33] used this empirical relation to reduce the scatter in the Hubble diagram to $\sigma < 0.2$ mag in V for a sample of nearly 30 SNIa from the *CTSS* search.

Impressed by the success of the Δm_{15}(B) parameter, Riess et al. [79] developed the *multi-color light curve shape method (MLCS)*, which parameterized the shape of SN light curves as a function of their absolute magnitude at maximum. This method also included a sophisticated error model and fitted observations in all colors simultaneously, allowing a color excess to be included. This color excess, which we attribute to intervening dust, enabled the extinction to be measured. Another method that has been used widely in cosmological measurements with SNIa is the "stretch" method described in Perlmutter et al. [74,77]. This method is based on the observation that the entire range of SNIa light curves, at least in the B and V-bands, can be represented with a simple time stretching (or shrinking) of a canonical light curve. The coupled stretched B and V light curves serve as a parameterized set of light curve shapes [26], providing many of the benefits of the MLCS method but as a much simpler (and constrained) set. This method, as well as recent implementations of Δm_{15}(B) [24,65], also allows extinction to be directly incorporated into the SNIa distance measurements. Other methods that correct for intrinsic luminosity differences or limit the input sample by various criteria have also been proposed to increase the precision of type Ia SNe as distance indicators [9,17,93,95]. While these latter techniques are not as developed as the Δm_{15}(B), *MLCS*, and stretch methods, they all provide distances that are comparable in precision, roughly $\sigma = 0.18$ mag about the inverse square law, equating to a fundamental precision of SNIa distances of $\sim 6\%$ (0.12 mag), once photometric uncertainties and peculiar velocities are removed. Finally, a "poor man's" distance indicator, the snapshot method [80], combines information contained in one or more SN spectra with as little as one night's multi-color photometry. This method's accuracy depends critically on how much information is available.

3 Cosmological Parameters

The standard model for describing the global evolution of the Universe is based on two equations that make some simple, and hopefully valid, assumptions. If the Universe is isotropic and homogenous on large scales, the Robertson-Walker Metric,

$$ds^2 = dt^2 - a(t)\left[\frac{dr^2}{1-kr^2} + r^2 d\theta^2\right]. \tag{4}$$

gives the line element distance(s) between two objects with coordinates r,θ and time separation, t. The Universe is assumed to have a simple topology such that, if it has negative, zero, or positive curvature, k takes the value $-1, 0, 1$, respectively. These models of the Universe are said to be open, flat, or closed, respectively. The dynamic evolution of the Universe needs to be input into the Robertson-Walker Metric by the specification of the scale factor $a(t)$, which gives the radius of curvature of the Universe over time – or more simply, provides the relative size of a piece of space at any time. This description of the dynamics of the Universe is derived from General Relativity, and is known as the Friedman equation

$$H^2 \equiv (\dot{a}/a)^2 = \frac{8\pi G\rho}{3} - \frac{k}{a^2}. \tag{5}$$

The expansion rate of our Universe (H), is called the Hubble parameter (or the Hubble constant, H_0, at the present epoch) and depends on the content of the Universe. Here we assume the Universe is composed of a set of components, each having a fraction, Ω_i, of the critical density

$$\Omega_i = \frac{\rho_i}{\rho_{crit}} = \frac{\rho_i}{\frac{3H_0^2}{8\pi G}}, \tag{6}$$

with an equation of state which relates the density, ρ_i, and pressure, p_i, as $w_i = p_i/\rho_i$. For example, w_i takes the value 0 for normal matter, +1/3 for photons, and -1 for the cosmological constant. The equation of state parameter does not need to remain fixed; if scalar fields are present, the effective w will change over time. Most reasonable forms of matter or scalar fields have $w_i \geq -1$, although nothing seems manifestly forbidden. Combining Eqs. (4–6) yields solutions to the global evolution of the Universe [13].

The luminosity distance, D_L, which is defined as the apparent brightness of an object as a function of its redshift z – the amount an object's light has been stretched by the expansion of the Universe – can be derived from Eqs. (4–6) by solving for the surface area as a function of z, and taking into account the effects of time dilation [25,26,50,82] and energy dimunition as photons get stretched traveling through the expanding Universe. D_L is given by the numerically integrable equation,

$$D_L = \frac{c}{H_0}(1+z)\kappa_0^{-1/2} S\{\kappa_0^{1/2} \int_0^z dz' [\sum_i \Omega_i (1+z')^{3+3w_i} - \kappa_0 (1+z')^2]^{-1/2}\}. \quad (7)$$

$S(x) = \sin(x)$, x, or $\sinh(x)$ for closed, flat, and open models respectively, and the curvature parameter κ_0, is defined as $\kappa_0 = \sum_i \Omega_i - 1$.

Historically, Eq. (7) has not been easily integrated and has been expanded in a Taylor series to give

$$D_L = \frac{c}{H_0}\{z + z^2 \left(\frac{1-q_0}{2}\right) + \mathcal{O}(z^3)\}, \quad (8)$$

where the deceleration parameter, q_0, is given by

$$q_0 = \frac{1}{2} \sum_i \Omega_i (1 + 3w_i). \quad (9)$$

From Eq. (9) we can see that, in the nearby Universe, the luminosity distances scale linearly with redshift, with H_0 serving as the constant of proportionality. In the more distant Universe, D_L depends to first order on the rate of acceleration/deceleration (q_0) or, equivalently, on the amount and types of matter that make up the Universe. For example, since normal matter has $w_M = 0$ and the cosmological constant has $w_\Lambda = -1$, a universe composed of only these two forms of matter/energy has $q_0 = \Omega_M/2 - \Omega_\Lambda$. In a universe composed of these two types of matter, if $\Omega_\Lambda < \Omega_M/2$, q_0 is positive and the universe is decelerating. These decelerating universes have D_L smaller as a function of z than their accelerating counterparts.

If distance measurements are made at a low-z and a small range of redshift at higher z (e.g., $0.3 > z > 0.5$), there is a degeneracy between Ω_M and Ω_Λ. It is impossible to pin down the absolute amount of either species of matter. One can only determine their relative dominance, which, at $z = 0$, is given by Eq. (9). However, Goobar and Perlmutter [27] pointed out that by observing objects over a larger range of high redshift (e.g., $0.3 > z > 1.0$) this degeneracy

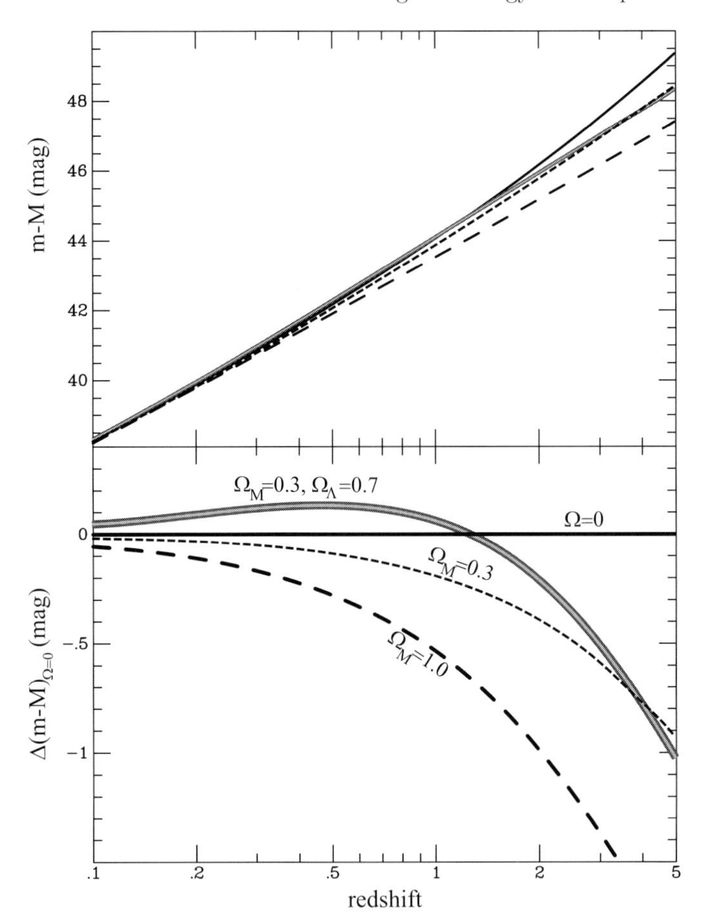

Fig. 1. D_L expressed as distance modulus $(m - M)$ for four relevant cosmological models; $\Omega_M = 0$, $\Omega_\Lambda = 0$ (empty Universe, *solid line*); $\Omega_M = 0.3$, $\Omega_\Lambda = 0$ (*short dashed line*); $\Omega_M = 0.3$, $\Omega_\Lambda = 0.7$ (*hatched line*); and $\Omega_M = 1.0$, $\Omega_\Lambda = 0$ (*long dashed line*). In the bottom panel the empty Universe has been subtracted from the other models to highlight the differences.

can be broken, providing a measurement of the absolute fractions of Ω_M and Ω_Λ.

To illustrate the effect of cosmological parameters on the luminosity distance, in Fig. 1 we plot a series of models for both Λ and non-Λ universes. In the top panel, the various models show the same linear behavior at $z < 0.1$ with models having the same H_0 being indistinguishable to a few percent. By $z = 0.5$ the models with significant Λ are clearly separated, with luminosity distances that are significantly further than the zero-Λ universes. Unfortunately, two perfectly reasonable universes, given our knowledge of the local matter density of the Universe ($\Omega_M \sim 0.2$), one with a large cosmological constant, $\Omega_\Lambda = 0.7$, $\Omega_M = 0.3$ and one with no cosmological constant, $\Omega_M = 0.2$, show differences of less

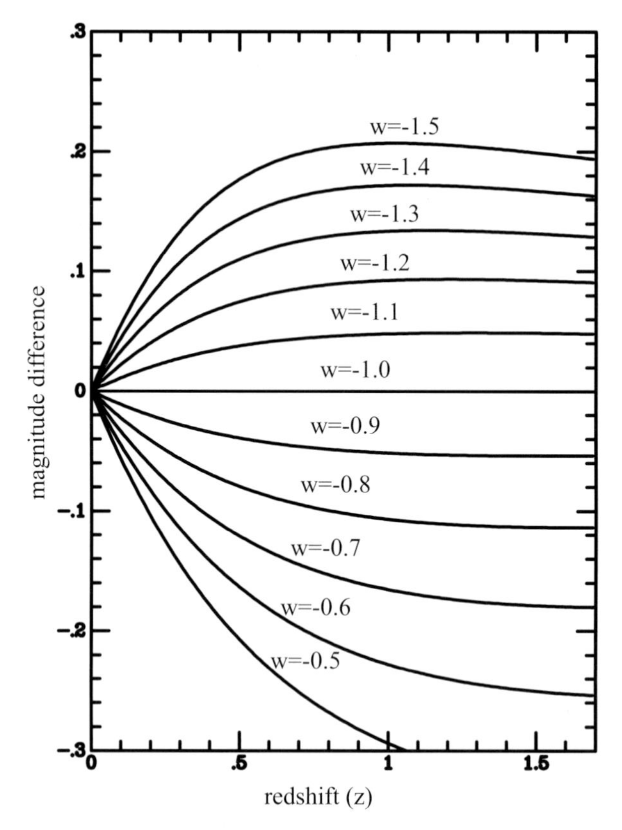

Fig. 2. D_L for a variety of cosmological models containing $\Omega_M = 0.3$ and $\Omega_x = 0.7$ with a constant (not time-varying) equation of state w_x. The $w_x = -1$ model has been subtracted off to highlight the differences between the various models

than 25%, even to redshifts of $z > 5$. Interestingly, the maximum difference between the two models is at $z \sim 0.8$, not at large z. Figure 2 illustrates the effect of changing the equation of state of the non-matter, dark energy component, assuming a flat universe, $\Omega_{tot} = 1$. If we are to discern a dark energy component that is not a cosmological constant, measurements better than 5% are clearly required, especially since the differences in this diagram include the assumption of flatness and also fix the value of Ω_M. In fact, to discriminate among the full range of dark energy models with time varying equations of state will require much better accuracy than even this challenging goal.

4 Measuring the Hubble Constant

Schmidt et al. [86], using a sample of 16 SNII, estimated H_0=73±6(statistical)±7 (systematic) using EPM. This estimate is independent of other rungs in the extragalactic distance ladder, the most important of which are the Cepheids, which currently calibrate most other distance methods (such as SNIa). The Cepheid and EPM distance scales, compared galaxy to galaxy, agree to within 5% and are consistent within the errors [16,52]. This provides confidence that both methods are providing accurate distances.

The current nearby SNIa sample [24,32,41,84] contains more than 100 objects (Fig. 3), and accurately defines the slope in the Hubble diagram from $0 < z < 0.1$ to 1%. To measure H_0, SNIa must still be externally calibrated with Cepheids, and this calibration is the major limitation to measuring H_0 with SNIa. Two separate teams have analyzed the Cepheids and SNIa but have obtained divergent values for the Hubble constant. Saha et al. [88] find $H_0 = 59 \pm 6$, whereas Freedman et al. [20] find $H_0 = 71 \pm 2 \pm$ (6 systematic). Of the 12 SNIa for which there are Cepheid distances to the host galaxy (SN1895B*, SN1937C*, SN1960F*, SN1972E, SN1974G*, SN1981B, SN1989B, SN1990N, SN1991T, SN1998eq, SN1998bu, and SN1999by), four were observed by non-digital means (marked by *) and are best excluded from analysis on the grounds that non-digital photometry routinely has systematic errors far greater 0.1 mag. Jha [41] has compared the SNIa distances using an updated version of *MLCS* to the Cepheid host galaxy distances measured by the two *Hubble Space Telescope (HST)* teams. Using only the digitally observed SNIa, he finds, using distances from the SNIa project of Saha et al. [88], $H_0 = 66 \pm 3 \pm$ (7 systematic) km s^{-1} Mpc^{-1}. Applying the same analysis to the Key Project distances by Freedman et al. [20] gives $H_0 = 76 \pm 3 \pm$ (8 systematic) km s^{-1} Mpc^{-1}. This

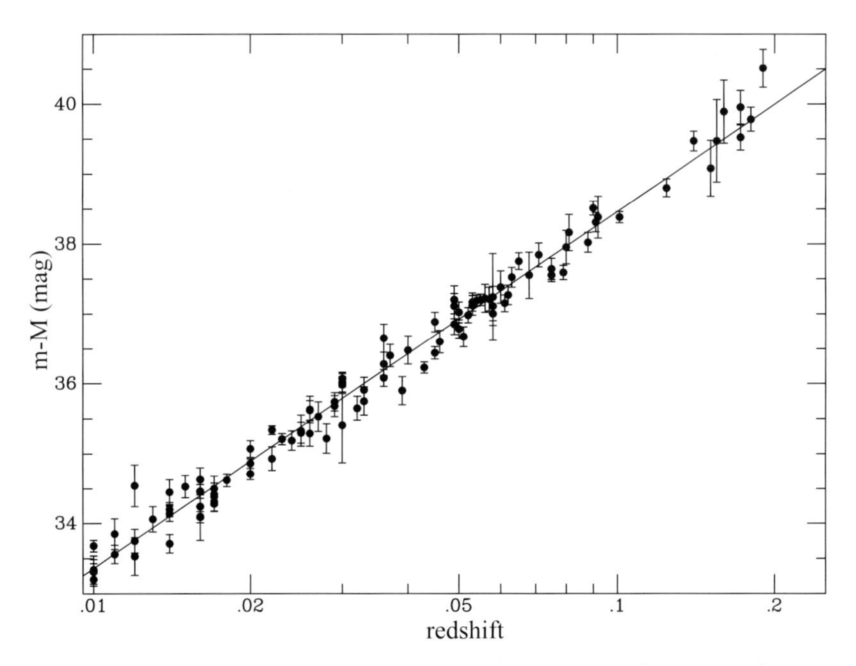

Fig. 3. The Hubble diagram for SNIa from $0.01 > z > 0.2$ [24,33,41,84]. The 102 objects in this range have a residual about the inverse square line of $\sim 10\%$.

difference is not due to SNIa errors, but rather to the different ways the two teams have measured Cepheid distances with *HST*. The two values do overlap when the systematic uncertainties are included, but it is still uncomfortable that the discrepancies are so large, particularly when some systematic uncertainties are common between the two teams.

At present, SNe provide the most convincing constraints with $H_0 \sim 70 \pm 10$ km s^{-1} Mpc^{-1}. However, future work on measuring H_0 lies not with the SNe but with the Cepheid calibrators, or possibly in using other primary distance indicators such as EPM or the Sunyaev-Zeldovich effect.

5 The Measurement of Acceleration

The intrinsic brightness of SNIa allow them to be discovered to $z > 1.5$ with current instrumentation (while a comparably deep search for type II SNe would only reach redshifts of $z \sim 0.5$). In the 1980s, however, finding, identifying, and studying even the impressively luminous type Ia SNe was a daunting challenge, even towards the lower end of the redshift range shown in Fig. 1. At these redshifts, beyond $z \sim 0.25$, Fig. 1 shows that relevant cosmological models could be distinguished by differences of order 0.2 mag in their predicted luminosity distances. For SNIa with a dispersion of 0.2 mag, 10 well observed objects should provide a 3σ separation between the various cosmological models. It should be noted that the uncertainty described above in measuring H_0 is not important in measuring the parameters for different cosmological models. Only the relative brightness of objects near and far is being exploited in Eq. (7) and the absolute value of H_0 scales out.

The first distant SN search was started by the Danish team of Nørgaard-Nielsen et al. [57]. With significant effort and large amounts of telescope time spread over more than two years, they discovered a single SNIa in a $z = 0.3$ cluster of galaxies (and one SNII at $z = 0.2$) [35,57]. The SNIa was discovered well after maximum light on an observing night that could not have been predicted, and was only marginally useful for cosmology. However, it showed that such high redshift SNe did exist and could be found, but that they would be very difficult to use as cosmological tools.

Just before this first discovery in 1988, a search for high redshift type Ia SNe using a then novel wide field camera on a much larger (4m) telescope was begun at the *Lawrence Berkeley National Laboratory (LBNL)* and the *Center for Particle Astrophysics*, at Berkeley. This search, now known as the *Supernova Cosmological Project (SCP)*, was inspired by the impressive studies of the late 1980s indicating that extremely similar type Ia SN events could be recognized by their spectra and light curves, and by the success of the *LBNL* fully robotic low-redshift SN search in finding 20 SNe with automatic image analysis [56,67].

The *SCP* targeted a much higher redshift range, $z > 0.3$, in order to measure the (presumed) deceleration of the Universe, so it faced a different challenge than the *CTSS* search. The high redshift SNe required discovery, spectroscopic confirmation, and photometric follow up on much larger telescopes. This precious telescope time could neither be borrowed from other visiting observers and staff nor applied for in sufficient quantities spread throughout the year to cover all SNe discovered in a given search field, and with observations early enough to establish their peak brightness. Moreover, since the observing time to confirm high redshift SNe was significant on the largest telescopes, there was a clear "chicken and egg" problem: telescope time assignment committees would not award follow-up time for a SN discovery that might, or might not, happen on a given run (and might, or might not, be well past maximum) and, without the follow-up time, it was impossible to demonstrate that high redshift SNe were being discovered by the *SCP*.

By 1994, the *SCP* had solved this problem, first by providing convincing evidence that SNe, such as SN1992bi, could be discovered near maximum (and K-corrected) out to $z = 0.45$ [73], and then by developing and successfully demonstrating a new observing strategy that could effectively guarantee SN discoveries on a predetermined date, all before or near maximum light [70–72,76]. Instead of discovering a single SN at a time on average (with some runs not finding one at all), the new approach aimed to discover an entire "batch" of half-a-dozen or more type Ia SNe at a time by observing a much larger number of galaxies in a single two or three day period a few nights before new Moon. By comparing these observations with the samc observations taken towards the end of dark time almost three weeks earlier, it was possible to select just those SNe that were still on the rise or near maximum. The chicken and egg problem was solved, and now the follow-up spectroscopy and photometry could be applied for and scheduled on a pre-specified set of nights. The new strategy worked – the *SCP* discovered batches of high redshift SNe,and no one would ever again have to hunt for high-redshift SNe without the crucial follow-up scheduled in advance.

The *High-Z SN Search (HZSNS)* was conceived at the end of 1994, when this group of astronomers became convinced that it was both possible to discover SNIa in large numbers at $z > 0.3$ by the efforts of Perlmutter et al.[70–72], and also use them as precision distance indicators as demonstrated by the *CTSS* group [32]. Since 1995, the *SCP* and *HZSNS* have both worked avidly to obtain a significant set of high redshift SNIa.

5.1 Discovering SNIa

The two high redshift teams both used this pre-scheduled discovery and follow-up batch strategy. They each aimed to use the observing resources they had available to best scientific advantage, choosing, for example, somewhat different exposure times or filters.

Quantitatively, type Ia SNe are rare events on an astronomer's time scale – they occur in a galaxy like the Milky Way a few times per millennium (see, e.g., [12,60,61] and the chapter by Cappellaro in this volume). With modern instruments on 4 meter-class telescopes, which observe 1/3 of a square degree to R = 24 mag in less than 10 minutes, it is possible to search a million galaxies to $z < 0.5$ for SNIa in a single night.

Since SNIa take approximately 20 days to rise from undetectable to maximum light [81], the three-week separation between observing periods (which equates to 14 rest frame days at $z = 0.5$) is a good filter to catch the SNe on the rise. The SNe are not always easily identified as new stars on the bright background of their host galaxies, so a relatively sophisticated process must be used to identify them. The process, which involves 20 Gigabytes of imaging data per night, consists of aligning a previous epoch, matching the image star profiles (through convolution), and scaling the two epochs to make the two images as identical as possible. The difference between these two images is then searched for new objects which stand out against the static sources that have been largely removed in the differencing process [73,74,76,87]. The dramatic increase in computing power in the 1980s was an important element in the development of this search technique, as was the construction of wide-field cameras with ever larger CCD detectors or mosaics of such detectors [104].

This technique is very efficient at producing large numbers of objects that are, on average, at or near maximum light, and does not require unrealistic amounts of large telescope time. It does, however, place the burden of work on follow-up observations, usually with different instruments on different telescopes. With the large number of objects discovered (50 in two nights being typical), a new strategy is being adopted by both the *SCP* and *HZSNS* teams, as well as additional teams like the *Canada France Hawaii Telescope (CFHT)* legacy survey, where the same fields are repeatedly scanned several times per month, in multiple colors, for several consecutive months. This type of observing program provides both discovery of new objects and their follow up, all integrated into one efficient program. It does require a large block of time on a single telescope – a requirement which was not politically feasible in years past, but is now possible.

5.2 Obstacles to Measuring Luminosity Distances at High-Z

As shown above, the distances measured to SNIa are well characterized at $z < 0.1$, but comparing these objects to their more distant counterparts requires great care. Selection effects can introduce systematic errors as a function of redshift, as can uncertain K-corrections and a possible evolution of the SNIa progenitor population as a function of look-back time. These effects, if they are large and not constrained or corrected, will limit our ability to accurately measure relative luminosity distances, and have the potential to reduce the efficacy of high-z type Ia SNe for measuring cosmology [74,77,83,87].

K-Corrections: As SNe are observed at larger and larger redshifts, their light is shifted to longer wavelengths. Since astronomical observations are normally made in fixed band passes on Earth, corrections need to be applied to account for the differences caused by the spectrum shifting within these band passes. These corrections take the form of integrating the spectrum of an SN over the relevant band passes, shifting the SN spectrum to the correct redshift, and re-integrating. Kim et al. [43] showed that these effects can be minimized if one does not use a single bandpass, but instead chooses the bandpass closest to the redshifted rest-frame bandpass, as they had done for SN1992bi [73]. They showed that the inter-band K-correction is given by

$$K_{ij}(z) = 2.5 \log \left[(1+z) \frac{\int F(\lambda) S_i(\lambda) d\lambda}{\int F(\lambda/(1+z)) S_j(\lambda) d\lambda} \frac{\int Z(\lambda) S_j(\lambda) d\lambda}{\int Z(\lambda) S_i(\lambda) d\lambda} \right], \quad (10)$$

where $K_{ij}(z)$ is the correction to go from filter i to filter j, and $Z(\lambda)$ is the spectrum corresponding to zero magnitude of the filters.

The brightness of an object expressed in magnitudes, as a function of z is

$$m_i(z) = 5 \log \left[\frac{D_L(z)}{\mathrm{Mpc}} \right] + 25 + M_j + K_{ij}(z), \quad (11)$$

where $D_L(z)$ is given by Eq. (7), M_j is the absolute magnitude of object in filter j, and K_{ij} is given by Eq. (10). For example, for $H_0 = 70$ km s^{-1} Mpc^{-1}, and $D_L = 2835$ Mpc ($\Omega_M = 0.3, \Omega_\Lambda = 0.7$), at maximum light a SNIa has $M_B = -19.5$ mag and a $K_{BR} = -0.7$ mag. We therefore expect an SNIa at $z = 0.5$ to peak at $m_\mathrm{R} \sim 22.1$ mag for this set of cosmological parameters.

K-correction errors depend critically on three uncertainties:

1. Accuracy of spectrophotometry of SNe. To calculate the K-correction, the spectra of SNe are integrated in Eq. (10). These integrals are insensitive to a grey shift in the flux calibration of the spectra, but any wavelength dependent flux calibration error will translate into erroneous K-corrections.
2. Accuracy of the absolute calibration of the fundamental astronomical standard systems. Eq. (10) shows that the K-corrections are sensitive to the shape of the astronomical band passes and to the zero points of these band passes.
3. Accuracy of the choice of SNIa spectrophotometry template used to calculate the corrections. Although a relatively homogenous class, there are variations in the spectra of SNIa. If a particular object has, for example, a stronger calcium triplet than the average SNIa, the K-corrections will be in error unless an appropriate subset of SNIa spectra are used in the calculations.

The first error should not be an issue if correct observational procedures are used on an instrument that has no fundamental problems. The second error is currently estimated to be small (~ 0.01 mag), based on the consistency of spectrophotometry and broadband photometry of the fundamental standards, Sirius and Vega [5]. To improve this uncertainty will require new, careful experiments to accurately calibrate a star, such as Vega or Sirius (or a White Dwarf or solar analog star), and to carefully infer the standard bandpass that defines the photometric system in use at telescopes. The third error requires a large database to match as closely as possible an SN with the spectrophotometry used to calculate the K-corrections. Nugent et al. [58] have shown that extinction and color are related and, by correcting the spectra to force them to match the photometry of the SN needing K-corrections, that it is possible to largely eliminate errors 1 and 3, even when using spectra that are not exact matches (in epoch or in fine detail) to the SNIa being K-corrected. Scatter in the measured K-corrections from a variety of telescopes and objects allows us to estimate the combined size of the effect for the first and third errors. These appear to be ~ 0.01 mag for redshifts where the high-z and low-z filters have a large region of overlap (e.g., R-band matched to B-band at $z = 0.5$).

Extinction: In the nearby Universe we see SNIa in a variety of environments, and about 10% have significant extinction [30]. Since we can correct for extinction by observing two or more wavelengths, it is possible to remove any first order effects caused by a changing average extinction of SNIa as a function of z. However, second order effects, such as possible evolution of the average properties of intervening dust, could still introduce systematic errors. This problem can also be addressed by observing distant SNIa over a decade or so of wavelength in order to measure the extinction law to individual objects. Unfortunately, this is observationally very expensive. Current observations limit the total systematic effect to < 0.06 mag, as most of our current data is based on two color observations.

An additional problem is the existence of a thin veil of dust around the Milky Way. Measurements from the *Cosmic Background Explorer (COBE)* satellite accurately determined the relative amount of dust around the Galaxy [89], but there is an uncertainty in the absolute amount of extinction of about $2-3\%$. This uncertainty is not normally a problem, since it affects everything in the sky more or less equally. However, as we observe SNe at higher and higher redshifts, the light from the objects is shifted to the red and is less affected by the Galactic dust. Our present knowledge indicates that a systematic error as large as 0.06 mag is attributable to this uncertainty.

Selection Effects: As we discover SNe, we are subject to a variety of selection effects, both in our nearby and distant searches. The most significant effect is the Malmquist Bias – a selection effect which leads magnitude limited searches to find brighter than average objects near their distance limit since brighter objects can be seen in a larger volume than their fainter counterparts. Malmquist Bias errors are proportional to the square of the intrinsic dispersion of the distance method, and because SNIa are such accurate distance indicators these errors are quite small, ~ 0.04 mag. Monte Carlo simulations can be used to estimate such selection effects, and to remove them from our data sets [74,76,77,87]. The total uncertainty from selection effects is ~ 0.01 mag and, interestingly, may be worse for lower redshift objects because they are, at present, more poorly quantified.

Gravitational Lensing: Several authors have pointed out that the radiation from any object, as it traverses the large scale structure between where it was emitted and where it is detected, will be weakly lensed as it encounters fluctuations in the gravitational potential [37,45,100]. On average, most of the light travel paths go through under-dense regions and objects appear de-magnified. Occasionally, the light path encounters dense regions and the object becomes magnified. The distribution of observed fluxes for sources is skewed by this process such that the vast majority of objects appear slightly fainter than the canonical luminosity distance, with the few highly magnified events making the mean of all light paths unbiased. Unfortunately, since we do not observe enough objects to capture the entire distribution, unless we know and include the skewed shape of the lensing a bias will occur. At $z = 0.5$, this lensing is not a significant problem: If the Universe is flat in normal matter, the large scale structure can induce a shift of the mode of the distribution by only a few percent. However, the effect scales roughly as z^2, and by $z = 1.5$ the effect can be as large as 25% [38]. While corrections can be derived by measuring the distortion of background galaxies near the line of sight to each SN, at $z > 1$, this problem may be one which ultimately limits the accuracy of luminosity distance measurements, unless a large enough sample of SNe at each redshift can be used to characterize the lensing distribution and average out the effect. For the $z \sim 0.5$ sample, the error is < 0.02 mag, but it is much more significant at $z > 1$ (e.g., for SN1997ff) [4,55], especially if the sample size is small.

Evolution: SNIa are seen to evolve in the nearby Universe. Hamuy et al. [29] plotted the shape of the SN light curves against the type of host galaxy. SNe in early hosts (galaxies without recent star formation) consistently show light curves which rise and fade more quickly than SNe in late-type hosts (galaxies with on-going star formation). However, once corrected for light curve shape the luminosity shows no bias as a function of host type. This empirical investigation provides reassurance for using SNIa as distance indicators over a variety of stellar population ages. It is possible, of course, to devise scenarios where some of the more distant SNe do not have nearby analogues, so as supernovae are studied at increasingly higher redshifts it can become important to obtain detailed spectroscopic and photometric observations of every distant SN to recognize and reject examples that have no nearby analogues.

In principle, it should be possible to use differences in the spectra and light curves between nearby and distant SNe, combined with theoretical modeling, to correct any differences in absolute magnitude. Unfortunately, theoretical investigations are not yet advanced enough to precisely quantify the effect of these differences on the absolute magnitude. A different, empirical approach to handle SN evolution [10] is to divide the SNe into subsamples of very closely matched events, based on the details of the their light curves, spectral time series, host galaxy properties, etc. A separate Hubble diagram can then be constructed for each subsample of SNe, and each will yield an independent measurement of the cosmological parameters. The agreement (or disagreement) between the results from the separate subsamples is an indicator of the total effect of evolution. A simple, first attempt at this kind of test has been performed by comparing the results for SNe found in elliptical host galaxies to SNe found in late spirals or irregular hosts, and the cosmological results from these subsamples were found to agree well [91].

Finally, it is possible to move to higher redshifts and see if the SNe deviate from the predictions of Eq. (7). At a gross level, we expect an accelerating Universe to be decelerating in the past because the matter density of the Universe increases with redshift, whereas the density of any dark energy leading to acceleration will increase at a slower rate than this (or not at all in the case of a cosmological constant). If the observed acceleration is caused by some sort of systematic effect, it is likely to continue to increase (or at least remain steady) with z, rather than disappear like the effects of dark energy. A first comparison has been made with SN1997ff at $z \sim 1.7$ [85], and it seems consistent with a decelerating Universe at that epoch. More objects are necessary for a definitive answer, and these should be provided by several large programs that have been discovering such type Ia SNe at the *W.M. Keck Telescope I (KECK I)*, *Subaru Telescope)*, and *HST* telescopes.

5.3 High Redshift SNIa Observations

The *SCP* [74] in 1997 presented their first results with 7 objects at a redshift around $z = 0.4$. These objects hinted at a decelerating Universe with a measurement of $\Omega_M = 0.88^{+0.69}_{-0.60}$, but were not definitive. Soon after, the *SCP* published a further result, with a $z \sim 0.84$ SNIa observed with the *KECK I* and *HST*added to the sample [75], and the *HZSNS* presented the results from their first four objects [22,87]. The results from both teams now ruled out a $\Omega_M = 1$ Universe with greater than 95% significance. These findings were again superceded dramatically when both teams announced results including more SNe (10 more *HZSNS* SNe, and 34 more *SCP* SNe) that showed not only were the SN observations incompatible with a $\Omega_M = 1$ Universe, they were also incompatible with a Universe containing only normal matter [77,83]. Figure 4 shows the Hubble diagram for both teams. Both samples show that SNe are, on average, fainter than would be expected, even for an empty Universe, indicating that the Universe is accelerating. The agreement between the experimental results of the two teams is spectacular, especially considering the two programs have worked in almost complete isolation from each other.

The easiest solution to explain the observed acceleration is to include an additional component of matter with an equation of state parameter more negative than $w < -1/3$; the most familiar being the cosmological constant ($w = -1$). Figure 5 shows the joint confidence contours for values of Ω_M and Ω_Λ from both experiments. If we assume the Universe is composed only of normal matter and a cosmological constant, then with greater than 99.9% confidence the Universe has a non-zero cosmological constant or some other form of dark energy.

Of course, we do not know the form of dark energy which is leading to the acceleration, and it is worthwhile investigating what other forms of energy are possible additional components. Figure 6 shows the joint confidence contours for the *HZSNS+SCP* observations for Ω_M and w_x (the equation of state of the unknown component causing the acceleration). Because this introduces an extra

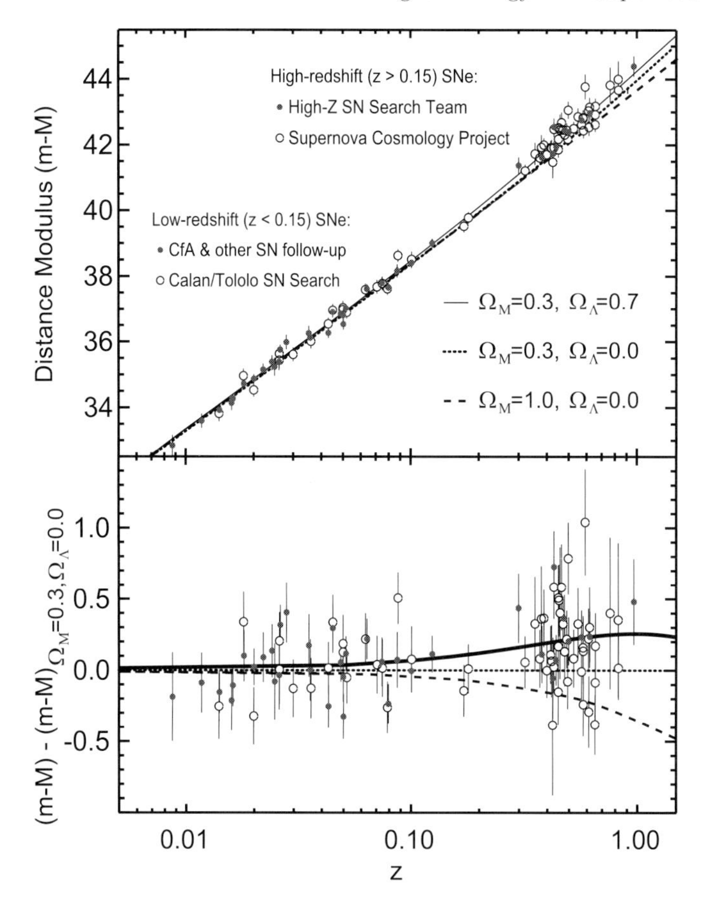

Fig. 4. *Upper panel:* The Hubble diagram for high redshift SNIa from both the *HZSNS* [83] and the *SCP* [77]. *Lower panel:* The residual of the distances relative to a $\Omega_M = 0.3$, $\Omega_\Lambda = 0.7$ Universe. The $z < 0.15$ objects for both teams are drawn from *CTSS* sample [32], so many of these objects are in common between the analyses of the two teams.

parameter, we apply the additional constraint that $\Omega_M + \Omega_x = 1$, as indicated by the *CMB* experiments [14]. The cosmological constant is preferred, but anything with a $w < -0.5$ is acceptable [23,77]. Additionally, we can add information about the value of Ω_M, as supplied by recent 2dF redshift survey results [98], as shown in the 2nd panel, where the constraint strengthens to $w < -0.6$ at 95% confidence [69].

6 The Future

How far can we push the SN measurements? Finding more and more SNe allows us to beat down statistical errors to arbitrarily small levels but, ultimately, systematic effects will limit the precision to which SNIa magnitudes can be

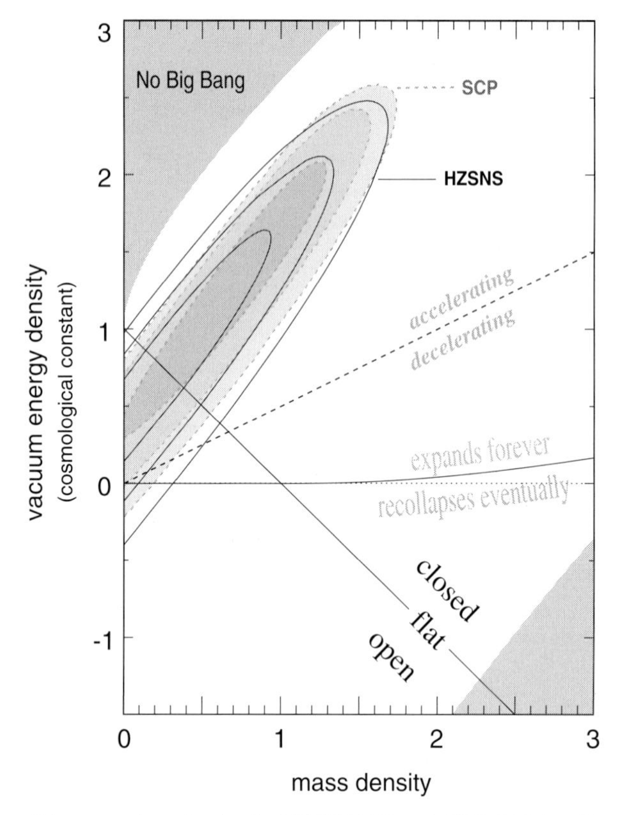

Fig. 5. The confidence regions for both *HZSNS* [83] and *SCP* [77] for Ω_M, Ω_Λ. The two experiments show, with remarkable consistency, that $\Omega_\Lambda > 0$ is required to reconcile observations and theory. The *SCP* result is based on measurements of 42 distant SNIa. (The analysis shown here is uncorrected for host galaxy extinction;see [77] for the alternative analyses with host extinction correction, which is shown to make little difference in this data set.) The *HZSNS* result is based on measurements of 16 SNIa, including 6 snapshot distances [80], of which two are *SCP* SNe from the 42 SN sample. The $z < 0.15$ objects used to constrain the fit for both teams are drawn from the *CTSS* sample [32], so many of these objects are common between the analyses by the two teams.

applied to measure distances. Our best estimate is that it will be possible to control systematic effects from ground-based experiments to a level of ~ 0.03 mag. Carefully controlled ground-based experiments on 200 SNe will reach this statistical uncertainty in $z = 0.1$ redshift bins in the range $z = 0.3 - 0.7$, and is achievable within five years. A comparable quality low redshift sample, with 300 SNe in $z = 0.02 - 0.08$, will also be achievable in that time frame [2].

The *SuperNova/Acceleration Probe (SNAP)* collaboration[1] has proposed to launch a dedicated cosmology satellite [1,68] – the ultimate SNIa experiment.

[1] See *http://snap.lbl.gov*

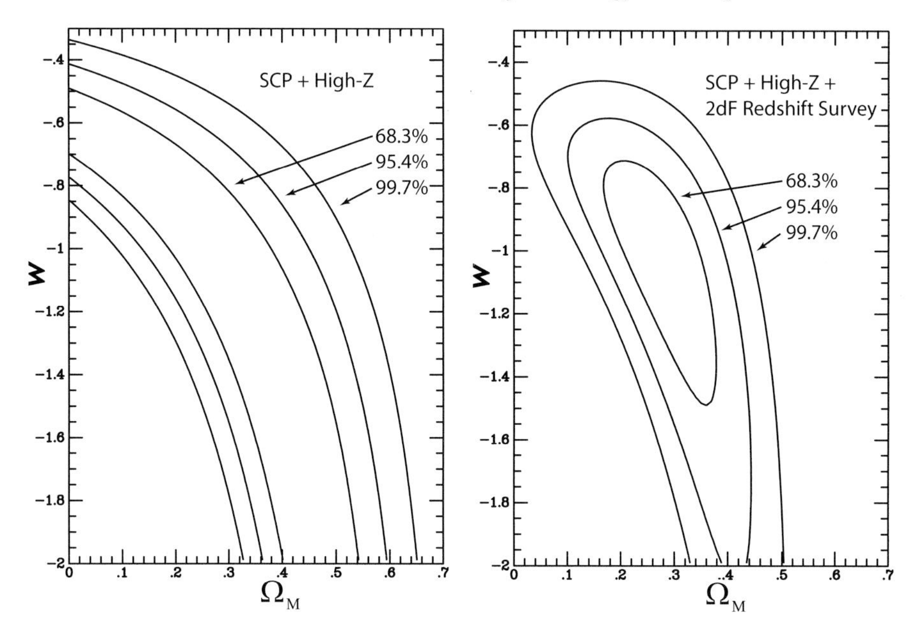

Fig. 6. Left panel: Contours of Ω_M versus w_x from current observational data. Right Panel: Contours of Ω_M versus w_x from current observational data, where the current value of Ω_M is obtained from the 2dF redshift survey. For both panels $\Omega_M + \Omega_x = 1$ is taken as a prior.

This satellite will, if funded, scan many square degrees of sky, discovering well over a thousand SNIa per year and obtain their spectra and light curves out to $z = 1.7$. Besides the large numbers of objects and their extended redshift range, space-based observations will also provide the opportunity to control many systematic effects better than from the ground [21,53]. Figure 7 shows the expected precision in the *SNAP* and ground-based experiments for measuring w, assuming a flat Universe. Perhaps the most important advance will be the first studies of the time variation of the equation of state w (see the right panel of Fig. 7 and [40,103]).

With rapidly improving *CMB* data from interferometers, the satellites *Microwave Anisotropy Probe (MAP)* and *Planck*, and balloon-based instrumentation planned for the next several years, *CMB* measurements promise dramatic improvements in precision on many of the cosmological parameters. However, the *CMB* measurements are relatively insensitive to the dark energy and the epoch of cosmic acceleration. SNIa are currently the only way to directly study this acceleration epoch with sufficient precision (and control on systematic uncertainties) that we can investigate the properties of the dark energy, and any time dependence in these properties. This ambitious goal will require complementary and supporting measurements of, for example, Ω_M from *CMB*, weak lensing, and large scale structure. The SN measurements will also provide a test of the cosmological results independent from these other techniques, which have their

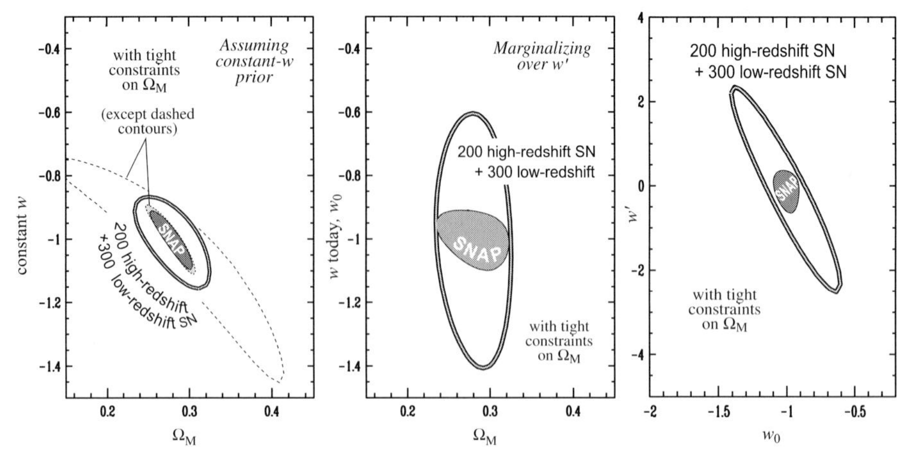

Fig. 7. Future expected constraints on dark energy: *Left panel:* Estimated 68% confidence regions for a constant equation of state parameter for the dark energy, w, versus mass density, for a ground-based study with 200 SNe between $z = 0.3 - 0.7$ (open contours), and for the satellite-based *SNAP* experiment with 2,000 SNe between $z = 0.3 - 1.7$ (filled contours). Both experiments are assumed to also use 300 SNe between $z = 0.02 - 0.08$. A flat cosmology is assumed (based on *Cosmic Microwave Background (CMB)* constraints) and the inner (solid line) contours for each experiment include tight constraints (from large scale structure surveys) on Ω_M, at the ± 0.03 level. For the *SNAP* experiment, systematic uncertainty is taken as $dm = 0.02(z/1.7)$, and for the ground-based experiment, $dm = 0.03(z/0.5)$. Such ground-based studies will test the hypothesis that the dark energy is in the form of a cosmological constant, for which $w = -1$ at all times. *Middle panel:* The same confidence regions for the same experiments not assuming the equation of state parameter, w, to be constant, but instead marginalizing over w', where $w(z) = w_0 + w'z$. (Weller and Albrecht [103] recommend this parameterization of $w(z)$ over the others that have been proposed to characterize well the current range of dark energy models.) Note that these planned ground-based studies will yield impressive constraints on the value of w today, w_0, even without assuming constant w. In fact, these constraints are comparable to the current measurements of w assuming it is constant (shown in the right panel of Fig. 6). *Right panel:* Estimated 68% confidence regions of the first derivative of the equation of state, w', versus its value today, w_0, for the same experiments.

own systematic errors Moving forward simultaneously on these experimental fronts offers the plausible and exciting possibility of achieving a comprehensive measurement of the fundamental properties of our Universe.

References

1. G. Aldering et al.: SPIE **4835**, 21 (2002)
2. G. Aldering et al.: SPIE **4836**, 93 (2002)
3. W.A. Baum: Astron. J. **62**, 6 (1957)
4. N. Benitez, A. Riess, P. Nugent, M. Dickinson, R. Chornock, A. Filippenko: Astrophys. J. Lett. **577**, L1 (2002)

5. M. Bessell: Pub. Astron. Soc. Pacific **102**, 1181 (1998)
6. D. Branch: In: *Encyclopedia of Astronomy and Astrophysics* (Academic, San Diego 1989) p. 733
7. D. Branch, G.A. Tammann: Ann. Rev. Astron. Astrophys. **30**, 359 (1992)
8. D. Branch, A. Fisher, P. Nugent: Astron. J. **106**, 2383 (1993)
9. D. Branch, A. Fisher, E. Baron, P. Nugent: Astrophys. J. Lett. **470**, L7 (1996)
10. D. Branch, S. Perlmutter, E. Baron, P. Nugent: In: *Resource Book on Dark Energy*, ed. by E.V. Linder (Snowmass 2001)
11. R. Cadonau: PhD Thesis, University of Basel (1987)
12. E. Cappellaro, M. Turatto, D.Yu. Tsvetkov, O.S. Bartunov, C. Pollas, R. Evans, M. Hamuy: Astron. Astrophys. **322**, 431 (1997)
13. P. Coles, F. Lucchin: In: *cosmology* (Wiley, Chicester 1995) p. 31
14. P. de Bernardis et al.: Nature **404**, 955 (2000)
15. R.G. Eastman, R.P. Kirshner: Astrophys. J. **347**, 771 (1989)
16. R.G. Eastman, B.P. Schmidt, R. Kirshner: Astrophys. J. **466**, 911 (1996)
17. A. Fisher, D. Branch, P. Hoeflich, A. Khokhlov: Astrophys. J. Lett. **447**, L73 (1995)
18. A.V. Filippenko: In: *SN1987A and Other Supernovae*, ed. by I.J. Danziger, K. Kjar (ESO, Garching 1991) p. 343
19. A.V. Fillipenko et al.: Astrophys. J. Lett. **384**, L15 (1992)
20. W.L. Freedman et al.: Astrophys. J. **553**, 47 (2001)
21. J. Frieman, D. Huterer, E.V. Linder, M.S. Turner: astro-ph 0208100 (2002)
22. P. Garnavich et al.: Astrophys. J. Lett. **493**, L53 (1998)
23. P. Garnavich et al.: Astrophys. J. **509**, 74 (1998)
24. L.G. Germany, A.G. Riess, B.P. Schmidt, N.B. Suntzeff: in preparation (2003)
25. G. Goldhaber et al.: In: *Thermonuclear Supernovae*, ed. by P. Ruiz-Lapuente, R. Canal, J. Isern (Aiguablava, June 1995; NATO ASI, 1997)
26. G. Goldhaber et al.: Astrophys. J. **558**, 359 (2001)
27. A. Goobar, S. Perlmutter: Astrophys. J. **450**, 14 (1995)
28. M. Hamuy, M.M. Phillips, J. Maza, M. Wischnjewsky, A. Uomoto, A.U. Landolt, R. Khatwani: Astron. J. **102**, 208 (1991)
29. M. Hamuy, M.M. Phillips, N.B. Suntzeff, R.A. Schommer, J. Maza, R. Aviles: Astron. J. **112**, 2391 (1996)
30. M. Hamuy, P.A. Pinto: Astron. J. **117**, 1185 (1999)
31. M. Hamuy et al.: Astron. J. **106**, 2392 (1993)
32. M. Hamuy et al.: Astron. J. **109**, 1 (1995)
33. M. Hamuy et al.: Astron. J. **112**, 2408 (1996)
34. M. Hamuy et al.: Astrophys. J. **558**, 615 (2001)
35. L. Hansen, H.E. Jorgensen, H.U. Nørgaard-Nielsen, R.S. Ellis, W.J. Couch: Astron. Astrophys. **211**, L9 (1989)
36. R.P. Harkness, J.C. Wheeler: In: *Supernovae*, ed. by A.G. Petschek (Springer-Verlag, New York 1990) p. 1
37. D.E. Holz, R.M. Wald: Phys. Rev. D **58**, 063501 (1998)
38. D.E. Holz: Astrophys. J. **506**, 1 (1998)
39. M.L. Humason, N.U. Mayall, A.R. Sandage: Astrophys. J. **61**, 97 (1956)
40. D. Huterer, M.S. Turner: Phys. Rev. D **64**, 123527 (2001)
41. S. Jha: PhD Thesis, Harvard University (2002)
42. C.T. Kowal: Astron. J. **272**, 1021 1968
43. A. Kim, A. Goobar, S. Perlmutter: Pub. Astron. Soc. Pacific **108**, 190 (1996)
44. R.P. Kirshner, J. Kwan: Astrophys. J. **193**, 27 (1974)

45. R. Kantowski, T. Vaughan, D. Branch: Astrophys. J. **447**, 35 (1995)
46. B. Leibundgut: PhD Thesis, University of Basel (1988)
47. B. Leibundgut, G.A. Tammann: Astron. Astrophys. **230**, 81 (1990)
48. B. Leibundgut, G.A. Tammann, R. Cadonau, D. Cerrito: Astron. Astrophys. Suppl. Ser. **89**, 537 (1991)
49. B. Leibundgut et al.: Astron. J. **105**, 301 (1993)
50. B. Leibundgut et al.: Astrophys. J. Lett. **466**, L21 (1996)
51. D.C. Leonard, A.V. Filippenko, D.R. Ardila, M.S. Brotherton: Astrophys. J. **553**, 861 (2001)
52. D.C. Leonard et al.: Pub. Astron. Soc. Pacific **114**, 35 (2002)
53. E. Linder, D. Huterer: astro-ph 0208138 (2002)
54. D.L. Miller, D. Branch: Astron. J. **100**, 530 (1990)
55. E. Mortsell, C. Gunnarsson, A. Goobar: Astrophys. J. **561**, 106 (2001)
56. R.A. Muller, H.J.M. Newberg, C.R. Pennypacker, S. Perlmutter, T.P. Sasseen, C.K. Smith: Astrophys. J. Lett. **384**, L9 (1992)
57. H.U. Nørgaard-Nielsen, L. Hansen, H.E. Jorgensen, A. Aragon Salamanca, R.S. Ellis: Nature **339**, 523 (1989)
58. P. Nugent, A. Kim, S. Perlmutter: Pub. Astron. Soc. Pacific **114**, 803 (2002)
59. N. Panagia: In: *Supernovae as Distance Indicators*, ed. by N. Bartel, (Springer-Verlag, Berlin 1985) p. 14
60. R. Pain et al.: Astrophys. J. **473**, 356 (1996)
61. R. Pain et al.: Astrophys. J. **577**, 120 (2002)
62. G. Pearce, B. Patchett, J. Allington-Smith, I. Parry: Astrophys. Space Sci. **150**, 267 (1988)
63. M.M. Phillips et al.: Pub. Astron. Soc. Pacific **99**, 592 (1987)
64. M.M. Phillips: Astrophys. J. Lett. **413**, L105 (1993)
65. M.M. Phillips, P. Lira, N.B. Suntzeff, R.A. Schommer, M. Hamuy, J. Maza: Astron. J. **118**, 1766 (1999)
66. P.J.E. Peebles: In: *Principles of Physical cosmology* (Princeton University Press, Princeton 1993)
67. S. Perlmutter, R.A. Muller, H.J.M. Newberg, C.R. Pennypacker, T.P. Sasseen, C.K. Smith: ASP Conf. Proc. **34**, 67 (1992)
68. S. Perlmutter, E. Linder: In *Dark Matter 2002, Proc. 5th International UCLA Symposium on Sources and Detection of Dark Matter and Dark Energy in the Universe*, ed. by D.B. Cline (Elsevier, Amsterdam 2003)
69. S. Perlmutter, M. Turner, M. White: Phys. Rev. Lett. **83**, 670 (1999)
70. S. Perlmutter et al.: IAUC 5956 (1994)
71. S. Perlmutter et al.: IAUC 6263 (1995)
72. S. Perlmutter et al.: IAUC 6270 (1995)
73. S. Perlmutter et al.: Astrophys. J. Lett. **440**, L41 (1995)
74. S. Perlmutter et al.: Astrophys. J. **483**, 565 (1997)
75. S. Perlmutter et al.: Nature **391**, 51 (1998)
76. S. Perlmutter et al.: In: *Thermonuclear Supernovae*, ed. by P. Ruiz-Lapuente, R. Canal, J. Isern (Aiguablava, June 1995; NATO ASI, 1997)
77. S. Perlmutter et al.: Astrophys. J. **517**, 565 (1999)
78. M.M. Phillips, L.A. Wells, N.B. Suntzeff, M. Hamuy, B. Leibundgut, R.P. Kirshner, C.B. Foltz: Astron. J. **103**, 1632 (1992)
79. A.G. Riess, W.H. Press, R.P. Kirshner: Astrophys. J. **473**, 88 (1996)
80. A.G. Riess, P. Nugent, A.V. Filippenko, R.P. Kirshner, S. Perlmutter: Astrophys. J. **504**, 935 (1998)

81. A.G. Riess, A.V. Filippenko, W. Li, B.P. Schmidt: Astron. J. **118**, 2668 (1999)
82. A.G. Riess et al.: Astron. J. **114**, 722 (1997)
83. A.G. Riess et al.: Astron. J. **116**, 1009 (1998)
84. A.G. Riess et al.: Astron. J. **117**, 707 (1999)
85. A.G. Riess et al.: Astrophys. J. **560**, 49 (2001)
86. B.P. Schmidt, R.P. Kirshner, R.G. Eastman, M.M. Phillips, N.B. Suntzeff, N.B. Hamuy, J. Maza, R. Aviles: Astrophys. J. **432**, 42 (1994)
87. B. Schmidt et al.: Astrophys. J. **507**, 46 (1998)
88. A. Saha, A. Sandage, G.A. Tammann, A.E. Dolphin, J. Christensen, N. Panagia, F.D. Macchetto: Astrophys. J. **562**, 313 (2001)
89. D.J. Schlegel, D.P. Finkbeiner, M. Davis: Astrophys. J. Suppl. **500**, 525 (1998)
90. A. Sandage, G.A. Tammann: Astrophys. J. **415**, 1 (1993)
91. M. Sullivan et al.: Mon. Not. R. Astron. Soc. , in press (2003)
92. G.A. Tammann, B. Leibundgut: Astron. Astrophys. **236**, 9 (1990)
93. G.A. Tammann, A. Sandage: Astrophys. J. **452**, 16 (1995)
94. A. Uomoto, R.P. Kirshner: Astron. Astrophys. **149**, L7 (1985)
95. S. van den Bergh: Astrophys. J. Lett. **453**, L55 (1995)
96. S. van den Bergh, J. Pazder: Astrophys. J. **390**, 34 (1992)
97. T.E. Vaughan, D. Branch, D.L. Miller, S. Perlmutter: Astrophys. J. **439**, 558 (1995)
98. L. Verde et al.: Mon. Not. R. Astron. Soc. **335**, 432 (2002)
99. R.V. Wagoner: Astrophys. J. Lett. **250**, L65 (1981)
100. J. Wambsgabss, R. Cen, X. Guohong, J. Ostriker: Astrophys. J. Lett. **475**, L81 (1997)
101. L. Wang, A.D. Howell, P. Höflich, J.C. Wheeler: Astrophys. J. **550**, 1030 (2001)
102. J.C. Wheeler, R. Levreault: Astrophys. J. Lett. **294**, L17 (1985)
103. J. Weller, A. Albrecht: Phys. Rev. D **65**, 103512 (2002)
104. D.M. Wittman, J.A. Tyson, G.M. Bernstein, R.W. Lee, I.P. dell'Antonio, P. Fischer, D.R. Smith, M.M. Blouke: SPIE **3355**, 626 (1998)

Supernova 1987A

Richard McCray

JILA, University of Colorado, Boulder CO 80309-0440, USA

Abstract. First, I summarize the main points that we have learned about the interior of SN1987A. Then, I describe in greater detail the rapidly developing impact of SN1987A with its inner circumstellar ring.

1 Introduction

At the present epoch of summer 2002, we are observing SN1987A approximately 15 years after its initial outburst. During the first 10 years, the radiation from SN1987A was dominated by energy deposited in the interior by the decay of newly synthesized radioisotopes. From observations of this radiation at many wavelengths, we learned a great deal about the dynamics and thermodynamics of the expanding debris. With the *Hubble Space Telescope (HST)*, we have also observed a remarkable system of three circumstellar rings, the origin of which still remains a mystery.

About 6 years ago, the blastwave from the supernova began to strike the inner circumstellar ring, resulting in the appearance of a rapidly brightening "hot spot" on the ring. Today, many more hot spots have appeared, and the radio, infrared, optical, and X-ray radiation from of the supernova is now dominated by the impact of the supernova debris with its circumstellar medium (CSM). This impact marks the birth of a supernova remnant, SNR1987A.

In this chapter, I will first summarize what we have learned about the interior structure of SN1987A and the dynamics and thermodynamics of the debris. Then I will discuss what we know about the CSM and rings, and what we are learning from observations of the interaction of the supernova debris with the CSM. Finally, I will hazard a few guesses about what we can expect to learn from SNR1987A during the next few decades.

2 Energetics

Before going into detail, it might be useful to summarize the main energy sources of SN1987A. These are listed in Table 1.

As Table 1 shows, SN1987A has three different sources of energy, each of which emerges as a different kind of radiation and with a different timescale. The greatest is the collapse energy itself, which emerges as a neutrino burst lasting a few seconds. The energy provided by radioactive decay of newly synthesized

Table 1. SN1987A Energetics

Source	Collapse	Radioactivity	Expansion
Definition	$\sim \frac{G\, M_{\odot}{}^{2}}{R_{N*}}$	$^{56}Ni \rightarrow^{56} Co \rightarrow^{56} Fe$ (0.07 $M_{\odot}$)	$\int_{debris} \frac{1}{2} V^2 dM$
Emerges as:	Neutrinos	O, IR	X-rays
	($kT \sim 4MeV$)	(+ X, γ)	(+ R, IR, O, UV)
Energy (erg)	10^{53}	10^{49}	10^{51}
Timescale	$\sim$ 10 seconds	$\sim$ 1 year	$\sim 10 - 1000$ years

elements is primarily responsible for the optical display. Most of this energy emerged within the first year after outburst, primarily in optical and infrared emission lines and continuum from relatively cool ($T \lesssim 5,000$ K) gas. Note that the radioactive energy is relatively small, $\sim 10^{-4}$ of the collapse energy.

The kinetic energy of the expanding debris can be inferred from observations of the spectrum during the first three months after explosion. We infer the density and velocity of gas crossing the photosphere from the strengths and widths of hydrogen lines in the photospheric spectrum. By tracking the development of the spectrum as the photosphere moves to the center of the debris, they can measure the integral defining the kinetic energy.7 Doing so, we find that $\sim 10^{-2}$ of the collapse energy has been converted into kinetic energy of the expanding debris [3]. Why this fraction is typically 10^{-2} and not, say, 10^{-1} or 10^{-3}, is one of the unsolved problems of supernova theory.

This kinetic energy will be converted into radiation when the supernova debris strikes the CSM. When this happens, two shocks always develop: the blastwave, which overtakes the CSM, and the reverse shock, which is driven inwards (in a Lagrangian sense) through the expanding debris. The gas trapped between these two shocks is typically raised to temperatures in the range $10^6 - 10^8$ K and will radiate most of its thermal energy as X-rays with a spectrum dominated by emission lines in the range $0.3 - 10$ keV.

Most of the kinetic energy of the debris will not be converted into thermal energy of shocked gas until the blastwave has overtaken a circumstellar mass comparable to that of the debris itself, $\sim 10 - 20$ $M_{\odot}$. Typically, that takes many centuries, and as a result, most galactic supernova remnants (*e.g.*, Cas A) reach their peak X-ray luminosities only after a few centuries and then fade.

As I shall describe, we believe that SN1987A is surrounded by a few solar masses of CSM within a distance of a parsec or two. Thus, a significant fraction of the kinetic energy of the debris will be converted into thermal energy within a few decades as the supernova blastwave overtakes this matter.

3 The Compact Object

The flash of 19 ± 1 neutrinos that was detected a few hours before the optical brightening of SN1987A provided compelling evidence that its core had collapsed to form a compact object [1]. The inferred energy ($\sim 3 \times 10^{53}$ erg), temperature (kT $\sim$ 4 MeV), and decay timescale ($\sim$ 4 s) of the flash were in remarkably good agreement with models in which a degenerate iron core collapsed to form a neutron star. But since then, and despite great effort, astronomers have been disappointed to find no firm evidence of a compact object near the center of the expanding debris of the supernova.

One thing is clear: if a compact object exists at the center of SN1987A, it must have bolometric luminosity $L_c \lesssim$ few$\times 100$ $L_\odot$. The bolometric (mostly far infrared) luminosity of the entire envelope is now $L_{env} \lesssim 1000$ $L_\odot$ and a brighter compact object at the center of the debris could not have escaped detection. This places a limit of $L_c \lesssim 10^{-2}$ times that of the 948 year old pulsar in the Crab Nebula. If the expected compact object in the center of SN1987A is a neutron star, it must be very faint – perhaps because it has an anomalously low magnetic field and/or spin rate. But that is not enough to escape detection. It also must not be radiating thermally above the detection limit, and it must accrete $\lesssim 10^{-11}$ $M_\odot$ yr^{-1}. The neutron star would probably cool fast enough so that its thermal radiation would barely escape detection [47]. Likewise, radiation pressure on the infal ling matter could suppress accretion if $L_c \gtrsim L'_E$, where $L'_E \sim$ few $\times$ 100 $L_\odot$ is a "modified Eddington limit" that takes into account the opacity due to resonance line scattering in addition to the electron scattering opacity [18].

Although the duration of the neutrino flash indicates that a neutron star formed during the supernova explosion, it is possible that enough matter fell back in the subsequent hours or days to cause it to collapse into a black hole [21]. The accretion rate need not be limited to the value $\dot{M}_E \approx 3\times10^{-8}$ $M_\odot$ yr^{-1} if the flow is dense and hot enough so that neutrinos can carry off the accretion luminosity. When it collapsed initially, the iron core is expected to have a mass at the Chandrasekhar limit, $M_c \approx 1.4$ $M_\odot$. After it formed, such a neutron star would have to accrete another $\Delta M_c \gtrsim 0.4$ $M_\odot$ in order to collapse to a black hole. Although we can't rule out this possibility, I regard it as unlikely, especially as I see no compelling argument to rule out a neutron star within the envelope of SN1987A today.

How might we detect a compact object in the debris of SN1987A if its bolometric luminosity is much less than that of the radioactive debris? There is little chance of doing so with radio ($\gtrsim$ 1 cm), UV, or X-ray observations, since the envelope will probably remain opaque in these bands for decades. Thus, our hopes turn to the submillimeter, infrared, optical, hard X-ray, and γ-ray bands.

We do see optical emission lines from the interior of SN1987A, but we also see clear evidence of extinction by dust in the debris. In fact, the dust appears to occult $\sim$ 50% of the emission from the far hemisphere of the supernova debris, so there is a good chance that a foreground dust cloud might completely obscure the compact object at optical wavelengths. Such obscuration could account for the absence of evidence of a central compact optical point source and the failure to confirm the purported detection [43] of a 2.1 ms optical pulsar.

We might detect the presence of a faint compact object in the envelope of SN1987A by observing optical or infrared emission lines from gas ionized by a central X-ray source [9]. The spectrum of the X-ray photoionized gas would display emission lines from atoms ionized twice or more and the emission lines would have widths FWHM $\lesssim 1000$ km s^{-1}. It should be easy to distinguish such a spectrum from that of the radioactive-illuminated debris, which displays only broad (FWHM ~ 3000 km s^{-1}) emission lines from neutral and once-ionized species. But this potentially sensitive method of detecting the compact object might also be confounded if the central X-ray nebula happens to be occulted by a foreground dust cloud in the debris.

4 The Debris

Most of us believe that supernovae make the heavy elements in the Universe, and this notion gains impressive support from theoretical calculations of supernova nucleosynthesis (see, e.g., [51]). Remarkably, this notion has little empirical support from observations of supernova spectra [39]. What have we learned about the debris of SN1987A from analysis of its spectrum?

4.1 Radioactivity

Certainly, the most important result has been the confirmation that the light of SN1987A comes from the radioactive decays of ^{56}Co and ^{57}Co, which were probably produced originally as ^{56}Ni and ^{57}Ni. To account for the bolometric light curve, we know that these isotopes must be present with masses $M(^{56}\mathrm{Co}) = 0.07\ \mathrm{M}_\odot$ and $M(^{57}\mathrm{Co}) = 0.003\ \mathrm{M}_\odot$, respectively, and this result is confirmed by direct observations of γ-ray and infrared emission lines from these isotopes [38]. Today the bolometric luminosity of SN1987A is dominated by recombination of hydrogen that was ionized at an earlier epoch [17] and heating by positrons from $\sim 2 \times 10^{-4}\ \mathrm{M}_\odot$ of ^{44}Ti [34].

4.2 Clumps and Voids

Another major result concerns the role of instabilities in mixing the debris of the explosion. The unexpectedly early emergence of X-rays and γ-rays showed that a fraction of the newly synthesized ^{56}Ni was expanding with radial velocities up to ~ 4000 km s^{-1}, far greater than would be possible in spherically symmetric explosion models in which the newly synthesized elements remain in concentric spherical layers. Moreover, the emission lines of hydrogen, helium, oxygen, and other elements did not have the flat-topped profiles that would be expected if these elements were excluded from the central regions of the envelope.

The mixing implied by these observations must be due to Rayleigh-Taylor instabilities that occur in the decelerating gas behind the blastwave. But the theory still has trouble in accounting quantitatively for the highest velocity clumps of ^{56}Co, even allowing for the fact that such clumps would be expected to penetrate further than indicated by 2-D simulations [2]. Since the Rayleigh-Taylor instabilities depend on the deceleration following the passage of the initial blast-wave, they only have a few e-folding timescales to grow. Therefore, they cannot account for the observed clumping unless substantial ($\sim 5-10\%$) perturbations already exist in the envelope before the explosion. Such perturbations are the natural consequence of the violent convection that occurs in the envelope of the progenitor.

The dynamical evolution of the envelope does not end when the Rayleigh-Taylor instabilities cease to grow. There is still one more mechanism that can alter the texture of the envelope: the radioactive energy released by the decay of ^{56}Ni. Since the ^{56}Ni clumps are opaque to γ-rays for several times the 8.8d mean lifetime of ^{56}Ni, they will become hotter than their surrounding substrate and will swell up during the first few weeks. As a result, the supernova debris develops a "foamy" texture, in which the iron group elements occupy $\gtrsim 30\%$ of the emitting volume, even though they comprise only $\sim 1\%$ of the mass. This scenario is confirmed by analysis of the light curves of emission lines from the iron group elements [30].

4.3 Thermal Evolution

The envelope of SN1987A cooled fast. By $t = 1$ yr after explosion, the emitting hydrogen had cooled to $T_H(1 \text{ yr}) \approx 4000$ K and was only about 1% ionized. Shortly thereafter, strong CO and SiO emission bands appeared in the infrared spectrum, indicating that the oxygen-rich gas containing these molecules had a temperature $T_{CO}(1 \text{ yr}) \approx 1000$ K [31]. On the other hand, the strength of the [OI] $\lambda\lambda$6300 lines indicates that $T_O(1 \text{ yr}) \approx 3000$ K, implying that the [OI] emission is coming from a warmer zone of oxygen that lacks CO (probably because it contains metastable He* which attacks CO [32]). The infrared emission lines of Fe, Co, and Ni indicate that these elements cooled from $T_{\rm Fe}(1 \text{ yr}) \approx 4000$ K to $T_{Fe}(3 \text{ yr}) \approx 200$ K. Evidently, the emission line spectrum is a composite coming from regions of distinct chemical composition, which have very different temperature histories owing to differences in radiative cooling efficiency. Because their line profiles are similar, we know that clumps having different compositions must share roughly the same volume. The fact that these clumps have different temperature evolution proves that the line-emitting region of SN1987A is mixed macroscopically but not microscopically.

Today, SN1987A is perhaps the coolest optical emission source known to astronomy. Even the hydrogen gas has cooled to $T \lesssim 350$ K [11,53]. In fact, after $t \approx 2$ years, most of the envelope was too cool to emit optically and the emission line spectrum was dominated by lines excited by nonthermal electrons resulting from radioactive energy deposition. The hydrogen-rich gas has ionized a fraction $\sim 10^{-3}$. Since the hydrogen recombination timescale is greater than the age of the supernova, the recombination lines reflect the ionization rate at earlier times and are insensitive to the present-day radioactive energy deposition.

The fact that the emission line spectrum at late times is produced primarily by nonthermal excitation offers the hope to infer the nucleosynthesis yields of elements from the observed line strengths. The idea is simple: at late times the emitting region is optically thin to γ-rays so that all elements are illuminated by a roughly uniform flux of γ-rays. Since the opacity to γ-rays is nearly independent of composition, the radioactive luminosity that is deposited (as nonthermal electrons) in regions of a given composition is nearly proportional to the net mass of gas having that composition. Kozma and Fransson [25] have calculated the nonthermal energy deposition fractions that appear in various emission lines. Therefore, it should be straightforward to infer the ratios of different element groups from the line ratios. Unfortunately, we have failed so far to realize the fruits of this idea. Kozma and Fransson [27,28] find that the theory under-predicts the strength of [OI] $\lambda\lambda$ 6300, 6364 doublet by nearly an order of magnitude if the oxygen-rich gas has a mass $\sim 2\ \mathrm{M}_{\odot}$. Evidently, the theory is still missing some crucial physical ingredient.

4.4 Internal Dust

Dust formed in the debris of SN1987A after $t \sim 450$ days [38]. The dust clouds evidently obscure about half of the far side of the supernova envelope and absorb roughly half of the luminosity emitted by the envelope, re-radiating it at far-infrared ($\gtrsim 5\ \mu$m) wavelengths. The red wings of near-infrared lines are absorbed almost as effectively as those of the optical lines, indicating that the dust clouds, where present, are very opaque.

What is this dust? Since the far infrared spectrum shows no spectral features, we can only make theoretical conjectures. Element abundances do not constrain the possible composition; even primordial gas of *Large Magellanic Cloud (LMC)* composition contains enough mass in refractory elements to make dust clouds of sufficient opacity if such elements could condense. Two conditions must be met for dust to condense: 1) the gas must be cold enough ($\lesssim 1200$ K for Fe_3O_4, $\lesssim 1300$ K for $MgSiO_3$, and $\lesssim 1800$ K for graphite), and 2) dense enough [23,24].

4.5 Lessons from the Emission Line Spectrum

Besides what we have learned about SN1987A itself, our efforts to interpret its emission line spectrum [26,38] have taught us some important lessons about interpreting the spectra of supernovae in general:

Infrared spectra are more useful than optical spectra because our most reliable information about the physical conditions in the envelope of SN1987A comes from its infrared spectrum. The main reason is simple. The luminosities of thermally excited forbidden lines from some ion X are proportional to

$$L_X \propto M_X \exp(-h\nu/kT)\,, \tag{1}$$

where M_X is the net mass of ion X. For optical lines, $h\nu/k \gtrsim 20,000$ K but, as we have seen, the supernova envelope has temperatures $T \lesssim 3000$ K. Since optical lines are so sensitive to temperature, they are very good thermometers but very poor indicators of anything else. Moreover, if the gas temperature is not uniform, optical line emission from cooler gas can be completely masked by that from hotter gas. In contrast, the Boltzmann factor for infrared lines is far less sensitive to temperature. Moreover, the infrared spectrum of SN1987A contains many spectral features that have been most revealing of the conditions in the envelope – for example, the CO bands and the [CoI] 10.2 μm line.

Be careful about inferring element abundances from emission line strengths as we learned from CaII [29]. [CaII] $\lambda\lambda$ 7293, 7324 is prominent in the spectrum of SN1987A (and many other supernovae), not because calcium is particularly abundant but because this line happens to be the most effective channel for hydrogen- and helium-rich gas in supernova envelopes to radiate thermal energy. As long as this is true, the strength of [CaII] $\lambda\lambda$ 7293, 7324 will be almost independent of the calcium abundance. If it is less than cosmic, the temperature of the hydrogen must rise enough so that the line will get rid of the thermal energy that is deposited there by γ-rays. Indeed, the [CaII] $\lambda\lambda$ 7293, 7324 line emission from primordial calcium in the hydrogen- and helium-rich gas completely masks the emission from a much greater mass of calcium that is probably present in Si and Ca-rich clumps in the envelope. The situation is similar for the near-infrared emission lines of Fe, Co, and Ni. For example, for $t \lesssim 2$ yr, the [FeII] 1.26 μm line comes mainly from newly synthesized iron. But thereafter, the iron-rich gas becomes too cool to emit this line, which then comes mainly from a much smaller mass of iron in the hydrogen- and helium-rich gas.

Internal extinction is a problem because even though we can see through the envelope of SN1987A at optical and near-infrared wavelengths, a good fraction of the envelope, especially the far side, is obscured by internal dust clouds. It's quite possible that clumps of some compositions are more likely to be obscured than others. Therefore, although we can construct models for the emission line spectrum evolution that conform fairly well to our preconceived notions of supernova nucleosynthesis, we may go astray if we use observations taken after the dust forms.

5 The Circumstellar Rings

The first evidence for CSM around SN1987A appeared a few months after the outburst in the form of narrow optical and UV emission lines seen with the *International Ultraviolet Explorer (IUE)* (see, e.g., [16] and the chapter by Panagia in this volume). Even before astronomers could image this matter, they could infer that:

- The gas was nearly stationary (from the linewidths).

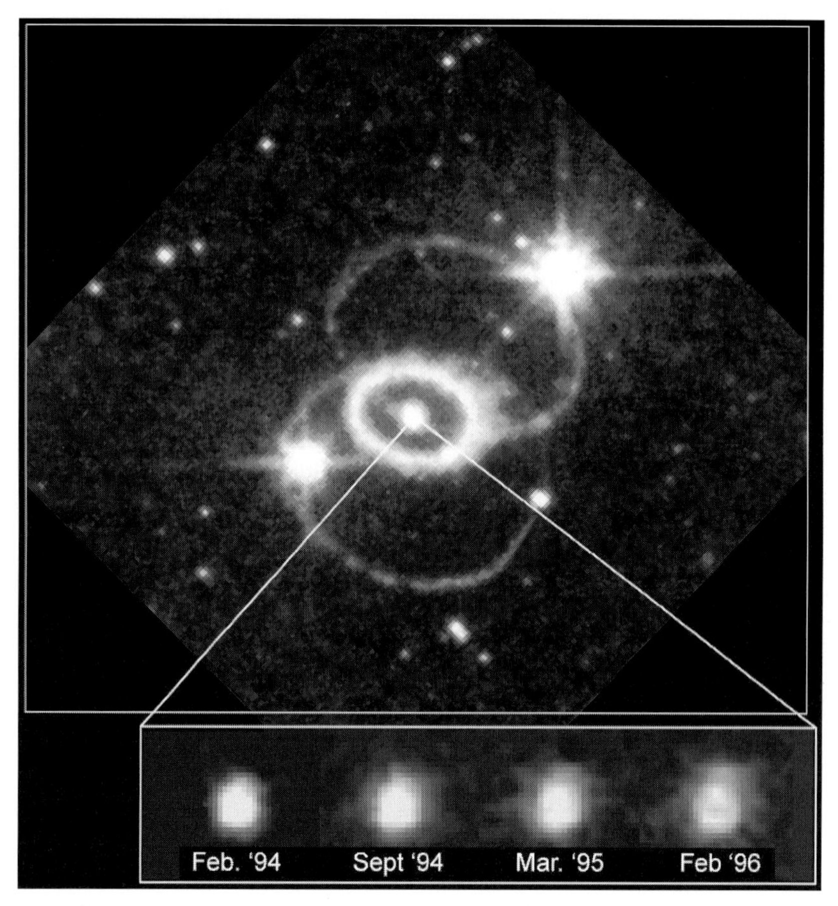

Fig. 1. *HST* image of SN1987A and its circumstellar rings. The inset at the bottom shows the evolution of the glowing center of the supernova debris.

- The gas was probably ejected by the supernova progenitor (because the abundance of nitrogen was elevated).
- The gas was ionized by soft X-rays from the supernova flash (from emission lines of NV and other highly ionized elements in the spectrum).
- The gas was located at a distance of about a light year from the supernova (from the rise time of the light curve of these lines).
- The gas had atomic density $\sim 3 \times 10^3 - 3 \times 10^4$ cm^{-3} (from the fading timescale of the narrow lines) [33].

Figure 1 shows an image of the circumstellar rings of SN1987A taken with the *HST Wide-Field Planetary Camera 2 (WFPC2)*[1]. Dividing the radius of the inner ring (0.67 lt-yr) by the radial expansion velocity of the inner ring (≈ 10 km s^{-1}) gives a kinematic timescale $\approx 20,000$ years since the gas in the ring

[1] See *http://oposite.stsci.edu/pubinfo/PR/97/03.html*

was ejected [13], assuming constant velocity expansion. The more distant outer loops are expanding more rapidly, consistent with the notion that they were ejected at the same time as the inner ring.

The rings observed by *HST* may be only the tip of the iceberg. They are glowing by virtue of the ionization and heating caused by the flash of extreme ultraviolet (EUV) and soft X-rays emitted by the supernova during the first few hours after outburst. But calculations [15] show that this flash was a feeble one. The glowing gas that we see in the triple ring system is probably only the ionized inner skin of a much greater mass of unseen gas that the supernova flash failed to ionize. For example, the inner ring has a glowing mass of only about ~ 0.04 $M_{\odot}$, just about what one would expect such a flash to produce.

In fact, ground-based observations of optical light echoes during the first few years after outburst provided clear evidence of a much greater mass of circumstellar gas within several light years of the supernova that did not become ionized [12,52]. The echoes were caused by scattering of the optical light from the supernova by dust grains in this gas. They became invisible a few years after outburst.

What accounts for this CSM and the morphology of the rings? I suspect that the supernova progenitor was originally a close binary system and that the two stars merged some 20,000 years ago. The inner ring might be the inner rim of a circumstellar disk that was expelled during the merger, perhaps as a stream of gas that spiraled out from the outer Lagrangian (L2) point of the binary system. Then, during the subsequent 20,000 years before the supernova event, ionizing photons and stellar wind from the merged blue supergiant (BSG) star eroded a huge hole in the disk. Finally, the supernova flash ionized the inner rim of the disk, creating the inner ring that we see today.

The binary hypothesis provides a natural explanation of the bipolar symmetry of the system, and may also explain why the progenitor of SN1987A was a BSG rather than a red supergiant (RSG) [45]. Unfortunately, we still lack a satisfactory explanation for the outer loops. If we could only see the invisible CSM that lies beyond the loops, we might have a chance of reconstructing the mass ejection episode.

Fortunately, SN1987A will give us another chance. When the supernova blast-wave hits the inner ring, the ensuing radiation will cast a new light on the CSM. As I describe below, this event is now underway.

6 The Crash Begins

The first evidence that the supernova debris was beginning to interact with CSM came from radio and X-ray observations. As Fig. 2 shows, SN1987A returned as a detectable source of radio and soft X-ray emission about 1200 days after the explosion and has been brightening steadily in both bands ever since. Shortly afterwards, astronomers imaged the radio source with the *Australia Telescope Compact Array (ATCA)* and found that the radio source was an elliptical annulus inside the inner circumstellar ring observed by *HST* (Fig. 7). From subse-

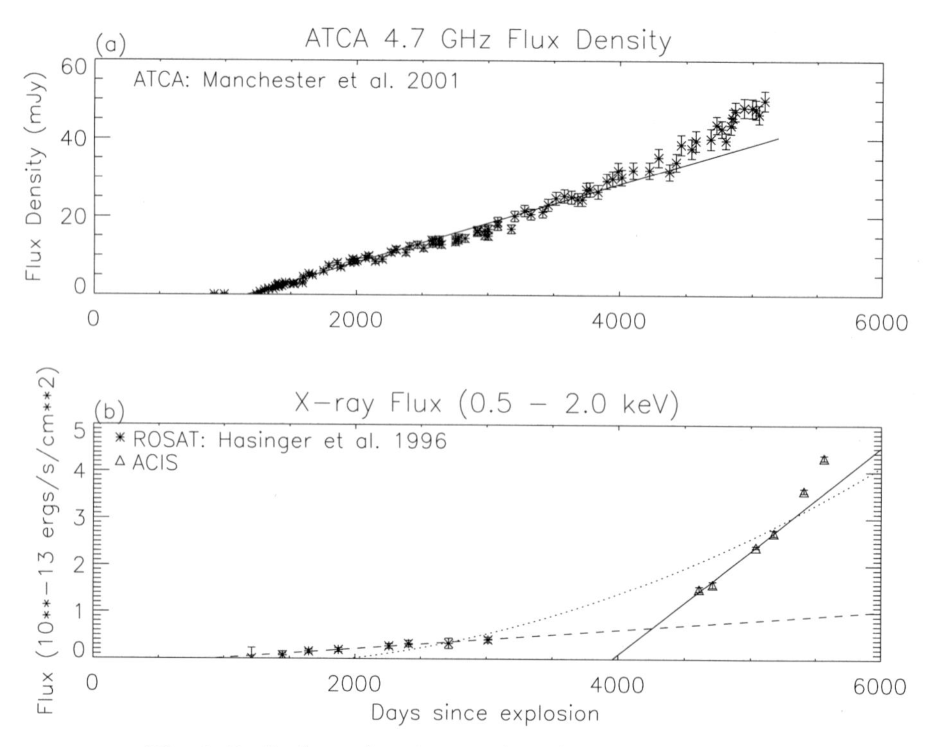

Fig. 2. Radio (upper) and X-ray (lower) light curves [20,37].

quent observations, they found that the annulus was expanding with a velocity $\sim 3,500$ km s^{-1} [19].

The radio emission most likely comes from relativistic electrons accelerated by shocks formed inside the inner ring where the supernova debris struck relatively low density ($n \sim 100$ cm^{-3}) CSM, and the X-ray emission comes from the shocked CSM and supernova debris [8]. Subsequently, Chevalier and Dwarkadas [10] suggested a model for the CSM in which the inner circumstellar ring is the waist of an hourglass-shaped bipolar nebula. The low-density CSM is a thick layer of photoionized gas that lines the interior of the bipolar nebula. The inner boundary of this layer is determined by a balance of the pressure of the hot bubble of shocked stellar wind gas and that of the photoionized layer. In the equatorial plane, the inner boundary of this layer is located at about half the radius of the inner ring. According to this model, the appearance of X-ray and radio emission at ~ 1200 days marks the time when the blastwave first entered the photoionized layer.

7 The Reverse Shock

Following Chevalier and Dwarkadas [10], Borkowski et al. [4] developed a more detailed model to account for the X-ray emission observed from SN1987A. They

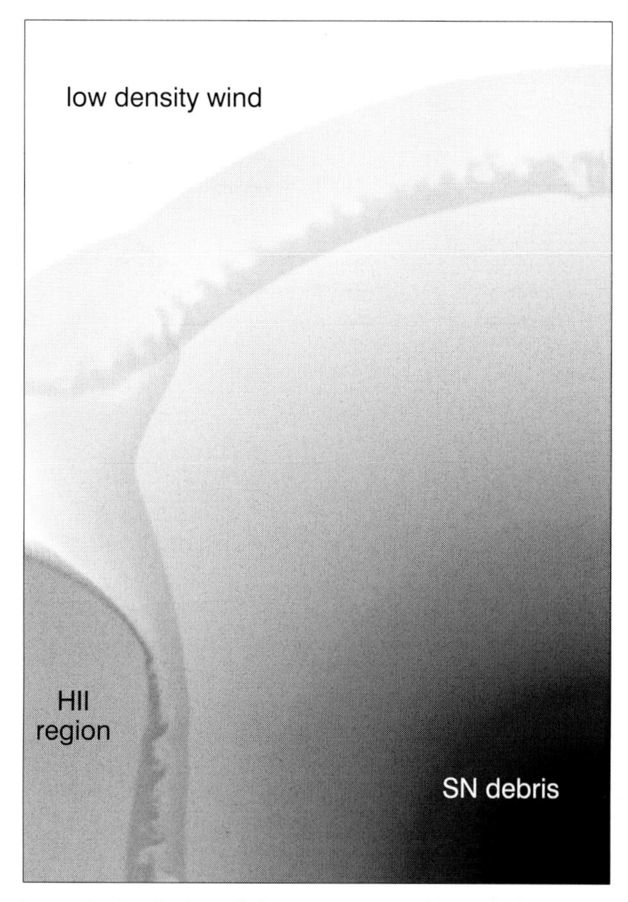

Fig. 3. Hydrodynamic simulation of the interaction of SN1987A debris with CSM [42]. The supernova center is at the lower right corner and the grey scale indicates density. The figure is cylindrically symmetric about the right boundary and the photoionized layer is assumed to be a thick torus (labeled HII region). A blastwave enters the photoionized layer (HII region), while a reverse shock arrests the free expansion of the supernova debris. Between the blastwave and the reverse shock are layers of shocked HII region and shocked supernova debris, separated by an unstable contact discontinuity.

used a 2-D hydro code to simulate the impact of the outer atmosphere of the supernova with an idealized model for the photoionized layer. They found a good fit to the *Röntgensatellit (ROSAT)* observations with a model in which the thickness of the photoionized layer was about half the radius of the inner ring and the layer had atomic density $n_0 \approx 150\ \mathrm{cm}^{-3}$.

The same model predicted that Lyα and Hα emitted by hydrogen atoms crossing the reverse shock should be detectable with the *HST Space Telescope Imaging Spectrograph (STIS)*. Then, in May 1997, only three months after these predictions were published, the first *STIS* observations of SN1987A were made and broad ($\Delta V \approx \pm 12,000$ km s^{-1}) Lyα emission lines were detected [49]. Within the observational uncertainties, the flux was exactly as predicted [40].

One might at first be surprised that such a theoretical prediction of the Lyα flux would be on the mark, given that it was derived from a hydrodynamical model based on very uncertain assumptions about the density distribution of circumstellar gas. But, on further reflection, it is not so surprising because the key parameter of the hydrodynamical model, the density of the circumstellar gas, was adjusted to fit the observed X-ray flux. Since the intensity of Lyα is derived from the same hydrodynamical model, the ratio of Lyα to the X-ray flux is determined by the ratio of cross sections for atomic processes, independent of the details of the hydrodynamics.

The broad Lyα and Hα emission lines are not produced by recombination because the emission measure of the shocked gas is far too low to produce detectable Lyα and Hα by recombination. Instead, the lines are produced by neutral hydrogen atoms in the supernova debris as they cross the reverse shock and are excited by collisions with electrons and protons in the shocked gas. Since the cross sections for excitation of the $n \geq 2$ levels of hydrogen are nearly equal to the cross sections for impact ionization, about one Lyα photon is produced for each hydrogen atom that crosses the shock. Thus, the observed flux of broad Lyα is a direct measure of the flux of hydrogen atoms that cross the shock. Moreover, since the outer supernova envelope is expected to be nearly neutral, the observed flux is a measure of the mass flux across the shock.

The fact that the Lyα and Hα lines are produced by excitation at the reverse shock gives us a powerful tool to map this shock. Since any hydrogen in the supernova debris is freely expanding, its line-of-sight velocity, $V_{\|} = \epsilon/t$, where ϵ is its depth measured from the mid-plane of the debris and t is the time since the supernova explosion. Therefore, the Doppler shift of the Lyα line will be directly proportional to the depth of the reverse shock: $\Delta\lambda/\lambda_0 = \epsilon/ct$ and by mapping the Lyα or Hα emission with *STIS*, we can generate a 3-dimensional image of the reverse shock.

Figure 4 illustrates this procedure. Panel (a) shows the location of the slit superposed on an image of the inner circumstellar ring, with the near (N) side of the tilted ring on the lower left. Panel (b) shows the actual *STIS* spectrum of Lyα from this observation. The slit is black due to geocoronal Lyα emission. The bright blue-shifted streak of Lyα extending to the left of the lower end of the slit comes from hydrogen atoms crossing the near side of the reverse shock, while the fainter red-shifted streak at the upper end of the slit comes from the far side of the reverse shock.

From this and similar observations with other slit locations we have constructed a map of the reverse shock surface, shown in panel (c). Note that the emitting surface is an annulus that lies inside the inner circumstellar ring, as would be expected from a hydrodynamic model such as that illustrated in Fig. 3. Presumably, the reverse shock in the polar directions lies at a greater distance from the supernova, where the flux of atoms in the supernova debris is too low to produce detectable emission. Panel (d) is a model of the *STIS* Lyα spectrum

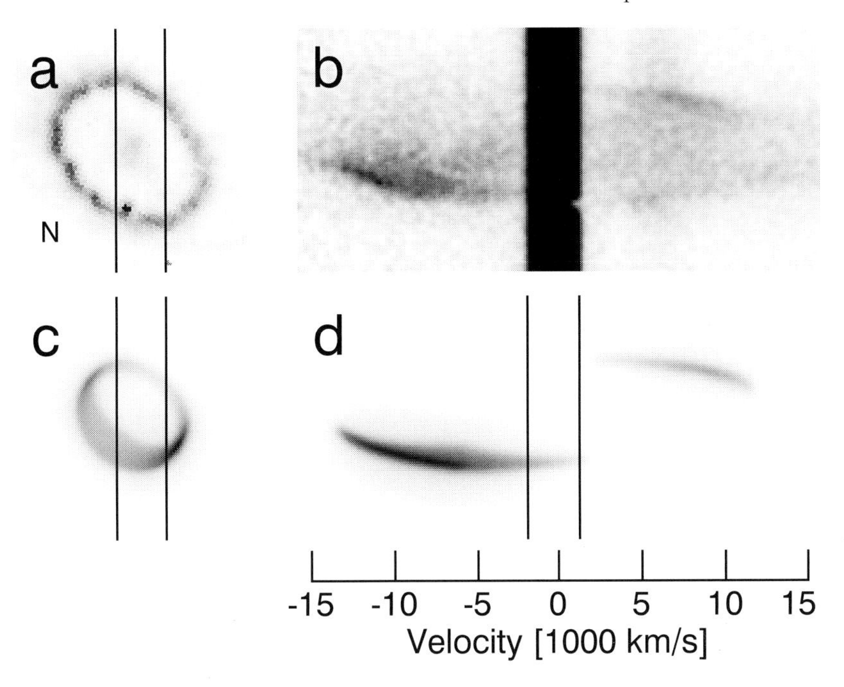

Fig. 4. *STIS* spectrum of Lyα emission from the reverse shock [40]

that would be expected from hydrogen atoms crossing the shock surface illustrated in panel (c). By comparing such model spectra with the actual spectra (e.g., panel (b)), we may refine our model of the shock surface [42].

Note that the broad Lyα emission is much brighter on the near (blue-shifted) side of the debris than on the far side and so is the reconstructed shock surface. There is one obvious reason why this might be so: the blue-shifted side of the reverse shock is nearer to us by several light-months, and so we see the emission from the near side as it was several months later than that from the far side. Since the flux of atoms across the reverse shock is increasing, the near side should be brighter. Unfortunately, this explanation fails quantitatively. The observed asymmetry is several times greater than can be explained by light-travel time delays. Moreover, the asymmetry is much greater in Lyα than it is in Hα. Recently, Michael et al. [42] have proposed that most of the asymmetry in Lyα can be attributed to resonant scattering of Lyα in the un-shocked supernova debris, which tends to reflect the Lyα emission in the radially outward direction. However, the Hα emission line from the reverse shock is not susceptible to resonant scattering, so the ($\sim 30\%$) asymmetry in Hα must be attributed in part to real asymmetry in the supernova debris. As we shall see, observations at radio and X-ray wavelengths also provide compelling evidence for asymmetry of the supernova debris.

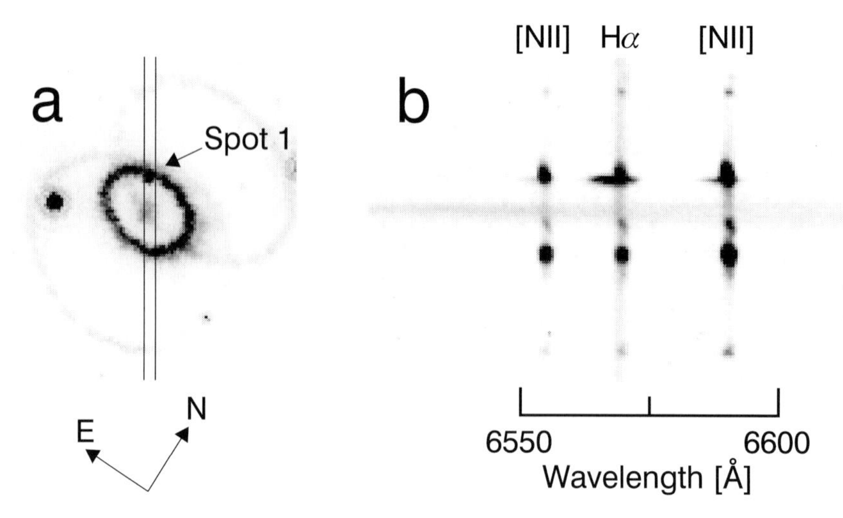

Fig. 5. Spectrum of Spot 1. (a) Slit orientation on the image of SN1987A's triple ring system; (b) Section of the *STIS* G750M spectrum.

8 The Hot Spots

One can estimate the time that the blastwave should strike the inner circumstellar ring from self-similar solutions for the hydrodynamics of a freely expanding stellar atmosphere striking a CSM [7]. Stellar atmosphere models give a good fit to the spectrum of SN1987A during the early photospheric phase with a stellar atmosphere having a power-law density law $\rho(r,t) = At^{-3}(r/t)^{-9}$ [14,48]. If this atmosphere strikes a CSM having a uniform density n_0, the blastwave will propagate according to the law $R_B(t) \propto A^{1/9} n_0^{-1/9} t^{2/3}$. For such a model, the time of impact can be estimated from the equation

$$t \propto A^{-1/6} n_0^{1/6} . \tag{2}$$

The coefficient, A, of the supernova atmosphere density profile can determined from the fit to the photospheric spectrum, leaving the density, n_0, of the gas between the supernova and the circumstellar ring as the main source of uncertainty. With various assumptions about the density distribution of this gas, predictions of the time of first contact ranged from 1996 to 2005 [10,35,36].

In April 1997, Sonneborn et al. [49] obtained the first *STIS* spectrum of SN1987A with the 2×2 arcsecond aperture. Images of the circumstellar ring were seen in several optical emission lines. No Doppler velocity spreading was evident in the ring images except at one point, located at position angle $PA = 29°$ (E of N), which we now call "Spot 1," where a Doppler-broadened streak was seen in Hα and other optical lines.

Figure 5 [46] shows a portion of a more recent (March 1998) *STIS* spectrum of Spot 1, where one can see vertical pairs of bright spots corresponding to emission from the stationary ring at Hα and [NII] $\lambda\lambda$ 6548, 6584. (One also sees three

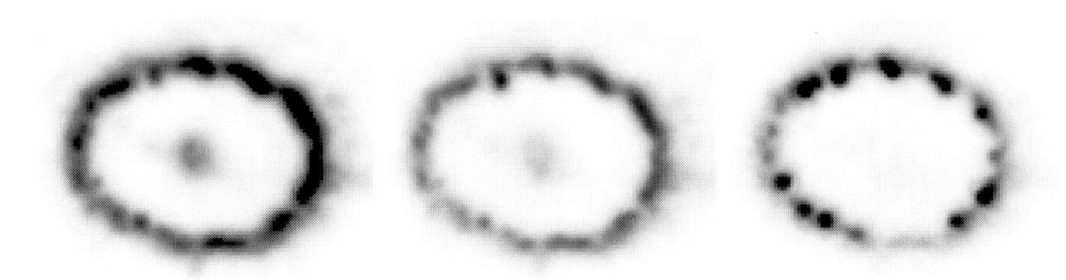

Fig. 6. *WFPC* narrow-band Hα images of the inner ring (left: February 1996; center: February 1998; right: May 2002). Both the inner ring and the central supernova debris have faded monotonically, while Spot 1, located at $PA = 29°$, has evidently brightened by February 1998. By May 2002, 16 hot spots have appeared.

more fainter spots at each wavelength, where the outer loops cross the slit, and a broad horizontal streak at the center due to the Hα emission from the rapidly expanding inner debris. The broad Hα from the reverse shock is invisible at this dispersion.) The emission lines are broadened (with FWHM $\approx$ 250 km s^{-1}) and blue-shifted (with $\Delta V \approx -80$ km s^{-1}) at the location of Spot 1, which is located slightly inside the stationary ring.

Spot 1 evidently marks the location where the supernova blastwave first touches the dense circumstellar ring. When a blastwave propagating with velocity $V_b \approx 4,000$ km s^{-1} through a CSM with density $n_0 \approx 150$ cm^{-3} encounters the ring, having density $n_r \approx 10^4$ cm^{-3}, one expects the transmitted shock to propagate into the ring with $V_r \approx (n_0/n_r)^{1/2} V_b \approx 500$ km s^{-1} if it enters at normal incidence, and more slowly if it enters at oblique incidence. Since Spot 1 is evidently a protrusion, a range of incidence angles, and hence of transmitted shock velocities and directions, can be expected. Obviously, the line profiles will be sensitive to the geometry of the protrusion. Since the protrusion is on the near side of the ring and is being crushed by the entering shocks, most of the emission will be blue-shifted, as is observed. But part of the emission is red-shifted because it comes from oblique shocks entering the far side of the protrusion.

Looking back at *HST* archival data, one finds that Spot 1 was first detected in March 1995. Spot 2, did not appear until November 1998, but shortly thereafter several more spots appeared [50]. By May 2002, 16 spots were evident. The hot spots are brightening rapidly, with doubling timescales ranging from a few months to 2 years.

The emission line spectrum of Spot 1 resembles that of a radiative shock in which the shocked gas has had time to cool from its post-shock temperature of $T_1 \approx 1.6 \times 10^5 [V_r/(100 \text{ km s}^{-1})]^2$ K to a final temperature of $T_f \approx 10^4$ K or less. As the shocked gas cools, it is compressed by a density ratio $n_f/n_r \approx (T_1/T_f) \approx 160[V_r/(100 \text{ km s}^{-1})]^2 \, [T_f/(10^4\text{K})]^{-1}$. We see evidence of this compression in the observed ratios of forbidden lines, such as [NII] $\lambda\lambda$ 6548, 6584 and [SII] $\lambda\lambda$ 6717, 6731, from which we infer electron densities in the range $n_e \sim 10^6$ cm^{-3}using standard nebular diagnostics [46].

The fact that the shocked gas in Spot 1 was able to cool and form a radiative layer within a few years sets a lower limit, $n_r \gtrsim 10^4$ cm^{-3}, on the density of unshocked gas in the protrusion. Given that limit, we can estimate an upper limit on the emitting surface area of Spot 1, from which we infer that Spot 1 should have an actual size no greater than about one pixel on the *WFPC2*. This result is consistent with the imaging observations.

The cooling timescale of shocked gas is sensitive to the postshock temperature, hence the shock velocity. For $n_r = 10^4$ cm^{-3}, shocks faster than 250 km s^{-1} will not be able to radiate and form a cooling layer within a few years. It is quite possible that such fast non-radiative shocks are present in the protrusions. For example, I estimated above that a blastwave entering the protrusion at normal incidence might have a velocity ~ 500 km s^{-1}. Faster shocks would be invisible in optical and UV line emission, but we are probably seeing evidence of such shocks in soft X-rays (see Sect. 9). We would still see the optical and UV line emission from the slower oblique shocks on the sides of the protrusion, however.

We have attempted to model the observed emission line spectrum of Spot 1 with a radiative shock code [46]. Our efforts to date have met with only partial success. A model that fits the UV emission line spectrum under-predicts the intensities of the optical emission lines by a factor $\sim$ 3. This is perhaps not surprising, given the complexity of the hydrodynamics. It is known, for example, that radiative shocks are subject to violent thermal instabilities [22], which we have not included in our initial attempts to model the shock emission.

9 The X-ray Source

As I have already mentioned in Sect. 6, we believe that the X-ray emission from SNR1987A seen by *ROSAT* (Fig. 2) comes from the hot shocked gas trapped between the supernova blastwave and the reverse shock. But, with its 10 arcsecond angular resolution, *ROSAT* was unable to image this emission; nor was *ROSAT* able to obtain a spectrum.

Our ability to analyze the X-rays from SNR1987A advanced dramatically with the launch of the *Chandra X-ray Observatory – AXAF (CHANDRA)* [6,44]. Figure 7 is a montage of images of SNR1987A, showing, for two epochs, the X-ray images in three bands. We see immediately that the X-ray images have brightened and changed substantially in the 18 month interval between these two observations. In the soft (0.3 – 0.8 keV) band, the locations of the bright X-ray emission correlate fairly well with the optical hot spots (see, e.g., Fig. 6). This fact suggests that the softer X-rays are coming from shocks where the blastwave is entering the inner ring. But we can be sure that the X-rays do not come from exactly the same shocks as the optical hot spots. As we discussed in Sect. 8, the shocks responsible for the optical emission are too slow to emit X-rays, but we would not be surprised if faster shocks are present in the same interaction region.

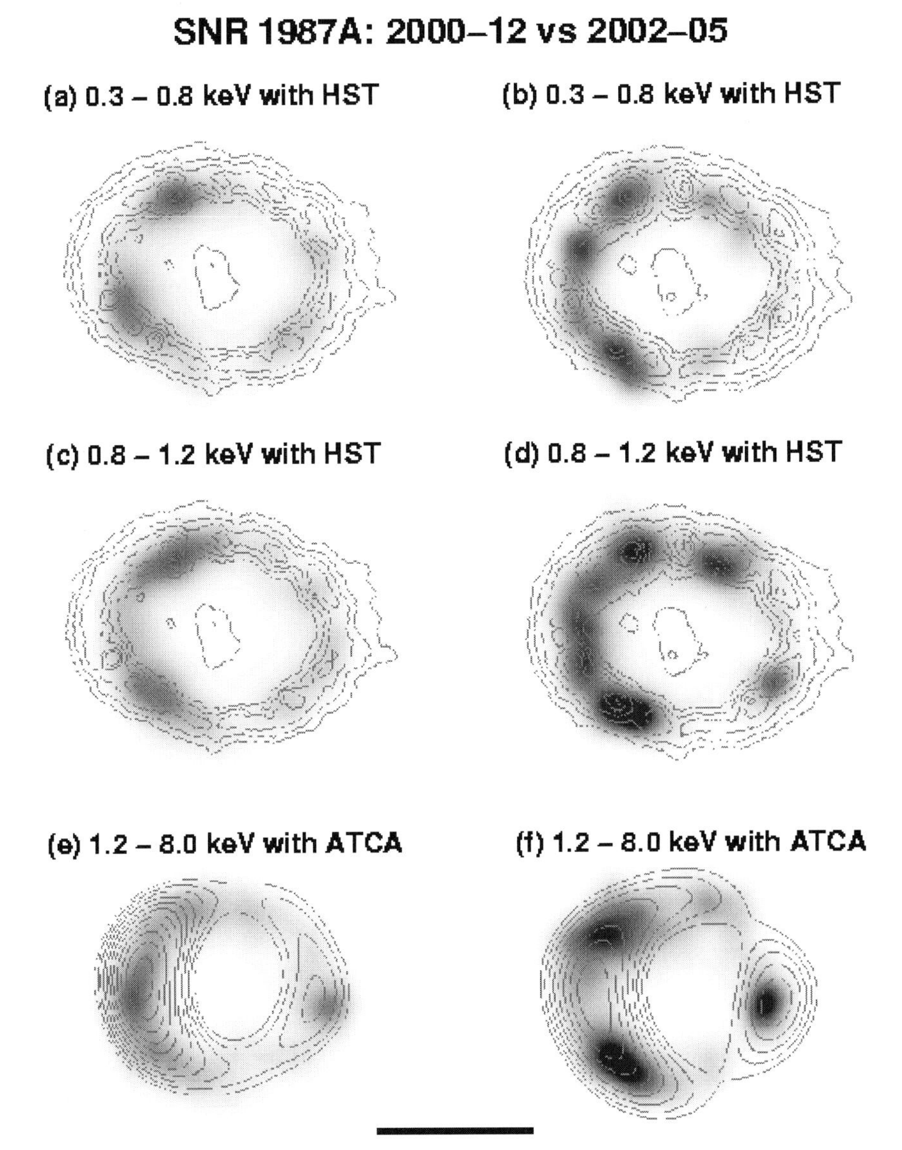

Fig. 7. Optical, radio and X-ray images of SNR1987A. North is to the top and East is to the left. The images on the left (a, c, e) represent observations of December 2000, while those on the right (b,d,f) represent observations of May 2002. The gray scale images represent the X-ray brightness in three bands. The contour lines in the top four images (a - d) represent the optical ring as seen by *HST*, while those in the bottom two represent the radio image as seen by *ATCA*.

In the hard (1.2–8 keV) band, the locations of bright X-ray emission correlate better with the images from the *Australia Telescope Compact Array (ATCA)* radio array than with the optical hot spots. This fact suggests that the harder X-rays are produced primarily by the hotter gas between the reverse shock and the blastwave (see, e.g., Fig. 3). We presume that the radio emission would correlate with the harder X-rays because the relativistic electrons responsible for the non-thermal radio emission have energy density proportional to that in the X-ray emitting gas and reside in roughly the same volume.

The fact that the X-ray and radio images are both brighter on the East (left) side than on the West could be explained by a model in which either: 1) the circumstellar gas inside the inner ring had greater density toward the East, or 2) the outer supernova debris had greater density toward the East. But the fact that most of the hot spots appeared first on the East side favors the latter hypothesis. If the circumstellar gas had greater density toward the East side and the supernova debris were symmetric, the blastwave would have propagated further toward the West side, and the hot spots would have appeared there first.

This conclusion is also supported by observations of Hα and LyαHα emission from the reverse shock (Sect. 7), which show that the flux of mass across the reverse shock is greater on the West side.

These observations highlight a new puzzle about SN1987A: why was the explosion so asymmetric? We might explain a lack of spherical symmetry by rapid rotation of the progenitor, but how do we explain a lack of azimuthal symmetry?

With the grating spectrometer on *CHANDRA*, we have also obtained a spectrum of the X-rays from SNR1987A, shown in Fig. 8 [41]. It is dominated by emission lines from helium- and hydrogen-like ions of O, Ne, Mg, and Si, as well as a complex of Fe-L lines near 1 keV, as predicted [5]. The characteristic electron temperature inferred from the spectrum, $kT_e \sim 3$ keV, is much less than the proton temperature, $kT_p \sim 30$ keV for a blastwave propagating with $V_b \approx 4,000$ km s^{-1} and that inferred from the widths of the X-ray emission lines. This result is a consequence of the fact that Coulomb collisions are too slow to raise the electron temperature to equilibrium with the ions.

The *CHANDRA* observations (Fig. 2) show that the current X-ray flux from SNR1987A is more than 3 times the value that would be estimated by extrapolating the *ROSAT* light curve to December 2001 (Fig. 2). The acceleration in the brightening rate suggests that most of the X-rays are now coming from the impact of the blastwave with the ring. The X-ray flux is expected to increase by another factor $\sim 10^2$ during the coming decade as the blastwave overtakes the inner circumstellar ring [5].

10 The Future

SNR1987A has been tremendously interesting so far, but the best is yet to come. During the next ten years, the blastwave will overtake the entire circumstellar ring. More hot spots will appear, brighten, and eventually merge until the entire

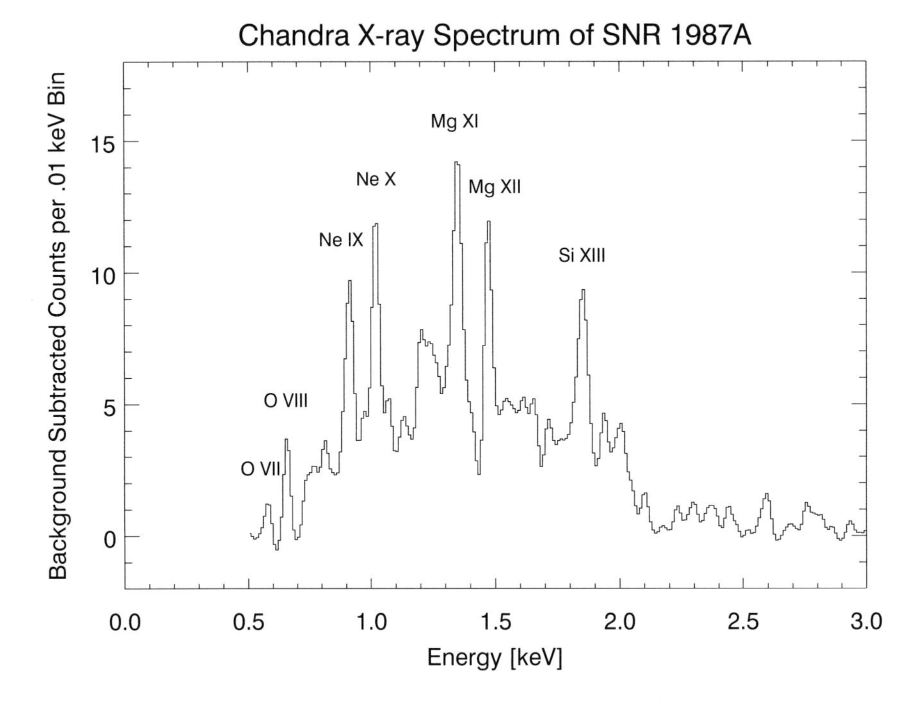

Fig. 8. X-ray spectrum of SNR1987A.

ring is blazing brighter than Spot 1. We expect that the Hα flux from the entire ring will increase to $F_{\mathrm{H}\alpha} \gtrsim 3 \times 10^{-12}$ erg cm^{-2} s^{-1}, or $\gtrsim$ 30 times brighter than it is today and that the flux of UV lines will be even greater [36].

As we have already begun to see, observations at many wavelength bands are needed to tell the entire story of the birth of SNR1987A. Fortunately, powerful new telescopes and technologies are becoming available just in time to witness this event.

Large ground-based telescopes equipped with adaptive optics will provide excellent optical and infrared spectra of the hot spots. We need to observe profiles of several emission lines at high resolution in order to unravel the complex hydrodynamics of the hot spots. These telescopes also offer the exciting possibility to image the source in infrared coronal lines of highly ionized elements (e.g., [Si IX] 2.58, 3.92 μm and [Si X] 1.43 μm) that may be too faint to see with the *HST*. Observations in such lines will complement X-ray observations to measure the physical conditions in the very hot shocked gas.

The observations with the *ATCA* have given us our first glimpse of shock acceleration of relativistic electrons in real time, but the angular resolution of *ATCA* is not quite good enough to allow a detailed correlation of the radio image with the optical and X-ray images. This correlation will become possible several years from now when the *Atacama Large Millimeter Array (ALMA)* is completed. Such observations will give us a unique opportunity to test our theories of relativistic particle acceleration by shocks.

Of course, we should continue to map the emission of fast Lyα and Hα from the reverse shock with *STIS*. Such observations give us a three-dimensional image of the flow of the supernova debris across the reverse shock, providing the highest resolution map of the asymmetric supernova debris. We expect this emission to brighten rapidly, doubling on a timescale $\sim$ 1 year. Most exciting, such observations will give us an opportunity to map the distribution of nucleosynthesis products in the supernova debris. We know that the debris has a heterogeneous composition. The early emergence of γ-rays from SN1987A showed that some of the newly synthesized ^{56}Co (and probably also clumps of oxygen and other elements) were mixed fairly far out into the supernova envelope by instabilities following the explosion [38]. When such clumps cross the reverse shock, the fast Hα and Lyα lines will vanish at those locations, to be replaced by lines of other elements. If we keep watching with *STIS*, we should see this happen during the coming decade.

The shocks in the hot spots are surely producing ionizing radiation, roughly half of which will propagate ahead of the shock and ionize heretofore invisible material in the rings. The effects of this precursor ionization will soon become evident in the form of narrow cores in the emission lines from the vicinity of the hot spots.

In Sect. 5 I pointed out that the circumstellar rings of SN1987A represent only the inner skin of a much greater mass of CSM, and that we obtained only a fleeting glimpse of this matter through ground-based observations of light echoes. The clues to the origin of the circumstellar ring system lie in the distribution and velocity of this matter, if only we could see it clearly. Fortunately, SNR1987A will give us another chance. Although it will take several decades before the blastwave reaches the outer rings, the impact with the inner ring will eventually produce enough ionizing radiation to cause the unseen matter to become an emission nebula. We have estimated [36] that the fluence of ionizing radiation from the impact will equal the initial ionizing flash of the supernova within a few years after the ring reaches maximum brightness. I expect that the circumstellar nebula of SNR1987A will be in full flower within a decade. In this way, SN1987A will be illuminating its own past.

References

1. W.D. Arnett, J.N. Bahcall, R.P. Kirshner, S.E. Woosley: Ann. Rev. Astron. Astrophys. **27**, 629 (1989)
2. M.M. Basko: Astrophys. J. **425**, 264 (1994)
3. H.A. Bethe, P. Pizzochero: Astrophys. J. Lett. **350**, L33 (1990)
4. K.J. Borkowski, J.M. Blondin, R. McCray: Astrophys. J. Lett. **476**, L31 (1997)
5. K. Borkowski, J.M. Blondin, R. McCray: Astrophys. J. **477**, 281 (1997)
6. D.N. Burrows et al.: Astrophys. J. Lett. **543**, L149 (2000)
7. R. Chevalier: Astrophys. J. Lett. **476**, L31 (1982)
8. R.A. Chevalier: Nature **355**, 617 (1992)

9. R.A. Chevalier, C. Fransson: Astrophys. J. **395**, 540 (1992)
10. R.A. Chevalier, V.I. Dwarkadas: Astrophys. J. Lett. **452**, L45 (1995)
11. N.N. Chugai, R.A. Chevalier, R.P. Kirshner, P. Challis: Astrophys. J. **483**, 925 (1997)
12. A.P.S. Crotts, W.E. Kunkel, P.J. McCarthy: Astrophys. J. Lett. **347**, L61 (1989)
13. A. Crotts, S.R. Heathcote: Nature **350**, 683 (1991)
14. R. Eastman, R.P. Kirshner: Astrophys. J. **347**, 771 (1989)
15. L. Ensman, A. Burrows: Astrophys. J. **393**, 742 (1992)
16. C. Fransson, A. Cassatella, R. Gilmozzi, R.P. Kirshner, N. Panagia, G. Sonneborn, W. Wamsteker: Astrophys. J. **336**, 429 (1989)
17. C. Fransson, C. Kozma: Astrophys. J. **408**, L25 (1993)
18. C.L. Fryer, S.A. Colgate, S. A., P.A. Pinto: Astrophys. J. **511**, 885 (1996)
19. B.M. Gaensler, R.N. Manchester, L. Staveley-Smith, V. Wheaton, A.K. Tzioumis, J.E. Reynolds, M.J. Kesteven: ASP Conf. Proc. **199**, 449 (2000)
20. G. Hasinger, B. Aschenbach, J. Truemper: Astron. Astrophys. **312**, L9 (1996)
21. J.C. Houck, R.A. Chevalier: Astrophys. J. **376**, 234 (1991)
22. D.E. Innes, J.R. Giddings, S.A.E.G. Falle: Mon. Not. R. Astron. Soc. **226**, 67 (1987)
23. T. Kosaza, H. Hasegawa, K. Nomoto: Astrophys. J. Lett. **346**, L81 (1989)
24. T. Kosaza, H. Hasegawa, K. Nomoto: Astron. Astrophys. **249**, 474 (1991)
25. C. Kozma, C. Fransson: Astrophys. J. **390**, 602 (1992)
26. C. Kozma, C. Fransson: Astrophys. J. **497**, 431 (1998)
27. C. Kozma, C. Fransson: Astrophys. J. **496**, 946 (1998)
28. C. Kozma, C. Fransson: Astrophys. J. **497**, 431 (1998)
29. H,-W. Li, R. McCray: Astrophys. J. **405**, 730 (1993)
30. H.-W. Li, R. McCray, R. Sunyaev: Astrophys. J. **419**, 824 (1993)
31. W. Liu, A. Dalgarno, S. Lepp: Astrophys. J. **396**, 679 (1992).
32. W. Liu, A. Dalgarno: Astrophys. J. **454**, 472 (1995)
33. P. Lundqvist, C. Fransson: Astrophys. J. **464**, 924 (1996)
34. P. Lundqvist, C. Kozma, J. Sollerman, C.A. Fransson: Astron. Astrophys. **374**, 629 (2001)
35. D. Luo, R. McCray: Astrophys. J. **372**, 194 (1991)
36. D. Luo, R. McCray, J. Slavin: Astrophys. J. **430**, 264 (1994)
37. R.N. Manchester, B.M. Gaensler, V.C. Wheaton, L. Staveley-Smith, A.K. Tzioumis, N.S. Bizunok, M.J. Kesteven, J.E. Reynolds: Pub. Astron. Soc. Australia **19**, 207 (2002)
38. R. McCray: Ann. Rev. Astron. Astrophys. **31**, 175 (1993)
39. R. McCray: ASP Conf. Proc. **99**, 273 (1996)
40. E. Michael, R. McCray, K.J. Borkowski, C.S.J. Pun, G. Sonneborn: Astrophys. J. Lett. **492**, L143 (1998)
41. E. Michael et al.: Astrophys. J. **574**, 166 (2002)
42. E. Michael et al.: Astrophys. J. , submitted (2003)
43. J. Middleditch et al.: New Astron. **5**, 243 (2000)
44. S. Park, D.N. Burrows, G.P. Garmire, J.A. Nousek, R. McCray, E. Michael, S. Zhekov: Astrophys. J. **567**, 314 (2002)
45. P. Podsiadlowski: Pub. Astron. Soc. Pacific **104**, 717 (1992)
46. C.S.J. Pun et al.: Astrophys. J. **572**, 906 (2002)
47. C. Schaab, F. Weber, M.K. Weigel, N.K. Glendenning: Nucl. Phys. A **605**, 531 (1996)
48. W. Schmütz, D.C. Abbott, R.S. Russell, W.-R. Hamann, U. Wessolowski: Astrophys. J. **355**, 255 (1990)

49. G. Sonneborn et al.: Astrophys. J. Lett. **492**, L139 (1998)
50. B.E.K. Sugerman, S.S. Lawrence, A.P.S. Crotts, P. Bouchet, S.R. Heathcote: Astrophys. J. **572**, 209 (2002)
51. F.-K. Thielemann, K. Nomoto, M. Hashimoto: Astrophys. J. **460**, 408 (1995)
52. L. Wang, E.J. Wampler: Astron. Astrophys. **262**, L9 (1992)
53. L. Wang et al.: Astrophys. J. **466**, 998 (1996)

Part II

Supernovae to γ-Ray Bursters

SN1998bw and Hypernovae

Koichi Iwamoto[1], Ken'ichi Nomoto[2], Paolo A. Mazzali[3], Takayoshi Nakamura[2], and Keiichi Maeda[2]

[1] Department of Physics, College of Science and Technology, Nihon University, Tokyo 101-8308, Japan
[2] Department of Astronomy and Research Center for the Early Universe, School of Science, The University of Tokyo, Tokyo 113-0033, Japan
[3] Osservatorio Astronomico di Trieste, Trieste, Italy

Abstract. SN1998bw is a peculiar type Ic supernova. Its peak brightness is almost ten times greater than typical type Ic in optical and even $10 - 100$ times brighter than other radio bright type Ib/c supernovae in radio bands. The extremely broad features in the optical spectra indicate high velocities of ejected matter $\sim 3 \times 10^4$ km s^{-1} which, combined with the breadth of light curves, lead to an estimated kinetic energy of explosion of $\sim 2 - 5 \times 10^{52}$ erg under the assumption of spherical symmetry. This class of hyperenergetic supernovae is termed "hypernova". There are some observational indications of possible asymmetry, which might reduce this energy requirement. However, it is still likely that SN1998bw is a hypernova. SN1998bw is the first supernova for which a possible link to a γ-ray burst was suggested due to its angular and time coincidence with GRB980425. The high radio luminosity and fast evolution of the radio light curves suggest the existence of a relativistic blast wave in SN1998bw, which might be related to the γ-ray burst directly or indirectly. Several other several hypernova candidates have been discovered or reinterpreted since SN1998bw. The unusual properties of these objects may require reconsideration of theories of stellar explosion mechanisms.

1 A Brief Summary of Discovery and Early Observations

The γ-ray burst (GRB) GRB980425 was detected on 25.91 April 1998 by the *Gamma-Ray Burst Monitor (GRBM)* on board *BeppoSAX* [91] and by the *Burst and Transient Source Experiment (BATSE)* of the *Compton Gamma-Ray Observatory (CGRO)* [45]. The burst position was quickly localized by the *Wide Field Cameras (WFCs)* of *BeppoSAX* and an afterglow was searched for in the *WFCs* error box of GRB980425. Subsequent observations by the *BeppoSAX Narrow Field Instruments (NFI)* led to detection of two unidentified X-ray sources in the error box [76,77]. Then, only 0.9 and 3 days later possible candidates for the optical and radio counterparts were discovered at a position consistent with one of the *NFI* sources by the *European Southern Observatory (ESO) New Technology Telescope (NTT)* [24–26] and by the *Australia Telescope Compact Array (ATCA)* [47,110,111], respectively.

The candidate for the optical counterpart did not resemble afterglows of other GRBs. The light curve had a maximum as broad as ~ 20 days and was followed by an exponential tail, which is different from GRB afterglows that show power-law declines, but quite similar to supernova (SN) light curves powered by radioactive decay of ^{56}Ni and ^{56}Co [24–26,61,75]. The optical spectra

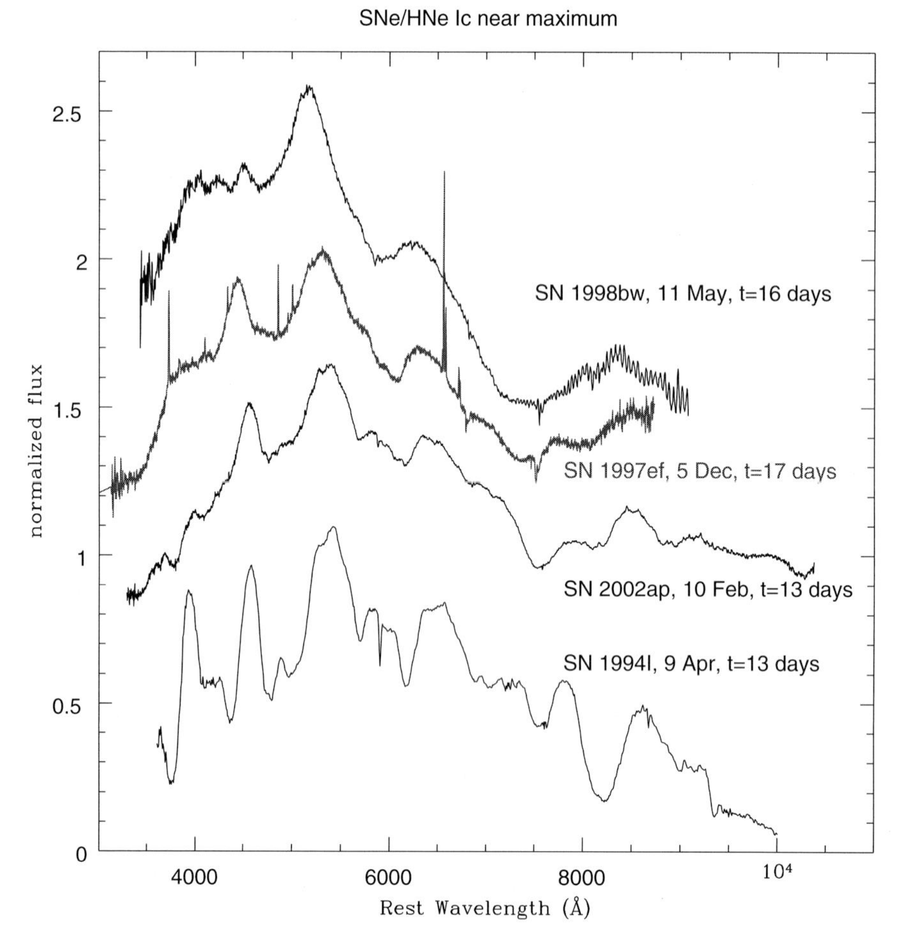

Fig. 1. The spectrum of SN1998bw at day 16 is compared with three other type Ic SNe, SN1994I, SN1997ef, and SN2002ap at comparable phases [60].

appeared to be thermal like in SNe, instead of the nonthermal synchrotron expected for GRB afterglows (Fig. 1). Therefore, the object was soon recognized as a supernova, SN1998bw [24–26,50]. In the spectra there were few lines, only some broad features at 3500 – 5000 Å and broad bumps around 5000, 6200 and 8000 Å [24–26,50]. The absence of any hydrogen lines and a strong Si II 6355 line indicated that it was neither a type II nor a type Ia SN, and thus that it should be either a type Ib [84] or a type Ic [50]. The lack of any clear He I features characteristic of type Ib eventually led to the conclusion that SN1998bw should be classified as a type Ic, despite some deviation from previously known type Ic characteristics [18,74]. SNIc are thought to be the result of the core-collapse induced explosion of a C+O star, which is a massive star that has lost all of its H-rich envelope and almost its entire of He layer [69].

SN1998bw is located in a spiral arm of the barred spiral galaxy ESO184-G82, for which a distance is measured as ~ 38 Mpc from its redshift velocity of 2550 km s^{-1} [24–26,50,101] and a Hubble constant $H_0 = 65$ km s^{-1} Mpc^{-1}. At such a distance, the peak absolute luminosity is estimated to be $\sim 10^{43}$ erg s^{-1}, which is about ten times brighter than typical core-collapse SNe (type Ib/c or type II), and even slightly brighter than the average thermonuclear SNe (type Ia). Provided that the GRB was at the same distance as SN1998bw, the γ-ray fluence of GRB980425 reported by the *BATSE* group of $(4.4 \pm 0.4) \times 10^{-6}$ erg cm^{-2}[45] corresponds to a burst energy of $\sim (7.2 \pm 0.65) \times 10^{47}$ erg in γ-rays, which is four orders of magnitude smaller than the average GRB. This implies that either GRB980425 was not a typical burst or, perhaps, it was not related to SN1998bw. The small probability ($\sim 10^{-4}$) of finding a SN and a GRB in the very small error box during such a short period of time [24–26] spurred heated discussions and Wang and Wheeler [106] have systematically studied the issue.

The peculiarity of SN1998bw is even more prominent at radio wavelengths. In general, SNIb/c are strong radio emitters (see [108] and the chapter by Sramek and Weiler in this volume). Among six known type Ib/c SNe with radio detection, SN1998bw is one of the most radio luminous. SN1998bw was also peculiar in the fast evolution of its radio light curves (see [109] and the chapter by Weiler et al. in this volume). The flux at 3.5, 6.3, and 12 cm reached the first maximum as early as ~ 10 days, then was followed by a dip, and again rose to a second maximum at around day $20 - 40$. Shorter wavelength bands evolved faster than longer ones, as expected for non-thermal synchrotron models. The existence of a sub-relativistic shock wave has been suggested from the lack of variability in early radio emissions [47,110,111] and from detailed modeling of radio light curves (see [49,109] and the chapter by Weiler et al. in this volume).

X-ray follow-up observations of GRB980425 by *BeppoSAX* have been reported by Pian et al. (see [77] and the chapter by Frontera in this volume) but neither of the two sources detected by *BeppoSAX NFI* in the *WFC* error box have shown behaviors expected for GRB X-ray afterglows. One source (S1) decayed slowly by a factor of 2 in 6 months, while the other (S2) quickly faded in a few days. The position of source S1 coincides with SN1998bw, and the slow light curve of S1 may be explained by interaction of the SN ejecta with a circumstellar medium (CSM). Thus, source S1 is likely associated with SN1998bw, while source S2 is not. Since the decay rate of source S2 was marginally consistent with a power-law decline, it could have been an afterglow of GRB980425. However, then SN1998bw would not be related to GRB980425, which appears unlikely. It should be kept in mind, however, that it is difficult to draw any definitive conclusions with the very limited X-ray data available [76–78].

Follow-up observations of SN1998bw have been continued for a long period; > 400 days in optical/IR, > 250 days in the radio, and ~ 6 months in the X-ray bands. Such observations have put rather strong constraints on theoretical models for the SN. This chapter is aimed at briefly reviewing these observations and the theoretical models for SN1998bw, placing particular emphasis on the optical data where the most detailed observations are available. In Sect. 2,

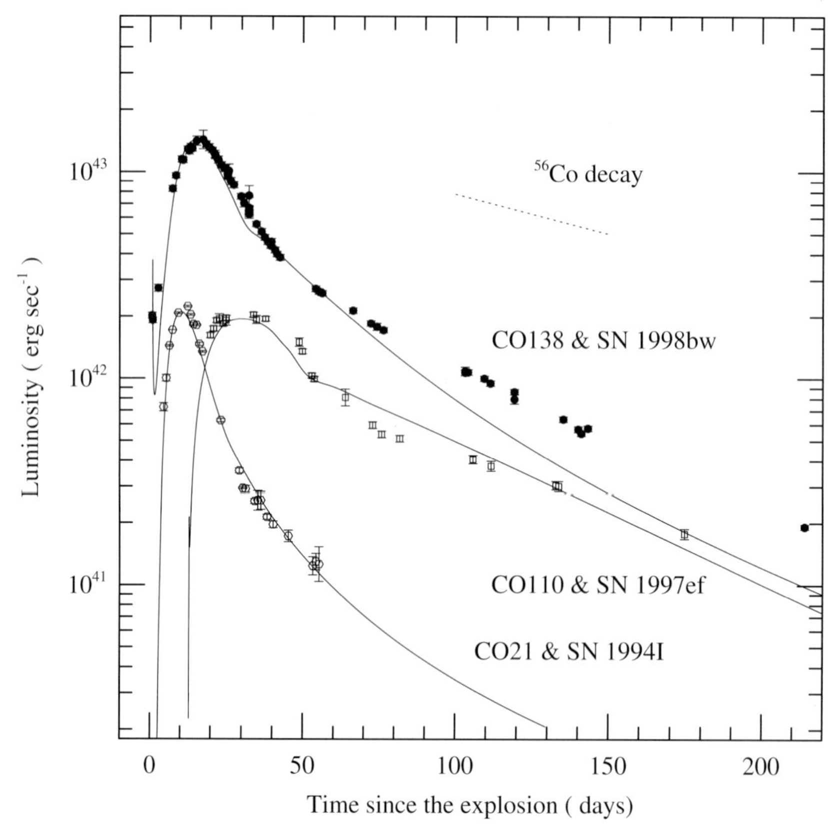

Fig. 2. Optical light curve of SN1998bw (filled circles [24–26]) compared with two other type Ic SNe, SN1997ef (open boxes [28]) and SN1994I (open circles [81]). Observed V light curves are transformed into bolometric light curves neglecting bolometric corrections. Solid lines are calculated light curves for models CO138 ($M_{\rm CO} = 13.8, E_{51} = 30, M_{\rm Ni} = 0.7$), CO100 ($M_{\rm CO} = 10, E_{51} = 8, M_{\rm Ni} = 0.15$), and CO21 ($M_{\rm CO} = 2.1, E_{51} = 1, M_{\rm Ni} = 0.07$), where $M_{\rm CO}$ is the mass of the progenitor C+O star in solar masses ($M_\odot$), E_{51} is the kinetic energy of the explosion in units of 10^{51} erg, and $M_{\rm Ni}$ is the mass of ^{56}Ni ejected in $M_\odot$ [37].

observations and theoretical models for the optical light curves and spectra are reviewed; nucleosynthesis in hypernova models is discussed in Sect. 3; and recent developments on aspherical explosion models and their implications for the explosion mechanism are briefly reviewed in Sect. 4. Analyses of other hypernovae are presented in Sect. 5. Finally, SN/GRB connections and possible evolutionary scenarios are discussed in Sect. 6.

2 Optical Light Curves and Spectra

SN1998bw has been extensively observed at optical/infrared (OIR) wavelengths both by photometry and by spectroscopy. Galama et al. [24–26] reported OIR light curves in U-, B-, V-, R-, and I-bands for the first 60 days observed at the *Australian National University (ANU) Mount Stromlo Observatory (MSO)*, the *ANU Siding Spring Observatory (SSO)*, the *Anglo-Australian Observatory (AAO)*, and the *European Southern Observatory (ESO)* at La Silla, Chile. McKenzie and Schaefer [61] reported photometry in B-, V-, and I-bands at intermediate times taken at the *Cerro Tololo Inter-American Observatory (CTIO)*. Patat et al. [75] published extensive coverage of spectroscopy and late-time optical photometry in the U-, B-, V-, R-, and I-bands and IR photometry in the J-, H-, and K-bands taken at *ESO*. Three early spectra from *ESO* (Fig. 4) were already presented by Iwamoto et al. [37]. Spectra taken by a joint campaign at the *SSO* were published in Stathakis et al. [93]. Patat et al. [75] also reported spectropolarimetry for two epochs. Sollerman et al. [92] reported photometric and spectroscopic observations at *ESO* at La Silla and by the *Very Large Telescope (VLT)* UT1 at Paranal, Chile.

2.1 Early Light Curves and Spectra

Light Curves: The early light curve of SN1998bw has been modeled by several authors [11,37,115]) based on spherically symmetric explosions. They reached similar conclusions that the early light curves can be successfully reproduced by the explosion of a massive C+O star with a kinetic energy of $> 10^{52}$ erg, more than ten times larger than the canonical SN explosion energy of $\sim 10^{51}$ erg. Figure 2 compares the light curve of SN1998bw with those of other SNIc and their model light curves. The model light curves are computed with a radiative transfer code [38] for a series of C+O star models given in Table 1.

SN1998bw showed a very early rise; it already had a luminosity of $\sim 10^{42}$ erg s^{-1} on day 1. This very early rise requires the presence of ^{56}Ni near the surface. Since ^{56}Ni is produced in deep inner layers of the ejecta, an extensive mixing must have occurred to dredge ^{56}Ni up to the outer layers [37]. Such mixing could take place due to a Rayleigh-Taylor instability at composition interfaces during shock propagation or due to an intrinsically non-spherical explosion. However, we can not rule out the possibility that the early rise was caused by an optical precursor somehow related to SN1998bw.

The light curve of SN1998bw around its maximum is quite similar to that of SNIa. It peaks at about day 16, which is close to the day 19.5 average time of maximum of SNIa [82]. The peak absolute magnitude is also comparable to that of normal SNIa. This encourages one to use the model light curves of SNIa to obtain a qualitative estimate of the model parameters such as mass of the ejecta, $M_{\rm ej}$, kinetic energy of the explosion, $E_{\rm K} = E_{51} \times 10^{51}$ erg, and the ^{56}Ni mass, $M_{\rm Ni}$. To illustrate this, we employ an analytic solution by Arnett [2]. The bolometric luminosity, $L_{\rm bol}$, of a "compact" SN that is powered by the radioactivity of ^{56}Ni is written as

$$L_{\rm bol} = \frac{\epsilon_{\rm Ni} M_{\rm Ni}}{\tau_{\rm Ni}} \Lambda(x, y) \qquad (1)$$

Table 1. Models and Their Parameters for SN1998bw

Model	$M_{\rm ms}$ ($M_\odot$)	$M_{\rm C+O}$ ($M_\odot$)	$M_{\rm ej}$ ($M_\odot$)	^{56}Ni mass ($M_\odot$)	$M_{\rm cut}$ ($M_\odot$)	$E_{\rm K}$ (10^{51} erg)
CO138E1[a]	~ 40	13.8	12	0.4	2	1
CO138E7[b]	~ 40	13.8	11.5	0.4	2.5	7
CO138E30[b]	~ 40	13.8	10.5	0.4	3.5	30
CO138E50[b]	~ 40	13.8	10	0.4	4	50

[a]CO138E1 is an ordinary, low-energy SNIc model in which a C+O star with a mass $M_{\rm CO} = 13.8\ M_\odot$ (which is the core of a 40 $M_\odot$ main sequence star [68]) explodes with $E_{\rm K} = 1.0 \times 10^{51}$ erg and $M_{\rm ej} = M_{\rm CO} - M_{\rm cut} \simeq 12\ M_\odot$, $M_{\rm cut}$ (2 $M_\odot$ in this case) denotes the mass cut, which corresponds to the mass of the compact stellar remnant, either a neutron star or a black hole.
[b]CO138E50, CO138E30, and CO138E7 are hypernova models in which the progenitor C+O star is the same as CO138E1 but explodes with different energies. The mass cut is chosen so that the ejected mass of ^{56}Ni is the value required to explain the observed peak brightness of SN1998bw. The compact remnant in these hypernova models may well be a black hole, because $M_{\rm cut}$ exceeds the maximum mass of a stable neutron star.

for early phases, where $\tau_{\rm Ni} = 8.8$ days and $\epsilon_{\rm Ni} = 2.96 \times 10^{16}$ erg g^{-1} are the decay time of ^{56}Ni and the energy deposited per gram of ^{56}Ni, respectively, and the function Λ is given by

$$\Lambda(x, y) = \exp(-x^2) \int_0^x \exp(-2zy + z^2) 2z dz. \tag{2}$$

It is assumed that ^{56}Ni is distributed homogeneously and that the γ-rays are all trapped in the ejecta, the latter being correct at the early phases we are considering here. The variable x is the dimensionless elapsed time,

$$x = t/\tau_c \tag{3}$$

normalized by the characteristic time of the light curve

$$\tau_c = (2\tau_h \tau_d)^{1/2}, \tag{4}$$

where τ_h and τ_d are the hydrodynamical time scale and the diffusion time scale of optical photons through the ejecta, respectively. The variable y is defined as

$$y = \tau_c/(2\tau_{\rm Ni}). \tag{5}$$

For the principal mode of diffusion (see [2,79]), τ_c turns out to be

$$\tau_c \sim 8 \text{ days} \left(\frac{\kappa}{0.05}\right)^{1/2} M_{\rm ej,\odot}^{3/4} E_{\rm K,51}^{-1/4} \tag{6}$$

where κ is the effective opacity. The symbol $M_{\mathrm{x},\odot}$ denotes $M_{\mathrm{x}}/M_{\odot}$ hereafter. The ejecta are assumed to be a sphere of constant density expanding homologously and, thus, M_{ej} and $E_{\mathrm{K}} = E_{\mathrm{K},51} \times 10^{51}$ erg are given by $M_{\mathrm{ej}} = (4\pi/3)R^3\rho$, $E_{\mathrm{K}} \sim (3/5)(1/2)M_{\mathrm{ej}}v_{\mathrm{ej}}^2$, where R, ρ, and v_{ej} are the radius, density, and surface velocity of the ejecta, respectively.

Differentiating Λ with respect to x, it is found that Λ has a maximum

$$\Lambda_{\mathrm{max}} = \exp(-2xy). \tag{7}$$

As given in Table 1 of Arnett [2], the maximum occurs at

$$x_{\mathrm{max}}y = t_{\mathrm{max}}/(2\tau_{\mathrm{Ni}}) \sim 0.42 + 0.48y \tag{8}$$

and

$$\Lambda_{\mathrm{max}} \sim 0.165/y \tag{9}$$

[3]. For SN1998bw, $t_{\mathrm{max}} \sim 16$ days corresponds to $y \sim 1$ and $\Lambda_{\mathrm{max}} \sim 0.165$, thus $L_{\mathrm{max}} \sim 1.3 \times 10^{43}\ \mathrm{M}_{\mathrm{Ni},\odot}$ erg s^{-1}. Given the fact that the ^{56}Co contribution doubles this luminosity at day ~ 16, the mass of ^{56}Ni is approximately given by

$$M_{\mathrm{Ni},\odot} \sim 0.38\ L_{\mathrm{max},43} \tag{10}$$

where $L_{\mathrm{max},43}$ is the peak luminosity in 10^{43} erg s^{-1}. The observed peak luminosity of SN1998bw, $L_{\mathrm{max},43} \sim 1 \times (d/37.8\mathrm{Mpc})^2$, translates into an estimated mass of ^{56}Ni $M_{\mathrm{Ni},\odot} \sim 0.4(d/37.8\mathrm{Mpc})^2$, which is much larger than that of typical core-collapse SNe with $M_{\mathrm{Ni},\odot} \lesssim 0.1$.

A constraint for $M_{\mathrm{ej},\odot}$ and $E_{\mathrm{K},51}$ can be obtained from $y \sim 1$ by the use of Eq. (3) such that

$$M_{\mathrm{ej},\odot}^3/E_{\mathrm{K},51} \sim 23 \left(\frac{\kappa}{0.05}\right)^{-2}. \tag{11}$$

Another constraint is necessary to resolve the degeneracy of masses and energies in this equation. One possibility is to use photospheric velocities or line velocities. An equivalent to this, but more quantitative, is to compare the observed and synthetic spectra.

In Fig. 3 the observed line velocities and photospheric velocities of SN1998bw are compared with calculated photospheric velocities for different models. The radius of the photosphere, r_{ph}, is defined by

$$\int_{r_{\mathrm{ph}}}^{\infty} \kappa\rho dr = 2/3. \tag{12}$$

For the constant density sphere model used in analytic light curves, the photosphere is located a fraction

$$r_{\mathrm{ph}} \sim \frac{80\pi E_{\mathrm{K}}}{27\kappa M_{\mathrm{ej}}^2} t^2 r_{\mathrm{ej}} \tag{13}$$

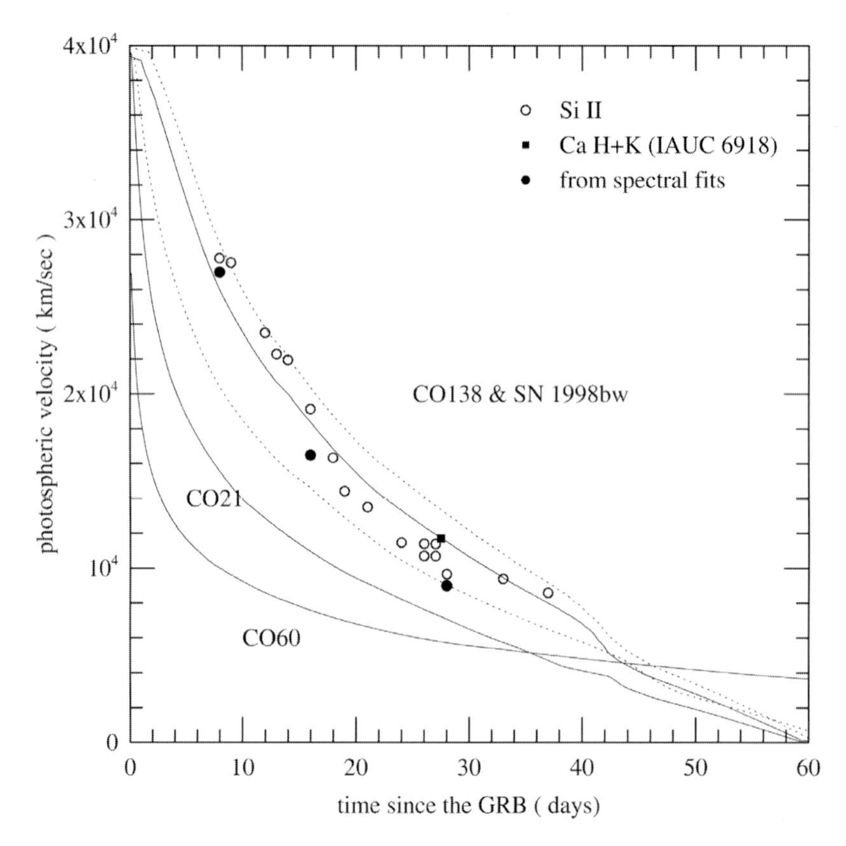

Fig. 3. The photospheric velocities obtained by spectral modeling (filled circles) and the observed velocities of Si II 6347 Å and 6371 Å lines measured at the absorption cores (open circles [75]), and that of the Ca II H+K doublet derived from the spectrum of 23 May 1998 (filled square) are compared with a series of C+O star explosion models [37].

of the ejecta radius from the surface. Then, the photospheric velocity $v_{\rm ph}$ at early phases is given by

$$v_{\rm ph} \sim v_{\rm ej}\left(1 - \frac{80\pi E_{\rm K}}{27\kappa M_{\rm ej}^2}t^2\right) = \left(\frac{10E_{\rm K}}{3M_{\rm ej}}\right)^{1/2}\left(1 - \frac{80\pi E_{\rm K}}{27\kappa M_{\rm ej}^2}t^2\right). \tag{14}$$

With Eq. (6), we obtain another constraint from $v_{\rm ph} \sim 27,000$ km s^{-1} at day 8 such that $E_{\rm K,51}/M_{\rm ej,\odot} \sim 4.3$. Combined with Eq. (5), this results in $M_{\rm ej,\odot} \sim 10$ and $E_{\rm K,51} \sim 43$.

This very large energy, exceeding 10^{52} erg, led us to refer to SN1998bw as a hypernova, an exceptional class of energetic SN explosions that have kinetic energy $E_{\rm K,51} \gtrsim 10$ [37,115]. The above estimate, despite its crudeness, provides model parameters in better agreement than expected with those given by detailed light curve modelings [37,64,115]. Nakamura et al. [64] picked a model

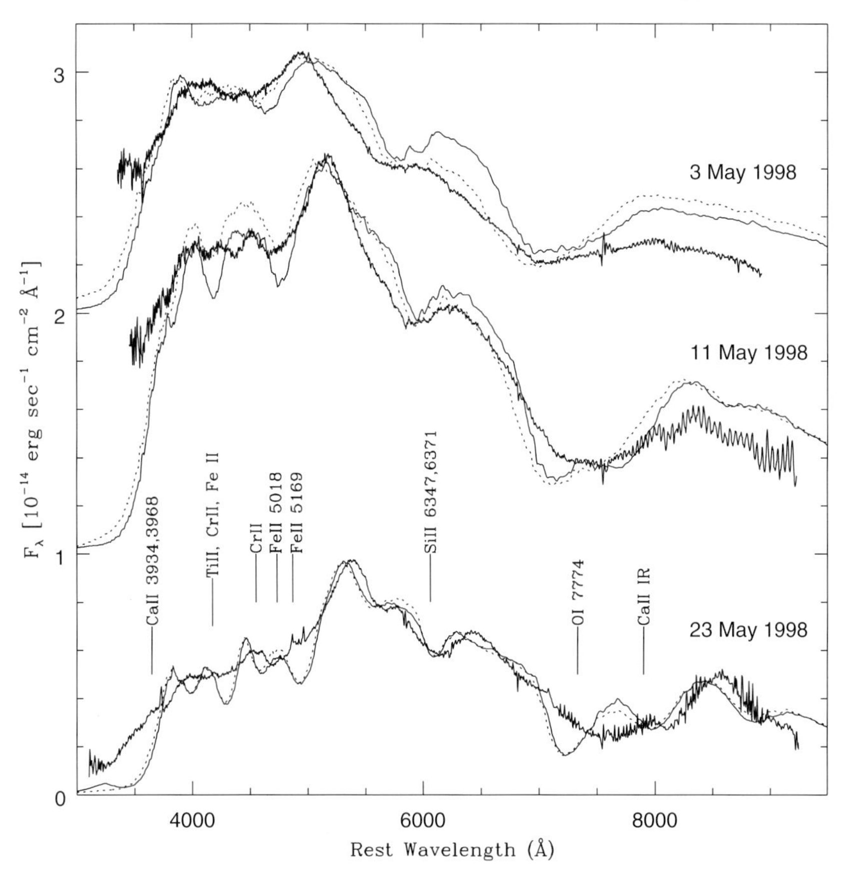

Fig. 4. Observed spectra of SN1998bw at three epochs: 3, 11, and 23 May 1998 (day 8, 16, and 28, respectively) [75] are compared with synthetic spectra (dashed lines) computed by a Monte Carlo code [56] using model CO138E50 [64]. A distance modulus of $(m-M) = 32.89$ mag and an extinction of $A_V = 0.05$ are assumed, which corresponds to a distance of 37.8 Mpc to SN1998bw for $H_0 = 65$ km s^{-1} Mpc^{-1}. The observed featureless spectra are the result of the blending of many metal lines having large velocities and large velocity dispersion. The apparent emission peaks are actually low opacity regions of the spectra in which photons can escape.

CO138E50 that has $M_{\mathrm{ej},\odot} = 10, E_{\mathrm{K},51} = 50, M_{\mathrm{Ni},\odot} = 0.4$ as the best model to reproduce the light curve and photospheric velocities for SN1998bw using a distance of 37.8 Mpc and an extinction of $A_V = 0.05$ (see Table 1 and Fig. 5).

Early Spectra: A more quantitative constraint on these basic model parameters and a crucial diagnosis of the chemical composition of the ejecta can be obtained by detailed spectral modeling [10,37,59,64,71]. Fig. 4 shows early spectra of SN1998bw for three epochs [64,75]. The spectra are dominated by broad absorption features at around 4700 – 5000 Å, 7000 – 6200 Å, and 8000 Å, the second and third of which are identified as the O I and Ca II IR triplet, re-

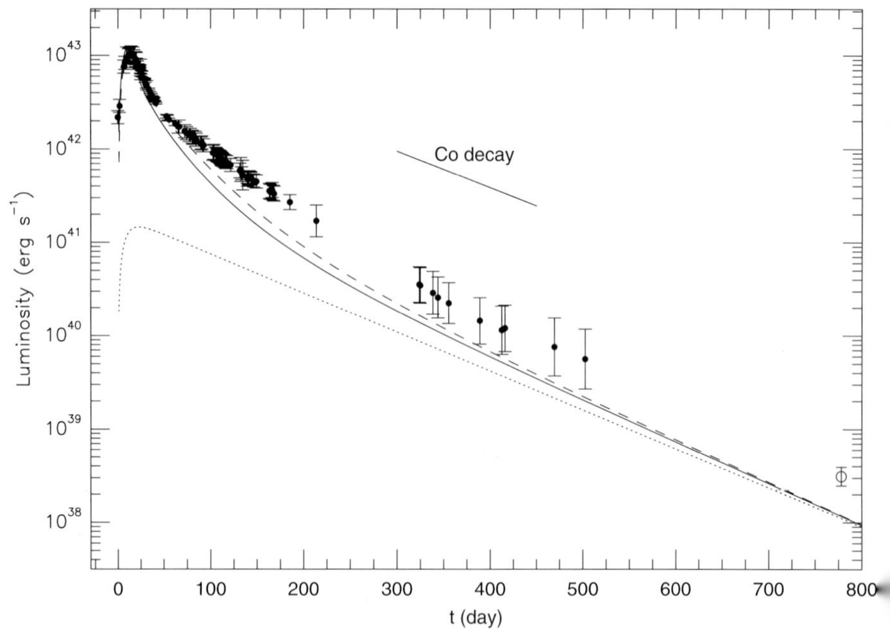

Fig. 5. The light curves of CO138E50 ($E_K = 5 \times 10^{52}$ erg, solid line) and CO138E30 ($E_K = 3 \times 10^{52}$ erg, long-dashed line) are compared with the bolometric light curve of SN1998bw [64,75]. A distance modulus of $(m - M) = 32.89$ mag and an extinction of $A_V = 0.05$ are adopted. The dotted line indicates the energy deposited by positrons for CO138E50. The *Hubble Space Telescope (HST)* observation on day 778 (open circle [21]) is shown assuming negligible bolometric correction.

spectively, and the first of which is likely due to Fe II lines. These line features are also seen in other type Ic SNe. However, for SN1998bw they are exceptionally broad and blueshifted. Stathakis et al. [93] found that the absorption line minima are shifted 10 – 50% blueward at day 15 in comparison with ordinary SNIc.

In Fig. 4 the observed spectra are compared with synthetic spectra computed with a hypernova model CO138E50 that has $M_{C+O,\odot} = 13.8$ and $E_{K,51} = 50$ [64]. The synthetic spectra were computed using a Monte Carlo code [56] improved by inclusion of photon branching and a new extended line list [51,57]. A distance modulus of $(m - M) = 32.89$ mag and an extinction of $A_V = 0.05$ were adopted. The assumption of low reddening is supported by the upper limit of 0.1 Å in the equivalent width of the Na I D-line obtained from high-resolution spectra [75].

The synthetic spectra of CO138E50 are in good agreement with the observed spectra. The Si II feature near 6000 Å and, in particular, the OI+CaII feature between 7000 and 8000 Å, are as broad as the observations. Nevertheless, the blue sides of those absorptions are still too narrow, indicating that the model CO138E50 may not contain enough mass in the high velocity ejecta. Nakamura et al. [64] tried a model CO138E50 with its density structure in the envelope artificially made shallower and examined the effect on the synthetic spectra. They found that a reduction of density gradient from $\rho \propto r^{-8}$ to $\rho \propto r^{-6}$ at $v > 30,000$ km s^{-1} leads to a significant increase of mass in the higher velocity region, making strong absorption features at $v \sim 60,000$ km s^{-1}. Similar conclusions are reached by Branch [10], who presented a parameterized synthetic spectrum for the early phases of SN1998bw. Branch [10] showed that the spectra are well reproduced by a spherically symmetric model with a kinetic energy of 60×10^{51} erg and with a very flat density distribution of $\rho \propto r^{-2}$ in the envelope. Spectral synthesis in 2- and 3-D general geometries will need to be done to draw further information from the observed spectra and obtain a more quantitative conclusion.

A P Cygni feature at 1.08 μm was discovered in the IR spectra of SN1998bw by Patat et al. [75] with high resolution spectroscopy. It is suggested to be the He I IR line at 10830 Å excited by non-thermal electrons produced by radioactive decay of ^{56}Co [12,75]. The presence of the He line in SNIc spectra and determination of the He abundance in their ejecta is important to constrain the evolution followed by progenitors of SNIc.

2.2 Late Time Light Curves and Spectra

The dominant sources of the light curve energy change with time. The light curve is first powered by γ-rays from ^{56}Co, then by positrons from ^{56}Co decay, and finally by γ-rays from ^{57}Co and ^{44}Ti.

The optical depth of the ejecta to γ-ray photons produced by ^{56}Co decay, τ_γ, is given by

$$\tau_\gamma = \kappa_\gamma \rho R = \frac{9\kappa_\gamma M_{\rm ej}^2}{40\pi E_{\rm K}} t^{-2} \sim 0.11 \left(\frac{\kappa_\gamma}{0.03}\right) \frac{M_{\rm ej,\odot}^2}{E_{\rm K,51}} \left(\frac{t}{100 \text{ days}}\right)^{-2} \tag{15}$$

where κ_γ is the effective γ-ray opacity for ^{56}Co γ-ray lines. In model CO138E50 for SN1998bw, the ejecta become optically thin in γ-rays ($\tau_\gamma < 1$) at around day 50. Using a deposition fraction $f_{\rm dep,\gamma} = 0.64\tau_\gamma$ for $\tau_\gamma < 0.25$ [14], we find

$$f_{\rm dep,\gamma} \sim 0.073 \left(\frac{\kappa_\gamma}{0.03}\right) \frac{M_{\rm ej,\odot}^2}{E_{\rm K,51}} \left(\frac{t}{100 \text{ days}}\right)^{-2} . \tag{16}$$

Assuming that the energy deposited by γ-rays is thermalized and subsequently radiated at OIR wavelengths, the bolometric luminosity $L_{\rm bol}$ is given by

$$L_{\rm bol} \sim f_{\rm dep,\gamma} \epsilon_{\rm decay} M_{\rm Ni},$$

where $\epsilon_{\rm decay}$ is the energy available per second per gram of ^{56}Co

$$\epsilon_{\rm decay} = \frac{\epsilon_{\rm Ni}}{\tau_{\rm Ni}} \exp\left(-\frac{t}{\tau_{\rm Ni}}\right) + \frac{\epsilon_{\rm Co}}{\tau_{\rm Co} - \tau_{\rm Ni}} \left[\exp\left(-\frac{t}{\tau_{\rm Co}}\right) - \exp\left(-\frac{t}{\tau_{\rm Ni}}\right)\right], \tag{17}$$

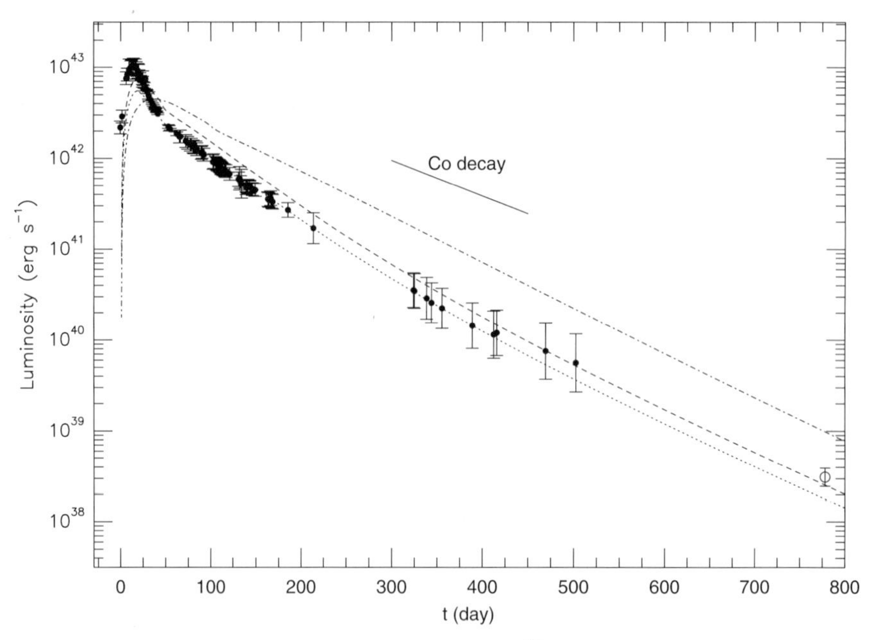

Fig. 6. The light curves of CO138E7 ($E_K = 7 \times 10^{51}$ erg, dashed line) and CO138E1 ($E_K = 1 \times 10^{51}$ erg, dash-dotted line) compared with the bolometric light curve of SN1998bw [64,75]. Also shown is the light curve of modified CO138E7 with a smaller ^{56}Ni mass of 0.28 $M_\odot$ (dotted line). The *HST* observation at day 778 (open circle [21]) is shown by assuming negligible bolometric correction.

where $\epsilon_{\rm Co} = 6.3 \times 10^{16}$ erg g^{-1} and $\tau_{\rm Co} = 111.3$ days are energy deposited per gram of ^{56}Co and the decay time, respectively. Since most of the energy deposition comes from ^{56}Co decay at late times, we have

$$L_{\rm bol} \sim \frac{f_{{\rm dep},\gamma}\epsilon_{\rm Co} M_{\rm Ni}}{\tau_{\rm Co} - \tau_{\rm Ni}} \exp\left(-\frac{t}{\tau_{\rm Co}}\right)$$
$$\sim 1.6 \times 10^{41} \text{ erg s}^{-1} \left(\frac{\kappa_\gamma}{0.03}\right) \left(\frac{M_{{\rm Ni},\odot}}{0.4}\right) \frac{M^2_{{\rm ej},\odot}}{E_{{\rm K},51}} \left(\frac{t}{111{\rm d}}\right)^{-2} \exp\left(-\frac{t}{111{\rm d}}\right). \quad (18)$$

This luminosity depends on $M_{\rm ej}$ and $E_{\rm K}$ as well as on $M_{\rm Ni}$. Thus we can use late-time light curves to distinguish different models that produce similar early light curves. For CO138E50, Eq. (11) gives $L_{\rm bol} \sim 2 \times 10^{40}$ erg s^{-1} at day 400, which is in agreement with the observations of SN1998bw (Fig. 5).

Figure 5 compares light curves calculated for hypernova models CO138E50 (solid line) and CO138E30 (long-dashed line) with the observed light curve of SN1998bw (filled circles). The light curve of CO138E50 is consistent with the observations until day 50 but declines at a faster rate afterwards. The light curve of CO138E30 shows a slower decline, in better agreement with SN1998bw, but it still declines too fast [61,64,75]. Fig. 6 shows light curves for the lower energy models CO138E7 (dashed-line) and CO138E1 (dash-dotted line), again with the observed light curve of SN1998bw (filled circles). The steady exponential decline of SN1998bw from day 60 to 400 is well reproduced by model CO138E7. This is presumably because CO138E7 is more efficient in trapping γ-rays emitted by ^{56}Co decay than higher-energy models CO138E50 and CO138E30. The observed low flux level at this phase requires a reduced ^{56}Ni mass of $0.28M_{\odot}$ (dotted line), which is too small to explain the maximum luminosity of SN1998bw. The expansion of CO138E1 is too slow because of its low energy and large mass and its light curve is too bright compared to the observation.

In summary, the early light curve of SN1998bw is well reproduced by higher-energy models, but they deviate from the observations at late times. On the other hand, the late-time light curve is more easily reproduced by a lower-energy model CO138E7, with a smaller ^{56}Ni mass, but this is too dim at early times. This difficulty may be overcome if there are multiple components of ejecta, higher velocity and lower velocity parts, implying that the density distribution or the ^{56}Ni distribution in the ejecta is not spherically symmetric as was proposed for SN1997ef [38,58].

Figures 5 and 6 show that the light curve of SN1998bw appears to flatten after about day 400. The model light curves do not reproduce this behavior. At $t \gtrsim 400$ days, the γ-ray deposition fraction $f_{\mathrm{dep},\gamma}$ decreases to below 1% in model CO138E50. About 3.5% of the decay energy of ^{56}Co is carried away by positrons (see, e.g., [4]) and they are effectively trapped in the ejecta because of the postulated weak magnetic fields (see, e.g., [13]). Therefore, the energy deposition from positrons becomes the dominant contribution to the light curve at $t \gtrsim 400$ days (dotted line in Fig. 5).

The luminosity by positrons is given as

$$L_{\mathrm{bol,e^+}} = 2 \times 10^{41} \left(\frac{M_{\mathrm{Ni},\odot}}{0.4} \right) \exp \left(- \frac{t}{111\mathrm{d}} \right) \ \mathrm{erg\ s^{-1}} \tag{19}$$

where the positrons are assumed to be completely trapped in the ejecta. Therefore, if the observed light curve exponential tail follows the positron powered model light curve, the ^{56}Co mass can be determined directly.

The light curve of SN1998bw showed a further flattening at around day 800. The last observed point in Figs. 5 and 6 is the *HST* observation on 11 June 2000 (day 778) [21]. The observed magnitude (V $= 25.41 \pm 0.25$ mag) is consistent with the prediction of CO138E7 (Fig. 6), but brighter than CO138E50 (Fig 5). As the density decreases, the recombination time scale becomes longer than the decay time and ionization freeze-out should make the bolometric light curve even flatter as is shown by Fransson and Kozma [20]. Another possible source of the excess luminosity is the emission due to the interaction of the ejecta with the CSM [92]. However, it is not clear whether this flattening is caused by SN1998bw itself or by a contribution from an underlying star cluster [21].

Late time spectra provide a wealth of information on the elemental abundances and their distribution in velocity space. SN1998bw seems to have entered the nebular phase between day 65 and day 115 (Fig. 7; [60,61]). The spectroscopic features at late times are very similar to those of type Ic SNe. Dominant

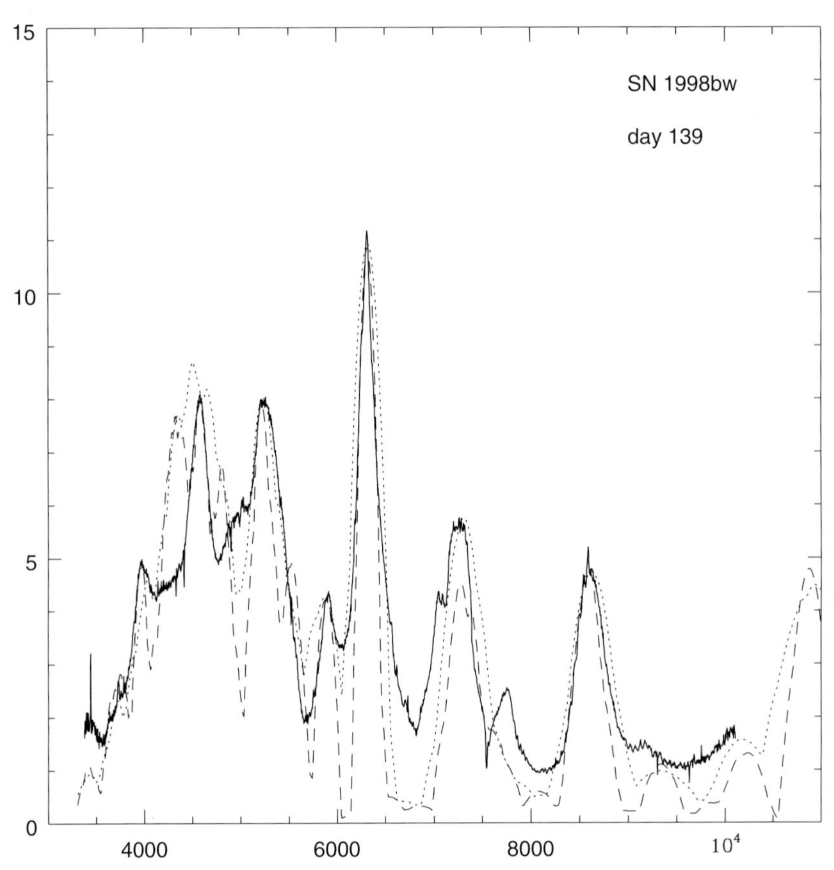

Fig. 7. A nebular spectrum of SN1998bw on 12 Sept 1998 (day 139) is compared to synthetic spectra obtained with a NLTE nebular model based on the deposition of γ-rays from ^{56}Co decay in a nebula of uniform density. Two models were computed. In one model (dotted line) the broad Fe II lines near 5300 Å are well reproduced. The derived ^{56}Ni mass is 0.65 $M_{\odot}$, the outer nebular velocity is 11,000 km s^{-1}, and the O mass is 3.5 $M_{\odot}$. The average electron density in the nebula is $\log n_e = 7.47$ cm^{-3}. In the other model (dashed line), only the narrow OI 6300 Å emission line is well reproduced. This model has smaller ^{56}Ni mass (0.35 $M_{\odot}$), O mass (2.1 $M_{\odot}$), and an outer velocity of 7500 km s^{-1}. The density is similar to that of the "broad-lined" model. The filling factor used is 0.1 for both models [59,71].

features include emission peaks of Mg I λ4571, Fe II blend around 5200 Å, O I $\lambda\lambda$ 6300, 6364 doublet, a feature around 7200 Å (identified as Ca II and O II by Mazzali et al. [59]), and Ca II IR triplet that is already visible in early spectra [59,75] (Fig. 7). What is different from ordinary SNIc, in particular, is the broadness of the line features. Patat et al. [75] estimated the expansion velocity of the Mg I emitting region and found a value $9,800 \pm 500$ km s^{-1} for day 201. Stathakis et al. [93] also found emission features 45 per cent broader than ordinary SNIc for day 94. Stathakis et al. [93] found it difficult to explain the extent of line blanketing by only using line broadening and suggested unusual relative abundances or physical conditions in the ejecta.

Mazzali et al. [60] analyzed the late time spectra of SN1998bw reported by Patat et al. [75] and found that both broad and relatively narrow lines are present in the spectra (Fig. 7). They interpreted the low ionization implied by the absence of Fe III nebular lines as a sign of clumpiness of the ejecta. Furthermore, some Fe lines are found to be broader than O lines, which is the opposite to what spherically symmetric models predict, implying that the explosion was aspherical [59]. Sollerman et al. [92] calculated late-time synthetic spectra and V, R, and I light curves for the models based on CO138 of Iwamoto et al. [37] and CO6 of Woosley et al. [115]. They found that the synthetic spectra and light curves are in general agreement with the observations of SN1998bw and concluded that roughly ~ 0.5 $M_\odot$ of ^{56}Ni is necessary, with an assumed distance of 35 Mpc, to power the light curve at $\sim$ 1 year after the explosion. The ^{56}Ni mass is very close to the values obtained by others. Woosley et al. [115] also stressed the need for mixing to account for the line features.

3 Explosive Nucleosynthesis

A strong shock wave traveling outwards heats up the hypernova ejecta, reaching temperatures high enough for thermonuclear burning to proceed. Explosive nucleosynthesis in the hypernova model for SN1998bw, and in general hypernova models, was calculated and compared with that of normal SN explosion models by Nakamura et al. [64,65]. They first carried out hydrodynamical simulations with a small-size nuclear reaction network and then recalculated detailed nucleosynthesis using a larger reaction network [34] with over 200 isotopes up to ^{71}Ge.

Burning products in each layer depend mostly on the maximum temperature that is experienced by the layer, $T_{\rm max}$. At $T_{\rm max} > 5 \times 10^9$ K, complete Si burning leaves products dominated by ^{56}Ni. The size of the sphere of dominant ^{56}Ni production is given approximately as

$$R_{\rm Ni} \sim 3700 \ \ (E/10^{51}{\rm erg})^{1/3} \ \ {\rm km}, \tag{20}$$

by assuming that

$$E = (4\pi/3)R_{\rm Ni}^3 a T_{\rm max}^4 \tag{21}$$

with $T_{\rm max} = 5 \times 10^9$ K (see, e.g., Thielemann et al. [99]). A higher E means a larger $R_{\rm Ni}$. The ^{56}Ni mass is determined by the mass enclosed within $R_{\rm Ni}$ in the progenitor, which is sensitive to the mass of the progenitor itself. From this argument, the large ^{56}Ni mass required to reproduce the light curve of SN1998bw rules out small-mass progenitor models for SN1998bw.

The top panel of Fig. 8 shows the composition in the ejecta of hypernova model CO138E50 as a function of the enclosed mass, M_r, and of the expansion velocity. The nucleosynthesis in SN model CO138E1 is also shown for comparison in the lower panel of Fig. 8. The yields of hypernova (CO138E7, 30, 50) and

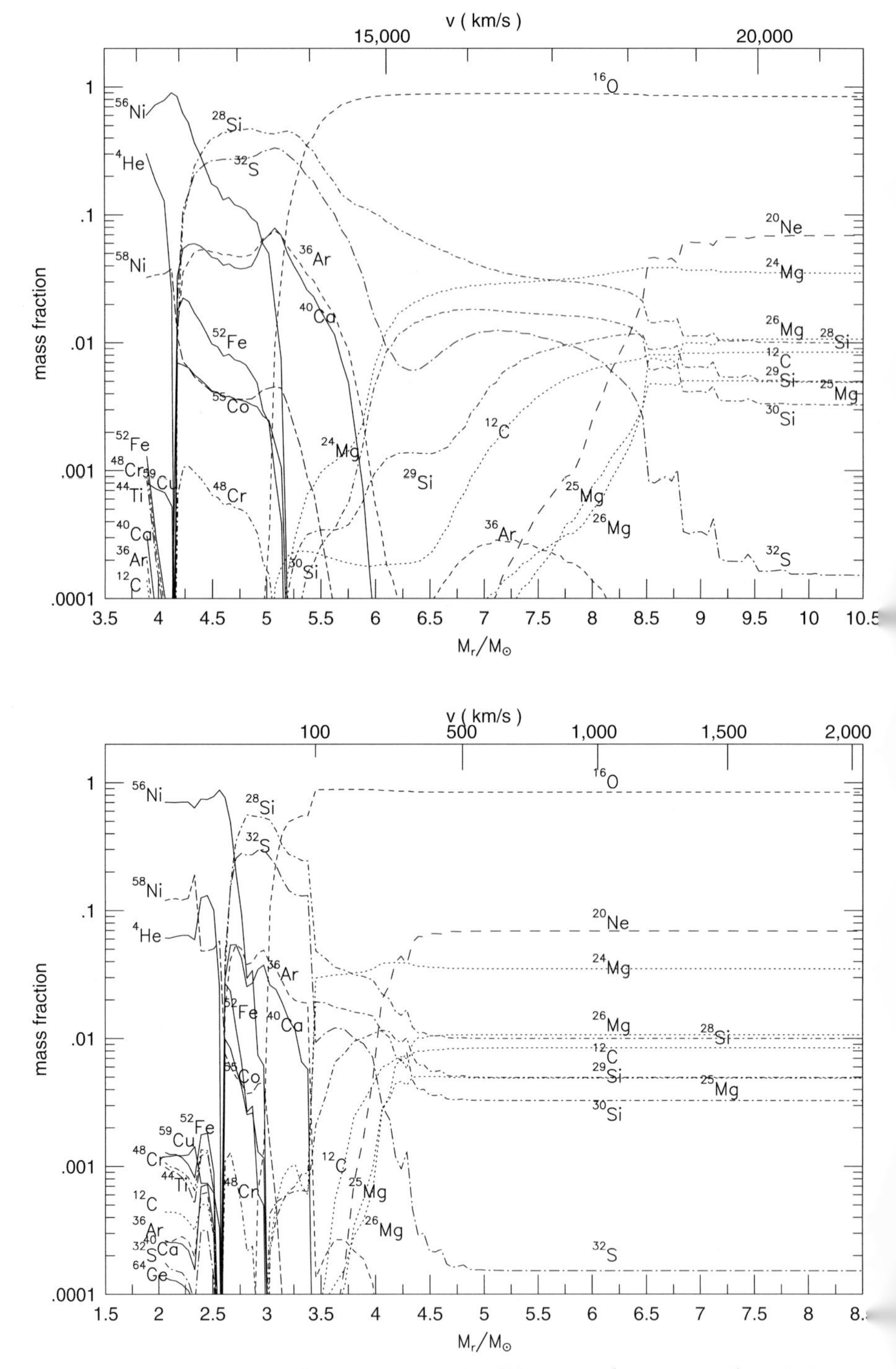

Fig. 8. Composition of the ejecta of hypernova model CO138E50 (upper panel) and of SN model CO138E1 (lower panel) [64].

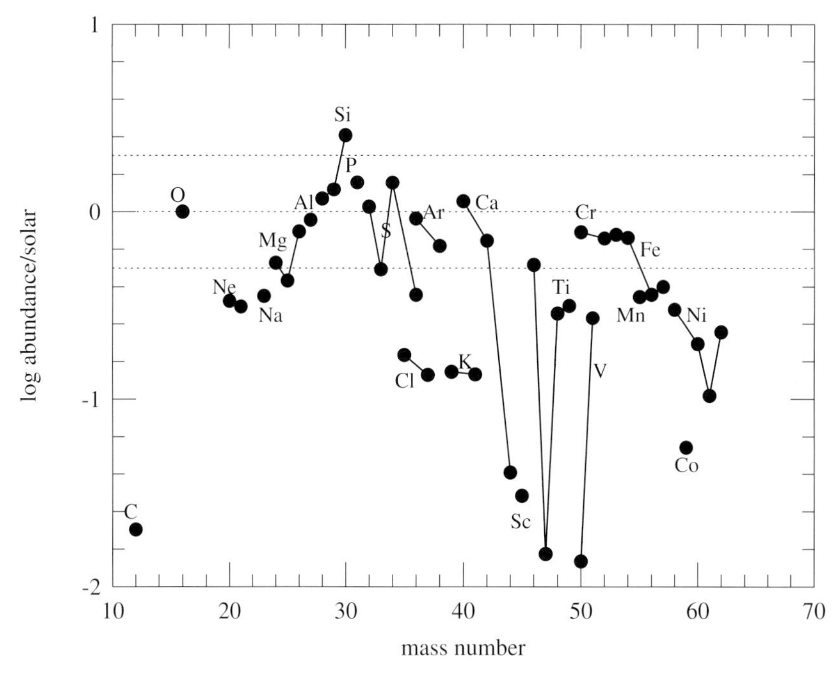

Fig. 9. Abundances of stable isotopes relative to the solar values for model CO138E50 [64].

SN (CO138E1) models are summarized in Table 2. Total abundances of stable isotopes for model CO138E50 relative to the solar abundances are shown in Fig. 9.

The complete Si burning region where ^{56}Ni is predominantly produced is much more extended (in mass) in the hypernova model (CO138E50) than in the ordinary SN model (CO138E1) simply because of the higher explosion energy. The large ^{56}Ni mass required for SN1998bw of $\sim 0.4 - 0.7$ $M_\odot$, implies that, in hypernova models for SN1998bw, the mass cut (the mass that collapses into a compact object) would be relatively small and thus isotopes such as ^{59}Cu, ^{63}Zn, and ^{64}Ge (which decay into ^{59}Co, ^{63}Cu, and ^{64}Zn, respectively) are ejected more abundantly than in ordinary SN models.

The α rich freeze-out is enhanced in the hypernova model because of the lower densities than in the SN model. Fig. 8 clearly shows that a larger number of α particles (^{4}He) are left behind in the hypernova model. As a result, the species produced by α capture such as ^{44}Ti and ^{48}Cr (which decay into ^{44}Ca and ^{48}Ti, respectively) are larger in CO138E50 than in CO138E1. Note that even in such an energetic hypernova model, He has a velocity much smaller than the value of 18,300 km s^{-1} which is implied if the feature observed in the near-IR is identified as He I 10830 Å [75]. However, a possible mixing might increase the He velocity above what the model predicts.

Table 2. Yields of Hypernova and Supernova Models (in $M_\odot$ [65]).

Model	C	O	Ne	Mg	Si	S	Ca	Ti	Fe	Ni
CO138E1	0.10	8.9	0.66	0.51	0.52	0.19	0.023	0.0007	0.44	0.065
CO138E7	0.09	8.5	0.58	0.47	0.52	0.19	0.029	0.0010	0.44	0.036
CO138E30	0.07	7.6	0.44	0.38	0.74	0.35	0.054	0.0009	0.45	0.023
CO138E50	0.06	7.1	0.37	0.34	0.83	0.41	0.066	0.0007	0.46	0.018

Model	^{44}Ti	^{56}Ni	^{57}Ni
CO138E1	3.4×10^{-4}	0.40	1.6×10^{-2}
CO138E7	4.5×10^{-4}	0.40	1.4×10^{-2}
CO138E30	2.3×10^{-4}	0.40	1.2×10^{-2}
CO138E50	5.5×10^{-5}	0.40	1.1×10^{-2}

The incomplete Si burning region is more extended in the hypernova model because of the higher explosion energy. The main products in this region are ^{28}Si, ^{32}S, and ^{56}Ni. Model CO138E50 produces 0.2 $M_\odot$ of ^{56}Ni in this region. Other important isotopes such as ^{52}Fe, ^{55}Co, and ^{51}Mn (decaying into ^{52}Cr, ^{55}Mn, and ^{51}V, respectively) are also produced more abundantly in the hypernova model. Oxygen burning takes place in more extended, lower density regions in the hypernova model. Finally, in the hypernova model more extensive burning of O, C, and Al result in more abundant production of isotopes such as Si, S and Ar, respectively.

4 Asphericity and Explosion Mechanism

The large explosion energy claimed for SN1998bw by light curve and spectral analyses is based on the assumption that the explosion was spherically symmetric. There is an alternative explanation for the large peak luminosity of SN1998bw. If the explosion was aspherical, the estimated explosion energy might be reduced considerably. There is growing evidence that SNIb/c show a small polarization ($\sim$ 1 percent), implying that their explosions might be somewhat aspherical. A similar degree of polarization was observed in SN1998bw at early photospheric phases [37,41,75].

Höflich et al. [35] studied such alternative possibilities for SN1998bw and found that an aspherical explosion of a compact C+O star (~ 2 $M_\odot$) with a conservative ^{56}Ni mass of ~ 0.2 $M_\odot$ and explosion energy of 2×10^{51} erg can reproduce the early light curve of SN1998bw. It is still unclear if such small mass aspherical models can explain the observed broad spectral features and high velocities. However, as discussed in Sect. 2.2, it is difficult to reproduce the entire light curve of SN1998bw using spherically symmetric models, implying

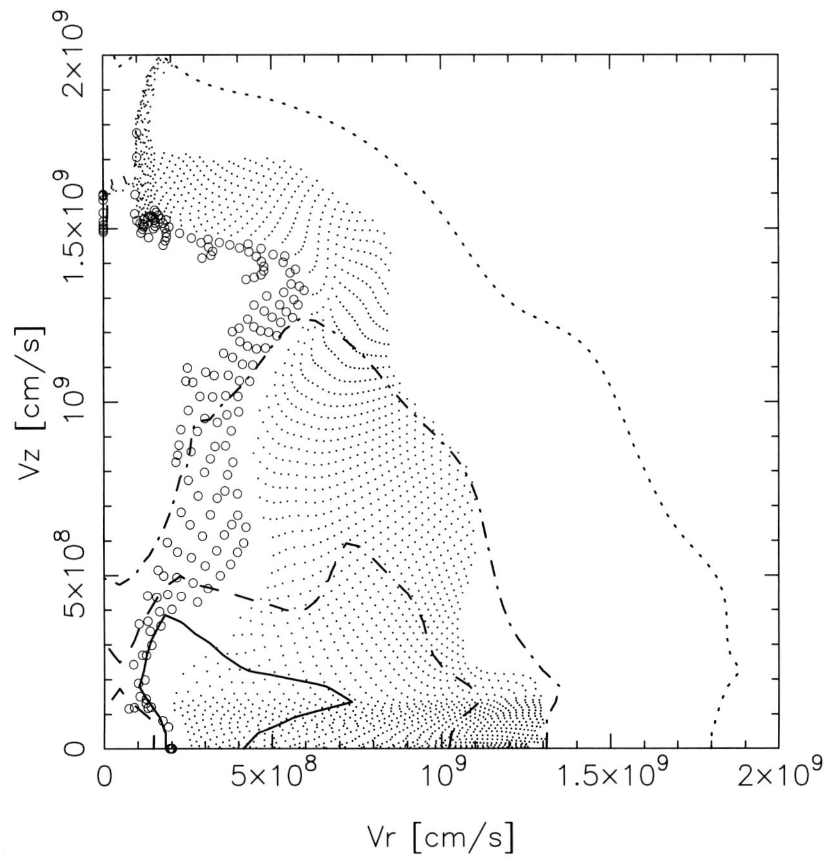

Fig. 10. Distribution of ^{56}Ni (open circles) and ^{16}O (dots) at the homologous expansion phase in an aspherical explosion model of SN1998bw [54]. Parameters are $E_{\rm exp} = 10^{52}$ erg, $v_z/v_r = 8$. Open circles and dots are test particles of ^{56}Ni and ^{16}O, respectively, indicating local volumes in which mass fractions of these elements are greater than 0.1. Lines are density contours of 0.5 (solid), 0.3 (dashed), 0.1 (dash-dotted), and 0.01 (dotted) of the maximum density.

that there exists some degree of asphericity in the explosion. An analysis of the late-time light curve suggests a ^{56}Ni mass $M_{\rm Ni,\odot} > 0.4$ [92], which is close to the values obtained by spherical symmetric models for the early light curves. Thus, the degree of possible asymmetry in SN1998bw seems to be moderate.

Maeda et al. [54] calculated nucleosynthesis in aspherical explosion models for SN1998bw with a 2D hydrodynamical code and a detailed nuclear reaction network. They used the same progenitor model, CO138, of Iwamoto et al. [37] and Nakamura et al. [64] and assumed aspherical initial velocity profiles. Figure 10 shows the composition in the ejecta at the homologous expansion phase for the case of $E_{\rm exp} = 10^{52}$ erg and an axial-to-radial velocity ratio $v_z/v_r = 8$. They found that the nebular line profiles in SN1998bw are reproduced better by such aspherical models if the explosion is seen at a small viewing angle from the jet axis.

Such aspherical initial velocity profiles may be the result of an aspherical SN explosion owing to the convection driven by neutrino heating [39]. Shimizu et al. [89] pointed out that anisotropy in the neutrino emission would increase the net energy gain by neutrino heating, which leads to a larger explosion energy than spherically symmetric models. A more speculative scenario, but one that may hold a better promise to produce a highly aspherical explosion, is the so-called "collapsar," the collapse of a rotating stellar core that forms a system consisting of a rotating black hole and an accretion torus around it [52]. Thermal neutrinos from the torus release a large amount of energy as electron-positron pairs. If the black hole has a strong magnetic field, the rotation energy of the hole might be extracted by the Blandford-Znajek mechanism [7]. Jet formation and propagation after energy deposition by the above processes can be studied by hydrodynamical simulations [1,53]. MacFadyen et al. [53] suggest that SN1998bw may be a case in which a black hole was produced by the fall back and the resulting jet was less collimated. Theoretical and empirical modeling is continuing and a better understanding of hypernovae can be expected in the future.

5 Other Hypernovae

Several other hypernovae have been discovered or recognized since SN1998bw. Such hypernovae include SN1997cy [30,103], SN1997ef [38,58], SN1997dq [55], SN1999E [83], SN1999as [46], and SN2002ap [60]. Among these, SN1997cy and SN1999E were SNe of type IIn, while the others were all SNIc. None of these seems to have been a candidate for a GRB counterpart. Figs. 1 and 11 show the near-maximum spectra and the absolute V-band light curves for these objects. The properties of these hypernovae can be deduced by analyzing their light curves and spectra, as illustrated for SN1998bw in the preceeding sections.

These objects all appear to fall in the class of hypernovae. However, they differ greatly in their properties. SN1999as was extremely bright ($M_V < -21.5$ mag) and was, in fact, the most luminous supernova ever discovered, while SN2002ap was no more luminous than a normal SN. The analysis also reveals that the kinetic energy, $E_{\rm K}$, may be related to the progenitor's main-sequence mass (Fig. 16). It seems that the higher energy hypernovae arise from the more massive progenitors. In general, hypernovae are at the high end of the range of SN progenitor masses, as their progenitors are more massive than normal core-collapse SNe ($\sim 15 - 20\ \mathrm{M}_\odot$) (Fig.16).

5.1 SN1997ef

This SN preceeded SN1998bw. Its broad-lined spectrum defied interpretation in the context of standard energy explosion models [38]. The realization, obtained from SN1998bw, that energies much larger than the supposedly standard value

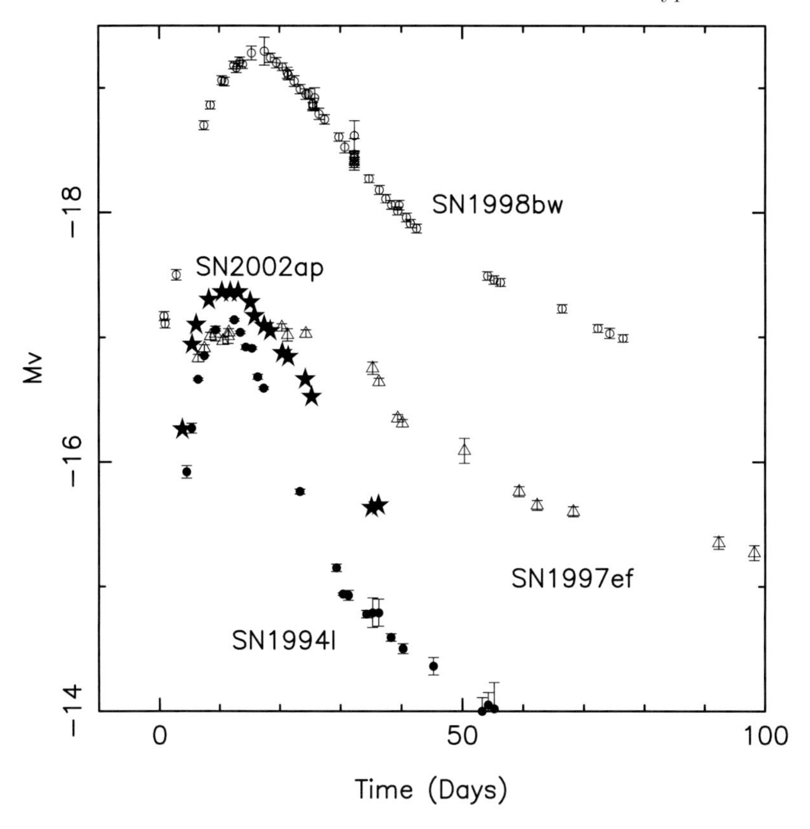

Fig. 11. The observed V-band light curves of type Ic SNe: 1998bw (*open circles*), 1997ef (*open triangles*), 2002ap (*stars*), and the normally-energetic 1994I (*filled circles*) [60].

of 10^{51} erg were possible, led to a reinterpretation of SN1997ef as a hypernova, although of smaller energy than SN1998bw.

Light Curve: The light curve of SN1997ef has a maximum as broad as ~ 25 days and enters its exponential tail at ~ 40 days after maximum, later than other SNIc. The light curve can be reproduced by different explosion models, CO60 (standard energy) and CO100 ($E = 10^{52}$erg) as seen in Fig. 12. A distance of 52.3 Mpc (i.e., distance modulus of $\mu = 33.6$ mag) is adopted for SN1997ef and a color excess of $E(B-V) = 0$ is assumed. The early rise of the light curve can be explained by the complete mixing of ^{56}Ni within the ejecta.

Model CO60 has the same kinetic energy ($E_{\rm K} = 1 \times 10^{51}$ erg) as model CO21 for SN1994I (see Table 1 for the model parameters), but the light curve of SN1997ef is much slower than that of SN1994I. Thus, the ejecta of CO60 should be more massive than CO21 by a factor of ~ 5. Model CO100 is about twice as massive as CO60, and less massive than CO138 for SN1998bw (Table 1) by only $\sim 20\%$. Thus, to reproduce the observed light curve, the explosion energy

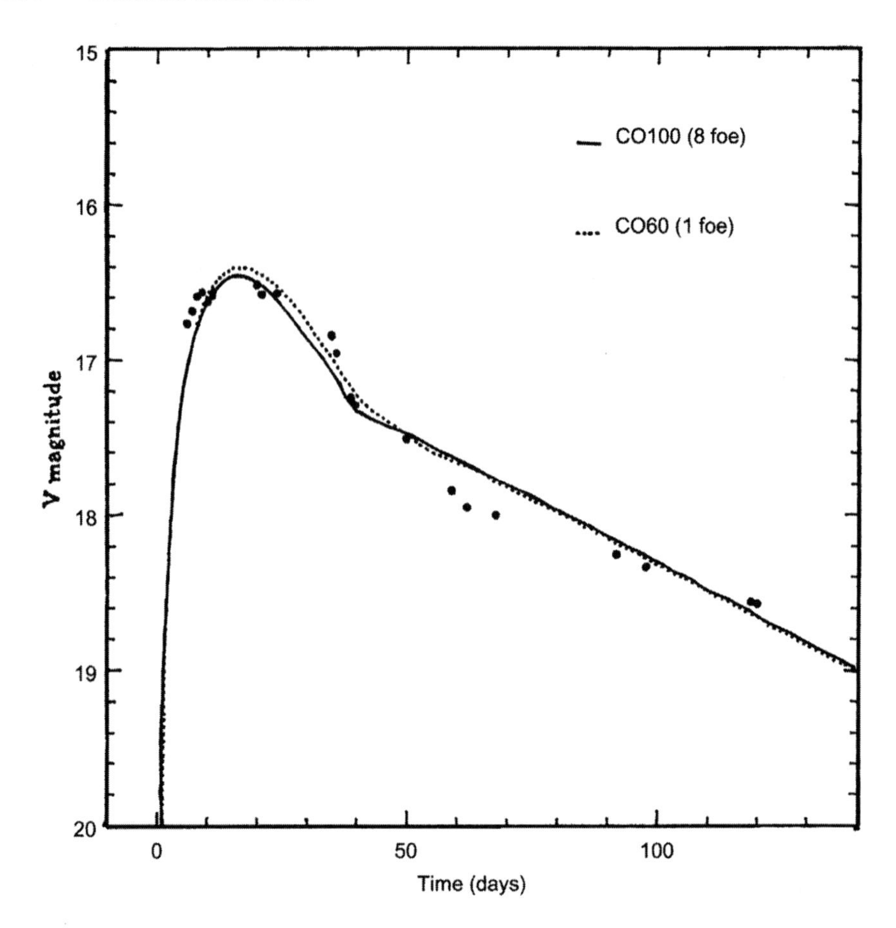

Fig. 12. Synthetic Visual light curves of models CO60 and CO100 [28] compared with the observations of SN1997ef [38].

of CO100 must be $\sim$ 10 times larger than that of CO60. This is a very energetic explosion, but still much weaker than CO138. The smaller $E_{\rm K}$ for a comparable mass is required for CO100 to reproduce the light curve of SN1997ef.

Entering the exponential tail around day 40, the V magnitude declines linearly with time at a rate of $\sim 1.1 \times 10^{-2}$ mag day^{-1}, which is slower than in other type Ic and is close to the ^{56}Co decay rate of 9.6×10^{-3} mag day^{-1}. Such a slow decline implies a more efficient γ-ray trapping in the ejecta of SN1997ef than in SN1994I. The ejecta of both CO100 and CO60 are massive enough to trap the γ-rays effectively, resulting in slower light curve exponential tails than CO21. However, the light curve exponential tail of both models declines somewhat faster than the observations.

A similar discrepancy has been noted for the type Ib supernovae SN1984L and SN1985F [5,95]. The late time light curve decline of these type Ib SNe is as slow as the ^{56}Co decay rate, so that the inferred value of ejecta mass, $M_{\rm ej}$, is significantly larger (and/or E_K is smaller) than those obtained from the early light curve shape. Baron et al. [5] suggested that the ejecta of these type Ib SNe must be highly energetic and as massive as ~ 50 $M_\odot$. Such a discrepancy between the early and late time light curves implies asphericity in the ejecta of SN1997ef, and perhaps also in other SNIb.

Spectra: As we have shown, light curve modeling provides direct constraints on M_{CO} and E_K. However, it is difficult to determine these values uniquely from the light curve shape alone, since models with different values of M_{ej} and E_K can yield similar light curves. Fortunately, spectral synthesis modeling can distinguish between these degenerate models because the spectrum contains much more information than a single band light curve.

Around maximum light, the spectra of SN1997ef show a few very broad features, and are quite different from those of ordinary type Ib/c, but similar to those of SN1998bw. However, at later epochs the spectra develop features that are easy to identify, such as the Ca II IR triplet at ~ 8200 Å, the O I absorption at 7500 Å, several Fe II features in the blue, and they look very similar to the spectrum of the ordinary SNIc SN1994I.

We computed synthetic spectra with a Monte Carlo spectrum synthesis code, using the density structure and composition of the hydrodynamic models CO60 and CO100, and compared them to three near-maximum epochs of SN1997ef: 29 November, 05 December, and 17 December 1997. These spectra are early enough to be very sensitive to changes in the kinetic energy. As in the light curve comparison, we adopted a distance modulus of $\mu = 33.6$ mag, and $E(B-V) = 0.0$.

In Fig. 13 (upper panel) we show the synthetic spectra computed with the ordinary SNIc model CO60. The lines in the spectra computed with this model are always much narrower than the observations. This clearly indicates a lack of material at high velocity in model CO60, and suggests that the kinetic energy of this model is much too small.

Synthetic spectra obtained with the hypernova model CO100 for the same 3 epochs are shown in Fig. 13 (lower panel). The spectra show much broader lines, and are in good agreement with the observations. In particular, the blending of the Fe lines in the blue, giving rise to broad absorption troughs, is well reproduced. The two "emission peaks" observed at ~ 4400 and 5200 Å are not real lines, but they correspond to the only two regions in the blue that are relatively line free so that photons can escape.

The spectra are characterized by a low temperature, even near maximum, because the rapid expansion combined with the relatively low luminosity (from the exponential tail of the light curve we deduce that SN1997ef produced about 0.15 $M_\odot$ of ^{56}Ni, compared to about 0.6 $M_\odot$ in a typical SNIa and 0.5 $M_\odot$ in SN1998bw) leads to rapid cooling. Thus, the Si II 6355 Å line is not very strong.

Model CO100 has $E = 10^{52}$ erg, $M_{ej} = 7.5$ $M_\odot$, $M(^{56}Ni) = 0.15$ $M_\odot$. From these values, we find $M_{CO} = 10$ $M_\odot$and $M_{rem} = 2.5$ $M_\odot$. A 10 $M_\odot$ CO core is formed in a ~ 30 $M_\odot$ star. Although model CO100 yields rather good

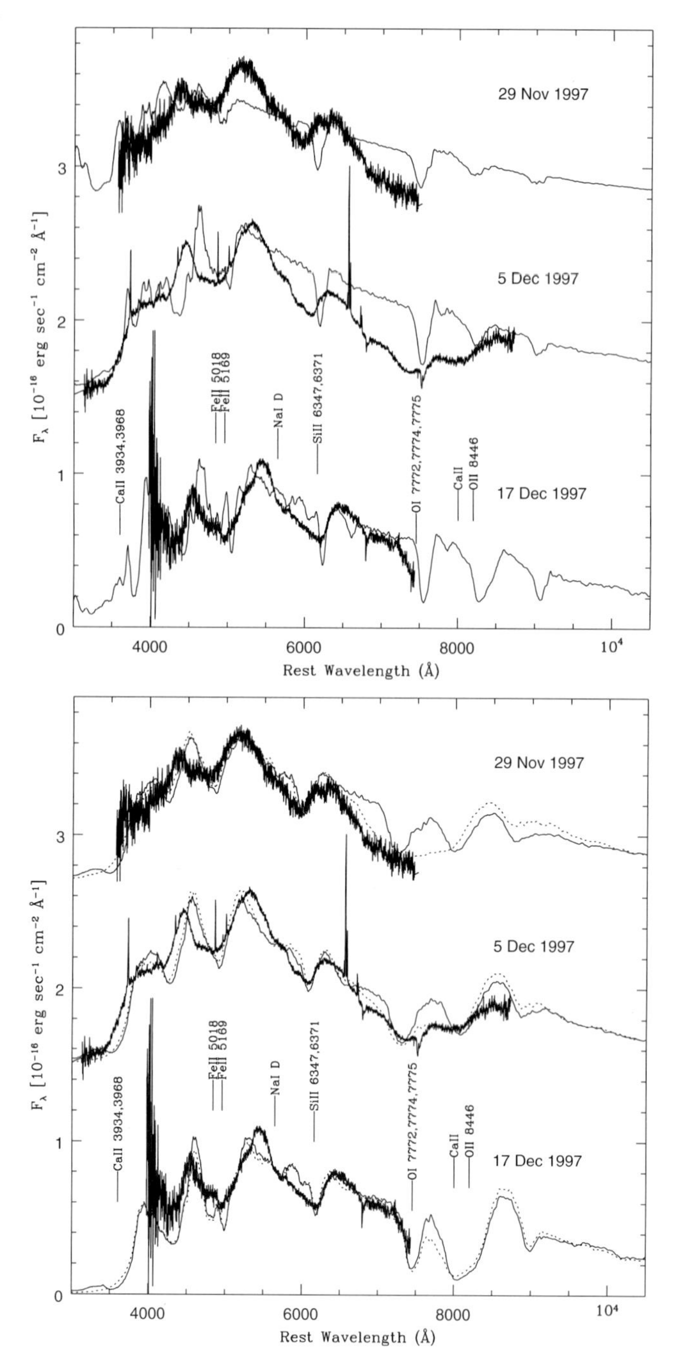

Fig. 13. Observed spectra of SN1997ef (bold lines) and synthetic spectra computed using models CO60 (upper panel) and CO100 (lower panel). The line features seen in the synthetic spectra for CO60 are much too narrow compared with observations, while the fits are greatly improved when model CO100 is used [38,58].

synthetic spectra, it still fails to reproduce the observed large width of the O I-Ca II feature in the only near maximum spectrum that extends sufficiently far to the red (05 Dec 1997). An improvement can be obtained by introducing an arbitrary flattening of the density profile at the highest velocities (shown as the dotted lines in Fig. 13 (lower panel) [10,58]. This modification leads to higher values of both $E_{\rm K}$ and $M_{\rm ej}$.

Possible Aspherical Effects: The light curve, the photospheric velocities, and the spectra of SN1997ef are better reproduced with the hyperenergetic model CO100 than with the ordinary SNIc model CO60. However, there remain several features that are still difficult to explain with model CO100.

- The observed velocity of Si II decreases much more rapidly than models predict. It is as high as $\sim 30,000$ km s^{-1} at the earliest phase, but it decreases to as little as $\sim 3,000$ km s^{-1} around day 50. Such a rapid drop of the photospheric velocity is difficult to explain by either CO100 or CO60, or by any other model that reproduces the light-curve shape.
- The observed light curve decline rate is slower than model CO100 in the exponential tail part, and it is also a bit flatter than the model around maximum (Fig. 12). Models with lower energies and/or larger masses give better fits to both the peak and the exponential tail of the light curve, but they fail to reproduce the large photospheric velocities observed at early times in SN1997ef.
- The photospheric epoch persists for at least 70 days [58] and line absorption is seen at very low velocities (~ 2000 km s^{-1}). This is in contradiction with the 1-D model CO100, which has a low velocity edge at ~ 5000 km s^{-1}, corresponding to the mass cut. These features may be a possible sign of asphericity in the ejecta. Some parts of the ejecta expand faster, forming the high velocity lines at early phases, while other parts expand more slowly to explain the slow decline of the exponential tail and the long duration of the photospheric phase, as well as the low velocity lines observed at epochs of a few months.
- Extensive mixing of ^{56}Ni is required to reproduce the short rise time of the light curve. One possibility to induce such mixing in velocity space is an asymmetric explosion. In extremely asymmetric cases, material ejection may take place in a jet-like form (see, e.g., [42,52]). The jet could easily mix some Ni from the deepest layers out to the high velocity surface. However, the lack of a coincidence with a GRB for SN1997ef suggests that if a jet was produced, it was either weak or it was not pointing towards us.

Unlike other hypernovae, SN1997ef doesn't seem to be a unique case. At least two other SNe, SN1997dq [55] and SN1999ey show very similar properties. Unfortunately, these two objects were not very well observed. Whether this is just a coincidence, or whether it really indicates that SN1997ef-like hypernovae are more frequent than other types remains to be resolved by further observations.

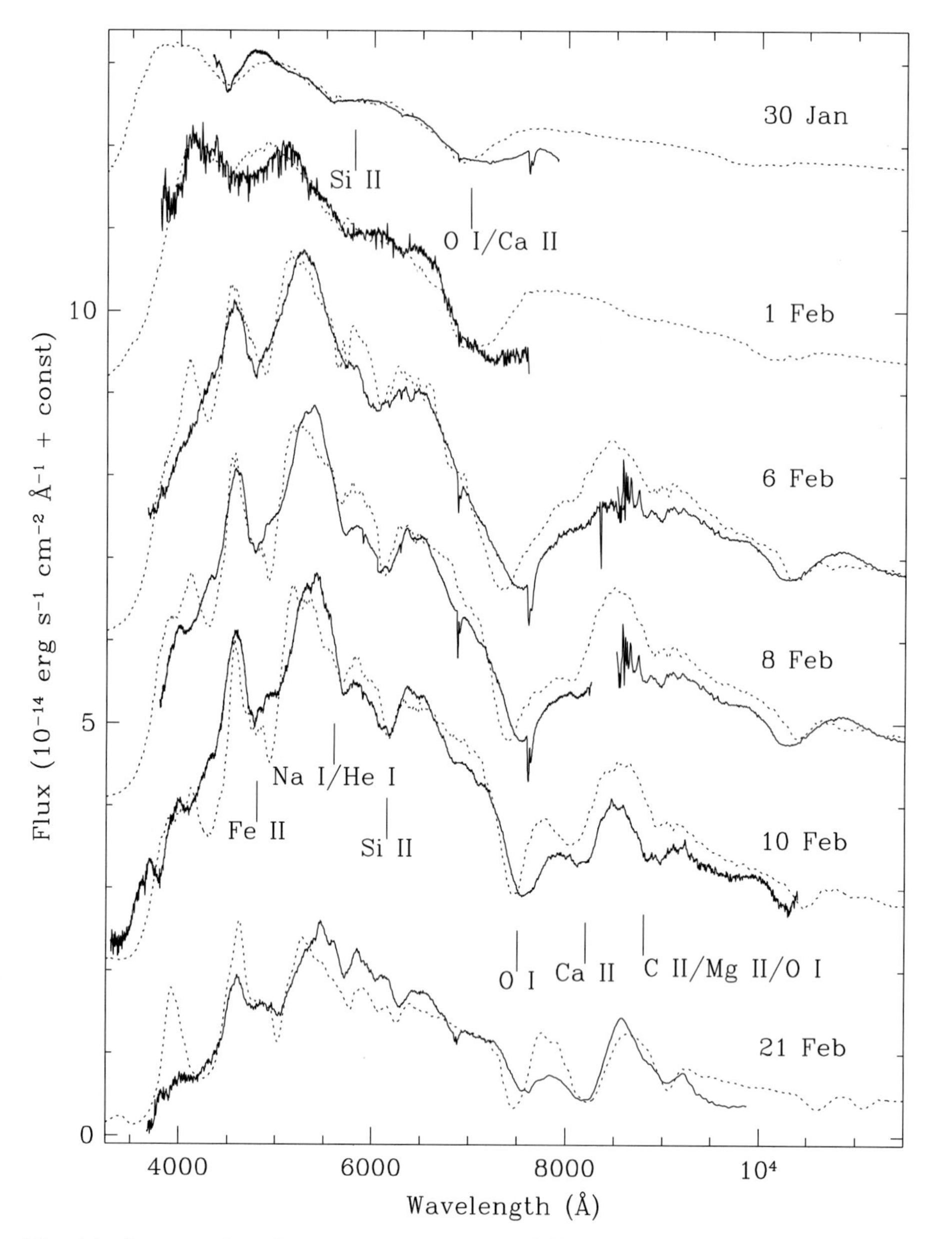

Fig. 14. A comparison between some spectra of SN2002ap observed in 2002 (*thick lines*: 30 January, *William Herschel Telescope (WHT)*; 01 February, *Gunma Astronomical Observatory (GAO)*; 06 February, *Beijing Astronomical Observatory (BAO)*; 08 February, *Subaru Telescope)*; 10 February, *Lick Observatory*; 21 February, *Asiago Observatory*) and synthetic spectra computed with model CO100/4 (*dashed lines*) [60].

5.2 SN2002ap

Type Ic SN2002ap was discovered in M74 on 30 January 2002 [33]. The SN was immediately recognized as a hypernova from its broad spectral features [19,22,44,62](Fig. 14), which indicated high velocities in the ejected material, the typical signature of a hypernova. It was therefore studied by several observatories and its relative proximity enabled observations with small telescopes. Luckily, the SN was discovered very soon after it exploded: the discovery date was 29 January, while observations of the area on 25 January showed no detectable emission from the supernova position [66]. This is among the earliest any SN has ever been observed, with the obvious exceptions of SN1987A and SN1993J.

Light Curve: SN2002ap reached V maximum on about 08 February at V = 12.3 mag (Fig. 11). SN2002ap peaked earlier than both hypernovae SN1998bw and SN1997ef, but later than the normal SN1994I, suggesting an intermediate value of the ejecta mass, M_{ej}.

Using a distance to M74 of 8 Mpc (μ = 29.5 mag [86]), and a combined Galaxy and M74 reddening of $E(B-V) = 0.09$ mag (estimated from a *Subaru Telescope) High Dispersion Spectrograph (HDS)* spectrum [96]), the peak absolute magnitude was $M_{\mathrm{V}} = -17.4$. This is comparable to SN1997ef and fainter than SN1998bw by almost 2 mag. Since the peak brightness depends on the ejected ^{56}Ni mass, SN2002ap, SN1997ef, and SN1994I appear to have synthesized similar amounts. Estimates were ~ 0.07 $\mathrm{M}_{\odot}$ for SN1994I [69] and 0.13 $\mathrm{M}_{\odot}$ for SN1997ef [58]. The ^{56}Ni mass for SN2002ap is estimated to be ~ 0.07 $\mathrm{M}_{\odot}$, which is similar to that of normal core-collapse SNe such as SN1987A and SN1994I.

Spectrum: If line width is the distinguishing feature of a hypernova, then clearly SN2002ap is a hypernova, as its spectrum resembles that of SN1997ef much more than that of SN1994I (Fig. 1). Line blending in SN2002ap and SN1997ef is comparable. However, some individual features that are clearly visible in SN1994I, but are completely blended in SN1997ef, can at least be discerned in SN2002ap (e.g., the Na I-Si II blend near 6000 Å and the Fe II lines near 5000 Å). Therefore, spectroscopically SN2002ap appears to be located just below SN1997ef on a "velocity scale," but far above SN1994I, which appears to confirm the evidence from the light curve.

The evolution of SN2002ap (Fig. 14) appears to follow closely that of SN1997ef, at a rate about 1.5 times faster. The spectra and the light curve of SN2002ap can be well reproduced by a model with ejected heavy element mass $M_{\mathrm{ej}} = 2.5 - 5$ $\mathrm{M}_{\odot}$ and $E_{51} = 4 - 10$ (Fig. 14 [61]). Both M_{ej} and E_{K} are much smaller than those of SN1998bw and SN1997ef (but they could be larger if a significant amount of He is present).

A Hypernova or a Supernova: Although SN2002ap appears to lie between normal core-collapse SNe and hypernovae, it should be regarded as a hypernova because its kinetic energy is distinctly higher than in normal core-collapse SNe. In other words, the broad spectral features that characterize hypernovae are the result of a high kinetic energy. However, SN2002ap was no more luminous than normal core-collapse SNe. This suggests that brightness alone should not be used to distinguish hypernovae from normal SNe. The criteria should be a high kinetic energy accompanied by broad spectral features. Further examples of hypernovae are necessary in order to establish whether a firm boundary exists between the two groups.

For the values of $M_{\rm ej}$, $M_{\rm ej}$, and $M(^{56}{\rm Ni})$ selected for SN2002ap, the progenitor main sequence mass, $M_{\rm ms}$, and the stellar remnant mass, $M_{\rm rem}$, can be constrained. Modeling the explosions of C+O stars with various masses, we obtain $M(^{56}{\rm Ni})$ as a function of the parameter set $E_{\rm K}$, $M_{\rm CO}$, $M_{\rm rem} = M_{\rm ej} - M_{\rm CO}$. The model which is most consistent with our estimates of $M_{\rm ej}$ and $M_{\rm ej}$ is one with $M_{\rm CO} \approx 5\ {\rm M}_\odot$, $M_{\rm rem} \approx 2.5\ {\rm M}_\odot$, and $E_{51} = 4.2$. The $5\ {\rm M}_\odot$C+O core forms in a He core of mass of $M_{\rm He} = 7\ {\rm M}_\odot$, corresponding to a main sequence mass of $M_{\rm ms} \approx 20 - 25\ {\rm M}_\odot$. The $M_{\rm ms} - M_{\rm He}$ relation depends on convection and metallicity (e.g. [67,104].

The estimated progenitor mass and explosion energy are both smaller than those of previous hypernovae such as SN1998bw and SN1997ef, but larger than those of normal core-collapse SNe such as SN1999em. This mass range is consistent with the non-detection of the progenitor in pre-discovery images of M74 [90].

Given the estimated mass of the progenitor, binary interaction, including the spiral-in of a companion star [71], is probably required in order for it to lose its hydrogen and some (or most) of its helium envelope. This would suggest that the progenitor was in a state of fast rotation. It is possible that a high rotation rate and/or envelope ejection are also necessary conditions for the birth of a hypernova.

Possible Aspherical Effects: SN2002ap was not apparently associated with a GRB. This may actually be not so surprising, since the explosion energy of SN2002ap was about a factor of $5 - 10$ smaller than that of SN1998bw, as was also indicated by the weak radio signature [6]. The present data show no clear signature of asymmetry, except perhaps for some polarization [40,48,107] which was smaller than that of SN1998bw. This suggests that the degree of asphericity is smaller in SN2002ap and that a possible jet may have been weaker, making GRB generation more difficult.

5.3 SN1999as

SN1999as was discovered on 08 February 1999, by the *Supernova Cosmological Project (SCP)* [46] in an anonymous galaxy at $z = 0.127$. The absolute magnitude was exceptionally bright, $M_{\rm V} < -21.5$, at least nine times brighter than SN1998bw.

SN1999as was spectroscopically classified as a SNIc because its photospheric phase spectra showed no conspicuous lines of hydrogen, He I, or Si II λ6355. The usual SNIc spectral lines such as Ca II and O I were very broad, like in other hypernovae. However, some narrow (~ 2000 km s^{-1}), but highly blueshifted ($\sim 11,000$ km s^{-1}), lines of Fe II were also present [32].

Fitting the observed light curve we have obtained the following constraints on the explosion model: ejected mass $M_{\rm ej} \simeq 10-20$ M$_\odot$, kinetic energy of the ejected matter $E_{\rm K} \simeq 10^{52}-10^{53}$ erg, and mass of ejected ^{56}Ni $M_{\rm Ni} > 4$ M$_\odot$ [16]. The progenitor may have been as massive as ~ 60 M$_\odot$, and the explosion almost certainly resulted in the formation of a black hole. Unfortunately, the spectral coverage of SN1999as is not very extensive, and a more accurate determination of the properties of this extraordinary supernova is therefore difficult.

The asymmetric hydrodynamical model of Maeda et al. [54] could describe this new class of hypernovae. In this model, a clump of freshly synthesized ^{56}Ni exists at high velocity ($\sim 15,000$ km s^{-1}) near the symmetry axis. Such a clump could produce narrow but high velocity absorption lines if the viewing angle with respect to the symmetry axis is small. On the other hand, the lack of a detected GRB may contradict this scenario.

5.4 Type IIn Hypernovae: SN1997cy and SN1999E

SN1997cy is a different kind of hypernova, as it is of SNIIn origin. However, the energy deduced from its light curve is extremely large. The energy reveals itself in the strength of the ejecta-CSM interaction. Furthermore, the SN may have a correlated GRB. SN1999E is very similar to SN1997cy in its spectrum and light curve [83].

SN1997cy displayed narrow Hα emission on top of broad wings, leading to its classification as a type IIn [30,103]. Assuming $A_V = 0.00$ for the galactic extinction we get an absolute magnitude at maximum of $M_{\rm V} \leq -20.1$. This makes SN1997cy the brightest SNII discovered to date. The light curve of SN1997cy does not conform to the classical templates of SNII, namely type IIL and type IIP, but resembles the slow evolution of the type IIn SN1988Z. As seen from the UVOIR bolometric light curve in Fig. 15, the SN light curve decline is much slower than the ^{56}Co decay rate between day 120 and 250 (the epoch is set assuming that the SN was coincident with GRB970514). This suggests that the source of the light output of the SN is circumstellar interaction.

In the interaction model [103], collision of the SN ejecta with the slowly moving CSM converts the kinetic energy of the ejecta into light, thus producing the observed intense light display of the SN. Our exploratory model considers the explosion of a massive star ($M = 25$ M$_\odot$) with parameterized kinetic energy $E_{\rm K}$. The collision between the SN ejecta and the CSM is assumed to begin near the stellar radius, at a distance r_1, where the density of the CSM is ρ_1. A power-law density profile $\rho \propto r^n$ is adopted to describe the CSM. The parameters $E_{\rm K}$, ρ_1, and n, are constrained through comparison with the observations.

The regions excited by the forward and reverse shock emit mostly X-rays. The density in the shocked ejecta is so high that the reverse shock is radiative and a dense cooling shell is formed (see, e.g., [94,97]). The X-rays are absorbed

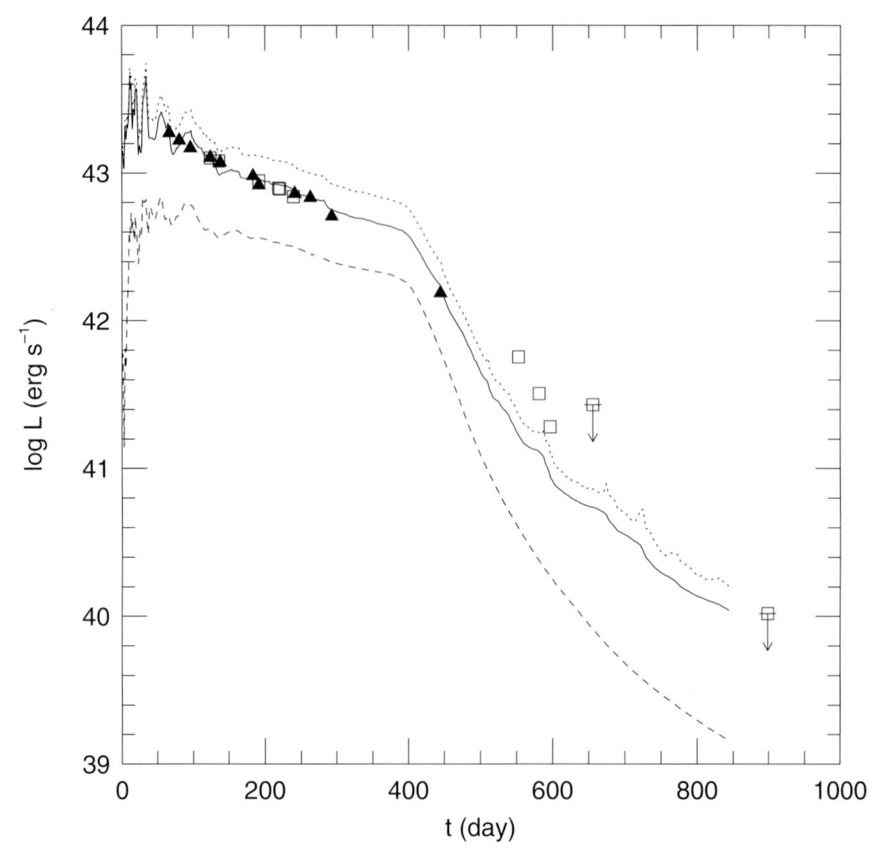

Fig. 15. The UVOIR bolometric light curve, L_{UVOIR}, of SN1997cy compared with the synthetic light curve obtained with the CSM interaction model. The solid and the dashed curves show L_{UVIOR} and the luminosity of the X-rays, L_{X}, escaping from the ejecta, respectively, and the dotted curve shows the total luminosity of the shocked ejecta, $L_{\mathrm{tot}}(= L_{\mathrm{UVIOR}} + L_{\mathrm{X}})$ [103].

by the outer layers and the core of the ejecta, and re-emitted as UV-optical photons.

Narrow lines are emitted from the slowly expanding, unshocked CSM photoionized by the SN UV outburst or by the radiation from the shocks; intermediate width lines come from the shock-heated CSM and broad lines come from the cooler region at the interface between the ejecta and the CSM.

Fig. 15 shows the model light curve which best fits the observations. The model parameters are: $E_{\mathrm{K}} = 3 \times 10^{52}$ erg, $\rho_1 = 4 \times 10^{-14}$ g cm^{-3} at $r_1 = 2 \times 10^{14}$ cm (which corresponds to a mass-loss rate of $\dot{M} = 4 \times 10^{-4}$ $\mathrm{M}_{\odot}$ yr^{-1} for a wind velocity of 10 km s^{-1}), and $n = -1.6$. The large mass-loss episode giving rise to the dense CSM is supposed to have occurred after the progenitor made a loop in the HR diagram from a blue supergiant (BSG) to a red supergiant (RSG). In this model, the mass of the low-velocity CSM is $\sim 5\ M_\odot$, which implies that the transition from BSG to RSG took place about 10^4 yr before the SN event.

The large CSM mass and density are necessary to have large shocked masses, and thus to reproduce the observed high luminosity, and so is the very large explosion energy. For models with low E_K and high ρ_1, the reverse shock speed is too low to produce a sufficiently high luminosity. For example, a model with $E_K = 10^{52}$ erg and ρ_1 as above yields a value of L_{UVOIR} lower than the observed luminosity by a factor of ~ 5. For high E_K or low ρ_1, the expansion of the SN ejecta is too fast for the cooling shell to absorb enough X-rays to sustain the luminosity. Thus, in our preferred model both E_K and $\dot{M}$ are constrained to be within a factor of ~ 3 of the quoted values.

The shape of the light curve also constrains the density structure of the CSM. For $n = -2$, the case of a steady wind, L_{UVOIR} decreases too rapidly around day 200. To reproduce the observed decrease after day ~ 300, the CSM density is assumed to drop sharply at the radius the forward shock reaches at day 300, so that the collision becomes weaker afterwards. (Such a change of the CSM density corresponds to the transition from BSG to RSG of the progenitor $\sim 10^4$ yr before the SN explosion.) This is consistent with the simultaneous decrease in the Hα luminosity and may be similar to that noted in Hα and the radio for type IIn SN1988Z (see [112] and the chapter by Sramek and Weiler in this volume).

The light curve drops sharply after day 550. The model reproduces such a light curve behavior (Fig. 15) by assuming that when the reverse shock has propagated through $\sim 5\ M_\odot$, it encounters an exceedingly low density region and dies. In other words, the model for the progenitor of SN1997cy assumes that most of the core material fell into a massive black hole of, say, $\sim 10\ M_\odot$, while the extended H/He envelope ($\sim 5\ M_\odot$) did not. Accretion leads to material being ejected from the massive black hole, possibly in jet-like form. The envelope is then hit by the jet and ejected at high velocity, which is the SN event.

In this model, the SN ejecta are basically the H/He layers of the progenitor, which therefore have the original solar abundance of heavy elements plus some heavy elements mixed out from the core (before fall back) or from the jet. This might explain the lack of oxygen and magnesium lines in the spectra, particularly in the nebular phases [103].

6 Properties of Hypernovae

It is possible to make some generalizations regarding the properties of hypernovae and the relation to their progenitor stars. In Table 3 we list the main properties of the objects studied.

6.1 The Explosion Kinetic Energy

In Fig. 16 we plot E_K as a function of the main sequence mass M_{ms} of the progenitor star derived from fitting the optical light curves and spectra of various hypernovae, of the normal SNe SN1987A, SN1993J, and SN1994I (see, e.g.,

Table 3. Properties of Supernovae and Hypernovae

	83N	94I	02ap	97ef	98bw
	Ib	Ic	Ic (Hypernovae)		
	Pre-explosion				
$M_{\rm ZAMS}$ ($M_\odot$)	15	15	21	34	40
$M_{\rm He}$ ($M_\odot$)	4	4	6.6	13	16
$M_{\rm CO}$ ($M_\odot$)	2	2.1	4.5	11.1	14
$M_{\rm exp}$ ($M_\odot$)	4	2.1	4.6	11.1	13.8
	Post-explosion				
$M_{\rm rem}$ ($M_\odot$)	1.25	1.2	2.1	1.6	2.9
$M_{\rm ej}$ ($M_\odot$)	2.75	0.9	2.5	9.5	10.9
$M_{\rm He}$ ($M_\odot$)	2.0	0	0.1	0	0
$M_{\rm CO}$ ($M_\odot$)	0.5	0.6	1.8	5.3	8
$M_{\rm IME}$ ($M_\odot$)	0.1	0.2	0.5	4	2
$M_{\rm Ni}$ ($M_\odot$)	0.15	0.07	0.1	0.13	0.7
$E_{\rm K}$ (10^{51} erg)	1	1	7	19	30

[36,68,69,87,88,113,116]), and of SN1997D [102] and SN1997br [117]. It appears that $E_{\rm K}$ increases with $M_{\rm ms}$, forming a "Hypernova Branch" reaching values much larger than the canonical 10^{51} erg for SNe. SN1997D and SN1999br, on the contrary, are located well below that branch, forming a "Faint SN Branch" [73].

This trend might be interpreted as follows: stars with $M_{\rm ms} \lesssim 20 - 25\ M_\odot$ leave a neutron star as a stellar remnant (SN1987A may be a borderline case between neutron star formation and black hole formation) while stars with $M_{\rm ms} \gtrsim 20 - 25\ M_\odot$ leave a black hole (see, e.g., [17]). Whether black hole formation generates a hypernova or a faint SN may depend on the angular momentum in the collapsing core, which in turn depends on the strength of stellar winds, metallicity, magnetic fields, and binarity.

Hypernovae may have rapidly rotating cores, owing possibly to the spiraling-in of a companion star in a binary system. The cores of faint type II SNe might not have a large angular momentum because the progenitor had a massive H-envelope which transported the angular momentum of the core to the envelope, possibly via a magnetic-field effect.

Between these two branches, there may be a variety of SNe. A dispersion in the properties of type IIP SNe has been reported [31].

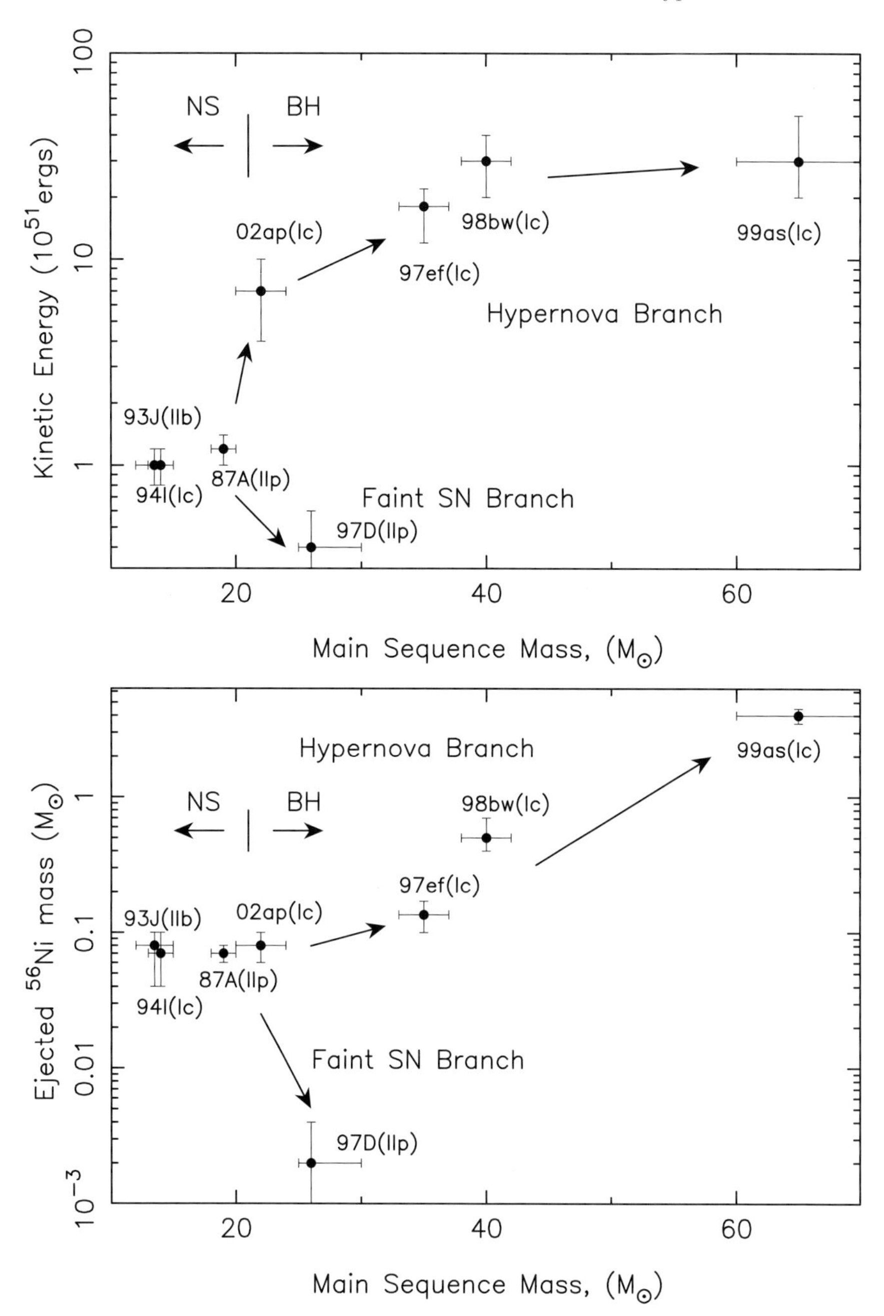

Fig. 16. Explosion energies and the ejected ^{56}Ni mass plotted versus the main sequence mass of the progenitor for several well studied supernovae/hypernovae [73].

6.2 The Mass of Ejected ^{56}Ni

A similar relation is observed between the mass of ^{56}Ni, $M(^{56}\mathrm{Ni})$, synthesized in core-collapse supernovae and M_{ms} (Fig. 16). This relation may be relevant for the study of the chemical evolution of galaxies.

Stars with $M_{\mathrm{ms}} \lesssim 20-25$ $\mathrm{M}_{\odot}$, forming a neutron star, produce $\sim 0.08\pm0.03$ $\mathrm{M}_{\odot}$ ^{56}Ni as in SN1993J, SN1994I, and SN1987A. For stars with $M_{\mathrm{ms}} \gtrsim 20-25$ $\mathrm{M}_{\odot}$, which form black holes, $M(^{56}\mathrm{Ni})$ appears to increase with M_{ms} in the hypernova branch, while SNe in the faint SN branch produce only very little ^{56}Ni. For faint SNe, because of the large gravitational potential, the explosion energy was so small that most ^{56}Ni fell back onto the compact remnant.

6.3 Asymmetry

All hypernovae of type Ic show some signatures of asymmetry, or at least of a departure from purely 1-dimensional spherically symmetric models. This may support the case for their connection with at least some GRBs.

Only for SN1998bw is the connection with a GRB well established. In the other cases, either a GRB was not generated or it must have been weak and/or not pointing towards us. The issue of directionality is very important. If hypernovae are aspherical, we expect to observe, at early times, a range of hypernova properties for the same Ni mass, the value of which can be established at late times independently of the shape of the ejecta. These objects may be very different at early phases, showing different light curves, velocities, and abundances. So far, however, this evidence is missing. Increasing the sample of observed hypernovae may change this situation.

6.4 The Gamma-Ray Burst/Hypernova Connection

For several hypernovae, a GRB connection has been suggested: GRB980425/ SN1998bw [23,37], GRB971115/ SN1997ef [106], GRB970514/ SN1997cy [30,103], GRB980910/SN1999E [100], and GRB991002/ SN1999eb [98].

Several GRBs are suggested to be associated with SNe: GRB980326 [8], GRB970228 [27,80], and GRB011121/SN2001ke [9,29]. The decline of the light curve of the optical afterglows of these GRBs slowed at late phases, and this can be reproduced if a red-shifted SN1998bw-like light curve is superposed on the power-law component.

Another question is whether the supernovae associated with GRBs have uniform peak luminosity, corresponding to the production of ~ 0.5 $\mathrm{M}_{\odot}$ of ^{56}Ni in SN1998bw. Fig. 16 shows that the ^{56}Ni mass, and thus the intrinsic maximum brightness of hypernovae, is highly variable. We thus need to increase the observational sample before we can define the luminosity function and the actual range of the mass of ^{56}Ni produced in supernovae/hypernovae.

There are several hypernovae, such as SN1998ey and SN2002ap, for which no GRB counterpart has been proposed. These were less energetic events than SN1998bw. It is possible that a weaker explosion is less efficient in collimating the γ-rays to give rise to a detectable GRB (GRB980425 was already quite weak compared to the average GRB), or that some degree of inclination of the beam axis with respect to the line-of-sight results in a seemingly weaker supernova and in the non-detection of a GRB. Again, only the accumulation of more data will allow us to address these questions.

The properties of hypernova nucleosynthesis suggest that hypernovae could make an important contribution to the Galactic (and cosmic) chemical evolution [63,72]. Hypernovae are probably the explosion of very massive stars. In view of the small frequency of GRBs, hypernovae may then be much more frequent than GRBs. In this case, only a special class of hypernovae gives rise to GRBs [73], or maybe all hypernovae have an associated GRB, but only a small fraction of very massive stars become hypernovae.

6.5 Possible Evolutionary Scenarios for Hypernovae

Here we discuss possible evolutionary paths leading to C+O star progenitors. In particular, we explore the paths to progenitors with rapidly rotating cores, because the explosion energy of hypernovae may be provided by rapidly rotating black holes [7].

Single Star Progenitor: If a star is as massive as $M_{\rm ms} \gtrsim 40\ \mathrm{M}_\odot$, it could lose its H and He envelopes in a strong stellar wind (see, e.g., [85]). This would be a Wolf-Rayet (W-R) star.

Close Binary Star Progenitor: Suppose we have a close binary system with a large mass ratio. In this case, the mass transfer from star 1 to star 2 inevitably takes place in a non-conservative way, and the system experiences a common envelope phase with star 2 spiraling into the envelope of star 1. If the spiral-in releases enough energy to remove the common envelope, we are left with a bare He star (star 1) and a main-sequence star (star 2), with a reduced separation. If the orbital energy is too small to eject the common envelope, the two stars merge to form a single star (see, e.g., [105]).

For the non-merging case, possible paths from the He stars to the C+O stars are as follows [70]:

- Small mass He stars tend to have large radii, so they fill their Roche lobes more easily and lose most of their He envelope via Roche lobe overflow.
- Larger mass He stars have radii too small to fill their Roche lobes, but have large enough luminosities to drive strong winds which can remove most of the He layer (see, e.g., [114]). Such mass-losing He stars would again be W-R stars.

Thus, from the non-merging scenario, we expect two different kinds of SNIc, fast and slow, depending on the mass of the progenitor. SNIc from smaller mass progenitors have a faster light curve and spectral evolution because the ejecta become quickly transparent to both γ-rays and optical photons. Slow SNIc, on the other hand, originate from W-R progenitors . The presence of both slow and fast SNe Ib/Ic was noted by Clocchiatti and Wheeler [15].

For the merging case, the merged star has a large angular momentum so that its collapsing core must be rapidly rotating. This would lead to the formation of a rapidly rotating black hole from which a hyperenergetic jet may emerge. If the merging process is slow enough that the H/He envelope is ejected, the star could become a rapidly rotating C+O star. Such stars are the candidate progenitors of type Ic hypernovae like SN1997ef and SN1998bw. If a significant fraction of the H-rich envelope is not lost upon merging, the star may become a hypernova of type IIn, possibly like SN1997cy, or a type Ib.

Acknowledgments

This work has been supported in part by the Grant-in-Aid for Scientific Research 07CE2002, 12640233, 14047206, 14540223 of the Ministry of Education, Science, Culture, Sports, and Technology in Japan.

References

1. M.A. Aloy, E. Müller, J.M. Ibáñez, J.M. Martí, A. MacFadyen: Astrophys. J. Lett. **531**, L119 (2000)
2. D. Arnett: Astrophys. J. **253**, 785 (1982)
3. D. Arnett: In: *Supernovae and Gamma-Ray Bursts*, ed. by M. Livio, N. Panagia, K. Sahu (Cambridge University Press, Cambridge 2001) p. 250
4. T.S. Axelrod: Ph.D. thesis, University of California, Berkeley, CA, USA (1980)
5. E. Baron, T.R. Young, D. Branch: Astrophys. J. **409**, 417 (1993)
6. E. Berger, S.R. Kulkarni, R.A. Chevalier: Astrophys. J. Lett. **577**, L5 (2002)
7. R.D. Blandford, R.L. Znajek: Mon. Not. R. Astron. Soc. **179**, 433 (1977)
8. J.S. Bloom et al.: Nature **401**, 453 (1999)
9. J.S. Bloom et al.: Astrophys. J. Lett. **572**, L45 (2002)
10. D. Branch: In: *Supernovae and Gamma-Ray Bursts*, ed. by M. Livio, N. Panagia, K. Sahu (Cambridge University Press, Cambridge 2001) p. 96
11. N. Chugai: Astron. Lett. **26**, 797 (2000)
12. A. Clocchiatti, J.C. Wheeler, M.S. Drotherton, A.L. Cochran, D. Wills, E.S. Barker: Astrophys. J. **462**, 462 (1996)
13. S.A. Colgate, A.G. Petschek: Astrophys. J. **229**, 682 (1979)
14. S.A. Colgate, A.G. Petschek, J.T. Kriese: Astrophys. J. Lett. **237**, L81 (1980)
15. A. Clocchiatti, J.C. Wheeler: Astrophys. J. **491**, 375 (1997)
16. J. Deng, K. Hatano, T. Nakamura, K. Maeda, K. Nomoto, P. Nugent, G. Aldering, D. Branch: In: *New Century of X-ray Astronomy, (ASP Conference Series 251)*, ed. by H. Inoue, H. Kunieda (Astron. Soc. of Pacific, San Francisco 2001) p. 238
17. E. Ergma, E.P.J. van den Heuvel: Astron. Astrophys. **331**, L29 (1998)
18. A. Filippenko: IAUC 6969 (1998)
19. A.V. Filippenko, R. Chornock: IAUC 7825 (2002)
20. C. Fransson, C. Kozma: Astrophys. J. Lett. **408**, L25 (1993)
21. J.U. Fynbo: Astrophys. J. Lett. **542**, L89 (2000)

22. A. Gal-Yam, O. Shemmer: IAUC 7811 (2002)
23. T. Galama et al.: Nature **395**, 670 (1998)
24. T. Galama et al.: Nature **395**, 670 (1998)
25. Galama et al.: GCN 60 (1998)
26. Galama et al.: IAUC 6895 (1998)
27. T.J. Galama et al.: Astrophys. J. **536**, 185 (2000)
28. P. Garnavich et al.: IAUC 6798 (1997)
29. P. Garnavich et al.: astro-ph 0204234 (2002)
30. L.M. Germany, D.J. Reiss, B.P. Schmidt, C.W. Stubbs, E.M. Sadler: Astrophys. J. **533**, 320 (2000)
31. M. Hamuy: astro-ph 0209174 (2002)
32. K. Hatano, D. Branch, K. Nomoto, J.S. Deng, K. Maeda, P. Nugent, G. Aldering: Bull. Am. Astron. Soc. **198**, 3902 (2001)
33. Y. Hirose: IAUC 7810 (2002)
34. R.W. Hix, F.-K. Thielemann: Astrophys. J. **460**, 869 (1996)
35. P. Höflich, J.C. Wheeler, L.Wang: Astrophys. J. **521**, 179 (1999)
36. K. Iwamoto, K. Nomoto, P. Höflich, H. Yamaoka, S. Kumagai, T. Shigeyama: Astrophys. J. Lett. **437**, L115 (1994)
37. K. Iwamoto et al.: Nature **395**, 672 (1998)
38. K. Iwamoto et al.: Astrophys. J. **534**, 660 (2000)
39. H.-T. Janka, E. Müller: Astron. Astrophys. **290**, 496 (1994)
40. K.S. Kawabata et al.: Astrophys. J. Lett. **580**, L39 (2002)
41. L.E. Kay, J.P. Halpern, K.M. Leighly, S. Heathcote, A.M. Magalhaes, A.V. Filippenko: IAUC 6969 (1998)
42. A.M. Khokhlov, P.A. Höflich, E.S. Oran, J.C. Wheeler, L. Wang, A.Yu. Chtchelkanova: Astrophys. J. Lett. **524**, L107 (1999)
43. K. Kinugasa, H. Kawakita, K. Ayani, T. Kawabata, H. Yamaoka, J.S. Deng, P.A. Mazzali, K. Maeda, K. Nomoto: Astrophys. J. Lett. **577**, L97 (2002)
44. K. Kinugasa, H. Kawakita, K. Ayani, T. Kawabata, H. Yamaoka: IAUC 7811 (2002)
45. R.M. Kippen et al.: GCN 67 (1998)
46. R. Knop et al.: IAUC 7128 (1999)
47. S. Kulkarni et al.: Nature **395**, 663 (1998)
48. D.C. Leonard, A.V. Filippenko, R. Chornock, R.J. Foley: astro-ph 0206368 (2002)
49. Z. Li, R.A. Chevalier: Astrophys. J. **526**, 716 (1999)
50. C. Lidman, V. Doublier, J.-F. Gonzalez, T. Augusteijn, O.R. Hainaut, H. Boehnhardt, F. Patat, B. Leibundgut: IAUC 6895 (1998)
51. L.B. Lucy: Astron. Astrophys. **345**, 211 (1999)
52. A.I. MacFadyen, S.E. Woosley: Astrophys. J. **524**, 262 (1999)
53. A.I. MacFadyen, S.E. Woosley, A. Heger: Astrophys. J. **550**, 410 (2001)
54. K. Maeda, T. Nakamura, K. Nomoto, P.A. Mazzali, F. Patat, I. Hachisu: Astrophys. J. **565**, 405 (2002)
55. T. Matheson, A.V. Filippenko, W. Li, D.C. Leonard, J.C. Shields: Astron. J. **121**, 1648 (2001)
56. P.A. Mazzali, L.B. Lucy: Astron. Astrophys. **279**, 447 (1993)
57. P.A. Mazzali: Astron. Astrophys. **363**, 705 (2000)
58. P.A. Mazzali, K. Iwamoto, K. Nomoto: Astrophys. J. **545**, 407 (2000)
59. P.A. Mazzali, K. Nomoto, F. Patat, K. Maeda: Astrophys. J. **559**, 1047 (2001)
60. P.A. Mazzali et al.: Astrophys. J. Lett. **572**, L61 (2002)
61. E.H. McKenzie, B.E. Schaefer: Pub. Astron. Soc. Pacific **111**, 964 (1999)

62. P. Meikle, L. Lucy, S. Smartt, B. Leibundgut, P. Lundqvist, R. Ostensen: IAUC 7811 (2002)
63. T. Nakamura, H. Umeda, K. Nomoto, F.-K. Thielemann, A. Burrows: Astrophys. J. **517**, 193 (1999)
64. T. Nakamura, P.A. Mazzali, K. Nomoto, K. Iwamoto: Astrophys. J. **550**, 991 (2001)
65. T. Nakamura, H. Umeda, K. Iwamoto, K. Nomoto, M. Hashimoto, W.R. Hix, F.-K. Thielemann: Astrophys. J. **555**, 880 (2001)
66. S. Nakano, R. Kushida, W. Li: IAUC 7810 (2002)
67. K. Nomoto, M. Hashimoto: Phys. Rep. **256**, 173 (1988)
68. K. Nomoto, T. Suzuki, T. Shigeyama, S. Kumagai, H. Yamaoka, H. Saio: Nature **364**, 507 (1993)
69. K. Nomoto et al.: Nature **371**, 227 (1994)
70. K. Nomoto, K. Iwamoto, T. Suzuki: Phys. Rep. **256**, 173 (1995)
71. K. Nomoto et al.: In: *Supernovae and Gamma-Ray Bursts*, ed. by M. Livio, N. Panagia, K. Sahu (Cambridge University Press, Cambridge 2001) p. 144
72. K. Nomoto, K. Maeda, H. Umeda, T. Nakamura: In: *The Influence of Binaries on Stellar Populations Studies*, ed. by D. Vanbeveren (Kluwer, Dordrecht 2001), p. 507
73. K. Nomoto, K. Maeda, H. Umeda, T. Ohkubo, J. Deng, P. Mazzali: astro-ph 0209064 (2002)
74. F. Patat and A. Piemonte: IAUC 6918 (1998)
75. F. Patat et al.: Astrophys. J. **555**, 917 (2001)
76. E. Pian, L.A. Antonelli, M.R. Daniele, S. Rebecchi, V. Torroni, G. Gennaro, M. Feroci, L. Piro: GCN 61 (1998)
77. E. Pian et al.: Astron. Astrophys. Suppl. Ser. **138**, 463 (1999)
78. E. Pian: In: *Supernovae and Gamma-Ray Bursts*, ed. by M. Livio, N. Panagia, K. Sahu (Cambridge University Press, Cambridge 2001) p. 85
79. P. Pinto, R. Eastman: Astrophys. J. **530**, 744 (2000)
80. D.E. Reichart: Astrophys. J. Lett. **521**, L111 (1999)
81. M.W. Richmond et al.: Astron. J. **111**, 327 (1996)
82. A.G. Riess, A.V. Filippenko, W. Li, B.P. Schmidt: Astron. J. **118**, 2668 (1999)
83. L. Rigon et al.: astro-ph 0211432 (2002)
84. E.M. Sadler, R.A. Stathakis, B.J. Boyle, R.D. Ekers: IAUC 6901 (1998)
85. G. Schaller, D. Schaerer, G. Meynet, A. Maeder: Astron. Astrophys. Suppl. Ser. **96**, 269 (1992)
86. M.E. Sharina, I.D. Karachentsev, N.A. Tikhonov: Astron. Astrophys. Suppl. Ser. **119**, 499 (1996)
87. T. Shigeyama, K. Nomoto: Astrophys. J. **360**, 242 (1990)
88. T. Shigeyama, T. Suzuki, S. Kumagai, K. Nomoto, H. Saio, H. Yamaoka: Astrophys. J. **420**, 341 (1994)
89. T.M. Shimizu, T. Ebisuzaki, K. Sato, S. Yamada: Astrophys. J. **552**, 756 (2001)
90. S.J. Smartt, P. Vreeswijk, E. Ramirez-Ruiz, G.F. Gilmore, W.P.S. Meikle, A.M.N. Ferguson, J.H. Knapen: Astrophys. J. Lett. **572**, L147 (2002)
91. P. Soffitta, M. Feroci, L. Piro: IAUC 6884 (1998)
92. J. Sollerman, C. Kozma, C. Fransson, B. Leibundgut, P. Lundqvist, F. Ryde, P. Woudt: Astrophys. J. Lett. **537**, L127 (2000)
93. R.A. Stathakis et al.: Mon. Not. R. Astron. Soc. **314**, 807 (2000)
94. T. Suzuki, K. Nomoto: Astrophys. J. **455**, 658 (1995)
95. D.A. Swartz, J.C. Wheeler: Astrophys. J. Lett. **379**, L13 (1991)

96. M. Takada-Hidai, W. Aoki, G. Zhao: Pub. Astron. Soc. Japan **54**, 899 (2002)
97. R. Terlevich, G. Tenorio-Tagle, J. Franco, J. Melnick: Mon. Not. R. Astron. Soc. **255**, 713 (1992)
98. R. Terlevich, A. Fabian, M. Turatto: IAUC 7269 (1999)
99. F.-K. Thielemann, K. Nomoto, M. Hashimoto: Astrophys. J. **460**, 408 (1996)
100. S.E. Thorsett, D.W. Hogg: GCN 197 (1999)
101. C. Tinney, R. Stathakis, R. Cannon, T.J. Galama: IAUC 6896 (1998)
102. M. Turatto et al.: Astrophys. J. Lett. **498**, L129 (1998)
103. M. Turatto et al.: Astrophys. J. **534**, L57 (2000)
104. H. Umeda, K. Nomoto: Astrophys. J. **565**, 385 (2002)
105. E.P.J. van den Heuvel: In: *Interacting Binaries*, ed. by H. Nussbaumer, A. Orr (Springer Verlag, Berlin 1994) p. 263
106. L. Wang, J.C. Wheeler: Astrophys. J. Lett. **504**, L87 (1998)
107. L. Wang, D. Baade, P.A. Höflich, J.C. Wheeler, C. Fransson, P. Lundqvist: astro-ph 0206386 (2002)
108. K.W. Weiler, N. Panagia, R.A. Sramek, S.D. Van Dyk, M.J. Montes, C.K. Lacey: In: *Supernovae and Gamma-Ray Bursts*, ed. by M. Livio, N. Panagia, K. Sahu (Cambridge University Press, Cambridge 2001) p. 198
109. K.W. Weiler, N. Panagia, M.J. Montes: Astrophys. J. **562**, 670 (2001)
110. M. Wieringa et al.: GCN 63 (1998)
111. M. Wieringa et al.: IAUC 6896 (1998)
112. C.L. Williams, N. Panagia, C.K. Lacey, K.W. Weiler, R.A. Sramek, S.D. Van Dyk: Astrophys. J. **(581)**, 396 (2002)
113. S.E. Woosley, R.G. Eastman, T.A. Weaver, P.A. Pinto: Astrophys. J. **429**, 300 (1994)
114. S.E. Woosley, N. Langer, T.A. Weaver: Astrophys. J. **448**, 315 (1995)
115. S.E. Woosley, R.G. Eastman, B. Schmidt: Astrophys. J. **516**, 788 (1999)
116. T. Young, E. Baron, D. Branch: Astrophys. J. Lett. **449**, L51 (1995)
117. L. Zampieri, A. Pastorello, M. Turatto, E. Cappellaro, S. Benetti, G. Altavilla, P. Mazzali, M. Hamuy: astro-ph 0210171 (2002)

Supernovae and γ-Ray Bursters

Titus J. Galama[1]

Astronomy Dept., California Institute of Technology, Pasadena, CA 91125, USA

Abstract. Since 1997 when the first counterparts to γ-ray bursts were discovered, evidence for a connection with supernovae has accumulated. I review the status of this possible connection between γ-ray bursts and supernovae, from a mostly observational perspective. In particular I will focus on the indirect evidence provided by the energies of γ-ray bursts, the duration of the events, the event rate, the locations and environments surrounding the bursts, the discovery of X-ray lines, the signatures of circumstellar winds and, finally, the association of SN1998bw with GRB980425 and the discovery of supernova-like features in the afterglows of some γ-ray bursts.

1 Introduction

Jan van Pardijs' last publication appeared on 22 October 1999 in the perspectives section of *Science*. In the article he discussed the status of the connection of γ-ray bursts (GRBs) to supernovae (SNe) at that time. Thus, I would like to dedicate this chapter to his memory.

In 1999 the status of the supernova connection was based mainly on the temporal and spatial coincidence of supernova SN1998bw with γ-ray burst GRB980425, on the many unusual characteristics of SN1998bw, and on the discovery of supernova-like features in the afterglows of GRB970228 and GRB980326. Since then, more evidence has been gathered in support of a possible connection of GRBs with SNe but, unfortunately, no cases as convincing as the aforementioned have been found. However, the localizations of GRBs by the detection of their counterparts have made it possible to study the environments in which GRBs arise. This has provided a number of indications of their association with massive stars and star formation. First, most GRBs appear to lie within regions of UV emission from massive stars in their host galaxies. Second, observations suggest that the progenitors of GRBs are surrounded by dense media. Thus, GRBs likely occur in the disks of galaxies and in, or close to, star-forming regions. Finally, evidence has emerged for the existence of collimated outflows (jets) in the explosions. The energies of GRBs, when corrected for the geometry of the explosion, are comparable to those of supernovae, further suggesting a relation between the two phenomena.

2 The Progenitors of GRBs

Observationally it is hard to distinguish between models of progenitors for GRBs. The GRB and the afterglow are produced when relativistic ejecta are decelerated, and no observable radiation emerges directly from the "hidden engine" that powers the GRB. Thus, in spite of all discoveries, the origin of GRBs has remained rather mysterious. Popular models for the origin of GRBs that (in principle) can provide the required energies, come in two classes: 1) compact object mergers such as the neutron star-neutron star (see, e.g., [9]) and neutron star-black hole mergers [20,22,23] and 2) the core-collapse of very massive stars (termed collapsars, "failed" supernovae, or hypernovae [25,39]).

The speculation is that compact object mergers give rise to short-duration GRBs (time-scales < 1 s), while collapsars produce long-duration GRBs (time-scales $\gg 1$ s) so that the short-duration GRBs are caused by merger systems, while the longer ones are due to the core-collapses of very massive stars. Note that no counterpart to a short duration GRB has yet been localized and identified.

In the following I will summarize observational constraints on progenitor models and what we have learned from such observations. Throughout this article please keep in mind that the nature of the evidence is suggestive rather than conclusive.

2.1 GRB Energetics

Jets have been discussed extensively in the context of GRBs. First, the similarity between some of the observed features of blazars and active galactic nuclei (AGNs) led to the speculation that jets also appear in GRBs. Second, the regions emitting the γ-rays as well as the afterglow must be moving relativistically. The emitted radiation is strongly beamed, and we can observe only a region with an opening angle $1/\Gamma$ off the line of sight, where Γ is the Lorentz factor of the outflow. Emission outside of this very narrow cone is not observed and we cannot tell whether we are looking at a spherical explosion or at a jet. These considerations have led to numerous speculations on the existence of jets and to attempts to search for their observational signatures. Finally, jets appear naturally in the context of several leading scenarios for the sources of GRBs.

Observationally it is found that the isotropic equivalent energies of GRBs range from 5×10^{51} to 1.4×10^{54} erg. However, transitions have been observed at optical and radio wavelengths which can be interpreted as being due to collimated (jetted) outflow (see Fig. 1 [16]). Frail et al. [10] have recently determined the jet break times for a sample of GRBs with known redshifts. From these, a wide range of jet opening angles is inferred in GRBs: from 3° to more than 25°, with a strong concentration near 4°. This relatively narrow collimation implies that the observed GRB rate has to be corrected for the fact that conical fireballs are visible to only a fraction of observers. Frail et al. [10] find that the "true" GRB rate is ~ 500 times larger than the observed GRB rate. Although the isotropic equivalent energies of GRBs span almost over three decades, when one corrects the observed energies for the geometry of the outflow, GRB energies appear narrowly clustered around 5×10^{50} erg (see Fig. 2 and [16,24]). These are energies that are very comparable to those of supernovae.

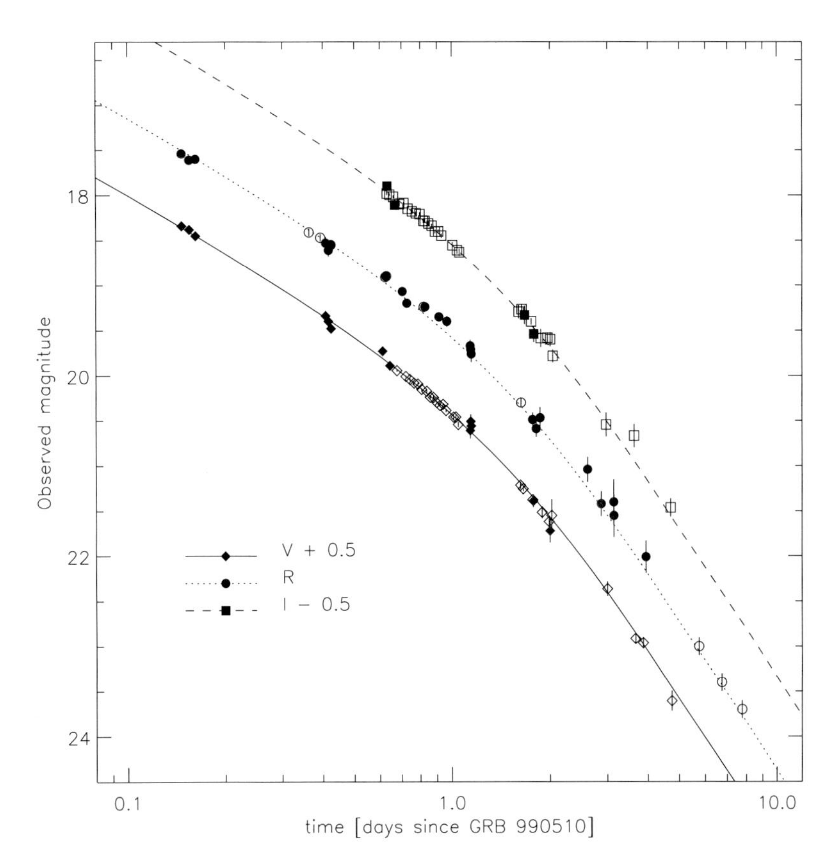

Fig. 1. GRB990510, the "classical" case for a jet. An achromatic break at optical and radio wavelengths at $t_{\rm jet} = 1.2$ days, implies a jet opening angle $\theta_0 = 0.08$. The temporal slopes before and after the break agree well with the theory if $p = 2.2$. For this burst the isotropic γ-ray energy $E_{\rm iso} = 2.9 \times 10^{53}$ erg, but the "true" total energy is only $E = 10^{51}$ erg [16].

2.2 Duration of the Event

Within the internal shock framework describing the prompt γ-ray emission of GRBs, the source is variable on < 1 s time-scales but lasts for tens of seconds. This rules out progenitor models that lead to simple single explosions, as the engine needs to persist for longer times. Both compact object mergers and collapsars can fulfill this requirement as the scenario for GRB production usually involves the formation of a disk that persists for a sufficiently long time and feeds a newly born black hole.

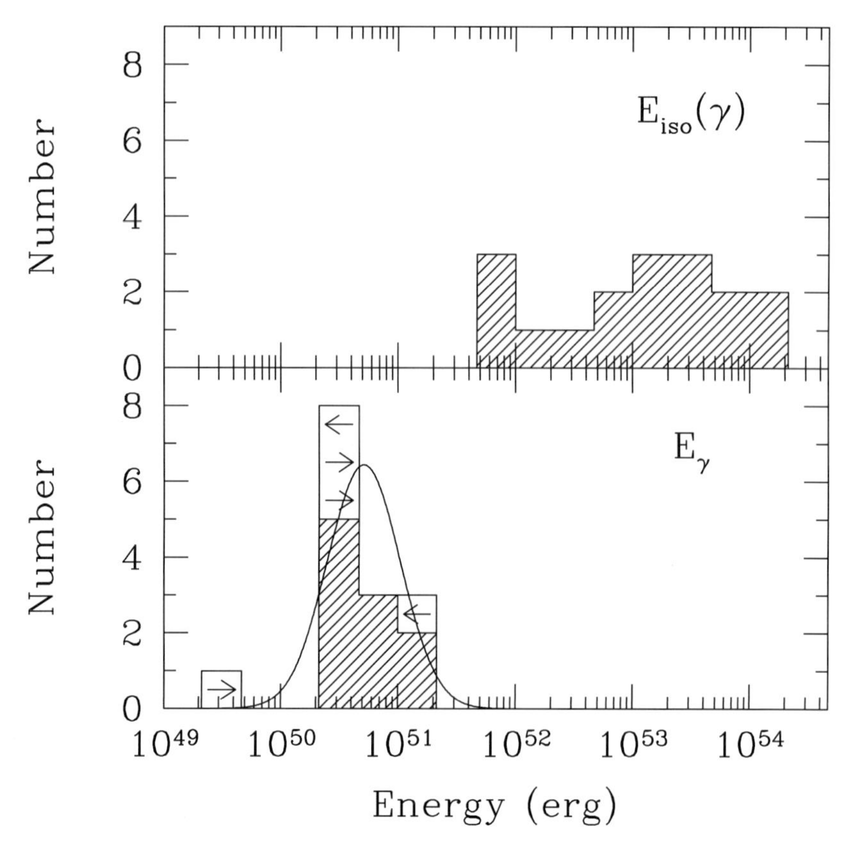

Fig. 2. The distribution of the apparent isotropic γ-ray burst energy of GRBs with known redshifts (top) versus the geometry corrected energy (bottom). While the isotropic energy E_{iso} spans three orders of magnitude, the geometrically corrected energy, $E_{\Gamma} = E_{iso}\theta^2/2$, is very narrowly distributed. This implies that the sources of GRBs produce roughly the same amount of energy, $\sim 5 \times 10^{50}$ erg, but that the energy is distributed over a variety of angles resulting in a wide distribution of isotropic energies [10].

2.3 The Event Rate

The true event rate is probably higher than the observed rate by about a factor of 500 as events appear collimated into narrow jets [10]. Frail et al. [10] estimate that this GRB event rate is about three times greater than the estimated rate of neutron star coalescence, and about 250 times smaller than the estimated rate of type Ib/c supernovae (the expected end product of a collapsar). Clearly, the collapsar scenario is capable of easily supplying a sufficient number of progenitors (including failed GRBs). Within the uncertainties of the estimates, the coalescence scenario is also (barely) capable of providing sufficient progenitors.

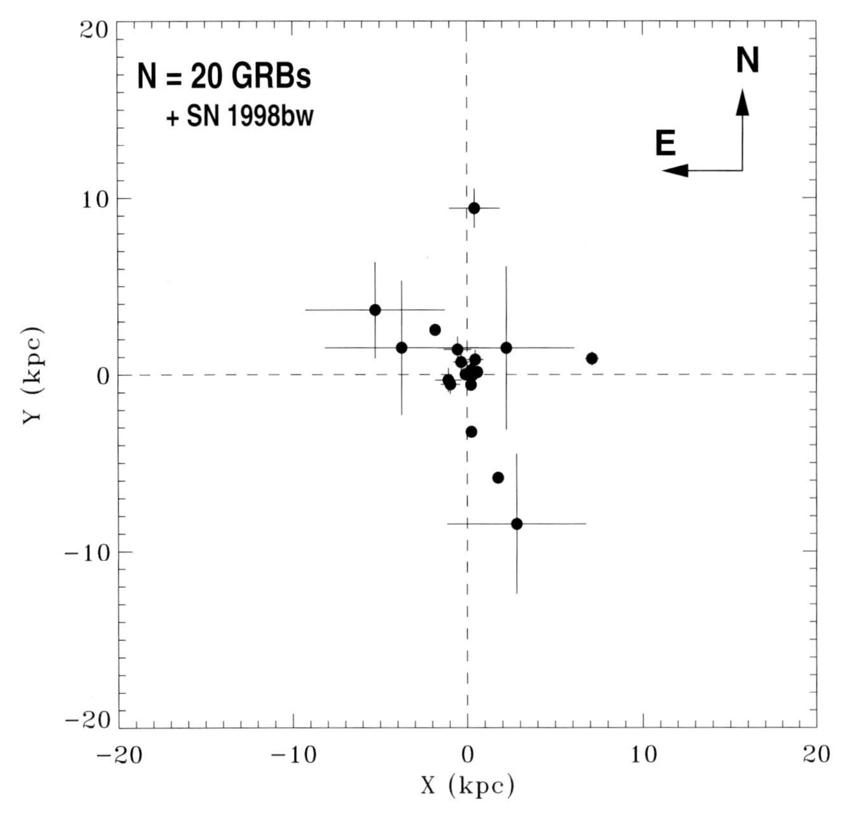

Fig. 3. The projected physical offset distribution of 20 γ-ray bursts, and SN1998bw/GRB980425, from the centers of their presumed host galaxies (1σ error bars) [4].

2.4 Offsets

The locations of GRBs in host galaxies may also put constraints on progenitor models. For instance, when a neutron star is formed in a supernova explosion it receives a substantial kick velocity of several hundred km s^{-1} (see, e.g., [34] and references therein). Taken together with the relatively long merger times for neutron stars ($\sim 10^8$ yr) one would expect some GRBs to be located outside of the host galaxy where the binary was formed [3]. Massive star collapses, on the other hand, are expected to occur in the star forming regions where they originated.

In a sample of *Hubble Space Telescope (HST)* observations it was found that GRBs consistently lie within the region of detectable rest frame UV light of their host galaxies [4]. Since UV light is predominantly produced by young and massive stars, this suggests an association with a young and massive stellar population. To be more quantitative, the median of the distribution of offsets of GRBs from their host galaxy's center is < 0.4 arcsec and a statistical comparison of the observed offset distribution with predicted distributions from progenitor models is consistent with a collapsar or promptly bursting binary scenario (see Fig. 3 and [4]). Slowly bursting binary mergers such as neutron star-neutron star or neutron star-black hole systems appear to be inconsistent with such a result.

2.5 The Circumburst Environment

The environments of at least some bursts agree well with ordinary interstellar medium (ISM) densities [24]. These bursts are thus not in their galaxies' halo (where one might expect a good number of slowly bursting binary mergers to reside).

Host galaxy absorption may also provide clues to the GRB progenitor (low extinction is expected for the merger models and high extinction for massive star collapses in star-forming regions). Some GRBs are found behind large columns of gas, $N_H = 10^{22} - 10^{23}$ cm^{-2} (see Fig. 4 and [15]), which are typical of the column densities of Galactic giant molecular clouds. In addition, GRB afterglows tend to have strong Mg I absorption lines, especially relative to Mg II, which indicates that they originate in denser regions than the normal diffuse ISM. In combination with the observed distribution of GRBs within their host galaxies (see Sect. 2.4), it thus appears that the long duration GRBs for which afterglows have been detected lie in the disks of galaxies and within, or at least close to, star forming regions.

The optical extinction toward GRBs, however, was found to be some $10 - 100$ times smaller than expected from the high column densities (see Fig. 4 and [15]). This would favor theoretical predictions that the early, hard radiation from GRBs and their afterglows destroys the dust in their environment, thus carving a path out of the molecular cloud through which the regular afterglow light travels relatively unobstructed [11,35].

On the other hand, early-time *W.M. Keck Telescope I (KECK I)* spectroscopic observations of GRB000926 (see Fig. 5 and [5]) reveal two absorption systems with a velocity separation of 168 km s^{-1}. They interpret this as being due to two individual clouds in the host galaxy. Dust destruction by the GRB and the afterglow is effective only to a certain distance from the GRB so that it is unlikely that the dust in both clouds would be destroyed. Further, the two clouds appear to have similar relative metal abundance and dust to gas ratio. If one cloud is associated with the GRB site, this would imply that the explosion did not significantly alter the surrounding environment. GRB010222 also shows a two-component optical line system similar to that of GRB000926 (S. Castro, private communication). Thus, the observed low optical extinctions could also be due GRB host galaxies typically having low dust to gas ratios.

2.6 X-Ray Lines

X-ray line features have been reported for various GRBs with statistical significance of $2 - 4\sigma$ [1,26,27,41]. However, strong upper limits were set in some other cases [42]. The relatively low statistical significance of these lines makes them somewhat uncertain, except, perhaps, in the afterglow of GRB991216 [27].

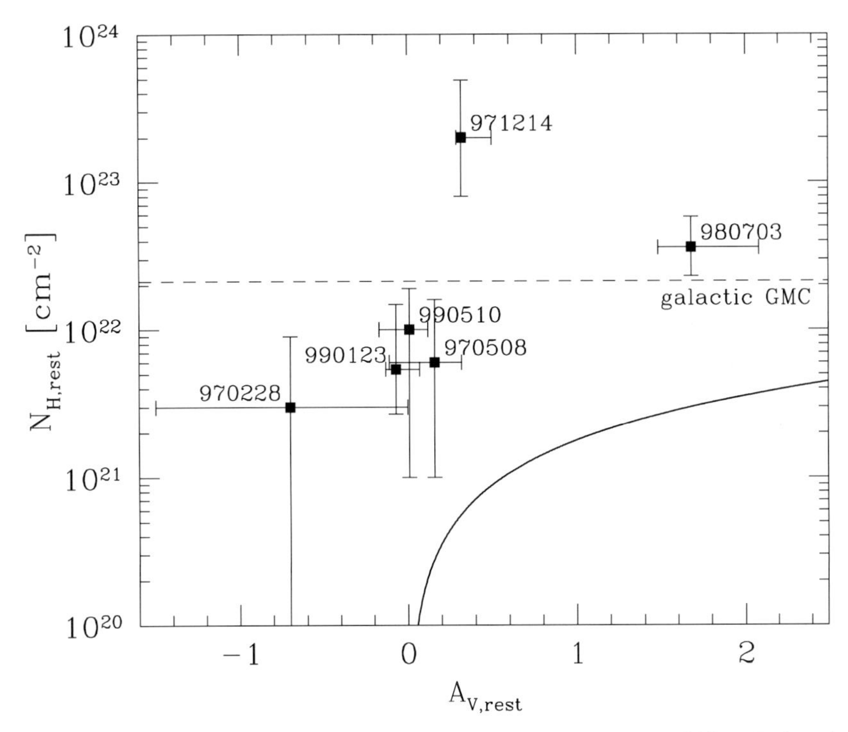

Fig. 4. Hydrogen column density N_H vs. the optical extinction at V-band, A_V, for a number of γ-ray burst afterglows; error bars are 68% confidence [15]. The solid curve is the Galactic A_V-N_H relation [28]. The dashed line shows the average column density of a giant molecular cloud of 170 $M_\odot$ pc^{-2} [31].

The detection of these features, if true, would provide strong constraints on the progenitors and their environment and, therefore, the detection of X-ray lines is of great importance.

The suspected lines are typically observed by a day after the event, and some of them are transient, lasting for about one day. There are two possible ways of explaining this timescale. First, if there is a large amount of iron located about a light day away from the explosion site, this iron is energized by the light from the burst, and from its afterglow, and then radiates either by fluorescence, recombination, or thermally [21]. This mechanism, however, almost precludes the existence of jets for, if the afterglow is narrowly collimated, the iron has to be located within the opening angle of the jet (to be energized by the burst), but not directly along the line of sight so as to not block the afterglow. Such a geometrical arrangement requires fine tuning for narrow jets. The requirement for a large amount of iron comes from the need to have a line optical depth ~ 1 at a relatively large distance (one light day) from the burst. Theoretical models trying to establish such a geometry require copious (but highly asymmetric) mass-loss [38], an preliminary supernova explosion prior to the ultimate collapse

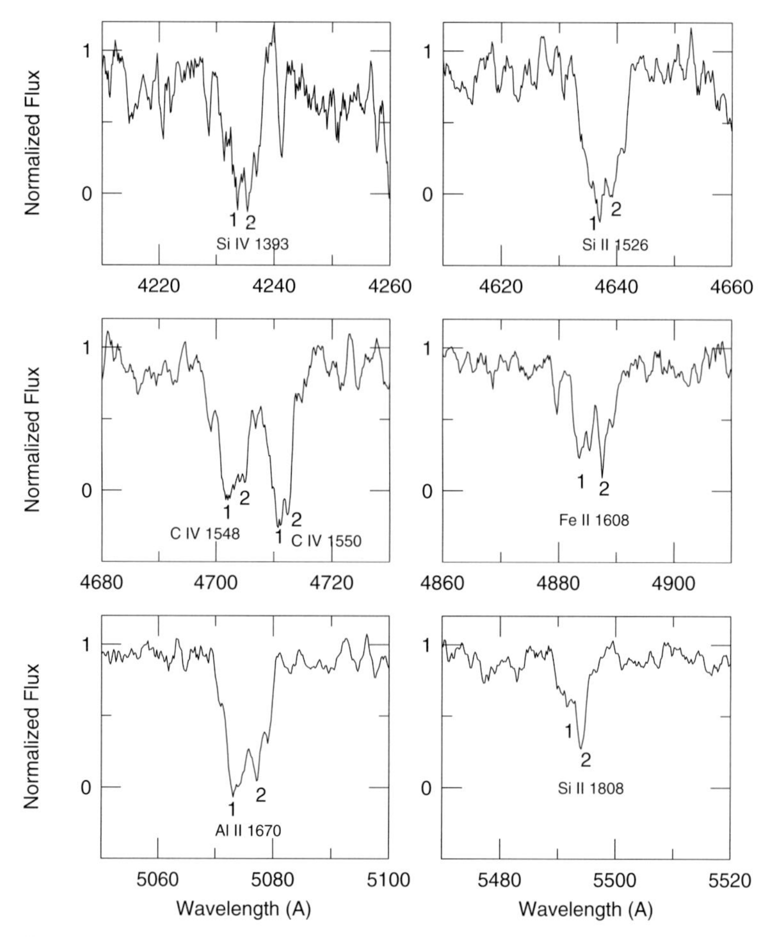

Fig. 5. A section of the spectrum of GRB000926 on 29.26 September 2000 UT showing the existence of two components with a mean separation of $\sim$ 4.05 Å at redshifts of $z_1 = 2.0370 \pm 0.0011$ (indicated by 1) and $z_2 = 2.0387 \pm 0.0011$ (indicated by 2). The spectrum has been smoothed by a boxcar of 7 pixels, equivalent to 1.12 Å.

of the central object [33], or reflection by a very dense local ISM whose geometry is constrained to allow the relativistic fireball to expand unimpeded [21].

A second possibility is that iron close to the source is producing the line emission. In this case, less iron is needed due to the smaller size, but the iron cannot be energized by the burst or the afterglow as both of these happen at large radii. Therefore, the source has to continue to operate, on a low level, for at least a day to energize its iron environment. The total additional energy required in this case is not large, as the overall energy observed in the line is small.

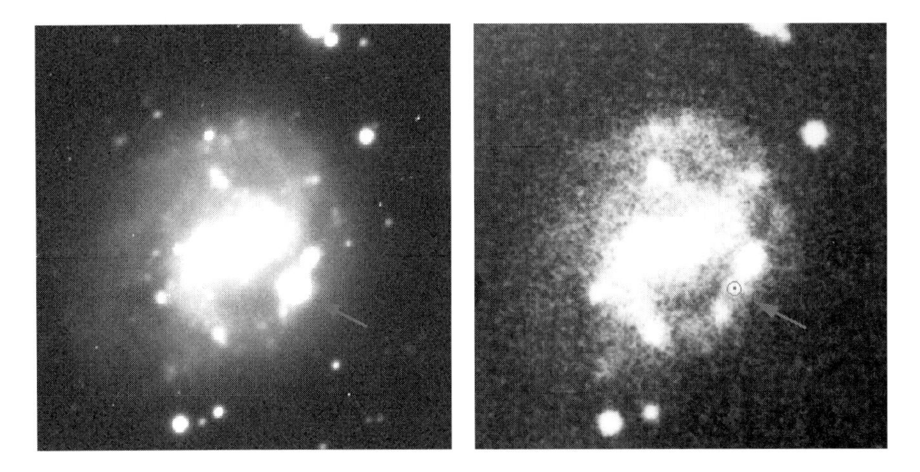

Fig. 6. Left: discovery R-band *New Technology Telescope (NTT)* image showing SN1998bw in the spiral galaxy ESO184-G82 on 01 May 1998. Right: pre-discovery (1976) *UK Schmidt Telescope* image [13].

2.7 Circumstellar Wind

If at least some GRBs are produced by the core-collapse of massive stars to black holes, then the circumburst environment will have been influenced by the strong wind of the massive progenitor star. For a constant mass-loss rate and constant wind speed, the circumstellar density falls as $\rho \propto r^{-2}$, where r is the radial distance from the progenitor. In this so-called circumstellar wind model, the afterglow evolution is different from that in a constant density ISM (see [6,7] for details). Observationally is has turned out to be rather hard to detect predicted differences in the afterglow's evolution for a constant density ISM and for a circumstellar wind (see, e.g., [7,30]).

2.8 The Supernova Connection

The first evidence for a possible GRB/SN connection was provided by the discovery of SN1998bw in the error box of GRB980425 (see Figs. 6 and 7 and [13]). The temporal and spatial coincidence of SN1998bw with GRB980425 suggest that the two phenomena are related [13,19]. But, the clearest indication that SN1998bw may be related to GRB980425 comes from the fact that the radio emitting shell in SN1998bw must be expanding at relativistic velocities, $\Gamma \gtrsim 2$ (see Fig. 8, [19], and the chapter by Weiler et al. in this volume). This has never previously been observed in a SN. From minimum energy arguments, it was estimated that this relativistic shock carried $\sim 5 \times 10^{49}$ erg, and could well have produced the GRB at early time [32]. Further, detailed analysis of the radio light curve [6,32] showed additional energy injection one month after the SN event – highly suggestive of a central engine (i.e. black hole versus neutron star

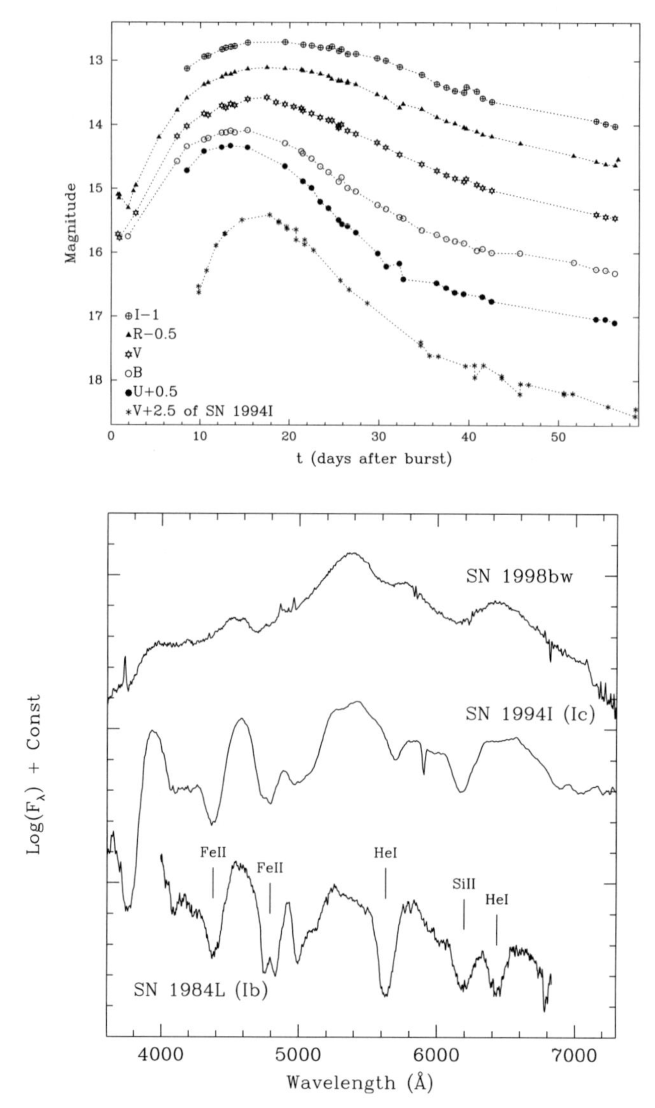

Fig. 7. Top: UBVRI light curves for SN1998bw/GRB980425. Time is in days since 25.91 April 1998 UT. For comparison the V-band light curve of the type Ic SN1994I is shown. Bottom: Representative spectra near maximum light of SN1998bw (type Ic), SN1994I (type Ic), and SN1984L (type Ib) are shown. The overall shape of the spectrum of SN1998bw is similar to that of SNIb/c, although the spectral features are less pronounced [13].

formation) rather than a purely impulsive explosion. However, the need for this additional energy injection has been disputed by Weiler et al. (see [36,37] and the chapter by Weiler et al. in this volume).

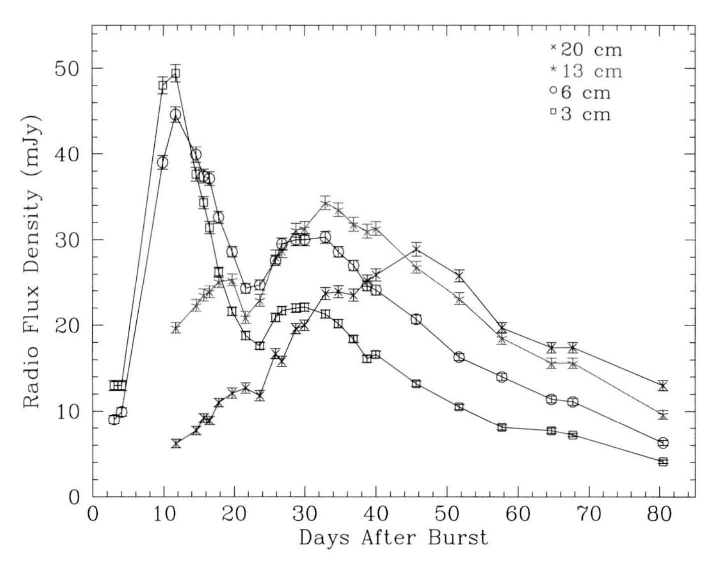

Fig. 8. The radio light curve of SN1998bw. The brightness temperature, T_B, is related to the specific intensity and is proportional to the flux divided by the solid angle of the source. Assuming that the supernova expands at the velocities indicated by optical observations Kulkarni et al. [19] find $T_B > 10^{12}$ K in the first month and exceeding 10^{13} K in the first week. Since this exceeds the 5×10^{11} K limit for catastrophic inverse Compton scattering losses, they conclude that the radio emitting sphere expanded at relativistic speeds, $\Gamma \gtrsim 2$ [19].

However, GRB980425 is most certainly not a typical GRB. The redshift of SN1998bw is $z = 0.0085$ and the corresponding γ-ray peak luminosity of GRB980425 and its total γ-ray energy budget are about a factor of $\sim 10^5$ smaller than those of "normal" GRBs [13]. Such low energy SN-GRBs may well be the most frequently occurring GRBs in the Universe, but they do not dominate the observed GRB population due to their faintness.

Evidence for possible supernova connections for normal, high luminosity GRBs comes from their late time red spectra and the late-time rebrightening of their afterglow light curves. GRB980326 shows possible evidence that, at late times, the emission was dominated by an underlying supernova [3]. A template supernova light curve, provided by the well-studied type Ib/c SN1998bw provides an adequate description of the observations (see Fig. 9).

Similarly, the optical afterglow of GRB970228 showed indications that the standard model was not sufficient to describe the observations in detail [12]. The early time decay of the optical emission is faster than that at later times and, as the source faded, it showed an unexpected reddening [12]. It was not until supernova-like emission accompanying GRB980326 was found that the behavior of GRB970228 could be understood. Also, for GRB970228 the late time light curve and reddening of the transient can be well explained by an initial power-law decay modified at late times by SN1998bw-like emission (see Figs. 10 and

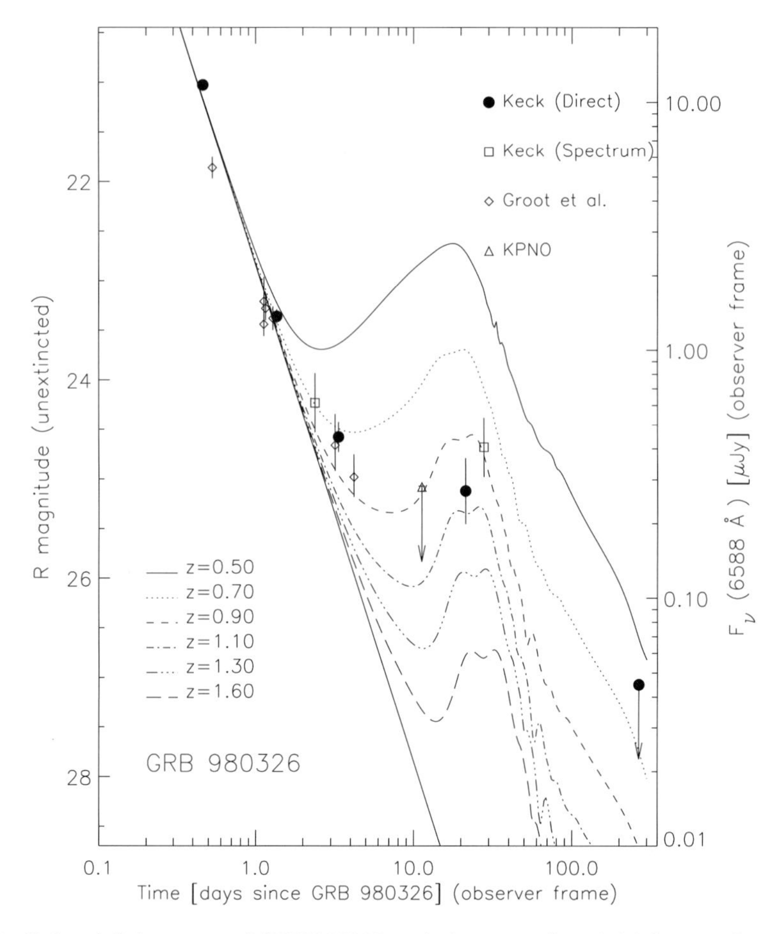

Fig. 9. R-band light curve of GRB980326 and the sum of an initial power-law decay plus a type Ic SN light curve for redshifts ranging from $z = 0.50$ to $z = 1.60$ [3].

11 and [14,29]). Theoretical work [18,40] has shown that these SNe most likely mark the birth events of stellar mass black holes as the final products of the evolution of very massive stars.

The relation between distant GRBs, like GRB980326, and GRB980425/SN1998bw is unclear. Is SN1998bw a different phenomenon or a more local and lower energy equivalent? Attempts have been made to unify the GRB980425/SN1998bw phenomenon with the more distant GRBs and it has been argued that the differences may be solely due to the viewing angle [8,17,31]. Are all afterglows consistent with such a phenomenon? Claims have been made for the existence of SN signatures in other afterglow light curves but, for a number of reasons, such as the presence of contaminating host galaxy light or the brightness of the regular afterglow itself, the evidence for cases other than GRB970228 and GRB980326 is less strong. A detailed study is provided by [8].

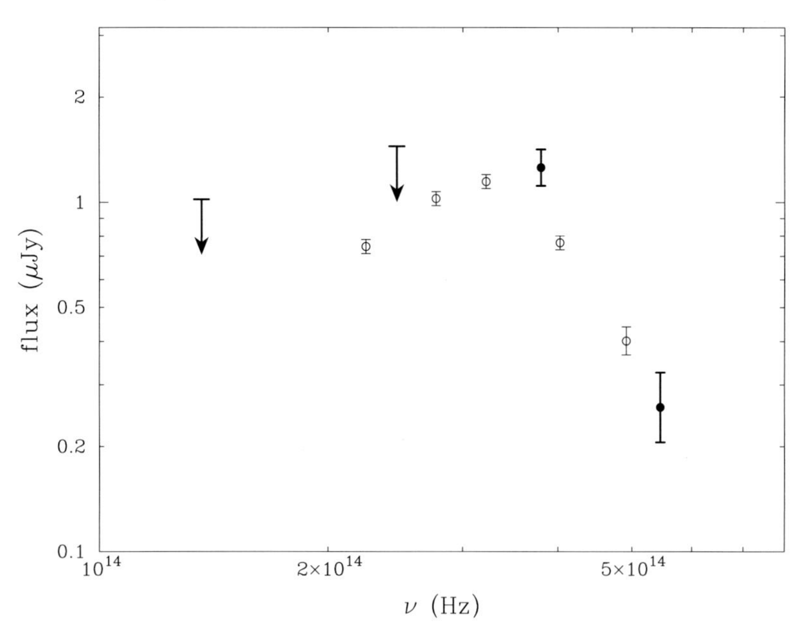

Fig. 10. The broad band spectrum of the optical transient (OT) of GRB970228 on 30.8 March 1997 UT (filled circle and upper-limit arrow). Also shown is the spectral flux distribution of SN1998bw (open circle) redshifted to the $z = 0.695$ of GRB970228. The similarity of the spectral flux distributions is remarkable [14].

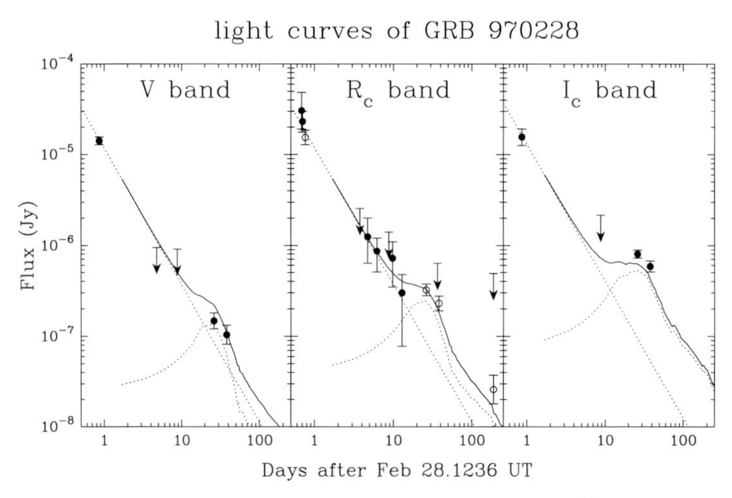

Fig. 11. The V-, R_c-, and I_c-band light curves of GRB970228 (flux versus time). The dotted curves indicate power-law decays, t^{β}, with $\beta = -1.5$, and redshifted SN1998bw light curves. The thick line is the resulting sum of SN and power-law decay light curves [14].

Although the light curves of other afterglows do not necessarily require a supernova component, many of them may be consistent with such an addition. However, the most convincing evidence may yet be provided by future observations of GRB afterglows around the time of the SN emission maximum. In particular, a direct detection of spectroscopic features associated with SNe in a GRB afterglow would be convincing and should be possible with $8 - 10$ m class telescopes such as the *Very Large Telescope (VLT)* or one of the *W.M. Keck Telescopes*.

References

1. L.A. Antonelli et al.: Astrophys. J. Lett. **545**, L39 (2000)
2. J.S. Bloom, S. Sigurdsson, O.R. Pols: Mon. Not. R. Astron. Soc. **305**, 763 (1999)
3. J.S. Bloom et al.: Nature **401**, 453 (1999)
4. J.S. Bloom, S.R. Kulkarni, S.G. Djorgovski: Astron. J. **123**, 1111 (2002)
5. S. Castro et al.: astro-ph 0110566 (2002)
6. R.A. Chevalier, Z. Li: Astrophys. J. Lett. **520**, L29 (1999)
7. R.A. Chevalier, Z. Li: Astrophys. J. **536**, 195 (2000)
8. S. Dado, A. Dar, A. De Rujula: Astron. Astrophys. **388**, 1079 (2002)
9. D. Eichler et al.: Nature **340**, 126 (1989)
10. D.A. Frail et al.: Astrophys. J. Lett. **562**, L55 (2001)
11. A. Fruchter, J.H. Krolik, J.E. Rhoads: astro-ph 0106343 (2001)
12. T.J. Galama et al.: Nature **387**, 479 (1997)
13. T.J. Galama et al.: Nature **395**, 670 (1998)
14. T.J. Galama et al.: Astrophys. J. **536,**, 185 (2000)
15. T.J. Galama, R.A.M.J. Wijers: Astrophys. J. Lett. **549**, L209 (2001)
16. F.A. Harrison et al.: Astrophys. J. Lett. **523**, L121 (1999)
17. K. Ioka, N.T. Takashi: Astrophys. J. Lett. **554**, L163 (2001)
18. K. Iwamoto et al.: Nature **395**, 672 (1998)
19. S.R. Kulkarni et al.: Nature **395**, 663 (1998)
20. J.M. Lattimer, D.N. Schramm: Astrophys. J. Lett. **192**, L145 (1974)
21. D. Lazzati, S. Campana, G. Ghisellini: Mon. Not. R. Astron. Soc. **304**, L31 (1999)
22. R. Mochkovitch et al.: Nature **361**, 236 (1993)
23. R. Narayan, B. Paczyński, T. Piran: Astrophys. J. Lett. **395**, L83 (1992)
24. A. Panaitescu, P. Kumar: Astrophys. J. **554**, 667 (2001)
25. B. Paczyński: Astrophys. J. Lett. **494**, L45 (1998)
26. L. Piro et al.: Astrophys. J. Lett. **514**, L73 (1999)
27. L. Piro et al.: Science **290**, 953 (2000)
28. P. Predehl, J.H.M.M. Schmitt: Astron. Astrophys. **293**, 889 (1995)
29. D.E. Reichart: Astrophys. J. **521**, 111 (1999)
30. R. Sari, T. Piran, J. Halpern: Astrophys. J. Lett. **519**, L17 (1999)
31. P.M. Solomon, A.R. Rivolo, J. Barrett, A. Yahil: Astrophys. J. **319**, 730 (1987)
32. J.C. Tan, C.D. Matzner, C.F. McKee: Astrophys. J. **551**, 946 (2001)
33. P. Valageas, J. Silk: Astron. Astrophys. **347**, 1 (1999)
34. E. Van den Heuvel, J. van Paradijs: Astrophys. J. **483**, 399 (1998)

35. E. Waxman, B.T. Draine: Astrophys. J. **537**, 796 (2000)
36. K.W. Weiler, N. Panagia, M.J. Montes: Astrophys. J. **562**, 670 (2001)
37. K.W. Weiler, N. Panagia, M.J. Montes, R.A. Sramek: Ann. Rev. Astron. Astrophys. **40**, 387 (2002)
38. C. Weth, P. Mészáros, T. Kallman, M.J. Rees: Astrophys. J. **534**, 581 (2000)
39. S.E. Woosley: Astrophys. J. **405**, 273 (1993)
40. S.E. Woosley, R.G. Eastman, B.P. Schmidt: Astrophys. J. **516**, 788 (1999)
41. A. Yoshida et al.: Astron. Astrophys. Suppl. Ser. **138**, 433 (1999)
42. D. Yonetoku et al.: Pub. Astron. Soc. Japan **52**, 509 (2000)

Part III

γ-Ray Bursters

Observational Properties of Cosmic γ-Ray Bursts

Kevin Hurley

University of California, Space Sciences Laboratory, Berkeley, CA 94720-7450, USA

Abstract. Cosmic γ-ray bursts are reviewed, with emphasis on the phenomenology in the X- and γ-ray ranges. Bursts last from around 10 ms to over 1000 s, with complex time histories which defy any coherent explanation, although there is good evidence for a bimodal distribution. Energy spectra span the range from below 10 keV to over 10 GeV with shapes that can be defined with a relatively simple function, although the ratio of the X-ray to γ-ray energy varies widely from one burst to another. The isotropy and inhomogeneity of the burster spatial distribution was the first indication that bursts were at cosmological distances; this has now been confirmed by direct redshift measurements. Some of the future uses of bursts for the study of the early Universe are discussed, as well as some of the remaining challenges in understanding their nature.

1 Introduction

A short Astrophysical Journal Letters article in 1973, when γ-ray astronomy was in its infancy, announced the discovery of 16 cosmic γ-ray bursts (GRBs) with the *VELA* satellites [41]. Today the literature on bursts alone has expanded to over 5000 articles, but two things have not changed. First, bursts are still the brightest objects we know of in the γ-ray sky, and second, their relation to supernovae, mentioned in the discovery paper, is still being discussed. Far fewer than one-half of the 10,000 or so bursts which have impinged on Earth since their discovery have actually been detected, and fewer than half of those have been studied in any detail. Still, despite the uncertainty and mystery surrounding their origin, it is possible to describe the phenomenology in detail. This chapter will be devoted to such a description. In the time domain, the discussion will be limited to the γ-ray burst proper; since the transition from a burst to its afterglow occurs gradually, the boundary between the two is not a sharp one, but by concentrating on energies above 25 keV or so, the distinction becomes clearer. However, it is also interesting to consider what happens during the burst at energies down to several keV and below, so the distinction between X- and γ-rays will not be adhered to strictly.

2 Gamma-Ray Burst Light Curves

GRB light curves, or time histories, have been observed from 1 keV to 18 GeV. In a "typical" γ-ray energy range, say $25 - 150$ keV, they may have durations

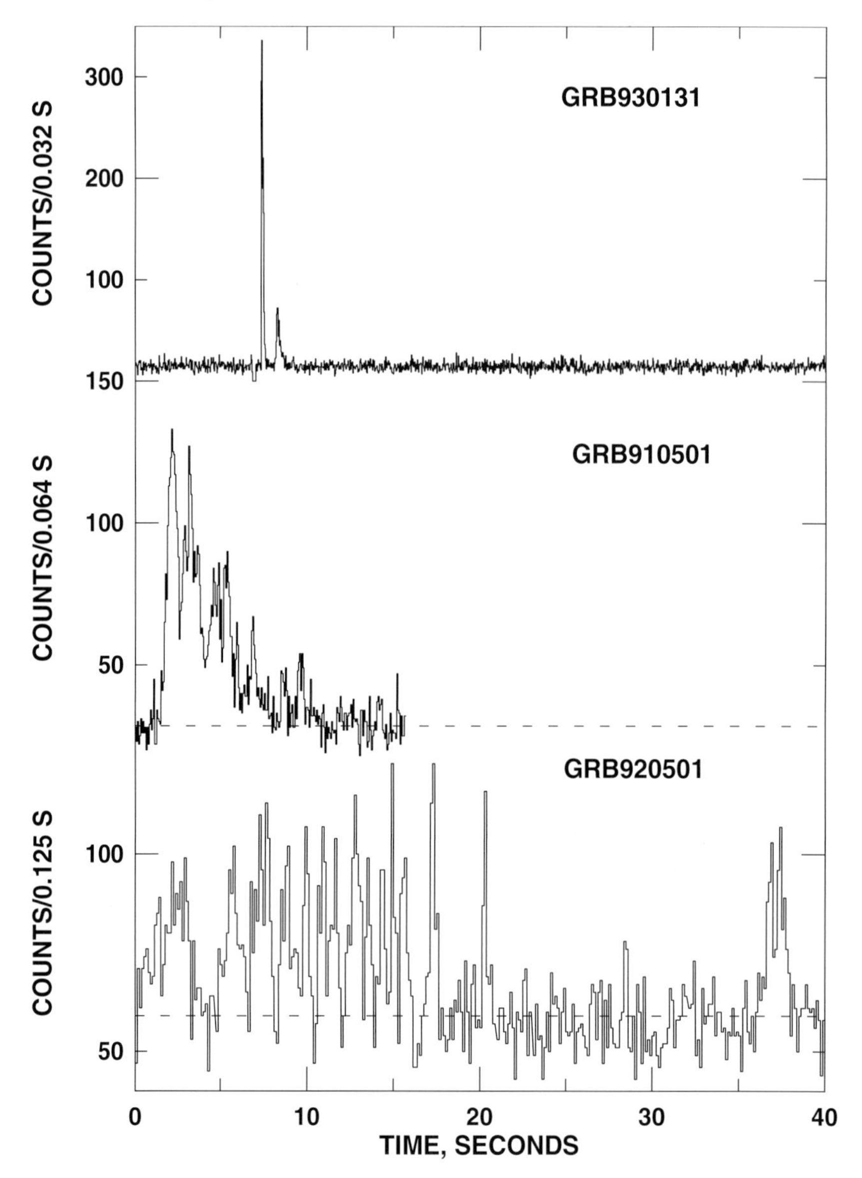

Fig. 1. Three γ-ray burst time histories. The naming convention is GRB followed by the year, month, and day of the burst.

anywhere from $\sim$ 10 ms to 1000 s. Figure 1 shows several examples from the *Ulysses* GRB experiment.

It has long been known that the durations of γ-ray bursts appear to exhibit a bimodal distribution, with the short bursts lasting $\sim$0.2 s and comprising $\sim$ 25% of the total, and the long bursts lasting $\sim$10 s [37,43] comprising the

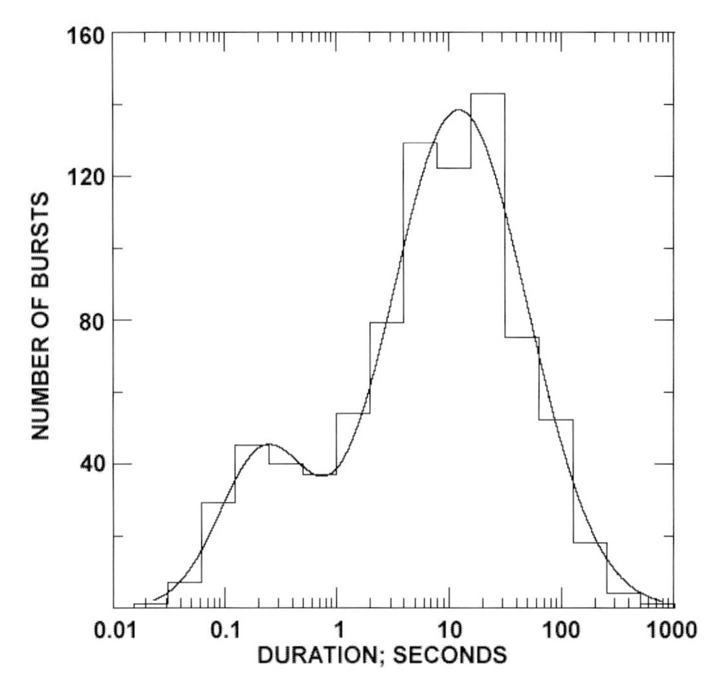

Fig. 2. Duration distribution of 836 γ-ray bursts (histogram) and a fit with the sum of two lognormal distributions. The burst durations were compiled from numerous sources of GRB data taken prior to the launch of the *CGRO*. Remarkably, although the data were taken with different instruments operating in different energy ranges, using different definitions of the duration, the bimodality of the distribution is still apparent. A more homogeneous data set was later analyzed by Kouveliotou et al. [43] with similar results.

remainder. An example is shown in Fig. 2. Other time history subclasses include "FRED" bursts (fast rise, exponential-decay)[13], bursts with precursors [42,47], and bursts with long quiescent episodes (e.g., GRB920501 in Fig. 1, between 22 and 35 s) [62]. It is not yet understood what causes these subclasses or the bimodality.

The power density spectra of the longer bursts behave as a $-5/3$ power-law over two decades in frequency, indicating a self-similar time structure [12].

Time histories are energy-dependent. The individual spikes which comprise a given time history tend to be longer at lower energies [26]. This trend continues to X-ray energies ($\sim$1-10 keV), where the light curves have less structure and are longer than at γ-ray energies. In some cases, X-ray precursors have been found [18,39,53]. Again, there is no generally agreed-upon explanation for these properties.

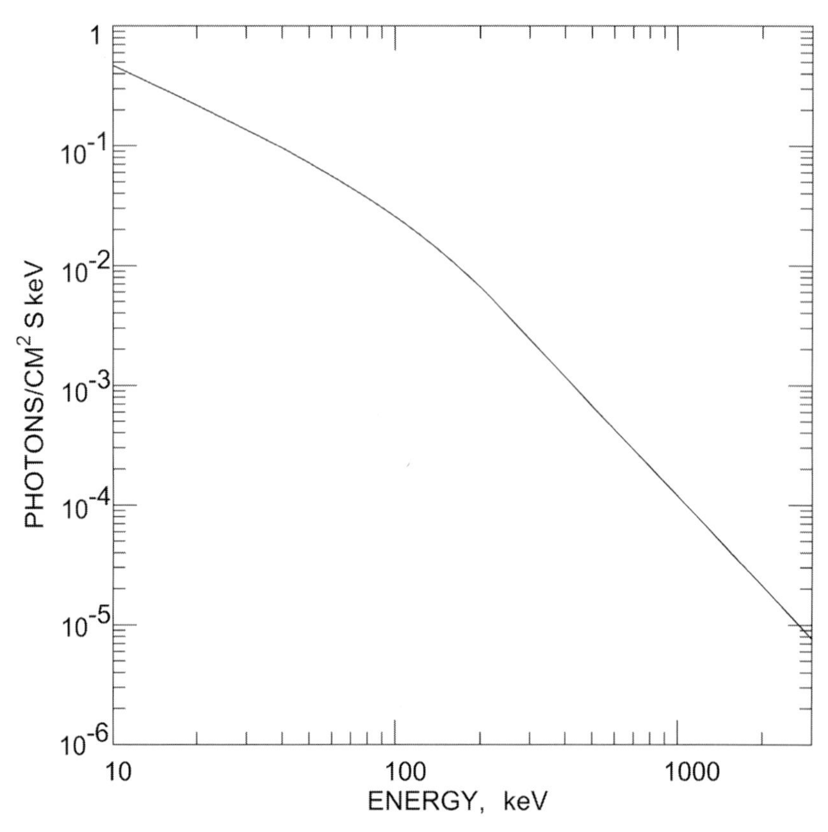

Fig. 3. A "typical" GRB spectrum as described by the Band function [6]. The function is $A \times (E/100\text{keV})^{\alpha}\exp(-E/E_o)$, $(\alpha - \beta)E_o > E$, and $A \times ((\alpha - \beta)E_o/100\text{keV}))^{\alpha-\beta}\exp(\beta - \alpha)(E/100\text{keV})^{\beta}$, $(\alpha - \beta)E_o < E$, where in this case $A = 0.05$, $\alpha = -1$, $\beta = -2$, and $E_o = 150$ keV

3 Gamma-Ray Burst Energy Spectra

Like the time histories, the energy spectra of GRBs have been measured from ~ 1 keV to 18 GeV. Also like the time histories, the spectra vary both from burst to burst, and are variable within a given burst. Unlike the time histories, they can be fit by a general function consisting of a power-law continuum with an exponential cutoff at low energies which smoothly joins a steeper power-law at high energies. This is called the Band spectrum, and it was originally used to fit the GRB energy spectra observed by *Burst and Transient Source Experiment (BATSE)* aboard *Compton Gamma-Ray Observatory (CGRO)* from energies around 10 keV or greater, up to several MeV or less [6]. More recently it has been used successfully to fit GRB spectra down to 2 keV [28]. The fitting parameters vary considerably from burst to burst, but a more or less typical spectrum is shown in Fig. 3.

Although the Band function does not represent any physical mechanism, there is general agreement that GRB spectra are produced by energetic electrons in a magnetic field, and that the actual production mechanisms may include synchrotron [19,45], synchrotron-self-Compton [15,19], and/or inverse Compton [31,58]. Apart from spectral breaks (e.g. the E_o parameter in the Band function) no other spectral features, such as annihilation or cyclotron lines, have been found in the *BATSE* data [7–10,57]. The extent to which these non-observations conflict with previous reports of line emission [25,32] has been discussed extensively.

Not all GRB spectra extend to MeV energies. The term "X-ray rich" has been used to describe bursts with a large ratio of X-ray ($\sim 2 - 10$ keV) to γ-ray ($> 20\,\mathrm{keV}$) flux or fluence [18,29,66]. These bursts appear to be true γ-ray bursts in every respect, except for their paucity of γ-rays. Perhaps related to them are the "NHE" (no high energy) bursts [60]. These are *BATSE* bursts in which no emission above 300 keV is observed.

Just as the time histories of γ-ray bursts vary, so do their energy spectra. Among the types of behavior found are hard to soft spectral evolution as the burst progresses [61], spectral evolution related to the burst fluence [44], and a correlation between the peak of the $\nu\mathrm{F}_\nu$ spectrum and the burst intensity [27].

4 Some Statistical Properties of γ-Ray Bursts

BATSE was the largest and most sensitive GRB experiment yet flown, and the data from it comprise the most uniform burst sample available. The Galactic distribution of *BATSE* bursts is shown in Fig. 4. It is isotropic, and the average value of $\mathrm{V}/\mathrm{V}_{\mathrm{max}}$ is 0.33 [56], which indicates that there is a true deficiency of weak bursts, i.e., a deficiency that is not related to detector efficiency or sensitivity. The all-sky rate of GRBs inferred from the *BATSE* data is slightly under one per day, and if burst sources emitted γ-rays isotropically, this would also reflect the Universal rate of bursts, at least within the *BATSE* energy window and down to its threshold. However, if the γ-radiation is beamed, as is now widely believed, the true Universal burst rate could be considerably higher, perhaps 50 per day.

When the distance scale to GRB sources was still uncertain, numerous studies were carried out to determine whether a given GRB source could emit more than one burst. Since the *BATSE* locations have uncertainties of several degrees, the investigations relied on various statistics to measure the clustering of sources [33]. No convincing evidence of repetition was ever found, and this is consistent with an origin in which the energy of the explosion destroys the system which generates it. However, at cosmological distances, GRBs may be gravitationally lensed, so it is also possible to find cases of apparent repetition; although no such cases have been found in clustering searches [49], lensing may in fact have been detected by other means [30].

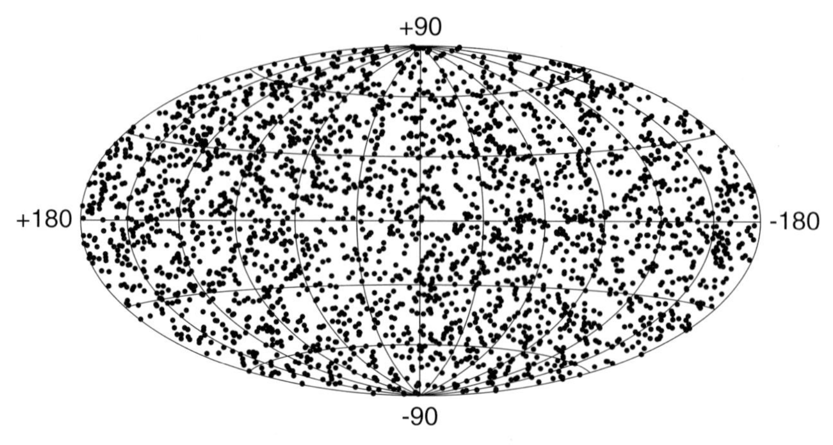

Fig. 4. The distribution of γ-ray bursts in Galactic coordinates observed by *BATSE*. This distribution was obtained from *http://www.batse.msfc.nasa.gov/batse/grb/skymap/*.

Source clustering might also indicate that nearby galaxies or clusters of galaxies are the sites of GRBs. However, coordinate system-independent statistical analyses have not revealed any such clustering [16].

5 Connections Between Various γ-Ray Burst Properties

Practically every GRB property is related to some other GRB property, usually in ways that are difficult to comprehend or explain.

5.1 Classes of Bursts

The long and the short bursts have different energy spectra on average. The short bursts tend to have harder energy spectra [22,43], suggesting that the two classes may originate from different processes. One possibility is that the short bursts are produced in neutron star-neutron star mergers, which proceed on short timescales, while the longer bursts may be produced in hypernovae, which proceed on longer timescales.

The clear distinction between short and long bursts has triggered a number of studies of burst properties to determine whether other classes of bursts exist. In one, 797 *BATSE* bursts were used to form a multivariate database. The GRB properties utilized were the duration (two definitions), fluence, peak flux, and hardness ratio (two definitions) [52]. Three classes were found: long/bright/soft spectrum bursts (59%), short/faint/hard bursts (23%), and intermediate duration ($2-5$ s) intermediate intensity/soft bursts (18%). As for the long and short classes, all three had isotropic distributions. The conclusions of this investigation agree with those of a similar, independent one [36].

Yet another class of GRB has been suggested: the very short (durations $<$ 100 ms) events [20]. This putative class, with only a dozen events in it so far, distinguishes itself not only on the basis of duration, but also on the fact that the time histories and energy spectra resemble one another. Also, the distribution is consistent with a Euclidean one; that is, it is both isotropic and consistent with a number-intensity distribution which follows a $-3/2$ power-law. An origin related to primordial black hole evaporation has been proposed for it.

The question has arisen from time to time whether there might be a class or classes of GRBs which have escaped detection so far by virtue of some combination of parameters such as duration, intensity, and spectral shape [35]. Only partial answers can be given, based on experiments which have explored regions of this parameter space. In one study, 9 years of *Solar Maximum Mission (SMM)* data were examined for bursts in the $0.8 - 10$ MeV range [34]. Only two bursts were found which were previously undetected because of their long durations. The conclusion was that there is no class of long (durations > 2 s), bright (peak fluxes > 0.04 photons $\mathrm{cm}^{-2}\ \mathrm{s}^{-1}$), hard spectrum bursts that *BATSE* has missed.

5.2 The Highest Energies

The upper limits to the energies of photons present in GRBs has increased steadily over the years. GRB940217 was detected by *Energetic Gamma Ray Experiment Telescope (EGRET)* aboard the *CGRO* up to 18 GeV [38]. However, the emission at 18 GeV appears to bear no relation to the GRB time history at lower energies (Fig. 5), as it arrives with a delay of $\sim$4500 s. Such behavior was actually anticipated in some models which invoked a collision between an expanding shell of debris and an external medium [40,50].

The high energy photon spectrum of GRB940217 may be fit by an $\mathrm{E}^{-2.83}$ power-law which, if extrapolated, indicates that higher energy γ-rays might be detectable by ground-based instruments. Moreover, studies of other *EGRET* bursts have revealed six more cases of high energy emission, in many cases delayed with respect to the time history at lower energies, and suggest that such emission might be a common feature of GRBs [24]. This has encouraged numerous searches, one of which, using the *Milagrito* (the *Milagro Gamma-Ray Observatory (MGRO)* prototype), has turned up evidence for high energy emission from GRB970417A [5]. In this case, the probability that the emission is due to background is 1.5×10^{-3}. Only a lower limit can be set for the photon energies of several hundred GeV. At these energies the opacity limit is $z \sim 0.3$ due to pair production on extragalactic starlight, so only detection of the closest bursts is expected. Searches for other examples are continuing with the *MGRO*.

5.3 The Longest Bursts

Bursts which last longer than several hundred seconds are rare, and they tend to be the most intense bursts. Fig. 6 indicates one reason why this is the case. The

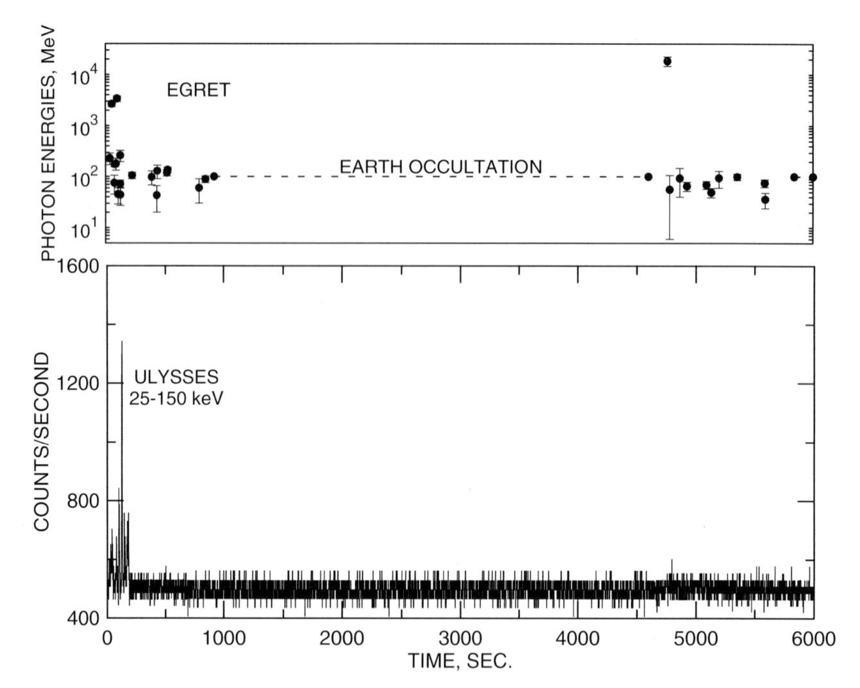

Fig. 5. The time history of GRB940217 at low energies, from the *Ulysses* experiment (bottom panel), and the energies of the photons detected by *EGRET* as a function of time (top panel). The single 18 GeV photon arrives long after the GRB has ended at lower energies. The probability that this photon is due to background is 5×10^{-6}.

long duration comes from a very weak tail which would probably be undetectable for less intense events. This constitutes a γ-ray afterglow which decays as a power-law and comprises $< 10\%$ of the burst energy. Another reason why some bursts appear to be very long, if they are observed at lower energies, is that the X-ray afterglow actually begins during the γ-ray burst [28].

5.4 GRB Distance Indicators

Given that GRB time histories are so poorly understood, and do not bear any resemblance to SNIa light curves, it comes as something of a surprise that they may be useful as distance indicators. The first studies concentrated on measuring the durations of bursts as a function of their intensity [54]. On the average, the dim bursts were found to have longer durations than the intense ones, as would be expected from cosmological time dilation. A subsequent study found that the times between different peaks in a GRB time history were also longer for the dimmer events [21]. However, a later study [51] using different methodology confirmed that the weaker bursts indeed appeared to be time stretched, but in a way which could not be attributed entirely to cosmology. Some stretching appeared to be inherent to the weaker events.

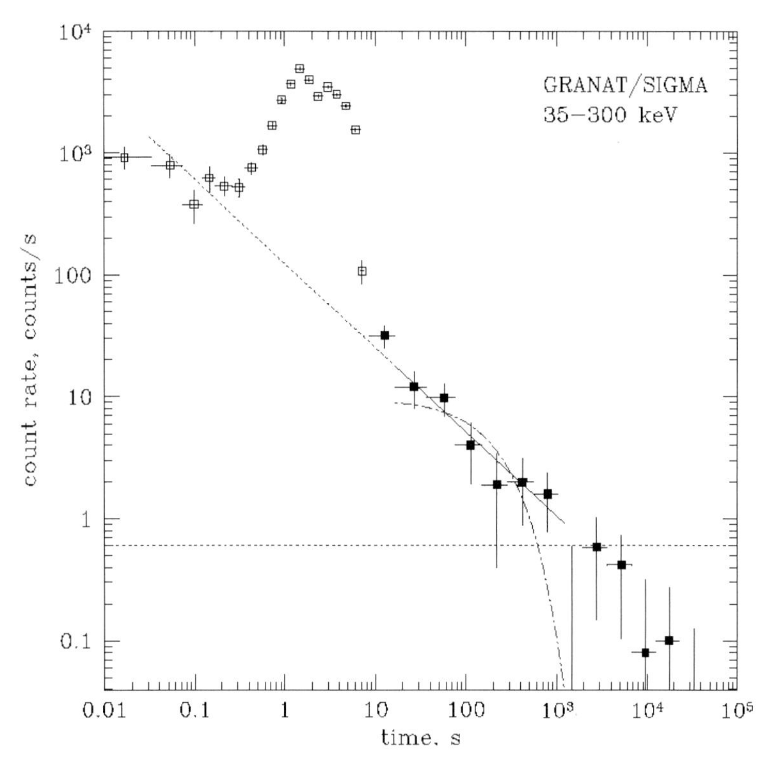

Fig. 6. The time history of GRB920723 in the 100-500 keV range as observed by the French *SIGMA* coded-mask hard X-ray telescope on the Russian *GRANAT* X-ray satellite [17]. Most of the ~1000 s duration is due to the γ-ray afterglow, which follows a power-law decay. With a small detector like the one on *Ulysses*, the burst duration appears to be < 10 s long.

Still, utilizing only the time histories and defining a variability measure for them, a correlation has been found between the variability and the luminosity for bursts with known redshifts. GRB980425, which is possibly related to SN1998bw, is included in the sample (Fig. 7) [63].

Utilizing the time histories at different energies and cross-correlating them to define a time lag between different energies in a given burst, an anticorrelation between peak intensity and lag has been found for a large sample of *BATSE* bursts [55]. Restricting the study to a much smaller sample of bursts with known redshifts, this has led to a luminosity-lag anticorrelation (Fig. 8). But in this case, GRB980425 does not fit the anticorrelation.

The effects of redshift have been searched for, and possibly found, in GRB energy spectra. A correlation between the intensity of a burst and the hardness of its spectrum has been noted [23,48] and interpreted as being due to the fact that bright, nearby bursts are expected to have harder energy spectra than dim, distant ones. However, in a subsequent study [46] it was demonstrated that at least some of the hardness-intensity correlation must be intrinsic to the sources.

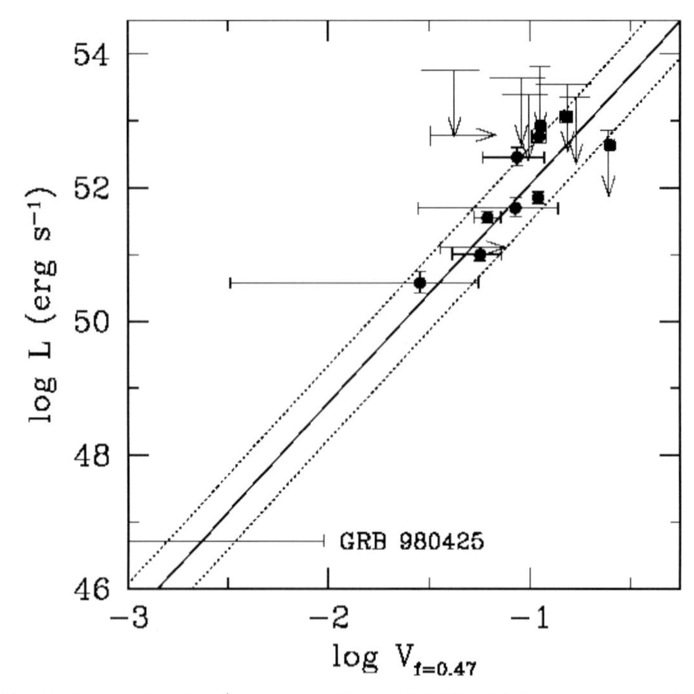

Fig. 7. The isotropic luminosity versus the variability V for a sample of bursts with known redshifts [63].

6 Long Wavelength and Non-electromagnetic Emission

Relatively little is known about emissions which accompany GRBs outside of the X- and γ-ray ranges. There have been many searches for prompt optical emission from GRB sources using robotic telescopes [59,65], but only one detection, from the *Robotic Optical Transient Search Experiment (ROTSE)* [2] has been obtained. The optical and γ-ray burst light curves are shown in Fig. 9. At its peak, the optical flash reached a V magnitude of 8.86, which is all the more remarkable considering that the burst source was at a redshift of 1.6 [14].

It is generally believed that GRBs may be accompanied by neutrino emission, but searches to date have given only negative results [1,3,4]. Gravitational radiation is expected to accompany neutron star-neutron star and neutron star-black hole mergers, so GRBs will be a target for the *Laser Interferometer Gravitational Wave Observatory (LIGO)* [11,64].

7 Remaining Challenges

When the distance scale to γ-ray bursts was completely unknown, there was really just one goal: to determine it. Today, the situation is more complex, and the goals are numerous. Among the challenges we face are the following:

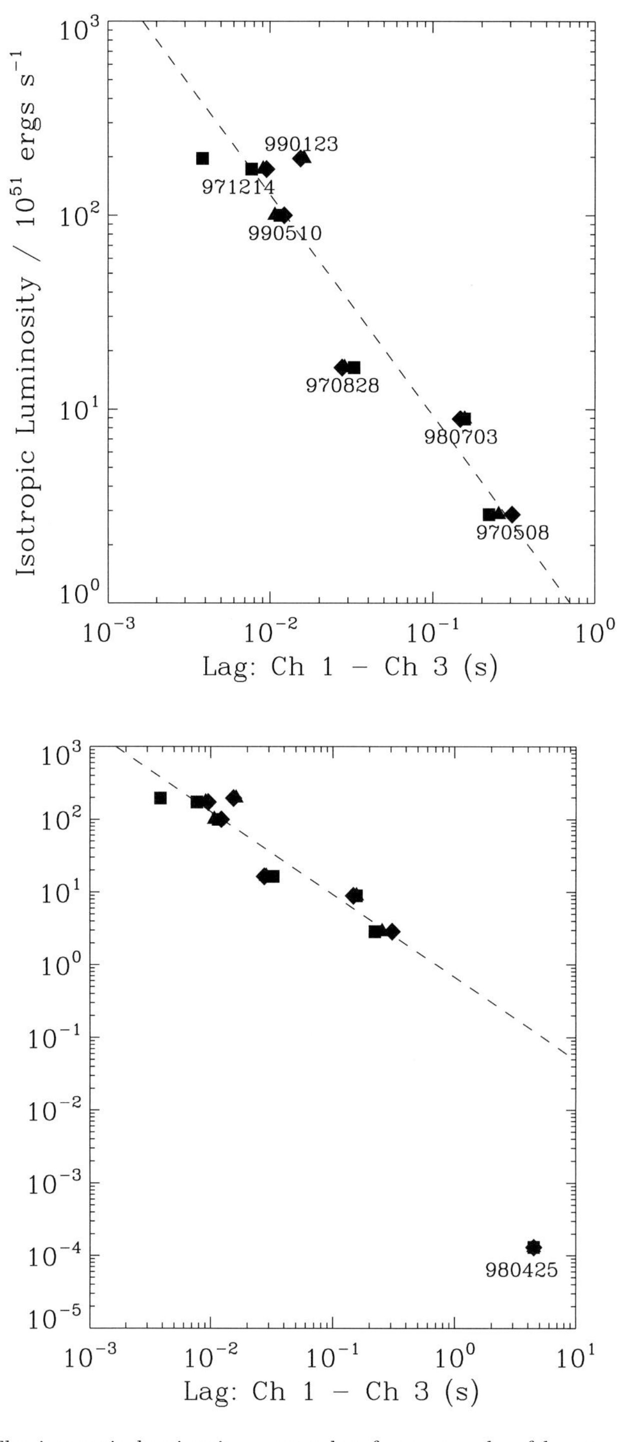

Fig. 8. Top: The isotropic luminosity versus lag for a sample of bursts with known redshifts [63]. Bottom: The expanded figure shows that GRB980425 does not fit the sample.

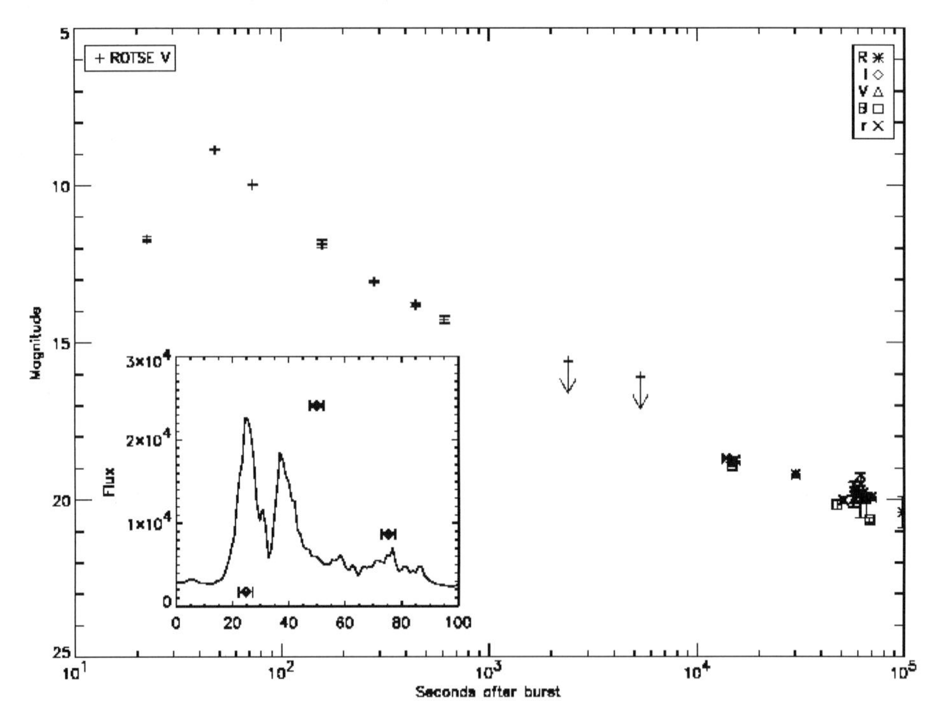

Fig. 9. Optical light curve for GRB990123, the only burst for which simultaneous optical emission has been found. Inset: The *ROTSE* optical fluxes compared to the *BATSE* γ-ray light curve. (Reprinted by permission from Nature [2], copyright 1999, Macmillan Publishing Ltd.)

- To understand the nature of the various classes of bursts and their relation to the production mechanism or mechanisms.
- To find the most distant GRB sources.
- To determine whether GRBs will be useful cosmological tools.
- To use GRBs to probe distant galaxies.
- To establish the relation between GRBs and SNe.
- To find the highest energies present in bursts.

All of these require that we detect and study many more bursts. In the near future, missions such as the *Interplanetary Network (IPN)* and *High Energy Transient Explorer (HETE-2)* will provide the data to attack these problems. In the longer term, the newer technologies of spacecraft such as the *Swift Gamma-Ray Burst Explorer (Swift)* and the *Gamma-Ray Large Area Space Telescope (GLAST)* will be needed to make progress.

8 Websites

Much of the data on γ-ray bursts is rapidly made public by posting it on websites. Similarly, many GRB missions are described and updated on the web long before descriptions appear in print. The following is a list of some useful websites.

8.1 Space-Based Missions

The *IPN* website *http://ssl.berkeley.edu/ipn3/index.html* contains localization data for several thousand GRBs and a table showing which spacecraft observed which bursts. It also has a bibliography of $\sim$5000 GRB publications and some maps showing the localizations of more recent, interesting bursts.

The *HETE-2* website is *http://space.mit.edu/HETE*. It contains information on the mission and its experiments.

The time histories, energy spectra, and sky maps for bursts detected by *BATSE* may be found on *http://www.batse.msfc.nasa.gov/batse/*, along with a description of the experiment and much more.

BeppoSAX, the mission which revolutionized the GRB field by first making it possible to obtain GRB positions both rapidly and with good precision, has its website at *http://www.asdc.asi.it/bepposax/*. Information on the mission, its experiments, and GRB data may be found there.

Swift a GRB *Medium-sized Explorer (MIDEX)* mission with γ-ray, x-ray, and optical telescopes, to be launched in 2003, is described at *http://swift.sonoma.edu*.

GLAST will detect GRBs up to the highest energies when it is launched around 2006. Its description is at *http://glast.gsfc.nasa.gov*.

8.2 Data

The *High Energy Astrophysics Science Archive Research Center (HEASARC)* at *http://heasarc.gsfc.nasa.gov* contains the data from many older missions, links to the websites of newer missions, an education and outreach facility, and much, much more.

Gamma-ray burst data is often made public by sending it to the *GRB Coordinates Network (GCN)* where it is sent out immediately to subscribers world-wide – in some cases while the burst is still in progress! The website, *http://gcn.gsfc.nasa.gov/gcn/*, contains an archive of the more than one thousand GRB messages that have been sent out over the years.

The site *http://www.ifa.au.dk/~hst/grb_hosts/index.html* is devoted to the host galaxies of GRBs. Images, redshifts, and other information are compiled there.

http://www.aip.de/~jcg/grb.html contains "a subjective collection of information and data on GRBs and their afterglows."

8.3 Ground-Based Instruments

The three main robotic telescopes dedicated to observations of simultaneous optical radiation from burst sources are the *Livermore Optical Transient Imaging System (LOTIS)*, *ROTSE*, and the *Télescope à Action Rapide pour les Objets Transitoires (TAROT – Rapid Action Telescope for Transient Objects)*. Their websites are *http://hubcap.clemson.edu/~ggwilli/LOTIS/*, *http://rotsei.lanl.gov/*, and *http://www.cesr.fr/~boer/tarot/*. Among other things they contain up-to-date sky patrol images taken the previous night.

References

1. M. Aglietta et al.: Astrophys. Space Sci. **231**, 355 (1995)
2. C. Akerloff et al.: Nature **398**, 400 (1999)
3. M. Ambrosio et al.: Astrophys. J. **546**, 1038 (2001)
4. E. Andres et al.: Nature **410**, 441 (2001)
5. R. Atkins et al.: Astrophys. J. Lett. **533**, L119 (2000)
6. D. Band et al.: Astrophys. J. **413**, 281 (1993)
7. D. Band et al.: Astrophys. J. **434**, 560 (1994)
8. D. Band et al.: Astrophys. J. **447**, 289 (1995)
9. D. Band et al.: Astrophys. J. **458**, 746 (1996)
10. D. Band, L. Ford, J. Matteson, M. Briggs, W. Paciesas, G. Pendleton, R. Preece: Astrophys. J. **485**, 747 (1997)
11. B. Barish, R. Weiss: Phys. Today **52**, 44 (1999)
12. A. Beloborodov, B. Stern: Astrophys. J. **535**, 158 (2000)
13. P. Bhat, G. Fishman, C. Meegan, R. Wilson, C. Kouveliotou, W. Paciesas, G. Pendleton: Astrophys. J. **426**, 604 (1994)
14. J. Bloom et al.: Astrophys. J. Lett. **518**, L1 (1999)
15. J. Brainerd: Astrophys. J. **538**, 628 (2000)
16. M. Briggs et al.: Astrophys. J. **459**, 40 (1996)
17. R. Burenin et al.: Astron. Astrophys. **344**, L53 (1999)
18. A. Castro-Tirado, S. Brandt, N. Lund, I. Lapshov, O. Terekhov, R. Sunyaev: 'WATCH Observations of Gamma-Ray Bursts During 1990-1992'. In: *Gamma-Ray Bursts - Second Workshop at Huntsville, AL, USA, October 20–22, 1994*, ed. by G. Fishman, J. Brainerd, K. Hurley (AIP Press, New York 1994) p. 17
19. J. Chiang, C. Dermer: Astrophys. J. **512**, 699 (1999)
20. D. Cline, C. Matthey, S. Otwinowski: Astrophys. J. **527**, 827 (1999)
21. M. Deng, B. Schaefer: Astrophys. J. Lett. **502**, L109 (1998)
22. J.-P. Dezalay, J. Lestrade, C. Barat, R. Talon, R. Sunyaev, O. Terekhov, A. Kuznetsov: Astrophys. J. Lett. **471**, L27 (1996)
23. J.-P. Dezalay et al.: Astrophys. J. Lett. **490**, L17 (1997)
24. B. Dingus, J. Catelli, E. Schneid: 'Bursts Detected and NOT Detected by EGRET Imaging Spark Chamber'. In: *Gamma-Ray Bursts, 4th Huntsville Symposium at Huntsville, AL, USA, September 15–20, 1997*, ed. by C. Meegan, R. Preece, T. Koshut (AIP Press, New York 1998), p. 349
25. E. Fenimore et al.: Astrophys. J. Lett. **335**, L71 (1988)
26. E. Fenimore, J. in't Zand, J. Norris, J. Bonnell, R. Nemiroff: Astrophys. J. Lett. **448**, L101 (1995)
27. L. Ford et al.: Astrophys. J. **439**, 307 (1995)

28. F. Frontera et al.: Astrophys. J. Suppl. **127**, 59 (2000)
29. F. Frontera et al.: Astrophys. J. **540**, 697 (2000)
30. P. Garnavich, A. Loeb, K. Stanek: Astrophys. J. Lett. **544**, L11 (2000)
31. G. Ghisellini, A. Celotti, D. Lazzati: Mon. Not. R. Astron. Soc. **313**, L1 (2000)
32. S. Golenetskii, E. Mazets, R. Aptekar, Yu. Guryan, V. Ilyinskii: Astrophys. Space Sci. **124**, 243 (1986)
33. J. Hakkila, C. Meegan, G. Pendleton, M. Briggs, J. Horack, D. Hartmann, V. Connaughton: 'GRB Repetition Limits from Current *BATSE* Observations'. In: *Gamma-Ray Bursts, 4th Huntsville Symposium at Huntsville, AL, USA, September 15–20, 1997*, ed. by C. Meegan, R. Preece, T. Koshut (AIP Press, New York 1998), p. 236
34. M. Harris, G. Share: Astrophys. J. **494**, 724 (1998)
35. J.C. Higdon, R.E. Lingenfelter: 'Selection Biases on the Spectral and Temporal Distribution of *BATSE* Gamma-Ray Bursts'. In: *Gamma-Ray Bursts, 4th Huntsville Symposium at Huntsville, AL, USA, September 15–20, 1997*, ed. by C. Meegan, R. Preece, T. Koshut (AIP Press, New York 1998), p. 40
36. I. Horvath: Astrophys. J. **508**, 757 (1998)
37. K. Hurley: 'Gamma-Ray Burst Observations: Past and Future'. In: *Gamma-Ray Bursts - Proceedings of the Gamma-Ray Burst Workshop at Huntsville, AL, USA, October 16–18, 1991*, ed. by W. Paciesas, G. Fishman (AIP Press, New York 1992) p. 3
38. K. Hurley et al.: Nature **372**, 652 (1994)
39. J. in't Zand, J. Heise, J. van Paradijs, E. Fenimore: Astrophys. J. Lett. **516**, L57 (1999)
40. J. Katz: Astrophys. J. Lett. **432**, L27 (1994)
41. R. Klebesadel, I. Strong, R. Olson: Astrophys. J. Lett. **182**, L85 (1973)
42. T. Koshut, C. Kouveliotou, W. Paciesas, J. van Paradijs, G. Pendleton, M. Briggs, G. Fishman, C. Meegan: Astrophys. J. **452**, 145 (1995)
43. C. Kouveliotou, C. Meegan, G. Fishman, N. Bhat, M. Briggs, T. Koshut, W. Paciesas, G. Pendleton: Astrophys. J. Lett. **413**, L101 (1993)
44. E. Liang, V. Kargatis: Nature **381**, 49 (1996)
45. N. Lloyd, V. Petrosian: Astrophys. J. **543**, 722 (2000)
46. N. Lloyd, V. Petrosian, R. Mallozzi: Astrophys. J. **534**, 227 (2000)
47. M. Lyutikov, V. Usov: Astrophys. J. Lett. **543**, L129 (2000)
48. R. Mallozzi, G. Pendleton, W. Paciesas: Astrophys. J. **471**, 636 (1996)
49. G. Marani, R. Nemiroff, J. Norris, K. Hurley, J. Bonnell: Astrophys. J. Lett. **512**, L13 (1999)
50. P. Meszaros, M. Rees: Mon. Not. R. Astron. Soc. **269**, L41 (1994)
51. I. Mitrofanov, M. Litvak, M. Briggs, W. Paciesas, G. Pendleton, R. Preece, C. Meegan: Astrophys. J. **523**, 610 (1999)
52. S. Mukherjee, E. Feigelson, G. Babu, F. Murtagh, C. Fraley, A. Raftery: Astrophys. J. **508**, 314 (1998)
53. T. Murakami, Y. Ogasaka, A. Yoshida, E. Fenimore: 'Ginga Observations of Gamma-Ray Bursts.' In: *Gamma-Ray Bursts - Proceedings of the Gamma-Ray Burst Workshop at Huntsville, AL, USA, October 16–18, 1991*, ed. by W. Paciesas, G. Fishman (AIP Press, New York 1992) p. 28
54. J. Norris, J. Bonnell, R. Nemiroff, J. Scargle, C. Kouveliotou, W. Paciesas, C. Meegan, G. Fishman: Astrophys. J. **439**, 542 (1995)
55. J. Norris, G. Marani, J. Bonnell: Astrophys. J. **534**, 248 (2000)
56. W. Paciesas et al.: Astrophys. J. Suppl. **122**, 465 (1999)

57. D. Palmer et al.: Astrophys. J. Lett. **433**, L77 (1994)
58. A. Panaitescu, P. Meszaros: Astrophys. J. Lett. **544**, L17 (2000)
59. H.-S. Park et al.: Astrophys. J. Lett. **490**, L21 (1997)
60. G. Pendleton et al.: Astrophys. J. **489**, 175 (1997)
61. R. Preece, G. Pendleton, M. Briggs, R. Mallozzi, W. Paciesas, D. Band, J. Matteson, C. Meegan: Astrophys. J. **496**, 849 (1998)
62. E. Ramirez-Ruiz, A. Merloni: Mon. Not. R. Astron. Soc. **320**, L25 (2001)
63. D. Reichart, D. Lamb, E. Fenimore, E. Ramirez-Ruiz, T. Cline, K. Hurley: Astrophys. J. **552**, 57 (2001)
64. M. van Putten, A. Levinson: Astrophys. J. Lett. **555**, L41 (2001)
65. G. Williams et al.: Astrophys. J. Lett. **519**, L25 (1999)
66. A. Yoshida, T. Murakami: 'A Spectral Study of an X-Ray Rich Gamma-Ray Burst.' In: *Gamma-Ray Bursts - Second Workshop at Huntsville, AL, USA, October 20–22, 1994*, ed. by G. Fishman, J. Brainerd, K. Hurley (AIP Press, New York 1994) p. 333

X-Ray Observations of γ-Ray Burst Afterglows

Filippo Frontera

Physics Department, University of Ferrara, Ferrara, Italy
and
Istituto Astrofisica Spaziale e Fisica Cosmica, CNR, Bologna, Italy

Abstract. The discovery by the *BeppoSAX* satellite of X-ray afterglow emission from the γ-ray burst which occurred on 28 February 1997 produced a revolution in our knowledge of the γ-ray burst phenomenon. Along with the discovery of X-ray afterglows, the optical afterglows of γ-ray bursts were discovered and the distance issue was settled, at least for long γ-ray bursts. The 30 year mystery of the γ-ray burst phenomenon is now on the way to solution. Here I review the observational status of the X-ray afterglow emission, its mean properties (detection rate, continuum spectra, line features, and light curves), and the X-ray constraints on theoretical models of γ-ray bursters and their progenitors. I also discuss the early onset afterglow emission, the remaining questions, and the role of future X-ray afterglow observations.

1 Introduction

The investigation into the nature and origin of celestial γ-ray bursts (GRBs) has been one of the major challenges of high energy astrophysics since they were discovered by chance about 30 years ago with the *VELA* satellites [77]. The GRB origin remained uncertain even when the *Burst and Transient Source Experiment (BATSE)* experiment aboard the *Compton Gamma-Ray Observatory (CGRO)* in the 1990s provided a much larger sample of GRBs (about 3000 events). Many outstanding results were obtained with *BATSE*, among them the completely isotropic distribution on the sky of GRBs and the inconsistency of the GRB number versus peak intensity relation with a homogeneous volume distribution in space, implying a deficit of faint bursts [18,96,114]. Two main explanations were given for these statistical properties: either GRBs have their origin in neutron stars distributed in a large Galactic halo at more than 100 kpc distance [17,19,61,83,136,156] or GRBs have an origin in sources at cosmological distances [96,98–100,108,115,137,155].

It was recognized (see, e.g., [39]) that the identification of a GRB counterpart at other wavelengths was needed in order to make a breakthrough in setting the distance scale to GRBs. Now the distance scale issue of at least the long (> 1 s) GRBs has been definitely settled thanks to the X-ray astronomy mission *BeppoSAX*, an Italian satellite with Dutch participation [12]. *BeppoSAX* has also provided the most exciting results of the last six years in GRB astronomy. The high performance of *BeppoSAX* for GRB studies is due to a particularly well-matched configuration of its payload, with both Wild Field and Narrow Field Instruments (WFIs and NFIs). The WFIs consist of a γ-ray ($40 - 700$ keV) all-sky monitor, the *Gamma-Ray Burst Monitor (GRBM)* [43], and two Wide-Field Cameras (WFCs) instruments covering the $2-28$ keV energy band [76]). The *NFI* include four focusing X-ray (0.1–10 keV) telescopes (one *Low-Energy Concentrator Spectrometer (LECS)* [120] and three *Medium-Energy Concentrator Spectrometers (MECS)* [13]). The *NFIs* also have two higher energy direct viewing detectors (the *High Pressure Gas Scintillation Proportional Counter (HPGSPC)* [88] and the *Phoswich Detector System (PDS)* [43]).

The *BeppoSAX* capability of precisely localizing GRBs was soon tested, only one month after first light, with the detection of GRB960720 [126]. Starting from December 1996, at the *BeppoSAX* Science Operation Center, an alert procedure for the search of simultaneous *GRBM* and *WFCs* detections of GRB events was implemented. Thanks to this procedure, the first X-ray afterglow of a GRB, GRB970228, was discovered on 28 February 1997 [23]. Simultaneously, an optical counterpart of the X-ray afterglow source was detected [164]. Less than two months later, the first optical redshift ($z = 0.835$) of GRB970508 was obtained thanks to the prompt localization provided by *BeppoSAX* [102]. The location of GRBs at cosmic distances was definitely established. At the same time, an exciting result on the GRB source properties came from radio observations of the same event: the radio afterglow exhibited rapid variations for about one month due to interstellar scintillations [40]. Thus, only a very compact source, expanding highly relativistically, could be the origin of GRBs [40].

Since the first detection of an X-ray afterglow, many other GRB afterglows have been detected, mainly with *BeppoSAX* and, less often, with other satellites, i.e., the *Röntgensatellit (ROSAT)*, the *Rossi X-ray Timing Explorer (RXTE)*, the *Advanced Satellite for Cosmology and Astrophysics (ASCA)*, the *Chandra X-ray Observatory – AXAF (CHANDRA)*, and the *XMM-Newton (XMM)*. In this chapter I review the status of our knowledge of GRB X-ray afterglows and discuss their implications. In Sect. 2, I discuss the measured afterglow properties (detection rate, continuum spectral properties, light curves, and X-ray line features); in Sect. 3, I discuss the early afterglow emission and its distinctive features with respect to the prompt emission; and in Sect. 4 I summarize the open questions and the prospects for future X-ray afterglow observations.

2 X-Ray Afterglow Observations and Properties

Tables 1, 2, and 3 summarize the main results of the X-ray searches for GRB afterglows. The following provides further elaboration.

2.1 X-Ray Afterglow Detection Rate

Table 1 lists the information on GRB follow-up observations of X-ray afterglows as of December 2001. Two X-ray rich events, GRB011030 [51] and GRB011130 [141], were not included in the list. GRB011030 was observed by *CHANDRA* [66] following the discovery of a radio transient in the GRB error box [162] and an X-ray source was detected but no fading observed [66]. GRB011130

Table 1. GRBs Observed for X-ray Afterglows

GRB	Instr. for GRB loc.[a]	Error box radius	Start of X-ray follow-up[b]	Instr. for follow-up[c]	X-ray counterpart flux at start[d] (erg cm^{-2} s^{-1})	Source name[e]	Error box radius	Ref.
970111	WFC	3′	16.5	NFI	$< 1.6\times10^{-13}$	1SAXJ1528.1+1937(?)		[35]
970228	WFC	3′	8.27	NFI	3.2×10^{-12}	1SAXJ0501.7+1146	50″	[23]
			...	HRI	$\sim 2\times10^{-12}$	RXJ050146+1149.9	10″	[44]
970402	WFC	3′	8.39	NFI	2.7×10^{-13}	1SAXJ1450.1-6920	50″	[109]
970508	WFC	1.9′	5.6	NFI	$\sim 7\times10^{-13}$	1SAXJ0653.8+7916	50″	[127,128]
970616[f]	BAT	2°	4.0	PCA	1.1×10^{-11}	01 18 57 −05 28 00	0.7′ × 18′	[90]
970815	ASM	6′ × 3′	76.8	HRI	$< 1.0\times10^{-13}$			[57]
970828	ASM	2.5′ × 1′	3.6	PCA	1.15×10^{-11}	18 08 32.2 +79 16 02	30″	[91,106,107,176]
971214	WFC	3.3′	6.55	NFI	$\sim 7\times10^{-13}$	1SAXJ1156.4+6513	60″	[30]
971227	WFC	8′	14.4	NFI	2.6×10^{-13}	1SAXJ1257.3+5924	90″	[6]
980329	WFC	3′	7.0	NFI	1.4×10^{-12}	1SAXJ0702.6+3850	50″	[177]
980425	WFC	8′	10.2	NFI	4.0×10^{-13}	1SAXJ1935.0-5248	90″	[121]
980515	WFC	4′	9.83	NFI	1.6×10^{-13}	1SAX J2116.8-6712	90″	[36]
980519	WFC	3′	4.74	NFI	6.0×10^{-13}	1SAXJ2322.3+7716	50″	[110,111]
980613	WFC	4′	8.63	NFI	3.3×10^{-13}	1SAXJ1017.9+7127	50″	[24,157]
980703	ASM	4′	22.0	NFI	4.7×10^{-13}	1SAXJ2359.1+0835	50″	[169]
980706[g]	COM	4°	2.7	PCA	4.0×10^{-11}			[92,166]
981226[h]	WFC	6′	11.3	NFI	$< 1.5\times10^{-13}$	1SAXJ2329.6-2356	60″	[47]
990123	WFC	2′	5.8	NFI	1.6×10^{-11}	1SAXJ1525.5+4446	50″	[67,69]
990217	WFC	3′	6.5	NFI	$< 1.0\times10^{-13}$			[129]
990506	BAT	9.6′	3.0	PCA	3.2×10^{-11}	11 54 46 −26 45 35	6′	[72]
990510	WFC	3′	8.0	NFI	5.2×10^{-12}	1SAXJ1338.1-8030	56″	[81]
990627	WFC	3′	8.0	NFI	3.5×10^{-13}	1SAXJ0148.5-7704	69″	[112]
990704	WFC	7′	8.3	NFI	8.0×10^{-13}	1SAXJ1219.5-0350	60″	[38]
990705	WFC	3′	11.38	NFI	1.9×10^{-13}	1SAXJ0509.9-7207	120″	[3]
990806	WFC	3′	7.77	NFI	4.0×10^{-13}	1SAXJ0310.6-6806	60″	[46,105]
990907[i]	WFC	6′	10.98	NFI	$1\text{–}2\times10^{-12}$	1SAXJ0731.2-6928	180″	[130]
991014	WFC	6′	11.76	NFI	4.0×10^{-13}	1SAXJ0651.0+1136	90″	[178]
991106[j]	WFC	3′	7.8	NFI	1.25×10^{-13} ?	1SAXJ2224.7+5423 ?	90″	[7]
991216	BAT	1.1°	4.03	PCA	1.24×10^{-10}	05 09 31.5 +11 17 05.7	1.5″	[131,161]
000115	BAT	1.7°	2.92	PCA	4.3×10^{-11}	08 03 12 −17 05 59	18′	[93]
000210	WFC	2′	7.2	NFI	4.5×10^{-13}	01 59 15.7 −40 39 32.3	1.6″	[25,54,134]
000214	WFC	6.5′	12.0	NFI	7.7×10^{-13}	1SAXJ1854.4-6627	50″	[8]
000528[k]	WFC	2′	8.3	NFI	1.8×10^{-13}	1SAXJ1045.1-3359	90″	[82]
000529[l]	WFC	4′	7.5	NFI	2.8×10^{-13}	1SAXJ0009.8-6128	120″	[37]
000615[m]	WFC	2′	10.04	NFI	6.7×10^{-14}	1SAXJ1532.4+7349	120″	[113]
000926	IPN	3′ × 10′	48.96	NFI	3.8×10^{-13}	17 04 01.6 +51 47 08.6	2″	[64,132]
001025[n]	IPN	5′	45.60	XMM	$1\text{-}2\times10^{-13}$	08 36 35.42-13 04 09.9	< 10″	[1,2]
001109	WFC	2.5′	16.5	NFI	7.1×10^{-13}	1SAXJ1830.1+5517	50″	[5]
010214	WFC	3′	6.28	NFI	1.0×10^{-12}	1SAXJ1741.0+4834	60″	[60]
010220	WFC	4′	16.0	NFI	$< 1.0\times10^{-13}$			[89]
010222	WFC	2.5′	8.0	NFI	1.2×10^{-11}	1SAXJ1452.2+4301	30″	[179]
011121	WFC	2′	21.25	NFI	$\sim 1.0\times10^{-12}$	1SAX J113426-7601.4	50″	[133,135]
011211	WFC	1′	11.13	XMM	1.9×10^{-3}	11 15 17.9 −21 56 57.5	< 10″	[139]

[a] *WFCs* on board *BeppoSAX*; *BAT*=*BATSE* on board *CGRO*; *ASM* on board *RXTE*; *COM*=*COMTEL* on board *CGRO*; *IPN*= Interplanetary Network.
[b] Earliest afterglow search, in hours from the GRB detection.
[c] Instrument used for the earliest afterglow search: *NFI* on board *BeppoSAX*; *HRI* on board *ROSAT*; *PCA* on board *RXTE*; *CHANDRA*; *XMM*.

Table 1. (Continued)

[d]Energy band: 2-10 keV. All upper limits are 3σ except for GRB981226 which is 2σ.
[e] The best available coordinates and error box radius. When no source name is available, we report here the equatorial coordinates α_{J2000}, δ_{J2000}, when available.
[f]No more sensitive follow-up of this source is available. No decay was observed. No detection of radio or near infrared (NIR) afterglow was obtained [56,58,171].
[g]It is unclear whether this flux was due to the afterglow. After 1.5 hrs from the first detection the source was no longer visible [92]. This would imply a decay index $\beta < -3$. Most likely, the *PCA* source is not related to the GRB [165].
[h]The source was not visible at the beginning of the observation campaign. Afterward, the source flux increased by more than a factor of 3 in about 10,000 s. Then the source started fading.
[i]Due to technical problems, the *BeppoSAX* observation had a duration of only 20 min. The probability of a serendipitous source is $\sim 3 \times 10^{-3}$ [130].
[j]Source fading was not observed. The probability of detecting an unrelated foreground source at this flux level or higher, in the GRB error box given by the *BeppoSAX WFCs* (see GCN 435) is about 10%.
[k]The source faded to a 2σ upper limit of 5.0×10^{-14} erg cm^{-2} s^{-1} after $78.86 - 99.06$ hrs from the GRB onset.
[l]The source was undetected (2σ upper limit of 1.3×10^{-13} erg cm^{-2} s^{-1}) 17 hrs after the main event.
[m]This flux was in the 1.6–4 keV band. The source was undetected (2σ upper limit of 3.0×10^{-14} erg cm^{-2} s^{-1}) by 20 hrs after the GRB event. This might be a field source unrelated to the GRB.
[n] This flux was in the 0.5-10 keV band.
[o] This flux was in the 0.2-10 keV band.

was also observed by *CHANDRA* and several X-ray sources were detected but no conclusions could be drawn as to their possible association with the GRB [142]. As can be seen, most of the 43 GRBs listed in Table 1 were promptly localized with the *BeppoSAX WFCs*, with the *RXTE Proportional Counter Array (PCA)*, the *All Sky Monitor (ASM)*, and the 3rd *Interplanetary Network (IPN)* contributing some additional examples. The minimum starting time of the follow-up observations ranges from $2-3$ hr for *RXTE*, to $5-6$ hr for *BeppoSAX*, to longer times for the other satellites. Given its limited sensitivity, the *RXTE PCA* has provided positive results only for very strong afterglow fluxes (few mCrab) and for a limited time interval. Unfortunately most of these PCA events were not followed up with more sensitive instruments.

The criterion for establishing the afterglow nature of an X-ray source is its fading on time scales of a few hours or days. Unfortunately, for the weakest sources this fading could not be well established, as the detected fluxes were very close to the instrumental sensitivity limit. In these cases the source found in the GRB error box could be a serendipitous field source unrelated to an X-ray afterglow, even though the probability of such a chance coincidence is low (see the notes to Table 1). With this caveat in mind, 37 of the followed 43 GRBs

(86%), show X-ray afterglows at a level $> 10^{-13}$ erg cm^{-2} s^{-1} in the 2–10 keV band. Thus, the detection rate of X-ray afterglows is about two times higher than that for optical afterglows (see the chapter by Pian in this volume). Therein lies the mystery of the so-called "dark bursts", i.e., bursts which show an X-ray afterglow but no detectable optical emission (see Sect. 4).

2.2 X-Ray Afterglow Continuum Spectra

The spectral shape of the X-ray afterglow emission gives important information about establishing the emission mechanisms. An initial question was whether the X-ray continuum emission was thermal or nonthermal. The X-ray afterglow of GRB970228 showed [45] that, in the X-ray range $(0.1 - 10$ keV), a blackbody model was not acceptable. The photon spectrum was well fit with a simple photoelectrically absorbed power-law (see Table 2). This spectral shape has been found in all X-ray afterglows known, although for some GRBs (GRB980329 [176] and GRB981226 [47]), in which thermal models were tested, it was found that the data do not have sufficient signal-to-noise to discriminate between power-law and blackbody models.

Table 2 reports the best fit power-law photon indices I (where the photon index I is related to the flux density F spectral index α, $F_\nu \propto \nu^{+\alpha}$, by $I = \alpha - 1$) and absorbing hydrogen column densities n_H. The values reported in square brackets were kept fixed during the fit. For comparison, the Galactic hydrogen column densities $n_{\rm H}^{\rm G}$ values along the GRB directions are also reported [32]. The best fit values of I and $n_{\rm H}$ are available for only the brightest of the detected afterglows. Most of the spectra have been determined up to 10 keV, and in only one case (GRB990123) was the source detected up to 60 keVwith the *BeppoSAX* PDS instrument, also confirming the power-law shape.

As can be seen from Table 2, due to the low statistical quality of the X-ray data, in many cases the Galactic $n_{\rm H}^{\rm G}$ was assumed (in square brackets) in the spectral fitting. In the cases for which estimates of $n_{\rm H}$ were derived, they are consistent, within the uncertainties, with the Galactic values, except for seven GRBs (GRB970828, GRB980329, GRB980703, GRB990510, GRB000926, GRB001109, and GRB010222) for which the derived $n_{\rm H}$ is significantly higher than the galactic value. Since the reported $n_{\rm H}$ values have been derived assuming only local, Galactic absorption ($z = 0$), the true $n_{\rm H}$ may be even higher, if it is in the host galaxy or in the ambient medium surrounding the burst. For example, Weiler et al. [170] (see also the chapter by Weiler et al. in this volume) found that significant amounts of optical extinction can be ascribed to the burst host galaxy or its local surroundings.

The best fit photon indices I range from ~ -1.5 to ~ -2.8 with a roughly Gaussian distribution with standard deviation $\sigma_I \approx 0.4$ and mean value $< I >= -1.95 \pm 0.03$, which is very close to the power law index of the well known synchrotron source, the Crab Nebula ($I = -2.1$). The only evidence of spectral variability is for GRB970508 and GRB000926 listed in Table 2. For comparison, the prompt $2 - 700$ keV emission from GRBs with detected X-ray afterglows is generally well described [48] by a photoelectrically absorbed, smoothly broken

Table 2. X-ray Afterglow Properties

GRB	Temporal index β	Photon index I	n_H/n_H^G ($\times 10^{21}$ cm^{-1})	2 – 10 keV flux at 10^5 s[a] (erg cm^{-2} s^{-1})	Redshift z	Ref.
970111	< -1.5 (3σ)	...	...	$< 1.0\times10^{-13}$	...	[35]
970228	$-1.33^{+0.11}_{-0.13}$	-2.06 ± 0.24	$3.5^{+3.3}_{-2.3}/1.6$	$\sim 6.8\times10^{-13}$	0.695	[23,45]
	$-1.50^{+0.35}_{-0.23}$	-2.1 ± 0.3	$3.5^{+3.3}_{-2.3}/1.6$	$\sim 1.1\times10^{-12}$	...	[44]
970402	$-1.3 - -1.6$	-1.7 ± 0.6	$<20/2.0$	$\sim 4.5\times10^{-14}$	...	[109]
970508	-1.1 ± 0.1[b]	-1.5 ± 0.8[c]	$6.0^{+13}_{-5.5}/0.5$	3.5×10^{-13}[d]	0.835	[127,128]
		-2.2 ± 0.7[e]	$0.5^{+5}_{-0.5}/0.5$	1.5×10^{-12}[f]		
970828	-1.44 ± 0.07	$[-2]$[g]	$4.1^{+2.1}_{-1.6}/0.34$[h]	6.0×10^{-13}	0.9758	[34,90,91]
971214	$-0.85^{+0.17}_{-0.15}$	-1.6 ± 0.2	$1.0^{+2.3}_{-1.0}/0.6$	...	3.42	[30]
971227	$-1.12^{+0.05}_{-0.08}$	$[-2.1]$	$[0.13]/0.13$	$\sim 1.4\times10^{-13}$	...	[6]
980329	-1.35 ± 0.03	-2.4 ± 0.4	$10 \pm 4/0.9$	2.0×10^{-13}	<3.9	[33,71,176]
980425[i]	-0.16 ± 0.04	-2.0 ± 0.3	$[0.39]/0.39$	$\sim 4.0\times10^{-13}$	0.0085	[121]
980519	-1.83 ± 0.30	$-2.8^{+0.5}_{-0.6}$	3–20/1.73	8.0×10^{-14}	...	[111]
980613	-1.19 ± 0.17	$> -2.4?$	...	$\sim 2.3\times10^{-13}$	1.097	[24,157]
	-1.05 ± 0.37[j]	...	...	...	...	[14,70]
980703	< -0.91 (3σ)	-2.51 ± 0.32	$36^{+22}_{-13}/0.34$[k]	4.5×10^{-13}	0.966	[168]
981226	$-1.31^{+0.39}_{-0.44}$[l]	-1.92 ± 0.47	$[0.18]/0.18$	$\sim 2.0\times10^{-13}$	...	[47]
990123	-1.44 ± 0.07	-1.86 ± 0.1[m]	$0.51^{+0.7}_{-0.1}/0.21$	1.25×10^{-12}	1.600	[67,69]
990510	-1.42 ± 0.07[n]	-2.03 ± 0.08	$2.1 \pm 0.6/0.94$	9.6×10^{-13}	1.619	[81,122]
990627	$-1.741^{+0.55}_{-0.63}$	-2.8 ± 0.7	$[0.52]/0.52$	$< 1.6\times10^{-13}$	...	[5]
990704	-0.83 ± 0.16	$-1.691^{+0.34}_{-0.60}$	$[0.3]/0.3$	$\sim 3.3\times10^{-13}$	...	[38]
990705[o]	-1.58 ± 0.06[p]	...	...	$< 1.2\times10^{-13}$	0.86	[3]
990806	-1.2 ± 0.3	$-2.161^{+0.50}_{-0.61}$	$[0.35]/0.35$	$\sim 2.0\times10^{-13}$	...	[105]
991014[q]	< -0.4 (3σ)	-1.53 ± 0.25	$[2.5]/2.5$	$\sim 3.0\times10^{-13}$	...	[177]
991216	-1.62 ± 0.07	-2.2 ± 0.2[r]	$3.51^{+1.5}_{-1.5}/2.1$	$\sim 5.0\times10^{-12}$	1.02	[62,131,160]
000115	< -1.0	...	...	$< 1.0\times10^{-11}$	...	[93]
000214	-0.8 ± 0.5	-2.0 ± 0.3[g]	$0.071^{+0.75}_{-0.07}/0.55$	$\sim 5.0\times10^{-14}$[s]	0.37–	[8]
	-1.41 ± 0.03[p]	...	...	...	0.47[t]	
000926	$-1.89^{+0.16}_{-0.19}$[u]	-1.85 ± 0.15[v]	$4.0^{+3.5}_{-2.5}$[w]$/0.27$	9.0×10^{-13}	2.06	[64,132]
	-1.70 ± 0.16[x]	-2.23 ± 0.3[y]	$[4.0]$[w]$/0.27$	...	...	
001109	-1.18 ± 0.05[p]	-2.4 ± 0.3	$8.7 \pm 0.4/0.42$	$\sim 8.0\times10^{-13}$	...	[5]
010214	$-2.1^{+0.6}_{-1.0}$ for $t > 8$ hr	$-1.3^{+0.6}_{-0.8}$	$[0.27]/0.27$	...	...	[60]
	$-0.99^{+0.09}_{-0.04}$ for $t < 8$ hr					
010222	-1.33 ± 0.04[z] for $t > 8.0$ hr	-1.97 ± 0.05	$1.5 \pm 0.3/0.16$	2.4×10^{-12}	1.477	[178]
	~ 0.8[p] for $t < 8.0$ hr					

Table 2. (Continued)

[a] All upper limits are 3σ except for GRB990627 and GRB990705 which are 2σ.
[b] From 6×10^4 s to 5.8×10^5 s an X-ray outburst occurred with integrated energy excess with respect to the power-law decay equal to $\sim 5\%$ that of the GRB primary event. The outburst was also visible at optical wavelengths (see references in [127]).
[c] Index measured before the outburst. During the outburst the slope was variable [127].
[d] Interpolated from the data before and after the outburst.
[e] Index measured soon after the outburst.
[f] Interpolated taking into account the outburst data.
[g] Evidence of an emission feature (see Sect. 2.4).
[h] $n_{\rm H}$ measured in the latest observation period.
[i] Assuming that the source S1 is coincident with the SN1998bw position (see Sect. 2.3).
[j] Result reported by [14].
[k] $n_{\rm H}$ value corrected for redshift.
[l] Temporal index during the fading phase of the source. See note h of Table 1.
[m] The afterglow is visible up to 60 keV with a power-law spectrum.
[n] Comparing the prompt emission with the X-ray afterglow emission, Pian et al. [122] derive the presence of a break in the afterglow light curve which tends, at early and late epochs, to a power-law of index $\beta \sim -1$ and $\beta \sim -2$, respectively, with a break time of ~ 0.5 days, compatible with that determined from the optical light curve [63,74,158].
[o] The field is highly contaminated by stray radiation.
[p] Derived by comparing the prompt emission and late afterglow data.
[q] The shortest event followed by *BeppoSAX* (3.29 s at 40–700 keV).
[r] Also evidence of two emission lines (see Table 3).
[s] Extrapolated from the measured light curve.
[t] Based on X-ray emission feature interpretation (see text).
[u] *BeppoSAX* plus *CHANDRA* data. Evidence of a discrepancy between optical and X-ray decline rates [132].
[v] *BeppoSAX* plus *CHANDRA* data $2-3$ days after the main event.
[w] Corrected for redshift [132]. This $n_{\rm H}^z$ value was added to the Galactic column density $n_{\rm H}^{\rm G}$. According to Harrison et al. [64], the $n_{\rm H}^z$ is not needed by the *CHANDRA* data.
[x] From *CHANDRA* data only.
[y] From *CHANDRA* data 13 days after the GRB.
[z] The extrapolation of the power-law light curve, derived from the *BeppoSAX* data, is above the X-ray flux measured 9 days after the burst with *CHANDRA* [65,178] (see text).

power-law [9] with low-energy index α_1 below the break, high-energy index α_2 above the break, and peak energy E_p of the $EF(E)$ spectra, all of which evolve rapidly with time. One relevant feature of these spectra is that, at the GRB onset, E_p is generally above the energy passband of the *BeppoSAX GRBM* [48] (see the example in Fig. 1).

Several hints, such as the nonthermal shape of the spectra and detection of polarized emission from the optical counterparts of two afterglows (GRB990510

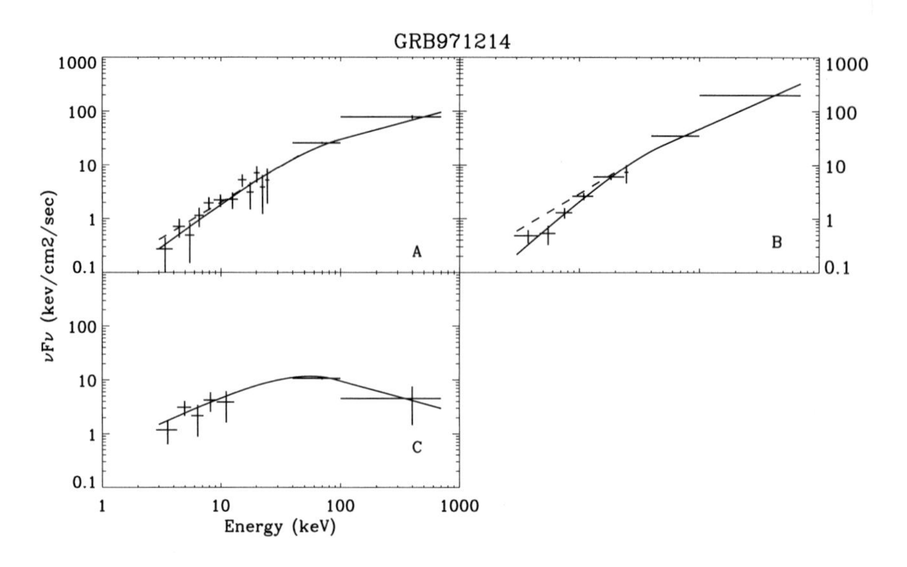

Fig. 1. Spectral evolution of GRB971214 in three contiguous time intervals, A (8 s), B (12 s) and C (18 s), into which the GRB light curve was subdivided [48].

[173,27] and GRB990712 [143]) point to synchrotron radiation as the basic emission mechanism of the electromagnetic radiation from GRB afterglows. Also, in the case of the prompt emission, synchrotron radiation appears to play an important role, even if other emission processes such as Inverse Compton (IC) may contribute at the earliest times (see, e.g., [48]). Synchrotron radiation is expected within the framework of the fireball model (see, e.g., the reviews by Piran [124,125] and references therein) as a result of the interaction between a relativistically expanding fireball and the external medium ("external shock"). Detailed calculations of the expected synchrotron spectral shape and its time evolution have been carried out for various external medium conditions (homogeneous medium [150], or radially declining density wind [21]). Also the effect of IC scattering of low-energy synchrotron photons on shock-accelerated electrons has been investigated (see, e.g., [118,153]).

Following Sari et al. [150], for the case of a homogeneous medium, the instantaneous synchrotron energy spectrum $F(E)$ is schematically described by a multi-broken power-law, with four different slopes. At low frequencies, the flux density $F_\nu \propto \nu^2$ holds up to the synchrotron self-absorption (SSA) frequency $\nu_{\rm SSA}$, then $F_\nu \propto \nu^{1/3}$ holds from $\nu_{\rm SSA}$ up to the peak frequency ν_m (ν_m corresponds to the minimum energy of the shock accelerated electrons). At still higher frequencies, the spectrum is related to the energy distribution of the electrons, which is assumed to exhibit a power law shape $N(E_e) \propto E_e^{-p}$ (E_e is the electron energy) from ν_m up to the cooling frequency ν_c (ν_c corresponds to the highest energy electrons that cool more rapidly than the expansion time of the source), by means of $F_\nu \propto \nu^{-(p-1)/2}$, while above ν_c, $F_\nu \propto \nu^{-p/2}$. Depending on the shock evolution, fully adiabatic or fully radiative, the characteristic frequencies, ν_m and ν_c, and the maximum flux density, F_m, vary in different ways with time. For the fully adiabatic case, $\nu_m \propto t^{-1/2}$ and $\nu_c \propto t^{-3/2}$. For the fully radiative case, $\nu_m \propto t^{-12/7}$ and $\nu_c \propto t^{-2/7}$. In the latter case, F_m also changes with time as $F_m \propto t^{-3/7}$. The different time behavior of the characteristic frequencies is thus expected to affect not only the spectral evolution of the emission but also the time behavior of the afterglow light curves (see Sect. 2.3).

The multiwavelength spectrum of the GRB970508 afterglow fits this scheme very nicely [49]. Unfortunately, broad band spectra are available for only a very limited number of GRBs. An attempt to collect the available data to estimate the multiwavelength spectra of GRB afterglows was performed [94], but the data are still scarce. Most of the few available spectra are consistent with the synchrotron shock model (in addition to GRB970508, GRB971216 [30], GRB980703 [168], and GRB991216 [41]). However, in two cases (GRB000926 [64] and GRB010222 [178]), the multiwavelength spectra clearly deviate from the shape expected from the synchrotron model. In both cases, the radio and optical data fit a synchrotron spectrum very well, but the extrapolation of the spectrum to X-rays is below the measured X-ray flux density. Harrison et al. [64] interpret this mismatch as due to the presence of a self-Inverse Compton component. The ambient medium density derived is $\sim 30\ \mathrm{cm}^{-3}$, which is higher than the average interstellar medium (ISM) density in a typical galaxy ($\sim 1\ \mathrm{cm}^{-3}$), but is consistent with a diffuse interstellar cloud such as those found in star forming regions.

Assuming a synchrotron origin, the photon indices I listed in Table 2 permit an estimate to be made of the power-law index of the energy distribution of the electrons accelerated in the external shocks. Given that most of the X-ray afterglow measurements were performed considerably after the main event (see Table 1 for the start times), these can be reasonably assumed to be taken in the regime of $\nu > \nu_m$, when the peak of the spectrum had moved toward the optical range. Thus, the electron index p is related to the photon index by one of two equations: either $p = -2I - 1$ or $p = -2I - 2$, depending whether the X-ray band is between ν_m and ν_c (fast cooling) or beyond these frequencies (slow cooling). From the mean value of I, we can derive a mean value of the electron index $< p >$ in the range from ~ 2 to ~ 3, depending on the cooling regime.

2.3 Time Behavior of the X-Ray Afterglow Light Curves

One of the most peculiar properties of the X-ray afterglow light curves is their power-law decay ($F(t) \propto t^{+\beta}$). The power-law fading was already noted in the first X-ray afterglow source 1SAX J0501.7+1146 associated with GRB970228 [23]. The X-ray fading slope ($\beta_X = -1.33^{+0.11}_{-0.13}$) was consistent with that ($\beta_{opt} = -1.46 \pm 0.16$) of the simultaneously discovered optical counterpart [164]. The power-law decay index and the photon index of the GRB970228 afterglow (see values in Table 2) were also found to be consistent with the relationship predicted by the fireball model and gave strong support to that model (see, e.g., [172]). Almost all other X-ray afterglow sources discovered so far show a similar power-law decay or, in a few cases, a broken power-law. Table 2 reports the best fit

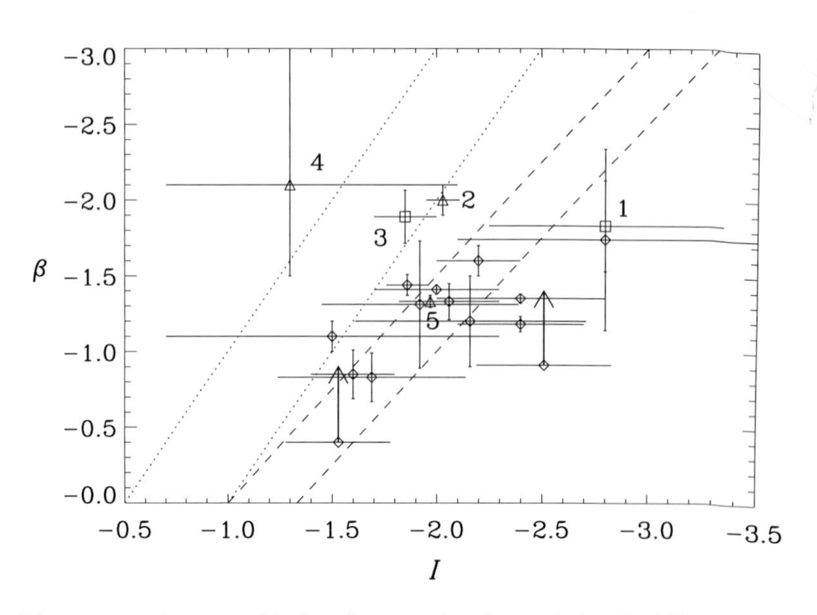

Fig. 2. The measured temporal index β versus the photon index I of GRB afterglows. The GRBs with evidence of a break in their X-ray light curves are indicated with triangles, while those with high temporal index ($\beta < -1.8$) are indicated with squares. 1: GRB980519, 2: GRB990510, 3: GRB000926, 4: GRB010214, 5: GRB010222. For GRB990510 the temporal index estimated by [122] is quoted. The region between the dashed lines is for a spherical expansion while the region between the dotted lines is for an expanding jet.

temporal indices, β, measured. Apart from GRB980425, which will be discussed separately below, the power-law indices are distributed like a Gaussian with mean value $< \beta >= -1.30 \pm 0.02$ and standard deviation $\sigma_\beta \approx 0.35$. A relevant feature of the light curves is that their extrapolation back to the time of the GRB event is, in general, consistent with the tail of the prompt X-ray light curve. This fact is, in some cases, directly measured (GRB970228 [23] and GRB970508 [127]) and in other cases is inferred from a different approach (see Sect. 3).

In a few cases (GRB990510 [121], GRB010222 [178] and GRB010214 [60]) comparison of the intensity level of the tail of the prompt emission with the late afterglow light curve (start time several hrs after the GRB event), shows that a broken power-law may be more consistent with the data. In Table 2 the power-law indices and break times of these GRBs are reported. The X-ray breaks are found to be consistent with those seen in the optical light curves (see [63,74,158] for GRB990510 and [95] for GRB010222).

The interpretation of achromatic breaks in the light curves of GRB afterglows has been discussed by various authors. In the framework of the fireball model, a break is expected if the relativistically expanding material is concentrated within a cone with angular width θ_c. As long as the Lorentz factor Γ of the outflowing material is larger than $1/\theta_c$, due to relativistic beaming, the radiation is emitted within an angle $\theta_b = 1/\Gamma$ from the cone axis, with $\theta_b < \theta_c$. When Γ drops below $1/\theta_c$, the observer begins to see the edge of the cone and then the effect of the collimated outflow, which causes a light curve steepening. At the same time (t_j), the jet begins to expand sideways, increasing the rate of decrease of the emitted radiation even more. This model has been discussed and detailed for various external conditions by several authors (see, e.g., [80,117,119,140,151] and references therein).

Assuming synchrotron radiation, adiabatic expansion, and a homogeneous ambient medium, following Sari et al. [151], the energy flux density, if $\nu_m < \nu < \nu_c$, is given by

$$F_\nu(t) \propto t^{-3(p-1)/4} \tag{1}$$

for $t \leq t_j$ and

$$F_\nu(t) \propto t^{-p} \tag{2}$$

for $t \geq t_j$. However, if $\nu > \nu_c > \nu_m$,

$$F_\nu(t) \propto t^{-3p/4+1/2} \tag{3}$$

for $t \leq t_j$ and

$$F_\nu(t) \propto t^{-p} \tag{4}$$

for $t \geq t_j$. In both cases, after the break the light curve has the same slope. In this framework, the temporal decay index, β, and the photon index, I, are strictly related: in the case of spherical expansion

$$\beta = 3I/2 + 3/2 \tag{5}$$

for $\nu < \nu_c$ and

$$\beta = 3I/2 + 2 \tag{6}$$

for $\nu > \nu_c$. In the case of an expanding jet

$$\beta = 2I + 1 \tag{7}$$

for $\nu < \nu_c$ and

$$\beta = 2I + 2 \tag{8}$$

for $\nu > \nu_c$. The above predictions can be easily tested with the available data.

In Fig. 2, the values of (I, β) for each afterglow source are shown (see also earlier test [159]). Even though the statistical uncertainties do not allow us to draw firm conclusions from this diagram, a clustering of the data within the region expected from a spherical expansion is apparent. The above derived mean values of I and β confirm this. Indeed, assuming $I = -1.95$, in the hypothesis of spherical expansion, β is expected to be in the range from -0.9 to -1.4, depending on whether $\nu < \nu_c$ or $\nu > \nu_c$, while in the case of an expanding jet, β is in the range from -1.9 to -2.9. This result seems to disagree with the results of Frail et al. [42] who, by examining the multiwavelength light curves of a sample of 17 GRB afterglows, find a break in a large fraction of them which they interpret as evidence for an expanding jet. They derive the jet opening angle for each of them. Even though both samples are still too small to infer general properties, our result could still be consistent with those of Frail et al. [42]. Only 10 of their 17 GRBs appear in Table 2 and, for most those 10, either the derived X-ray parameters refer to epochs before the breaks reported by Frail et al. or the X-ray light curves are of poor statistical quality.

As can be seen from Fig. 2, the position of three events is consistent with the jet region between the dotted lines. Two show evidence of a break in their X-ray light curves (GRB010510 and GRB010214), and one (GRB000926) exhibits a high temporal index (but it shows a break in the optical light curve [132]). The X-ray light curve of GRB990510 was discussed by Pian et al. [122], who found it consistent with the jet model and derived the following parameters of the jet and circumburst medium: ambient density of 0.13 cm^{-3}, electron energy distribution index of $p = 2$, isotropic equivalent energy of $E_0 = 5 \times 10^{53}$ erg, and opening angle of $\theta_c = 2.7°$.

In the case of GRB010222, which is within the spherical expansion region of Fig. 2 but shows a break consistent with that observed in the optical band [95], in 't Zand et al. [178] interpret it as probably due to the transition of a spherically expanding fireball from a relativistic to a non-relativistic phase (NRP) [28]. Also in this case a break is expected when the shock wave has swept up a rest-mass energy equal to its initial energy [87,132,172]. Assuming synchrotron radiation from an adiabatically cooling shock as above, the change in the temporal index is now shallower than for the case of an expanding jet. After the transition to NRP, we have

$$\beta = 3I + 18/5 \tag{9}$$

for $\nu < \nu_c$ and

$$\beta = 3I + 5 \tag{10}$$

for $\nu > \nu_c$. In the case of GRB010222, in spite of the fact that the measured values of β and I after the break are still inconsistent with each other assuming NRP, this model predicts values of the electron index p, when derived from β or I, which are much more consistent with each other than in the case of an expanding jet (-1.94 ± 0.10 versus -2.22 ± 0.03 for transition to NRP against -1.33 ± 0.03 versus -1.94 ± 0.10 an expanding jet) [178]. Assuming the NRP model, from the break time estimate, a high ambient gas density is inferred ($\sim 10^6$ cm^{-3} [178]), which is consistent with the fact that a Comptonized component is needed by the X-ray afterglow spectral data (see Sect. 2.2).

The existence of high temporal indices ($\beta \leq -1.8$, see Table 2) has also been taken as evidence for beamed emission (GRB980519 [151]) or for a transition to NRP (GRB000926 [132]). However, in the case of GRB980519, the X-ray data are consistent with a spherical expansion (see Fig. 2), and in the case of the GRB000926 afterglow, the observational scenario appears more complex. By combining together optical and X-ray observations, two breaks are apparent [132]. Piro et al. [132] interpret the early break as due to a relativistic transition from a spherical to a jet-like phase, and the later break as due to the transition of

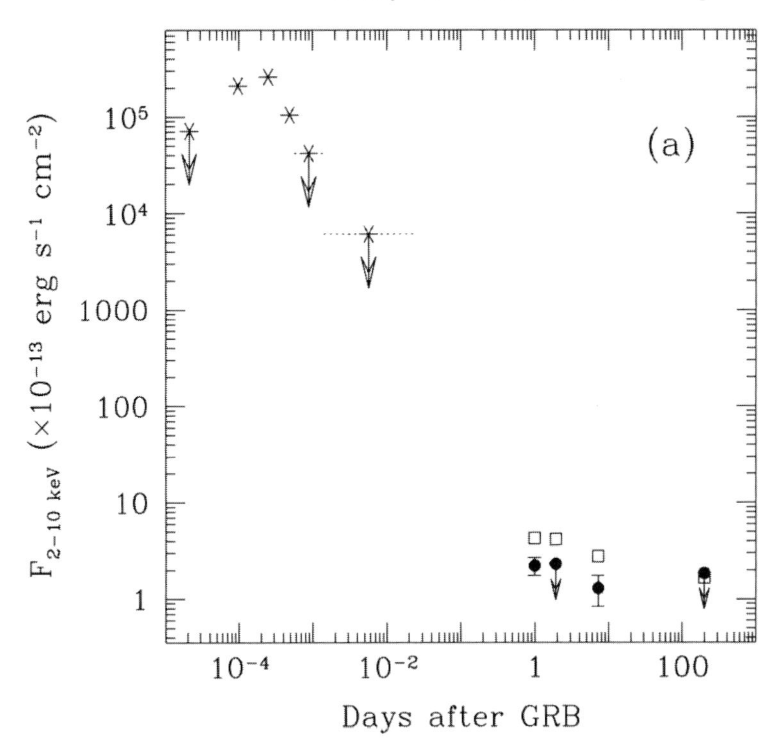

Fig. 3. The 2–10 keV light curve of the prompt emission from GRB980425 (stars) and from the two candidate afterglow sources S1 (open squares) and S2 (filled circles) in the 2–10 keV energy band. The arrows give 3σ upper limits [121].

the jet to a NRP [87]. This interpretation implies the presence of a moderately collimated outflow ($\theta_c \approx 27°$) that expands in a dense medium ($n \approx 4 \times 10^4$ cm^{-3}). Unfortunately the derived medium density is much higher than that ($n \sim 30$ cm^{-3}) inferred from the broad band spectrum [64], as discussed in Sect. 2.2. Also, in the case of GRB010222 a late break could be the explanation of the inconsistency of the *BeppoSAX* light curve with the *CHANDRA* flux point taken 9 days after the GRB [178].

The presence of beamed emission would help to reduce the dramatic GRB energy problem. While the isotropic γ-ray energy $E_{iso}(\gamma)$ ranges from 5×10^{51} to 3×10^{54} erg (see, e.g., [4,11,154]), the E_γ for a jet with opening angle θ_c would be a factor of $2/\theta_c^2$ lower. Frail et al. [42] attribute the range of variation of $E_{iso}(\gamma)$ to a range of jet cone angles and give an estimate of the mean value of $E_\gamma \sim 5 \times 10^{50}$ erg, where a conversion efficiency of the jet kinetic energy into electromagnetic radiation of 0.2 is assumed [59]. Rossi et al. [144] assume a jet as well, but with a variation in the output energy which depends on the offset angle from the jet axis ($\propto \theta^{-2}$) and attribute the dispersion in the $E_{iso}(\gamma)$ energy to differences in the jet axis-to-observer viewing angles.

The afterglow of GRB980425 deserves a separate discussion, because of its large impact on the GRB-supernova connection issue. The GRB occurred at a time, and in a position, which are both consistent with those of the peculiar type Ic supernova SN1998bw in the nearby galaxy ESO184-G82 ($z = 0.0085$) [50,75,79]. Following the detection with *BeppoSAX WFCs* and *GRBM* (see Table 1), a prompt X-ray observation of the GRB error box led to the discovery of two weak sources, S1 and S2, within the *WFCs* error box, one of which, S1, was consistent with the SN1998bw position [121]. The 2–10 keV flux level of S2, about two times lower than that of S1, was at the sensitivity limit of the *BeppoSAX MECS* telescopes so that even very small variations of this source could lead to a positive detection or an upper limit. Indeed, the source was no longer detected one day after the first observation and it was barely detectable again 6 days later. Finally, it was not detected during an observation in November 1998 (see Fig. 3). The source S1 was stronger, but showed a very weak power-law fading with index $\beta = -0.2$. Neither S1 nor S2 showed "typical" fading behavior, illustrating the peculiar nature of the burst and afterglow. Further observations of the GRB980425 field with more sensitive X-ray instrumentation are of key importance. An observation performed with XMM-Newton [123] has shown that the X-ray flux from S1 has further faded from the last *BeppoSAX* observation according to an exponential law, while S2, still visible, is the superposition of more sources, likely AGNs.

2.4 Emission/Absorption Features in the X-Ray Spectra

Evidence of emission features has been reported in the X-ray afterglow spectra of GRB970508 [128], GRB970828 [175], GRB991216 [131], and GRB000214 [8].[1] Table 3 summarizes the major properties of these features and Fig. 4 shows one of the two features discovered in the GRB991216 X-ray afterglow. A feature at 3.5 keV (97% confidence level) was observed from GRB970508 only during the first part (10 hrs) of the follow-up observation of this burst, before a delayed outburst occurred in both the X-ray and optical bands [127]. Also in the case of GRB970828 an emission feature at 5 keV (confidence level of 98.3%) was observed for a limited time duration ($\sim$ 5.5 hrs) during a flare activity of the GRB afterglow. The two emission features from the GRB991216 afterglow (4.7σ excess above the continuum emission at 3.5 keV, 2σ excess for the feature at 4.4 keV) were detected for the entire observation duration (3.4 hrs) performed 37 hrs after the main event with *CHANDRA*. In the last case (GRB000214), the continuum level was observed to fade while the emission feature at 4.7 keV (3σ significance level) remained stable for the entire duration of the observation ($\sim$ 29 hrs). Taking into account that three of these GRBs have host galaxies with known redshifts ($z = 0.835$ for GRB970508 [102], $z = 0.9578$ for GRB970828 [34], and $z = 1.02$ for GRB991216 [169]), the line features in their respective comoving frames are consistent with Fe fluorescence lines (Fe I-Fe XX Kα line in the

[1] After this chapter was written, evidence of 5 emission features from Mg XI, Si XIV, S XVI, Ar XVIII, Ca XX, were reported [139], but their reality is questioned [16,146].

Table 3. Properties of the Detected X-ray Emission Features

Quantity	GRB970508	GRB970828	GRB991216	GRB000214
E_l (keV)	3.4 ± 0.3	$5.04^{+0.23}_{-0.31}$	3.49 ± 0.06	4.7 ± 0.2
			4.4 ± 0.5	
$FWHM_l$ (keV)	< 0.5	$0.73^{+0.89}_{-0.73}$	0.54 ± 0.16	< 0.4
			> 2.35	
F_l^{a} (10^{-13} erg cm^{-2} s^{-1})	3 ± 1	1.5 ± 0.8	1.7 ± 0.5	0.67 ± 0.22
			2.7 ± 1.4	
EW_l (keV)	~ 1.5	~ 1.8	0.5 ± 0.013	~ 2.1
			~ 0.8	
F_ν^{b} (10^{-12} erg cm^{-2} s^{-1})	~ 7	~ 0.4	~ 2.3	~ 0.3
Redshift z	0.835	0.9578	1.02	0.37–0.47[c]
E_l^0 (keV)[d]	6.24 ± 0.55	$9.86^{+0.45}_{-0.61}$	7.05 ± 0.12	$6.4 - 6.97$[b]
			8.9 ± 1.0	
L_l^{e} (10^{44} erg s^{-1})	6.7 ± 2.7	4.8 ± 2.5	5.7 ± 1.6	0.45 ± 0.15[f]
			9.0 ± 4.7	
hr from GRB[g]	6–16	32–38	37–40	12–41
Mission	*BeppoSAX*	*ASCA*	*CHANDRA*	*BeppoSAX*
Instrument	*NFI*	*SIS+GIS*	Gratings+*ACIS-S*	*NFI*
References	[128]	[175]	[131]	[8]

[a] Line flux.
[b] Continuum flux in the $2 - 10$ keV band.
[c] Depends on the Fe ionization state (Fe K X-ray fluorescence assumed).
[d] Energy line corrected for cosmological redshift.
[e] Derived assuming $H_0 = 65$ km s^{-1} Mpc^{-1} and $q_0 = 0.5$.
[f] Assuming $z = 0.47$.
[g] Time interval during which the line was visible.

case of GRB970508, Fe K line of H-like ions in the case of the lower energy line from GRB991216) or Fe recombination edges (recombination of Fe H-like ions for both the line feature from GRB970828 and the higher energy feature from GRB991216). In the case of GRB000214, for which the redshift is not known, if the feature is interpreted as an Fe K fluorescence line, the implied redshift of the GRB source ranges from $z = 0.37$ to 0.47, depending on the ionization status of iron, from neutral to hydrogen-like Fe, respectively.

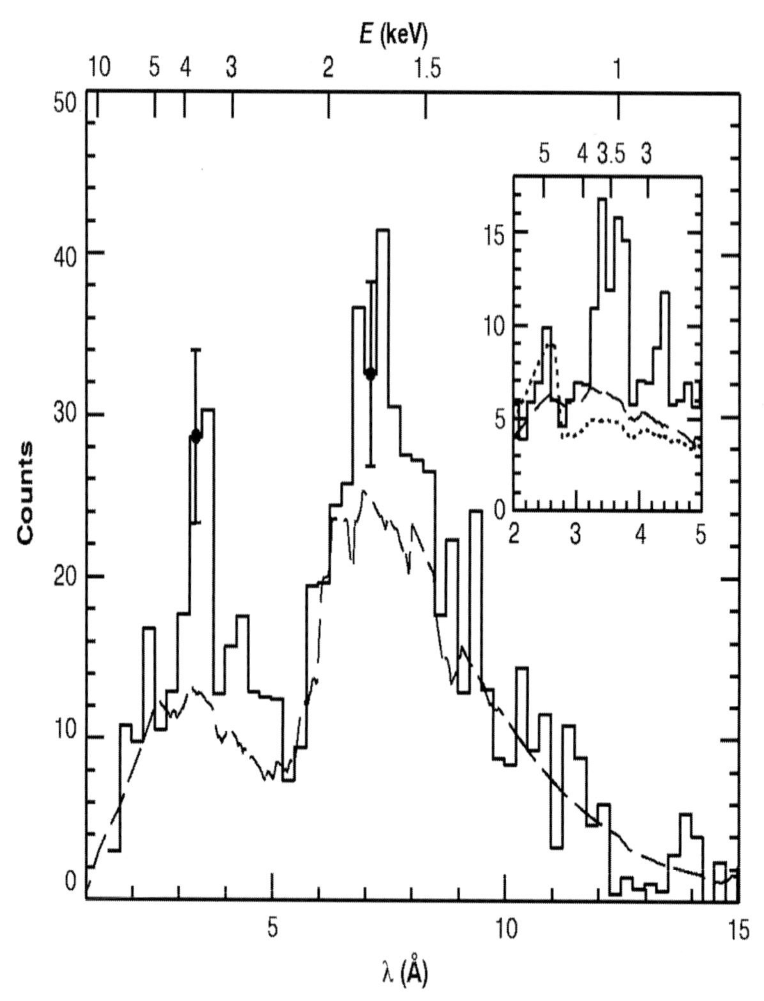

Fig. 4. Spectrum of the X-ray afterglow of GRB991216 obtained with the *CHANDRA* high energy gratings. The 4.7σ excess at 3.5 keV is apparent. In the inset the fine structure of the line is shown [131].

Independent of the specific identification, all the detected lines point to the presence of ionized Fe at the time of the afterglow measurements, with the ionizing radiation likely due to the GRB flash. The line widths (see Table 3) are not accurately determined except in the case of the 3.5 keV line from GRB991216 (see Fig. 4). In this case the line width has been interpreted [131] as being of kinematic origin, implying velocities of the line emitting material of $\sim 0.1c$. This velocity would also be that of the material ejected by the GRB progenitor (if a supernova) a few months before the main GRB event. The large equivalent widths of the lines favor a reflection geometry for the line emitting material [167]. Indeed, for a reflection process the ionizing flux is always efficiently reprocessed into line photons by the outermost Fe layer (optical depth $\tau_{Fe} \sim 1$), while for a transmission process an efficient re-processing of the ionizing flux would require a particular tuning of the density of both the scattering material and free electrons. The amount of Fe needed mainly depends on the density of the scattering cloud and on its distance from the GRB site. In the reflection model worked out by Vietri et al. [167], the required iron mass to obtain the observed line fluxes is significant (a sizable fraction of a solar mass in the case of GRB991216) which is not expected even for the largest known stars. Only a supernova remnant could contain so much iron. However in the model by Rees and Mészáros [138], in which a continuing (hrs to days) power output after the burst in a magneto-hydrodynamic (MHD) wind is required, it is the interaction of this wind with the walls of a funnel previously created by the GRB itself (assumed to be produced in a collapsing star) that gives rise to the emission features. They result from a reprocessing of part of the wind luminosity by a thin layer of the funnel. In this case, the iron mass needed to obtain the line flux is much lower ($\sim 10^{-8}$ $M_\odot$ for GRB991216), thanks to the high material density and the high recombination rate of the Fe ions. In this model the GRB environment is not required to be enriched in Fe by a previous supernova explosion. In a later paper, Mèszàros and Rees [104] proposed a different model for the emission line in which the total amount of Fe is significantly larger ($\sim 10^{-5}$ $M_\odot$). Here, a hypernova progenitor releases a relativistic shock when the inner black hole is sufficiently powerful to drive the shock progress into the outlying stellar layers. The heated matter (heavily contaminated by the Fe nuclear matter), which is not accelerated to relativistic energies, expands outside the main body of the star in a cocoon and produces the Fe line thermally.

In this context, the discovery of a transient absorption feature (13 s duration) at 3.8 keV in the prompt emission of GRB990705 [3] could provide an important datum for settling the issue of the Fe abundance and its mass and location within the environment in which the emission lines are produced. Assuming that the observed absorption feature was due to a redshifted Fe K-edge [3] (consistent with the optical redshift $z_{\rm opt} = 0.8424 \pm 0.0002$ of the GRB host galaxy [86]), the derived Fe abundance ($A_{Fe} \sim 75$) is typical of a supernova remnant (SNR) and its transient nature could be a consequence of its photoionization by the GRB photons. From the estimated column density, supposing a uniform distribution of SNR material, its distance from the GRB site was estimated to be ~ 0.1 pc. Such a short distance makes it very likely that the GRB site was coincident with that of a supernova explosion. The time for the remnant to grow to this size was estimated as ~ 10 yrs. A drawback of this scenario [15,84] is the huge amount of iron (tens of solar masses) required, unless the absorbing iron is clumped and a clump is, by chance, along the line of sight.

Lazzati et al. [84] propose an alternative model that does not require a large Fe mass. In this model the assumption is that the quasi-complete ionization of iron is achieved in a much shorter time than the duration of the feature (13 s), which is due to resonant scattering of the GRB photons off H-like iron (transition 1s-2p, $E_{\rm rest} = 6.927$ keV), while its disappearance is due to a halt of the electron recombination by an electron temperature increase. In this model, to fit the data, a Fe relative abundance ~ 10 times the solar value and a Fe mass of ~ 0.2 $M_\odot$ are required. Also in this model, an iron rich environment, typical of an SNR, is required and the distance to the absorbing iron is low ($\sim 2 \times 10^{16}$ cm). As a consequence, independent of the specific model, a SN explosion preceding the GRB event is required, which is likely to be associated with the GRB event itself. In this framework, the afterglow emission features appear when the strong ionizing flux has decreased and electron recombination restarts.

3 The Early X-Ray Afterglow

A peculiar property of the X-ray afterglow light curves is their very smoothly fading behavior, except in a few cases in which an outburst (GRB970508), a flaring activity (GRB970828), and a rise from below the *BeppoSAX MECS* sensitivity (GRB981226), were observed [47,127,175]. By comparison, the time profile of the prompt γ-ray emission is erratic and sometimes even spiky. In order to explain the complex time profiles of GRBs, it has been hypothesized [78,147,148] that the GRB itself could be due to internal shocks, with fast-rise exponential-decay (FRED) events perhaps also involving external shocks [31]. However, there is a general consensus that the late afterglow emission is likely due to external shocks (see, e.g., [101,150]). In this scenario, the two phenomena can evolve in different ways and can involve energetics which are not necessarily correlated. It has also been suggested [149,152] that the early afterglow could start already only a few tens of seconds after the burst. In fact, in a few cases like GRB970228 [23,26] and GRB970508 [127], the extrapolation of the $2-10$ keV afterglow fading law back to the time of the GRB main event gives a flux consistent with that measured during the last portion of the event. Using the ratio between the $2-10$ keV fluence of the last portion of the prompt emission and that of the late afterglow, Frontera et al. [48] demonstrated that this property is valid also for an entire sample of GRBs with known afterglows; approximately the second half of the GRB prompt X-ray emission is strictly correlated with the afterglow emission. This suggests that at least the tail of the GRBs can be due to afterglow emission, while at the onset of the primary event this contribution is negligible. Once the rise of the afterglow has been established, the framework of the fireball model permits the initial Lorentz factor Γ_0 of the shocked ambient medium to be evaluated [152]. From the above results $\Gamma_0 \sim 150$ for all GRB events investigated, except for GRB980425 in which a much lower value ($\Gamma_0 < 50$) is found [48].

At higher energies (> 10 keV), as mentioned in Sect. 2.2, late afterglow emission has been discovered only from GRB990123 with the *BeppoSAX PDS* instrument [69]. However, early afterglow emission is also expected from, e.g., electron-synchrotron self-Inverse Compton and proton-synchrotron processes [153,179] and some evidence of a possible early, high energy afterglow has been reported [20,22,55]. The most suggestive result is that found by Giblin et al. [55] with the strong *BATSE* event GRB980923 in the $25-300$ keV range. The *BATSE* data show a GRB light curve which, after 40 s of erratically variable flux, smoothly decays with a power-law shape of index $\beta \sim -1.8$ for about 400 s until the event becomes undetectable. Correspondingly, the energy spectrum shows a power-law shape with a photon index I, which, after 40 s from the GRB onset, exhibits a jump from -1 to -1.7, which is typical of the X-ray afterglow photon indices (see Sect. 2.2).

4 Open Issues and Future Prospects

In spite of the tremendous advances in the years since the discovery of the first X-ray afterglows, many questions about the GRB phenomenon are still open which can only be answered with further X-ray observations. Analysis of the afterglow radiation seems to indicate a synchrotron origin, but in some cases multiwavelength spectra (see Sect. 2.2) show the presence of an additional component in the X-ray band interpreted as electron self-Inverse Compton radiation. It is important to obtain good quality, multiwavelength spectra for a large sample of GRBs to determine the relevant emission mechanisms of the afterglow radiation.

Breaks in the X-ray afterglow light curves have been observed for a few GRB afterglows, while at longer wavelengths such breaks are more frequent [42]. The jet model predicts that these breaks are achromatic if the jet propagates into a medium with uniform density. However, if the jet moves into a wind-like circumstellar environment ($\rho \propto r^{-s}$ with $s \approx 2$), the break is expected to be less relevant in X-rays than in the optical band [80]. In addition, other models have been proposed for interpreting such breaks (e.g., transition to NRP, see Sect. 2.3). Thus it is of key importance to measure X-ray afterglow light curves with high statistical quality in order to test the jet model and to derive the ambient medium properties to be compared with the properties derived from the multiwavelength spectra.

The ambient medium density, along with other GRB environmental properties (composition, velocity, temperature, metal abundance) can also be inferred from the detection of X-ray lines or edges in the spectra of the prompt and delayed X-ray emission. This information is important for determining the nature of the GRB progenitor, which is one of the most debated issues in astrophysics today. For the case of collapse of a massive progenitor, the GRB environment is expected to be Fe polluted and line features are more likely. On the contrary, in a clean environment, such as that expected for neutron star mergers, Fe lines are not likely. The X-ray lines detected so far point toward the presence of a medium highly ionized by the GRB main event, with iron abundance higher than solar. It is likely that this iron is produced and ejected by the GRB progenitor before the burst, which supports the hypothesis of a supernova-like explosion (hypernova, collapsar, or supranova [116,166,174]). However many questions are still open even in this scenario. For example:

- Why do we not see any absorption feature in the prompt emission of the GRBs which show an emission feature during the afterglow phase (GRB970508 or GRB000214)?
- Why are the emission lines in the afterglow spectra of GRB970508 and GRB970828 observed only for a limited amount of time, while in the case of GRB000214 the line is stable?
- Why, in many other GRBs observed with *BeppoSAX* do we not see any features?
- Is this due to the limited sensitivity of the *BeppoSAX* instrumentation or is it due to an intrinsic physical process?

Only more sensitive and continuous observations of GRBs from the main event through the late afterglow can answer these questions.

High quality light curves of the GRB prompt emission and afterglows, with no interruption, could also resolve the issue of the onset time of the afterglow emission phase, its time profile and its spectral evolution. This information is of key importance to study the feedback of the intense γ-ray flux on the dynamics of the fireball and on the afterglow emission properties [10,103,163]. High quality light curves at multiple wavelengths could definitely settle the question of the rate of GRBs with single or multiple broken power-law temporal behavior. If the breaks in the light curves imply the presence of jets, the GRB energy problem could be solved, or mitigated, and ordinary supernovae could be energetic enough to be the origin of GRB events [42]. High quality light curves would also permit the study of short time variations in the afterglow fading, and thus of the degree of homogeneity of the circumburst medium [47].

Another open issue concerns the origin of short (< 1 s) GRBs. Until now, no prompt (within a few hrs) follow-up of these events has been possible and no afterglow source has ever been identified [73]. A claim for the detection of a transient (100 s duration) and fading hard X-ray emission, obtained by summing the *BATSE* light curves of a sample of 76 short GRBs has recently been reported (see [85], but also [22]). Lazzati et al. [85] interpret this emission as possible evidence for afterglows from short GRBs. This claim still has to be confirmed by other, more sensitive observations and its origin is not clear. It could also be prompt emission from truly long GRBs which have a spectrally hard, short and strong pulse at their onset, and a long tail which is not individually observable because it is below the sensitivity limit of the instrumentation used. More sensitive GRB observations with immediate position identification are needed to disentangle the short GRB origin issue and the afterglow properties of these events.

The cause of "dark" GRBs (see also the chapter by Pian in this volume) is still a mystery. A possible explanation is the presence of dust in the environment around GRB sites that obscures the optical emission, even if the GRB output may be enough to destroy the dust in the immediate surroundings [33]. Observations of X-ray lines from these GRBs could be crucial for establishing the environmental properties. However other possibilities cannot be excluded, such as a rapid decline of the optical light or a high redshift of the object. Only fast afterglow searches, starting minutes after the main event with multiple wavelength coverage of the afterglow emission, can definitely settle this issue.

The X-ray flashes detected with *BeppoSAX* [68] with no γ-ray counterparts are still a mystery. They could be GRBs at very high redshifts (z >10) or with very soft spectra, or from fireballs with a low Lorentz factor. More sensitive observations of these events are needed.

The fireball model of GRBs is currently the most successful for interpreting the X-ray data. Other models have been proposed, such as the cannonball (see, e.g., [29]) and only future observations will be able to distinguish between, and determine the validity of, these alternative models.

One of the major limitations of the current investigations of GRB afterglows is the delay from the main event to the follow-up observations. Many of the issues discussed herein will be solved when future missions, now under development (in particular, *Swift Gamma-Ray Burst Explorer (Swift)* [53] and *Gamma-Ray Large Area Space Telescope (GLAST)*) [52] are operating. *Swift* will perform immediate ($\sim$ 1 min) follow-up observations in the X-ray and optical bands and will trigger very prompt optical, near-infrared, and radio observations with ground-based telescopes. The γ-ray mission *GLAST* will allow a sensitive and high time resolution study of the highest energies (GeV) emitted by GRBs and other missions, like the European *International Gamma-Ray Astrophysics Laboratory (INTEGRAL)* (see, e.g., [97]) and the Italian *Astro-rivelatore Gamma a Immagini LEggero (AGILE)* [161] are expected to make key contributions to the resolution of the GRB mystery.

Acknowledgments

I wish to thank L. Amati for his generous contribution to the preparation of the Tables 1 and 2 of this chapter, Mario Vietri for his very useful comments and suggestions for improving the chapter, and N. Masetti and John Stephen for their careful reading of the manuscript. I also wish to thank the colleagues of the *BeppoSAX* team (L. Amati, C. Guidorzi, L. Nicastro, L. Piro, and P. Soffitta), who have permitted the publication of preliminary results. Many thanks especially to my wife for her sympathy for my days of work at home for this chapter, when I should have been available for family needs. I also like to acknowledge financial support from the *Italian Space Agency (ASI)* and from the Ministry of the University of Italy (COFIN funds).

References

1. B. Altieri, N. Schartel, M. Santos, L. Tomas, M. Guainazzi, L. Piro, A. Parmar: GCN 869 (2000)
2. B. Altieri, N. Schartel, D. Lumb, L. Piro, A. Parmar: GCN 884 (2000)
3. L. Amati et al.: Science **290**, 953 (2000)
4. L. Amati et al.: Astron. Astrophys. **390**, 81 (2002)
5. L. Amati et al.: in preparation (2003)
6. L.A. Antonelli et al.: Astron. Astrophys. Suppl. Ser. **138**, 435 (1999)

7. L.A. Antonelli et al.: GCN 445 (1999)
8. L.A. Antonelli et al.: Astrophys. J. Lett. **545**, L39 (2000)
9. D. Band et al.: Astrophys. J. **413**, 281 (1993)
10. A.M. Beloborodov: Astrophys. J. **565**, 808 (2002)
11. J.S. Bloom, D.A. Frail, R. Sari: Astron. J. **121**, 2879 (2001)
12. G. Boella, R.C. Butler, G.C. Perola, L. Piro, L. Scarsi, J.A.M. Bleeker: Astron. Astrophys. Suppl. Ser. **122**, 299 (1997)
13. G. Boella et al.: Astron. Astrophys. Suppl. Ser. **122**, 327 (1997)
14. M. Boer, B. Gendre: Astron. Astrophys. **361**, L21 (2000)
15. M. Boettcher, C.D. Dermer, L. Amati, F. Frontera: In: *Gamma-Ray Bursts in the Afterglow Era, Proc. International Workshop at Rome, Italy, 17–20 October 2000*, ed. by E. Costa, F. Frontera, J. Hjorth (Springer, Heidelberg 2001) p. 160
16. K.N. Borozdin and S.P. Trudolyubov: ApJL, submitted (2003) (astro-ph/ 0205208)
17. J.J. Brainerd: Nature **355**, 552 (1992)
18. M.S. Briggs et al.: Astrophys. J. **459**, 40 (1996)
19. T. Bulik, D.Q. Lamb, P.S. Coppi: Astrophys. J. **505**, 666 (1998)
20. R.A. Burenin et al.: Astron. Astrophys. **344**, L53 (1999)
21. R.A. Chevalier, Z.-Y. Li: Astrophys. J. Lett. **520**, L29 (1999)
22. V. Connaughton: Astrophys. J. **567**, 1028 (2001)
23. E. Costa et al.: Nature **387**, 783 (1997)
24. E. Costa et al.: IAUC 6939 (1998)
25. E. Costa et al.: GCN 553 (2000)
26. E. Costa: in *AIP Conference Proc.*, ed. R.M. Kippen, R.S. Malozzi, G. J. Fishman, vol. 526, p. 365 (2000)
27. S. Covino et al.: Astron. Astrophys. **348**, L1 (1999)
28. Z.G. Dai, T. Lu: Astrophys. J. Lett. **519**, L155 (1999)
29. A.Dar, A. De Rújula: astro-ph 0012227 (2000)
30. D. Dal Fiume et al.: Astron. Astrophys. **355**, 454 (2000)
31. C.D. Dermer, M. Boettcher, J. Chiang: Astrophys. J. Lett. **515**, L49 (1999)
32. J.M. Dickey, F.J. Lockman: Ann. Rev. Astron. Astrophys. **28**, 215 (1990)
33. S.G. Djorgovski et al.: In: *Proc. IX Marcel Grossmann Meeting*, ed. V. Gurzadyan, R. Jantzen, R. Ruffini (World Scientific, Singapore 2001)
34. S.G. Djorgovski, D.A. Frail, S.R. Kulkarni, J.S. Bloom, S.C. Odewahn, A. Diercks: Astrophys. J. **562**, 654 (2001)
35. M. Feroci et al.: Astron. Astrophys. **332**, L29 (1998)
36. M. Feroci, L. Piro, M.R. Daniele, G. Gennaro, S. Rebecchi, G. Celidonio, L.A. Antonelli, J. Heise: IAUC 6909 (1998)
37. M. Feroci et al.: GCN 685 (2000)
38. M. Feroci et al.: Astron. Astrophys. **378**, 441 (2001)
39. G.J. Fishman, C.A. Meegan: Ann. Rev. Astron. Astrophys. **33**, 415 (1995)
40. D.A. Frail, S.R. Kulkarni, S.R. Nicastro, M. Feroci, G.B. Taylor: Nature **389**, 261 (1997)
41. D.A. Frail et al.: Astrophys. J. Lett. **538**, L129 (2001)
42. D.A. Frail et al.: Astrophys. J. Lett. **562**, L55 (2001)
43. F. Frontera, E. Costa, D. dal Fiume, M. Feroci, L. Nicastro, M. Orlandini, E. Palazzi, G. Zavattini: Astron. Astrophys. Suppl. Ser. **122**, 357 (1997)
44. F. Frontera et al.: Astron. Astrophys. **334**, L69 (1998)
45. F. Frontera et al.: Astrophys. J. Lett. **493**, L67 (1998)

46. F. Frontera, M. Capalbi, M.R. Daniele, E. Montanari, A. Paolino, A. Tesseri, C. Pastor, G. Gandolfi: IAUC 7235 (1999)
47. F. Frontera et al.: Astrophys. J. **540**, 697 (2000)
48. F. Frontera et al.: Astrophys. J. Suppl. **127**, 59 (2000)
49. T.J. Galama, R.A.M.J. Wijers, M. Bremer, P.J. Groot, R.G. Strom, C. Kouveliotou, J. van Paradijs: Astrophys. J. Lett. **500**, L97 (1998)
50. T.J. Galama et al.: Nature **395**, 670 (1998)
51. G. Gandolfi: GCN 118 (2001)
52. N.A. Gehrels, I. Michelson: Astropart. Phys. **11**, 277 (1999)
53. N.A. Gehrels: 'Swift Gamma-Ray Burst MIDEX'. In: *X-Ray and Gamma-Ray Instrumentation for Astronomy XI*, ed. K.A. Flanagan, O.H.W. Siegmund (SPIE 4140, 2000) p. 42
54. G. Garmire, L. Piro, G. Stratta, M. Garcia, J. Nichols: GCN 782 (2000)
55. T. Giblin, J. Van Paradijs, C. Kouveliotou, V. Connaughton, R.A.M.J. Wijers, M.S. Briggs, R.D. Preece, G.J. Fishman: Astrophys. J. Lett. **524**, L47 (1999)
56. J. Gorosabel et al.: Astron. Astrophys. Suppl. Ser. **138**, 455 (1999)
57. J. Greiner: IAUC 6742 (1997)
58. P.T. Groot, T.J. Galama, K. Hurley, C. Kouveliotou: IAUC 6723 (1997)
59. D. Guetta, M. Spada, E. Waxman: Astrophys. J. **557**, 399 (2001)
60. C. Guidorzi et al.: Astron. Astrophys. in press (2003)
61. J. Hakkila, C.A. Meegan, G.N. Pendleton, G.J. Fishman, R.B. Wilson, W.S. Paciesas, M.N. Brock, J.M. Horack: Astrophys. J. **422**, 659
62. J.P. Halpern et al.: Astrophys. J. **543**, 697 (2000)
63. F.A. Harrison et al.: Astrophys. J. Lett. **523**, L121 (1999)
64. F.A. Harrison et al.: Astrophys. J. **559**, 123 (2001)
65. F.A. Harrison, A.A. Yost, S.R. Kulkarni: GCN 1023 (2001)
66. F.A. Harrison, S. Yost, D. Fox, J. Heise, S.R. Kulkarni, P.A. Price, E. Berger: GCN 1143 (2001)
67. J. Heise et al.: IAUC 7099 (1999)
68. J. Heise, J. in 't Zand, R.M. Kippen, P.M. Woods: In: *Gamma-Ray Bursts in the Afterglow Era, Proc. International Workshop at Rome, Italy, 17–20 October 2000*, ed. by E. Costa, F. Frontera, J. Hjorth (Springer, Heidelberg 2001) p. 16
69. J. Heise et al.: in preparation (2003)
70. J. Hjorth et al.: Astrophys. J. **576**, 113 (2002)
71. S. Holland et al.: GCN 778 (2000)
72. K. Hurley, S. Barthelmy: GCN 290 (1999)
73. K. Hurley et al.: Astrophys. J. **567**, 447 (2002)
74. G.L. Israel et al.: Astron. Astrophys. **348**, L51 (1999)
75. K. Iwamoto et al.: Nature **395**, 672 (1998)
76. R. Jager et al.: Astron. Astrophys. Suppl. Ser. **125**, 557 (1997)
77. R.W. Klebesadel, I.B. Strong, R.A. Olson: Astrophys. J. Lett. **182**, L85 (1973)
78. S. Kobayashi, T. Piran, R. Sari: Astrophys. J. **490**, 92 (1997)
79. S.R. Kulkarni et al.: Nature **395**, 663 (1998)
80. P. Kumar, A. Panaitescu: Astrophys. J. Lett. **541**, L9 (2000)
81. E. Kuulkers et al.: Astrophys. J. **538**, 638 (2000)
82. E. Kuulkers et al.: GCN 700 (2000)
83. D.Q. Lamb: Pub. Astron. Soc. Pacific **107**, 1152 (1995)
84. D. Lazzati, G. Ghisellini, L. Amati, F. Frontera, M. Vietri, L. Stella: Astrophys. J. **556**, 471 (2001)
85. D. Lazzati, E. Ramirez-Ruiz, G. Ghisellini: Astron. Astrophys. **379**, L39 (2001)

86. E. Le Floc'h et al.: Astrophys. J. **581**, L81 (2002)
87. M. Livio, E. Waxman: Astrophys. J. **538**, 187 (2000)
88. G. Manzo, S. Giarrusso, A. Santangelo, F. Ciralli, G. Fazio, S. Piraino, A. Segreto: Astron. Astrophys. Suppl. Ser. **122**, 341 (1997)
89. R. Manzo et al.: GCN 956 (2001)
90. F.E. Marshall, T. Takeshima, S.D. Barthelmy, C.R. Robinson, K. Hurley: IAUC 6683 (1997)
91. F.E. Marshall, J.K. Cannizzo, R.H.D. Corbet: IAUC 6727 (1997)
92. F.E. Marshall, T. Takeshima, P. Woods: GCN 138 (1998)
93. F.E. Marshall, T. Takeshima, T. Giblin, R.M. Kippen: GCN 519 (2000)
94. N. Masetti et al.: Astron. Astrophys. Suppl. Ser. **138**, 453 (1999)
95. N. Masetti et al.: Astron. Astrophys. **374**, 382 (2001)
96. C.A. Meegan, G.J. Fishman, R.B. Wilson, J.M., Horack, M.N. Brock, W.S. Paciesas, G.N. Pendleton, C. Kouveliotou: Nature **355**, 143 (1992)
97. S. Mereghetti, D.I. Cremonesi, J. Borkowski: In: *Gamma-Ray Bursts in the Afterglow Era, Proc. International Workshop at Rome, Italy, 17–20 October 2000*, ed. by E. Costa, F. Frontera, J. Hjorth (Springer, Heidelberg 2001) p. 363
98. P. Mészáros, M.J. Rees: Astrophys. J. **405**, 278 (1993)
99. P. Mészáros, P. Laguna, M.J. Rees: Astrophys. J. **415**, 181 (1993)
100. P. Mészáros, M.J. Rees, H. Papathaniassou: Astrophys. J. **432**, 181 (1994)
101. P. Mészáros, M.J. Rees: Astrophys. J. **476**, 2 (1997)
102. M.R. Metzger, S.G. Djorgovski, S.R. Kulkarni, C.C. Steidel, K.L. Adelberger, D.A. Frail, E. Costa, F. Frontera: Nature **387**, 878 (1997)
103. P. Mészáros, E. Ramirez-Ruiz, M.J. Rees: Astrophys. J. **554**, 660 (2001)
104. P. Mèszàros, M.J. Rees: Astrophys. J. Lett. **556**, L37 (2001)
105. E. Montanari et al.: In: *Gamma-Ray Bursts in the Afterglow Era, Proc. International Workshop at Rome, Italy, 17–20 October 2000*, ed. by E. Costa, F. Frontera, J. Hjorth (Springer, Heidelberg 2001) p. 195
106. T. Murakami et al.: IAUC 6729 (1997)
107. T. Murakami, Y. Ueda, A. Yoshida, N. Kawai, F.E. Marshall, R.H.D. Corbet, T. Takeshimaet al.: IAUC 6732 (1997)
108. R. Narayan, B. Paczyński, T. Piran: Astrophys. J. Lett. **395**, L83 (1992)
109. L. Nicastro, L. Amati et al.: Astron. Astrophys. **338**, L17 (1998)
110. L. Nicastro, L.A. Antonelli, G. Celidonio, M.R. Daniele, C. de Libero, G. Spoliti, L. Piro, E. Pian: IAUC 6912 (1998)
111. L. Nicastro, et al.: Astron. Astrophys. **138**, 437 (1999)
112. L. Nicastro, L.A. Antonelli, M. Dadina, M.R. Daniele, E. Costa, E. Pian: IAUC 7213 (1999)
113. L. Nicastro et al.: In: *Gamma-Ray Bursts in the Afterglow Era, Proc. International Workshop at Rome, Italy, 17–20 October 2000*, ed. by E. Costa, F. Frontera, J. Hjorth (Springer, Heidelberg 2001) p. 198
114. W.S. Paciesas et al.: Astrophys. J. Suppl. **122**, 465 (1999)
115. B. Paczyński: Astrophys. J. **363**, 218 (1990)
116. B. Paczyński: Astrophys. J. Lett. **494**, L45 (1998)
117. A. Panaitescu, P. Mészáros: Astrophys. J. **526**, 707 (1999)
118. A. Panaitescu, P. Kumar: Astrophys. J. **543**, 66 (2000)
119. A. Panaitescu, P. Kumar : Astrophys. J. **554**, 667 (2001)
120. A.N. Parmar et al.: Astron. Astrophys. Suppl. Ser. **122**, 309 (1997)
121. E. Pian et al.: Astrophys. J. **536**, 778 (2000)
122. E. Pian et al.: Astron. Astrophys. **372**, 456 (2001)

123. E. Pian: Talk presented at the 3rd International Workshop on *Gamma-Ray Bursts in the Afterglow Era, held in Rome, 17-20 September 2002*
124. T. Piran: Phys. Rep. **314**, 575 (1999)
125. T. Piran: Phys. Rep. **333**, 529 (2000)
126. L. Piro et al.: Astron. Astrophys. **329**, 906 (1998)
127. L. Piro et al.: Astron. Astrophys. **331**, L41 (1998)
128. L. Piro et al.: Astrophys. J. Lett. **514**, L73 (1999)
129. L. Piro et al.: IAUC 7111 (1999)
130. L. Piro et al.: GCN 409 (1999)
131. L. Piro et al.: Science **290**, 955 (2000)
132. L. Piro, G. Garmire et al.: Astrophys. J. **558**, 442 (2001)
133. L. Piro et al.: GCN 1172 (2001)
134. L. Piro et al.: Astrophys. J. **577**, 680 (2002)
135. L.Piro et al.: in preparation (2003)
136. P. Podsiadlowski, M.J. Rees, M. Ruderman: Mon. Not. R. Astron. Soc. **273**, 755 (1995)
137. M.J. Rees, P. Mészáros: Mon. Not. R. Astron. Soc. **258**, 41P (1992)
138. M.J. Rees, P. Mészáros: Astrophys. J. Lett. **545**, L73 (2000)
139. J.N. Rooves et al.: Nature **416**, 512 (2002)
140. J.E. Rhoads: Astrophys. J. **525**, 737 (1999)
141. G. Ricker et al.: GCN 1165 (2001)
142. G. Ricker, P. Ford, G. Monnelly, N. Butler, R. Vanderspek, D. Lamb: GCN 1185 (2001)
143. E. Rol et al.: Astrophys. J. **544**, 707 (2000)
144. E. Rossi, D. Lazzati, M.J. Rees: Mon. Not. R. Astron. Soc. **332**, 945 (2002)
145. M. Santos-Lleo, N. Loiseau, P. Rodriguez, B. Altieri, N. Schartel: GCN 1192 (2001)
146. R.E. Rutledge and M. Sako: MNRAS in press (2003)(astro-ph/0206073)
147. R. Sari, T. Piran: Astrophys. J. **485**, 270 (1997)
148. R. Sari, T. Piran: Mon. Not. R. Astron. Soc. **287**, 110 (1997)
149. R. Sari: Astrophys. J. Lett. **489**, L37 (1997)
150. R. Sari, T. Piran, R. Narayan: Astrophys. J. Lett. **497**, L17 (1998)
151. R. Sari, T. Piran, J.P. Halpern: Astrophys. J. Lett. **519**, L17 (1999)
152. R. Sari, T. Piran: Astron. Astrophys. Suppl. Ser. **138**, 537 (1999)
153. R. Sari, A.A. Esin: Astrophys. J. **548**, 787 (2001)
154. M. Schmidt: Astrophys. J. Lett. **523**, L117 (1999)
155. A. Shemi, T. Piran: Astrophys. J. Lett. **365**, L55 (1990)
156. I.S. Shklovskii, I.G. Mitrofanov: Mon. Not. R. Astron. Soc. **212**, 545 (1985)
157. P. Soffitta et al.: In: *Gamma-Ray Bursts in the Afterglow Era, Proc. International Workshop at Rome, Italy, 17–20 October 2000*, ed. by E. Costa, F. Frontera, J. Hjorth (Springer, Heidelberg 2001) p. 201
158. K.Z. Stanek, P.M. Garnavich, J. Kaluzny, W. Pych, I. Thompson: Astrophys. J. Lett. **522**, L39 (1999)
159. G. Stratte et al.: in *Gamma-Ray Bursts in the Afterglow Ere, Proc. of the International Workshop held in Rome, Italy, 17-20 October 2000*, ed. by E. Costa, F. Frontera, J. Hjorth (Springer, Heidelberg 2001) p. 118
160. T. Takeshima, C. Markwardt, F. Marshall, T. Giblin, R.M. Kippen: GCN 478 (1999)
161. M. Tavani et al.: In: *Gamma-Ray Astrophysics 2001, Proc. Symposium at Baltimore, MD, USA, 4–6 April 2001* (AIP 2003) in press

162. G.B. Taylor, D.A. Frail, S.R. Kulkarni: GCN 1136 (2001)
163. C. Thompson, P. Madau: Astrophys. J. **538**, 105 (2000)
164. J. Van Paradijs et al.: Nature **386**, 686 (1997)
165. J. Van Paradijs, C. Kouveliotou, R.A.M. Wijers : Ann. Rev. Astron. Astrophys. **38**, 379 (2000)
166. M. Vietri, L. Stella: Astrophys. J. Lett. **507**, L45 (1998)
167. M. Vietri, G. Ghisellini, D. Lazzati, F. Fiore, L. Stella: Astrophys. J. Lett. **550**, L43 (2001)
168. P.M. Vreeswijk et al.: Astrophys. J. **523**, 171 (1999)
169. P. Vreeswijk et al.: GCN 496 (1999)
170. K.W. Weiler, N. Panagia, M.J. Montes: Astrophys. J. **562**, 670 (2001)
171. J.C. Wheeler: IAUC 6697 (1997)
172. R.A.M.J. Wijers, M.J. Rees, P. Mészáros: Mon. Not. R. Astron. Soc. **288**, L51 (1997)
173. R.A.M.J. Wijers et al.: Astrophys. J. Lett. **523**, L33 (1999)
174. S. Woosley: Astrophys. J. **405**, 273 (1993)
175. A. Yoshida et al.: Astron. Astrophys. Suppl. Ser. **138**, 439 (1999)
176. J.J.M. in 't Zand et al.: Astrophys. J. Lett. **505**, L119 (1998)
177. J.J.M. in 't Zand et al.: Astrophys. J. **545**, 266 (2000)
178. J.J.M. in 't Zand et al.: Astrophys. J. **559**, 710 (2001)
179. B. Zhang, P. Mészáros: Astrophys. J. **559**, 110 (2001)

Optical Observations of γ-Ray Burst Afterglows

Elena Pian

INAF, Astronomical Observatory of Trieste, Via G.B. Tiepolo 11, I-34131 Trieste, Italy

Abstract. Optical, IR and UV observations of GRB fields have allowed detection of counterparts and host galaxies of the high energy transients, thus crucially contributing to our present knowledge of the GRB phenomenon. Measurements of afterglow variable emission, polarized light, redshifted absorption and emission line spectra, as well as host galaxy brightnesses and colors have clarified many fundamental issues related to the radiation mechanisms and environments of GRBs, setting the background towards discovering the nature of their progenitors.

1 Introduction

Gamma-ray burster (GRB) counterparts are bright multiwavelength emitters, unlike supernovae (SNe) which emit most of their power at the optical frequencies. However, observations in the optical band have had the biggest impact in the study of the GRB phenomenon by sampling the time histories of the GRB counterparts from a few seconds up to a few years after the explosion, by determining the nature of the afterglow emission (synchrotron radiation), and by unambiguously establishing the extragalactic origin of GRBs.

GRB optical counterparts initially can be extremely bright. However, the delayed emission, even though it is long lived with respect to the prompt γ-ray event, still fades quite rapidly and the host galaxies of GRBs are very small and faint, because of their cosmological distances. Therefore, telescopes of all aperture sizes have been involved in the investigation of the optical afterglows. The smaller telescopes are more flexible for a rapid follow-up and the larger ones (including the *Hubble Space Telescope (HST)*) play a leading role in obtaining early spectroscopy and polarimetry, in the photometric monitoring at late epochs, and in the study of the host galaxies. The most important observations of optical afterglows and their host galaxies, which have led to the current understanding of the physics of these sources, are discussed below. Previous reviews on this subject include [36,51,80,115,128,135,139,185,224,231]. Although the present review focuses on the optical observations, information from the scanty infrared (IR) and ultraviolet (UV) data is also included, because of the proximity of these bands to the optical. This chapter is organized as follows: in Sect. 2 the steps which led to the first detection of optical afterglow emission are summarized and the observational problems related to the afterglow investigation are described; in Sect. 3 the temporal behavior of the optical prompt and afterglow emission is reviewed; in Sect. 4 the observed spectral continuum at IR, optical and UV wavelengths is compared to the standard afterglow model based on the propagation of an external shock, and the characteristics of the afterglow absorption spectra are illustrated. A discussion of those aspects of GRB host galaxies which seem to be relevant to the afterglow investigation is included in Sect. 5. Future prospects for the exploration of GRB optical counterparts are sketched in Sect. 6.

2 Optical Searches of GRB Error Boxes

The search for optical counterparts of GRBs started soon after the first GRB discovery in the belief that they held the key to the physical origin of the GRB phenomenon. The first searches in the 1970's were based on inspection, on archival plates or on new photographic images of the large error boxes yielded by the early high energy missions (see [231] for a review of these pioneering attempts). The reasons of their failure, as we now understand, reside in the inadequacy of the methods. Since GRB afterglows fade quickly, only timely and sensitive surveys of accurate, rapidly disseminated GRB error boxes can be effective in detecting optical counterparts.

The first identification of an optical afterglow was made possible on 28 February 1997 by the *BeppoSAX* satellite [24], whose *Wide Field Cameras (WFCs)* [122] determined the position of the GRB with a 3 arcmin uncertainty radius and disseminated it to the community within a few hours [39]. Optical observations made 21 hours and $\sim$1 week after the GRB allowed the detection of a variable source in the GRB error box, which had been refined and reduced to ~ 1 arcmin^2. Variability and positional coincidence suggested association of the optical transient with the GRB [230]. Five years later, from about 100 GRBs accurately and rapidly localized by *BeppoSAX* or by other spacecraft, and quickly observed by optical telescopes (delays of a few hours to a few days after the GRB explosion), ~ 30 have detected optical afterglows.

These are listed in Table 1, together with those GRBs for which, despite the lack of an optical afterglow, detection at other wavelengths has allowed identification of a host galaxy (in these cases the limit on the magnitude of the optical transient is reported). Col. 2 of Table 1 reports the instrument, or suite of instruments, which have localized the GRB: "*BeppoSAX*" means that the event triggered the *Gamma-Ray Burst Monitor (GRBM)* of the *BeppoSAX* satellite [58], and was localized with arcmin precision by the onboard *WFCs*. In two cases the *BeppoSAX WFCs* detected the afterglow in the X-rays and localized the event which triggered the *Burst and Transient Source Experiment (BATSE)*, but not the *GRBM*. For almost all GRBs localized by the *Rossi X-ray Timing Explorer (RXTE)* satellite, the accurate positioning was obtained with its *All Sky Monitor (ASM)*, the exception being GRB990506, which was localized precisely only after the *RXTE Proportional Counter Array (PCA)* detected its X-ray afterglow [26]. Many GRBs have been localized by the *Interplanetary Network (IPN)* of spacecraft, whose synergy allows triangulation of the GRB position, yielding, in many cases, arcmin-sized error boxes, although with delays seldom shorter than 12 hours, and often longer than 24 hours [117,118]. For most cases, a refinement of the localization error box has been possible after detection of the X-ray afterglow by the *BeppoSAX Narrow Field Instruments (NFI)*, by the *RXTE PCA*, or by the *Chandra X-ray Observatory – AXAF (CHANDRA)* (see also the chapter by Frontera in this volume).

Optical afterglows are generally identified for being previously unknown sources and from their variability. The former is done by comparing deep images of the GRB field, acquired soon after the GRB detection, with the *Space Telescope Science Institute (STScI) Digitized Sky Survey (DSS)*, and the latter by comparing early images with those obtained at later epochs. An alternative technique for the selection of GRB counterpart candidates is based on the characteristic power-law spectral shape of afterglows, and can be successfully employed when sufficient color information is available [87,206]. This method may suffer from redshift-dependent biases and possible contamination by other classes of sources; however, it can be advantageously applied to a single set of images instead of requiring two (or more) sets of images taken at different epochs.

It should be noted that, so far, optical counterparts have been detected only for long duration GRBs, which represent one of the two populations into which GRBs have been subdivided (see, however, [6,167] for the possible existence of a third class). According to their duration, the GRBs of the *BATSE* sample are divided into long (75% of the total) and short (25%) events, with average durations of $\sim$20 s and $\sim$0.2 s, respectively. This bimodality is reflected also in the spectral hardness, with long GRBs tending to be softer than sub-second GRBs [130], and may be due to a different origin of the two classes of sources. Sub-second GRBs could not be accurately localized by the *BeppoSAX WFCs* [82], however, four of them have been localized with arcmin precision by the *IPN*, and followed up in the optical with delays no shorter than $\sim$20 hours. No afterglows have been detected and the upper limits show that, within the limited statistics, the optical afterglow behavior of sub-second GRBs may not differ from that of long ones [88,119]. The GRB000301C, which was detected by the *IPN* with a 2 s duration and exhibited a bright, variable counterpart (see Sect. 3.2), cannot be unambiguously classified as a long, sub-second hard, or intermediate GRB [124].

Considering only GRBs for which the angular localization is better than $\sim$30 arcmin^2 and which were disseminated within 24 hr, the statistics of detected optical afterglows is $\sim$40% of the total. Therefore, many GRBs are optically "dark," although nearly all of these have X-ray and/or radio afterglows. Many processes can cause optical searches to be unsuccessful (see also [49]): 1) optical afterglows can be intrinsically faint, rapidly decaying, or dim because of a large redshift causing the Lyα break to affect the optical spectrum, or 2) the lack of an optical detection may be due to insufficient sensitivity of the search [55,73,228]. The different decay rates of the optical afterglows on the one hand, and the rather wide range of magnitudes measured, at equal intervals after the GRB, for optical counterparts with comparable decay rates on the other (see Table 1), indicate that the chance of detection may be very different from case to case for similarly prompt and deep exposures. This, together with the lack of a straightforward correlation between optical emission and γ-ray brightness of

Table 1. Parameters of GRB Optical Counterparts[a]

GRB	Instrument[b]	z	β_1^{c}	β_2^{c}	R_{OT}^{d} (mag)	R_{host}^{e} (mag)	A_R^{f} (mag)	Refs.
970228	*BeppoSAX*	0.695	−1.7		20.3±0.2	24.6±0.2	0.63	[17,60,79]
970508	*BeppoSAX*	0.835	−1.3		21.0±0.1 [g]	25.0±0.2	0.13	[14,62,163]
970828	*BATSE, RXTE*	0.9578	...	...	> 23.7 (4hr)	25.1±0.3	0.10	[22,49,91]
971214	*BeppoSAX*	3.418	−1.4		23.0±0.1	26.2±0.2	0.04	[97,132,168]
980326	*BeppoSAX*	∼ 1?[h]	−2.0		22.8±0.1	29.0±0.3 (V)	0.21	[15,66,92]
980329	*BeppoSAX*	3–5	−1.2		23.7±0.2	27.8±0.3	0.19	[59,112,137,171,200]
980425	*BeppoSAX*	0.0085	...	...	15.60±0.05	14.11±0.05	0.17	[72,77,229]
980519	*BATSE, WFCs*	...	−1.73	−2.22	20.64±0.03	∼ 25.5	0.69	[106,108,123]
980613	*BeppoSAX*	1.097	∼ −0.2	∼ −1	23.1±0.1	24.3±0.2	0.23	[48,50,98,105,115]
980703	*BATSE, RXTE*	0.966	−1.4		21.00±0.05	22.4±0.1	0.15	[32,47,113]
981226	*BeppoSAX*	...	...	...	> 23 (10hr)	24.2±0.1	0.06	[111,149]
990123	*BeppoSAX*	>1.6004	−1.13	−1.8	20.4±0.1	23.9±0.1	0.04	[33,61,134]
990308	*BATSE, RXTE*	...	∼ −1.2		20.7±0.1	> 28.4	0.07	[109,217]
990506	*BATSE, RXTE*	1.3	...	...	> 19 (1hr)	24.8±0.3	0.18	[18,110,246]
990510	*BATSE, WFCs*	>1.619	−0.82	−2.18	19.00±0.05	27.4±0.3 (V)	0.53	[64,102,235]
990705	*BeppoSAX*	∼0.86	−1.7	< −2.6	$H \sim 19$	22.0±0.1	0.20	[3,154,210]
990712	*BeppoSAX*	0.4331	−0.97		21.25±0.05	21.90±0.15	0.08	[63,209,235]
991208	*IPN*	0.706	−2.3[i]	−3.2	18.5±0.1 (2d)	24.2±0.2	0.04	[35,65]
991216	*BATSE, RXTE*	1.02	−1.0	−1.8	18.0±0.1	25.3±0.2	1.64	[100,192,236,237]
000131	*IPN*	4.5	−2.3		23.0±0.1 (3d)	> 25.6	0.14	[4]
000210	*BeppoSAX, CHANDRA*	0.846	...	...	> 22 (12.4hr)	23.5±0.1	0.05	[194]
000301C	*IPN*	2.04	−1.1	−2.9	20.1±0.1 (2d)	28.0±0.3	0.13	[67,83,124,155,222]
000418	*IPN*	1.118	−1.22		21.9±0.1 (3d)	23.8±0.2	0.08	[16,129,164]

Table 1. (Continued)

GRB	Instrument[b]	z	β_1^c	β_2^c	R_{OT}^d (mag)	R_{host}^e (mag)	A_R^f (mag)	Refs.
000630	*IPN*	...	−1.0		23.2±0.2	26.7±0.2	0.03	[73,126]
000911	*IPN*	1.058	−1.4		20.40±0.08 (1.4d)	∼ 25	0.31	[51,146,173]
000926	*IPN*	2.037	−1.5	−2.3	19.50±0.02	∼ 25	0.06	[28,74,104,196]
001007	*IPN*	...	−2.05		20.2 (3.5d)	24.73±0.15	0.11	[30,195]
001011	*BeppoSAX*	...	−1.33		22.4±0.1	25.1±0.3	0.26	[87]
001018	*IPN*	...	...	...	> 22.6 (>2d)	24.50±0.09	0.06	[19,20]
010222	*BeppoSAX*	1.476	−0.65	−1.7	20.15±0.05	25.7±0.2	0.06	[68,125,156,226]
010921	*HETE-2*	0.45	−1.59		19.29±0.06	21.40±0.05	0.38	[197]
011030[j]	*BeppoSAX*	...	...	...	> 21 (8hr)	∼ 25 (V)	1.00	[71,166]
011121	*BeppoSAX*	0.36	−1.7		19.47±0.05	24.70±0.05	1.26	[23,120]
011211	*BeppoSAX*	2.14	−0.83	−1.7	20.9±0.1	25.0±0.3	0.11	[27,70,114]

[a] Detected before 31 December 2001.
[b] Instrument, or suite of instruments, which localized the GRB.
[c] Fitted temporal decay index in the R-and, $F(t) \propto t^{+\beta}$. The subscripts 1 and 2 refer to the decay rate before and after the light curve steepening, respectively (see Sect. 3.2).
[d] R-band magnitude of the optical transient (OT) one day after explosion (unless otherwise noted). Magnitudes are in the Cousins system and are corrected for the host galaxy contribution and for Galactic extinction.
[e] Host galaxy magnitude, corrected for Galactic extinction. Magnitudes are in the Cousins system.
[f] Galactic extinction derived from the dust maps of Schlegel et al. [219], except for GRB970228 [60].
[g] At 1 day after the explosion, the afterglow was still brightening. This magnitude was observed and not derived from a fit.
[h] Estimate based on decomposition of the optical transient light curve [15].
[i] Castro-Tirado et al. [35] argue that the slope may be flatter at times earlier than 2 days based on an upper limit obtained from sky patrol films soon after the GRB.
[j] This event belongs to the class of X-ray flashes; see the chapter by Frontera in this volume.

the prompt event, makes it difficult to predict the optical detection level and to devise an optimal observing strategy.

In addition, optical afterglows may be affected by extinction in the plane of our Galaxy (which is transparent to γ- and hard X-rays and to radio radiation), and by absorption in their host galaxies (see, e.g., non-Galactic absorption estimates in the chapter by Weiler et al. in this volume), which makes their detection all the more challenging. By simulating the absorption experienced at the center of a dense dust clump, similar to those found in star forming regions in our Galaxy and in external galaxies, in a number of directions randomly distributed, Lamb [141] determined that along only 35% of the lines of sight is the optical depth $\tau \lesssim 1$, while along the remainder it is $\tau \gg 1$. This statistic is consistent with the percentage of dark GRBs, supporting the idea that dust absorption local to the burst may hamper, or completely prevent, optical detection of the GRB afterglow and strengthening the importance of IR observations (see Sect. 4.1). The presence of substantial quantities of dust at the burst explosion site favors, in turn, the association of GRBs with star forming regions (see Sect. 5). Although all dark GRBs may be the result of dust extinction, perhaps this is only one of the possible causes for failed optical detections. An alternative view [144] is that dark GRBs can be heavily absorbed only if dust sublimation by the strong UV/optical [239] and X-ray radiation [69] following the explosion does not play a significant role. If dust destruction around the burst site is important[81], then dark GRBs may be a distinct population differing significantly from GRBs with detected optical afterglows.

3 Temporal Characteristics

3.1 The Optical Flash

Prompt emission at optical and near infrared (NIR) wavelengths simultaneous with a GRB, or delayed by a few seconds, is expected to take place as a consequence of a reverse shock propagating into the explosion ejecta, and is therefore distinct from the afterglow which is produced by the interaction of the forward shock (blastwave) with the interstellar medium [158,159,174,212]. This low energy prompt emission can be very bright, in principle up to ~ 5 mag for an intense GRB at $z \sim 1$, but is expected to last only a few tens of seconds.

Since current instrumentation disseminates accurate GRB localizations with delays of a few hours, GRB fields are usually imaged in the optical no earlier than several hours after the explosion. Even if dissemination occurred in real time (as was the case for the large *BATSE* error boxes), the initiation of observations at a traditional optical telescope would still take at least several minutes. Small robotic telescopes, which can slew automatically and rapidly to the GRB position in response to a localization delivered in real time are therefore the most efficient instruments for optical GRB follow-up at the earliest epochs.

On 23 January 1999, an optical flare was detected by the *Robotic Optical Transient Search Experiment (ROTSE)* telescope (consisting of a two-by-two array of 35 mm lenses), starting 22 seconds after the onset of a GRB which triggered both *BeppoSAX* and *BATSE*. At maximum, the optical transient reached $V \simeq 9$ [1], implying a power output in the optical of about ~1% of that at the γ-ray energies [78], in agreement with the reverse shock interpretation and allowing an estimate of the plasma initial Lorentz factor [213]. In Fig. 1 the *ROTSE* data are reported (from immediately after the light maximum, to ~ 10 minutes after the GRB, before the decreasing flux level became undetectable), along with the measurements of the successive afterglow taken at larger telescopes. The *ROTSE* points are fitted by a steeper temporal power-law than the afterglow points, indicating two different radiation mechanisms. Owing possibly to the exceptional brightness of the event, to the rapidity of the *ROTSE* slew, to the precise *BeppoSAX* localization (which allowed identification of the transient in the $16° \times 16°$ *ROTSE* CCD image), and to the limited sensitivity of state of the art robotic telescopes, GRB990123 still remains the unique case of detection of an optical flash simultaneous with the GRB event itself. Searches for prompt optical transients in other GRB error boxes, both with *ROTSE* and with other robotic systems, have yielded no detections to limiting magnitudes from ~4 to ~15 at epochs between 10 seconds and 30 minutes after the GRB [2,25,35,127,181,182].

3.2 Optical Afterglow Emission

At epochs between a few hours and ~1 day after the GRB990123, the afterglow decays follow approximately a temporal power-law, $t^{+\beta}$, with an index β ranging from ~ -0.7 to ~ -3 (see Table 1, Cols. 4 and 5, and Fig. 1). This behavior was predicted before the detection of the first afterglow as a consequence of the simplest version of the fireball model [159]. According to this scenario (see the chapter by Waxman in this volume), the GRB990123 explosion triggers a relativistic forward shock which plows into the interstellar medium (or into a pre-explosion wind from the GRB990123 stellar progenitor [37,190]) and accelerates the particles. These interact with the magnetic field and radiate at all frequencies through the synchrotron mechanism (a significant contribution from inverse Compton scattering emission is also predicted and observed in some cases [103,177,215]). Linear polarization, measured in the optical afterglows of GRB990510 and GRB990712 at the level of a few percent [40,207,242], represents a good test of the synchrotron mechanism [93,151] and of collimated emission [86] (see below). NIR polarimetry of GRB000301C yielded an upper limit of 30%. Although not very constraining, this is also consistent with a synchrotron origin of the continuum in a relativistic jet [227].

On average, the initial luminosities of GRB optical afterglows are two orders of magnitude larger than maximum supernova luminosities, and obviously outshine their parent galaxies. At longer intervals after the GRB, when the flux of the transient subsides to less than the brightness of the host galaxy, it is possible to measure the magnitude of the latter. In some cases (notably GRB970228, GRB980326 and GRB011121), the light curve of the optical afterglow exhibits

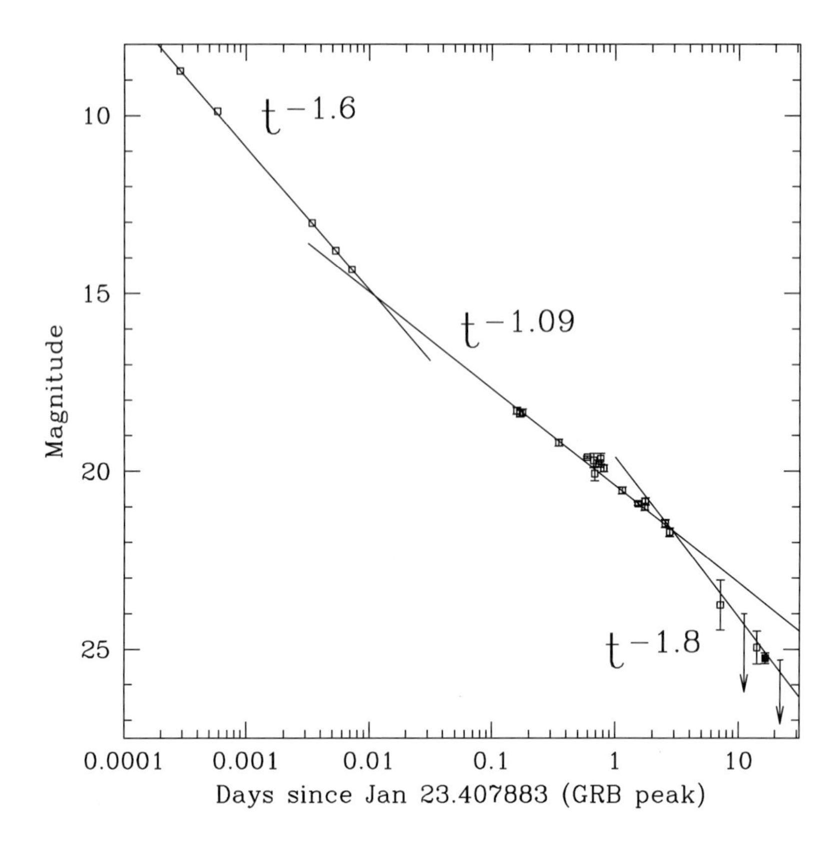

Fig. 1. R-band light curve of the GRB990123 afterglow. All points, except for the *HST* point (rightmost filled square), represent measurements taken from the ground (see [61] for references) and are reduced to a common flux standard with the host galaxy flux subtracted. Error bars (1σ) are shown where available, and arrows indicate 95% confidence upper limits (from [61]).

a re-brightening a few weeks after the GRB with respect to a power-law behavior. Following claims for the possible association of GRB980425 with the close-by ($z = 0.0085$) SN1998bw [77,133] (see also the chapter by Galama in this volume), the light curves of those optical transients have been decomposed into a non-thermal, pure afterglow contribution and a supernova profile, using SN1998bw as a template which has been appropriately redshifted. In some cases the results are convincing [15,23,35,79,201], while in other cases they are not decisive [13,123,209]. Systematic decomposition into non-thermal and supernova emission components has also been attempted by Dado et al. [41–43], for the afterglows of GRBs with known redshifts, in the framework of the "Cannonball Model" and good results have been obtained in most cases. The afterglow magnitudes reported in Table 1 (Col. 6) have been obtained from a fit with a single or double power-law after subtraction of the host galaxy flux, and in some cases after decomposition of a possibly underlying supernova (see references for individual cases in Col. 9). In those cases where the few optical measurements did not allow a proper fit, the optical magnitude measured closest to 1 day after explosion was used (e.g., GRB980613).

In a large fraction of the best monitored optical and/or NIR afterglows the initial power-law decline steepens at times ranging from ~0.5 to ~5 days after the GRB explosion. The effect is clearly seen as a smooth increase of the flux decay rate [12,33,61,74,100,102,103,121,124,134,155,156,196,205,225,226], and is suggested also by the X-ray data in a few cases [103,189,193,243]. In Table 1 (Cols. 4 and 5) the early and late temporal indices are reported as determined via empirical fits to the optical light curves with double power-laws (see also Fig. 1). The change in the temporal slope is thought to be due to the presence of a decelerating jet. Collimation of the radiation in a jet structure would reduce the huge energy requirements ($\sim 10^{52}$–10^{54} erg) for isotropic emission of the observed γ-ray brightnesses at the measured GRB distances and thus help resolve the paradox of the energy conversion efficiency [56,162,190,220]. When the aperture of the radiation cone (beaming angle), which progressively increases as the relativistic plasma decelerates, becomes larger than the jet opening angle, the observer perceives a faster light dimming, independent of wavelength, due to the jet edge becoming visible and/or to jet sideways expansion [161,204,214]. The change in fading rate is smooth, due to light travel time effects at the end surface of the jet [165,176]. Under a jet model, the steepening of the afterglow light curve is a probe of the GRB and of the afterglow emitting geometry. Stanek et al. [226] note an anti-correlation between the slope change $\Delta\beta$ and the isotropic γ-ray energy of the burst, suggesting that different jet opening angles may be responsible for it. Specific jet models for individual cases have been proposed [9,10,179], and afterglow emission from jets has been modeled as a function of the viewing angle [90,208,245], as was also suggested by Dado et al. [41].

If jets are unavoidable to relax the energy crisis in GRBs, and a light curve steepening is their signature, one may wonder why not all observed optical afterglows exhibit a detectable steepening in their light curves. This may simply be due to under sampling: when not detected, the steepening may have occurred at early, poorly sampled epochs (many afterglows are described by power-laws with temporal indices steeper than −2, see Table 1), or at late epochs, when the afterglow behavior is significantly contaminated by the emerging host galaxy emission or that of a possible supernova, so that discerning a decay rate variation is more difficult (see, e.g., [62]). On the other hand, light curve steepening cannot be unequivocally ascribed to a decelerating jet, but may instead (or additionally) be caused by the transition of a spectral break through the observing frequency band [211] (see Sect. 4.1) or by the propagation of the external shock in a inhomogeneous medium [37,123,160,175] (although in these cases the steepening would be frequency dependent; see however [136] for detectability of a jet in a stratified medium), or by the transition of the plasma kinematic conditions from relativistic to Newtonian in a dense medium [44,45,116,156]. In some cases the interpretation is not unique [99,148], although simultaneous multiwavelength observations may resolve the ambiguity [38,103,178,193]. We finally note that some months after the GRB, a flattening of the afterglow light curve may be expected [150].

Optical inter-day or intra-day variations superimposed on the overall afterglow decline are rarely detected, because of the limited photometric precision of the measurements. Two cases where significant deviations from a steady decay have been observed during the optical monitoring are GRB970508, which exhibited an initial shallow decline, followed by a 2-day re-bursting of a factor $\sim$5 amplitude [31,76,184] correlated with an X-ray flare [191], and GRB000301C, which showed intra-day achromatic variability of $20-30\%$ amplitude [10,155]. For both events, an interpretation based on microlensing has been proposed [41,83,84]. Rapid variability can otherwise be produced by small scale inhomogeneities of the plasma flow, or irregularities of the external medium through which the blastwave propagates [114,238], or local re-energization episodes [175].

4 Spectral Properties

4.1 The IR-Optical Afterglow Continuum

The classical fireball model has specific predictions for the temporal evolution of the broad band spectral shape [211]. This has been modified to include the detectable effects of the presence of a jet [214]. Both the spectral slope and the temporal decay rate depend on the index p of the electron energy distribution $N(\Gamma) \propto \Gamma^{-p}$ (where Γ denotes the electron energy above a certain cutoff). The radio-to-X-ray spectral shape is characterized by smooth breaks at typical frequencies (self-absorption, peak and cooling frequencies), which evolve with time in a predictable way [89], so that simultaneous multiwavelength observations at various epochs during the evolution of the afterglow allow the measurement of the spectral slopes and breaks and an estimate of the relevant physical parameters of the afterglow (see e.g., [178,180,241] and the chapter by Frontera in this volume).

When the optical photometric observations are accurate and sufficiently extended in time to make a good signal-to-noise ratio measurement of the spectral and temporal slopes possible, they show that the spectra of some afterglows, corrected for Galactic extinction, are steeper (i.e., redder) than expected from the fireball theory based on comparison with the temporal decay rate. This has been commonly attributed to absorption intrinsic to the source or, especially for GRBs at very high redshift, intervening absorption along the line of sight [198,202,221]. A small amount of reddening by dust in the GRB host galaxy has been invoked in many cases to reconcile the observations with the theoretical scenarios, using extinction curves typical of our own Galaxy, of star-forming galaxies, or of the *Large Magellanic Cloud (LMC)* and *Small Magellanic Cloud (SMC)* [42,43,46,103,124,147,156,171,234]. While even a moderate quantity of dust at the GRB source redshift may significantly attenuate the observed optical spectrum (which corresponds at the average $z \sim 1$ to rest-frame UV wavelengths), or even completely obscure it (see Sect. 2), NIR data are less affected

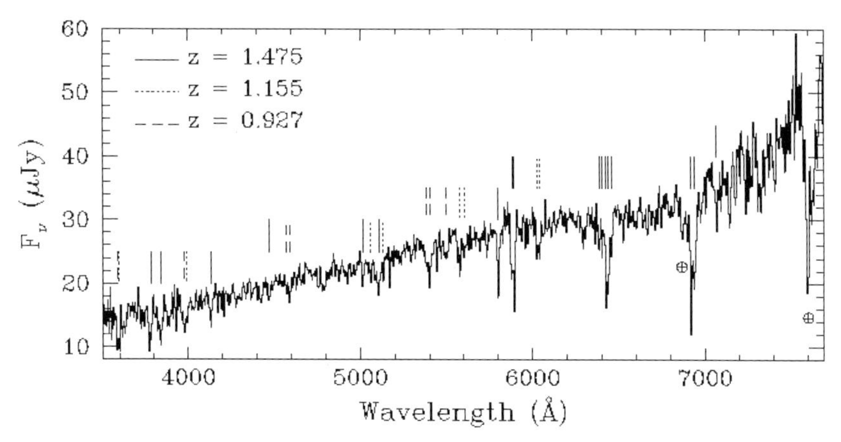

Fig. 2. Low resolution spectrum of the GRB010222 afterglow taken at the *Telescopio Nazionale Galileo*, corrected for Galactic extinction. Absorption lines from systems at three different redshifts are detected. Telluric features are marked with the symbol ⊕ (from [156]).

and may be more effective in determining the overall afterglow spectrum, when combined with data at other frequencies [46,137,171]. Observations in the NIR range are therefore critical for the study of afterglows.

4.2 Absorption Features

Low and medium resolution spectra of bright optical afterglows have allowed, in a number of cases, the detection of absorption lines of metallic species caused by intervening absorbers, and the measurement of lower limits to the GRB redshift (see Table 1 and Fig. 2). Frequently, the evidence of a low-ionization, high density medium (e.g., related to the detection of Mg I in absorption) suggests that the absorbing system is actually the host galaxy [156,163,235]. Spectroscopy of the likely host galaxy generally shows that the highest absorption redshift coincides with the redshift of the galaxy emission lines, confirming the association of the GRB with the proposed host. No variability of the absorption line equivalent widths is detected at the 20% level (which represents the 1σ uncertainty on the measurements) on time scales of some hours to few days [235].

For GRB000301C and GRB000131 the redshifts have been determined through identification of absorption edges with the Lyman limit and intervening Lyα forest, respectively. The afterglow of the former GRB is the only one which has been observed by *HST* at UV wavelengths: in the low-resolution spectrum taken 5 days after explosion with the *Space Telescope Imaging Spectrograph (STIS)* instrument equipped with the *NUV-MAMA* prism a discontinuity at $\sim$2800 Å is clearly detected, which has been identified with the hydrogen ionization edge [222], implying $z \simeq 2$ (see Fig. 3). The measurement was then confirmed and refined from optical spectroscopy at the *W.M. Keck Telescope I (KECK I)* and *Very Large Telescope (VLT)* [29,124]. The redshift of GRB000131, $z \simeq 4.5$,

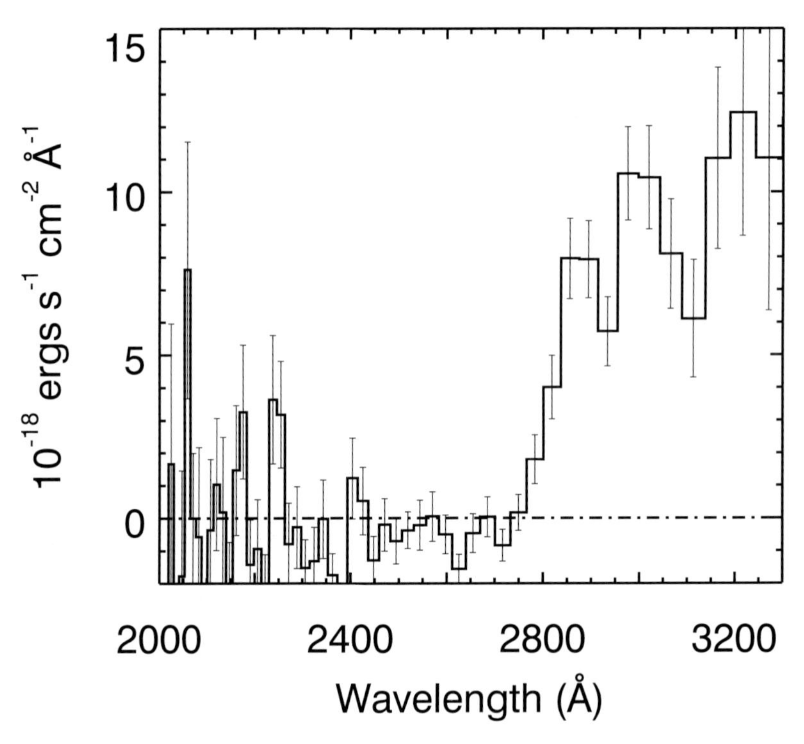

Fig. 3. Deconvolved, flux-calibrated, UV spectrum of GRB 000301C taken with the *HST STIS NUV-MAMA* prism. A break is clearly seen at 2797 Å. If caused by the onset of Lyman continuum absorption due to HI gas associated with the host galaxy, the redshift is $z = 2.067 \pm 0.025$. This matches the ground-based measurement of the redshift (from [222]).

was determined photometrically from simultaneous NIR and optical observations (see Fig. 4), and supported by optical spectrophotometry [4].

The distance of GRB980329 is controversial: the afterglow photometry suggested a redshift as large as $z \sim 5$ or lower, $3 \lesssim z \lesssim 4.4$, according to whether the observed continuum suppression shortward of ~6500 Å is identified with Lyα intervening absorption [59], or with molecular hydrogen dissociation by the strong initial UV flash [52]. A redshift of $z < 3.9$ is suggested by the absence of the Lyα break in the host galaxy spectrum [51,138].

The redshifts measured so far, either with spectroscopy or broad band photometry, span the range ~0.4 to ~4 (Table 1), excluding the peculiar case of GRB980425 (see the chapters by Galama and by Iwamoto et al. in this volume), and prove that long duration GRBs have an extragalactic, cosmological origin, which makes their early bright optical afterglows excellent probes of the high redshift universe.

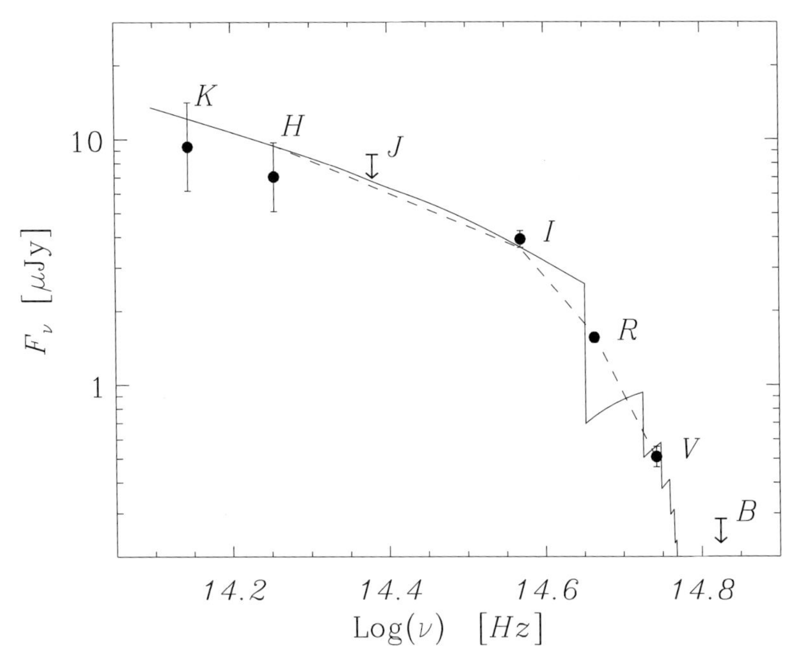

Fig. 4. Spectral energy distribution of the GRB000131 afterglow, corrected for Galactic reddening, as derived from broad band *VLT* and *New Technology Telescope (NTT)* photometry. The uncertainties of the H- and K-band fluxes include the formal error from the extrapolation of the light curves back to the epoch of the optical measurements, $t = 3.5$ days. A fit with a power-law spectrum with Lyα forest absorption and SMC reddening is shown as a dashed curve. This yields $A_V = 0.18$, when an intrinsic spectral slope $\beta = -0.70$ and a redshift of $z = 4.5$ is assumed. The solid curve shows the corresponding spectrum with its Lyman absorption edges (from [4]).

5 Host Galaxies

For almost every well studied optical afterglow, deep late epoch optical or NIR observations from the ground or with *HST* have detected a galaxy close to the point-like optical transient or at its position after it has faded away (see Table 1). Host galaxies have been detected also for some dark GRBs with arcsecond afterglow localizations from radio telescopes or from *CHANDRA*. Few observations of host galaxies have been made at longer wavelengths [11,57,101,223]). The host optical magnitudes (and upper limits) are consistent with those expected for a reasonable redshift distribution and galaxy population [107]. This solves the "no-host galaxy" problem [7,8,143,216], which, *a posteriori*, turns out to be clearly related to the very faint flux of the host galaxies. They are usually detected only with long exposures at telescopes larger than 2 m diameter.

HST observations taken at early stages of the afterglow evolution, when the transient is still bright (see Fig. 5), show that the GRB always lies within the stellar field of its host galaxy. If only late epoch *HST* images are available, their

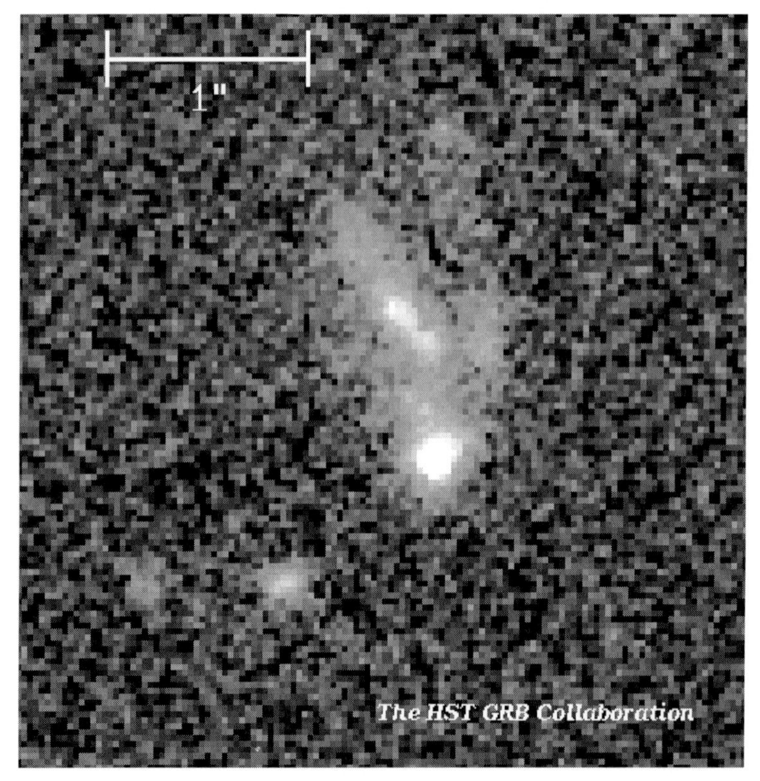

Fig. 5. *HST STIS* image of the host galaxy of GRB990123 in white light. North is up, and east is to the left. The galaxy morphology is irregular. The optical transient is the bright point-like source on the southern edge of the galaxy (from [61]).

comparison with accurate astrometry of the bright transient on early epoch ground-based images still yields projected angular offsets of a fraction of an arcsecond between the transient and the galaxy center. When normalized to the individual host half-light radii, the median offset is 0.98 [21]. This has led to the conclusion that GRBs are associated with star forming regions, which would support their origin as hypernovae [169] (or collapsars [152]) or as "supranovae" [233], as opposed to progenitor scenarios which envisage the explosion as taking place at many kiloparsecs from the parent galaxy in its halo (like binary neutron star systems, see the chapter by Waxman in this volume).

There is specific evidence that the host galaxies of GRBs undergo strong star formation:

- Their integrated colors are remarkably blue [60,61].
- They are usually underluminous (luminosities of a fraction of the characteristic luminosity L_* of the Schechter [218] luminosity function), small, and often have a compact morphology [62,103,188], which are characteristics common to galaxies hosting star formation at the typical GRB redshifts of $z \sim 1$ [5,96].

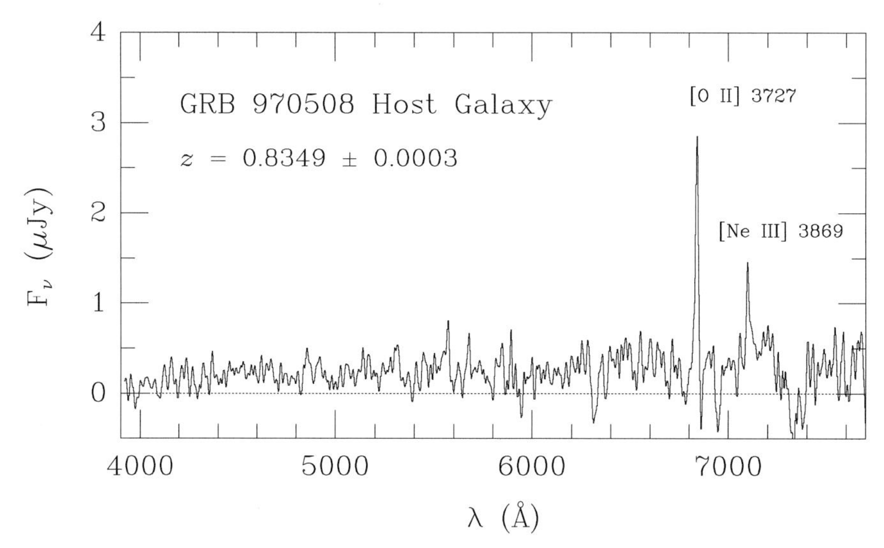

Fig. 6. Spectrum of the host galaxy of GRB970508, obtained at the *KECK I* telescope. Prominent emission lines are labeled (from [14]).

- Their spectra exhibit star formation emission lines like [O II], [Ne III] (Fig. 6), [O III], Lyα and Balmer series [14,47,75,132,235].
- Their star formation rates, derived either from the emission line intensities or from the UV rest frame galaxy continuum are high, compared with the galaxy size [47,49,51,60,113,115,235], although sometimes substantially obscured [49,51] and possibly measurable only at rest frame far-IR wavelengths [57].
- They sometimes have morphologies consistent with being mergers or interacting systems [49,50,61], where star formation is enhanced.

Furthermore, the imprints of local extinction on the afterglow spectra (see Sect. 4.1) point to dusty and likely star forming environments as the favored GRB explosion sites. Finally, the observed host galaxy magnitudes and redshifts are consistent with a model in which the comoving rate density of GRBs is proportional to the cosmic star formation rate density [107,131,153,199,240]. These suggestions are consistent with the fact that most measured GRB redshifts are around $z \sim 1$, where the cosmic star formation rate is one order of magnitude larger than locally [139]. Since this is predicted to increase monotonically back to $z \sim 10$, one may expect that, given the opportunity of detecting GRBs up to that redshift, it would be possible to select the youngest star forming galaxies in the Universe.

6 Conclusions: Open Problems and Future Prospects

The enormous progress achieved over the last few years through the optical afterglow follow-up studies has also indicated some fundamental, uncertain aspects of GRB and afterglow physics:

- the conversion of energy into radiation,
- the structure and geometry of the emitting regions,
- the nature, density and composition of the circumburst medium, and
- the cosmological evolution of the GRB population.

The solution of these problems will ultimately lead to unveiling the major unknown, i.e. the identity of GRB progenitors. While at the present time these issues remain a matter of investigation, current knowledge suggests an observational approach to tackle them most effectively.

The most serious limitation of present day observations is the substantial temporal delay between GRB trigger and follow-up of its field at lower frequencies (a few hours). This causes a lack of sampling of the initial portions of the light curves. Bridging this gap will be possible when real-time dissemination of positions becomes available from the coming generation of high energy satellites, and follow-up can be rapidly achieved by small, fast reacting telescopes with suitably large fields of view. Thanks to the *UltraViolet and Optical Telescope (UVOT)* camera onboard *Swift Gamma-Ray Burst Explorer (Swift)* mission (to be launched in 2003), and to the advent of ground-based robotic telescopes [25,34,172,182,244], the follow-up can start as early as a few tens of seconds after GRB detection. This will allow astronomers to catch the transient counterpart in its maximum emission state and to follow the temporal evolution of its optical to IR continuum. It is at these early epochs that the models differ most in their predictions [41,161,177,211,212], and the high signal to noise can discriminate among them with the greatest confidence. Early, simultaneous optical/NIR searches will either detect more counterparts, and reduce the number of dark GRBs, or put stronger constraints on the "darkness." In addition to robotic systems, all sky optical monitoring cameras [170,183] present the advantage that they will be independent of spacecraft triggers and, therefore, can detect possible precursors of GRBs and "burstless" afterglows[1], which would be missed by robotic telescopes.

Weeks after the GRB explosion, the light curve will result from the sum of different components of comparable brightness: the fading afterglow, the possible supernova rising to maximum light, and the host galaxy background. To disentangle these contributions we will need sensitive and densely sampled photometric observations at late epochs in the optical and IR. These will also possibly discern the presence of a light echo due to dust scattering [54,203] or sublimation [232,239]. This task is a prerogative of the sensitive, high resolution optical and NIR cameras in space, such as the *HST WFPC2, STIS, Near Infrared*

[1] Highly collimated jets misaligned with respect to the line of sight could prevent detection of the GRB, while the optical afterglow might be detectable after jet spreading [160].

Camera and Multi-Object Spectrometer (NICMOS), and the newly-deployed *Advanced Camera for Surveys (ACS)*. The future *Next Generation Space Telescope (NGST)* will build on the legacy of these instruments and push the research on GRB late afterglows and hosts toward even larger redshifts.

The measurement of a large number of redshifts, through early absorption spectra of optical afterglows as well as through emission line spectra of the host galaxies, is necessary to construct a luminosity function of GRBs to be compared with models of star forming rate evolution. This will allow us to test the link between GRBs and star formation history up to very high redshifts, many hints of which have been so far collected [139]. A primary role in early and late sensitive optical and IR spectroscopy will be played by the ground-based $4-8$ m class telescopes in both hemispheres. Early bright counterparts will be excellent targets for spectroscopic monitoring: variations in the equivalent widths of the absorption lines will be measured with good signal to noise ratio, thus making it possible to place constraints on the density and distribution of the circumburst medium [53,145,187], which is a critical diagnostic of the GRB progenitor. Furthermore, the intense initial optical flares associated with GRBs will also act as background "light bulbs" to probe the ionization state and metallicity of the intergalactic medium (IGM) through high resolution spectroscopy. More generally, multiwavelength observations of GRBs will allow a series of cosmological tests on a wide range of redshifts [140,142,186] (see also the chapter by Loeb in this volume).

Prompt, long, and intensive polarimetric monitoring of afterglows will detect possible changes of polarized light percentage and position angle and thereby set constraints on the most important open issue of afterglow physics, the generation of magnetic fields [85,93–95,157]. Moreover, knowledge of the magnetic field geometry and of the circumburst density profile will be instrumental to defining the structure of the jet and its interaction with the ambient medium.

Finally, real time dissemination of accurate localizations of short, sub-second GRBs and the prompt follow-up of these fields in the optical will hopefully afford detection of their elusive counterparts and will allow us to obtain clear insights into their genesis and physics.

Acknowledgments

I am grateful to the members of the GRACE Consortium led by E. van den Heuvel for a longstanding, fruitful collaboration, and in particular to F. Frontera, A. Fruchter, N. Masetti and E. Palazzi, who have been my closest collaborators in GRB optical counterpart research since the beginning of the "afterglow era." I thank S. Recchi for valuable input, and A. De Rújula, A. Fruchter, P. Mazzali and E. Waxman for a critical reading of the manuscript. I would like to dedicate this review to the memory of Jan van Paradijs.

References

1. C. Akerlof et al.: Nature **398**, 400 (1999)
2. C. Akerlof et al.: Astrophys. J. Lett. **532**, L25 (2000)
3. L. Amati et al.: Science **290**, 953 (2000)
4. M.I. Andersen et al.: Astron. Astrophys. **364**, L54 (2000)
5. A. Babul, M.J. Rees: Mon. Not. R. Astron. Soc. **255**, 346 (1992)
6. A. Balastegui, P. Ruiz-Lapuente, R. Canal: Mon. Not. R. Astron. Soc. **328**, 283 (2001)
7. D.L. Band, D.H. Hartmann: Astrophys. J. **493**, 555 (1998)
8. D.L. Band, D.H. Hartmann, B.E. Schaefer: Astrophys. J. **514**, 862 (1999)
9. E. Berger et al.: Astrophys. J. **556**, 556 (2001)
10. E. Berger et al.: Astrophys. J. **545**, 56 (2000)
11. E. Berger, S.R. Kulkarni, D.A. Frail: Astrophys. J. **560**, 652
12. S.G. Bhargavi, R. Cowsik: Astrophys. J. Lett. **545**, L77 (2000)
13. G. Bjornsson, J. Hjorth, P. Jakobsson, L. Christensen, S. Holland: Astrophys. J. Lett. **552**, L121 (2001)
14. J.S. Bloom, S.G. Djorgovski, S.R. Kulkarni, D.A. Frail: Astrophys. J. Lett. **507**, L25 (1998)
15. J.S. Bloom et al.: Nature **401**, 453 (1999)
16. J.S. Bloom, A. Diercks, S.G. Djorgovski, D. Kaplan, S.R. Kulkarni: GCN 661 (2000)
17. J.S. Bloom, S.G. Djorgovski, S.R. Kulkarni: Astrophys. J. **554**, 678 (2001)
18. J.S. Bloom, D.A. Frail, R. Sari: Astron. J. **121**, 2879 (2001)
19. J.S. Bloom, A. Diercks, S.R. Kulkarni, F.A. Harrison, B.B. Behr, J.C. Clemens: GCN 915 (2001)
20. J.S. Bloom, S.G. Djorgovski, F.A. Harrison, P. Mao,D. Stern: GCN 918 (2001)
21. J.S. Bloom, S.R. Kulkarni, S.G. Djorgovski: Astron. J. **123**, 1111 (2002)
22. J.S. Bloom, S.R. Kulkarni, T.J. Galama, D.A. Frail, S.G. Djorgovski: GCN 1134 (2001)
23. J.S. Bloom et al.: Astrophys. J. **572**, 45 (2002)
24. G. Boella, R.C. Butler, G.C. Perola, L. Piro, L. Scarsi, J.A.M. Bleeker: Astron. Astrophys. Suppl. Ser. **122**, 299 (1997)
25. M. Boer, J.-L. Atteia, M. Bringer, B. Gendre, A. Klotz, R. Malina, J.A. de Freitas Pacheco, H. Pedersen: Astron. Astrophys. **378**, 76 (2001)
26. H. Bradt, A.M. Levine, F.E. Marshall, R.A. Remillard, D.A. Smith, T. Takeshima: In: *Gamma-Ray Bursts in the Afterglow Era, Proc. International Workshop at Rome, Italy, 17–20 October 2000*, ed. by E. Costa, F. Frontera, J. Hjorth (Springer, Heidelberg 2001) p. 26
27. I. Burud, J. Rhoads, A.S. Fruchter, J. Hjorth: GCN 1213 (2001)
28. S.M. Castro, S.G. Djorgovski, S.R. Kulkarni, J.S. Bloom, T.J. Galama, F.A. Harrison, D.A. Frail: GCN 851 (2000)
29. S.M. Castro, A. Diercks, S.G. Djorgovski, S.R. Kulkarni, T.J. Galama, J.S. Bloom, F.A. Harrison, D.A. Frail: GCN 605 (2000)
30. J.M. Castro Cerón et al.: astro-ph 0110049 (2001)
31. A.J. Castro-Tirado et al.: Science **279**, 1011 (1998)
32. A.J. Castro-Tirado et al.: Astrophys. J. Lett. **511,**, L85 (1999)
33. A.J. Castro-Tirado et al.: Science **283,**, 2069 (1999)
34. A.J. Castro-Tirado et al.: Astron. Astrophys. Suppl. Ser. **138**, 583 (1999)
35. A.J. Castro-Tirado et al.: Astron. Astrophys. **370**, 398 (2001)

36. A.J. Castro-Tirado: In: *Gamma-Ray Bursts in the Afterglow Era, Proc. International Workshop at Rome, Italy, 17–20 October 2000*, ed. by E. Costa, F. Frontera, J. Hjorth (Springer, Heidelberg 2001) p. 121
37. R.A. Chevalier, Z.-Y. Li: Astrophys. J. Lett. **520**, L29 (1999)
38. R.A. Chevalier, Z.-Y. Li: Astrophys. J. **536**, 195 (2000)
39. E. Costa et al.: Nature **387**, 783 (1997)
40. S. Covino et al.: Astron. Astrophys. **348**, L1 (1999)
41. S. Dado, A. Dar, A. De Rújula: Astron. Astrophys. **388**, 1079 (2002)
42. S. Dado, A. Dar, A. De Rújula: astro-ph 0203315 (2002)
43. S. Dado, A. Dar, A. De Rújula: Astrophys. J. Lett. **572**, L143 (2002)
44. Z. Dai, T. Lu: Astrophys. J. Lett. **519**, L155 (1999)
45. Z. Dai, T. Lu: Astron. Astrophys. **367**, 501 (2001)
46. D. Dal Fiume, L. Amati: Astron. Astrophys. **355**, 454 (2000)
47. S.G. Djorgovski, S.R. Kulkarni, J.S. Bloom, R. Goodrich, D.A. Frail, L. Piro, E. Palazzi: Astrophys. J. Lett. **508**, L17 (1998)
48. S.G. Djorgovski, S.R. Kulkarni, S.C. Odewahn, H. Ebeling: GCN 117 (1998)
49. S.G. Djorgovski, D.A. Frail, S.R. Kulkarni, J.S. Bloom, S.C. Odewahn, A. Diercks: Astrophys. J. **562**, 654 (2001)
50. S.G. Djorgovski, J.S. Bloom, S.R. Kulkarni: astro-ph 0008029 (2000)
51. S.G. Djorgovski et al.: In: *Gamma-Ray Bursts in the Afterglow Era, Proc. International Workshop at Rome, Italy, 17–20 October 2000*, ed. by E. Costa, F. Frontera, J. Hjorth (Springer, Heidelberg 2001) p. 218
52. B.T. Draine: Astrophys. J. **532**, 273 (2000)
53. B.T. Draine, L. Hao: Astrophys. J. **569**, 780 (2002)
54. A.A. Esin, R. Blandford: Astrophys. J. Lett. **534**, L151 (2000)
55. D.A. Frail et al.: Astrophys. J. Lett. **525**, L81 (1999)
56. D.A. Frail et al.: Astrophys. J. Lett. **562**, L55 (2001);
57. D.A. Frail et al.: Astrophys. J. **565**, 829 (2002)
58. F. Frontera, E. Costa, D. dal Fiume, M. Feroci, L. Nicastro, M. Orlandini, E. Palazzi, G. Zavattini: Astron. Astrophys. Suppl. Ser. **122**, 357 (1997)
59. A.S. Fruchter: Astrophys. J. Lett. **512**, L1 (1999)
60. A.S. Fruchter et al.: Astrophys. J. **516**, 683 (1999)
61. A.S. Fruchter et al.: Astrophys. J. Lett. **519**, L13 (1999)
62. A.S. Fruchter et al.: Astrophys. J. **545**, 664 (2000)
63. A.S. Fruchter, P.M. Vreeswijk, R. Hook, E. Pian: GCN 752 (2000)
64. A.S. Fruchter, R. Hook, E. Pian: GCN 757 (2000)
65. A.S. Fruchter, P.M. Vreeswijk, V. Sokolov, A. Castro-Tirado: GCN 872 (2000)
66. A.S. Fruchter, P.M. Vreeswijk, P. Nugent: GCN 1029 (2001)
67. A.S. Fruchter, P.M. Vreeswijk: GCN 1063 (2001)
68. A.S. Fruchter, I. Burud, J. Rhoads, A. Levan: GCN 1087 (2001)
69. A.S. Fruchter, J.H. Krolik, J.E. Rhoads: Astrophys. J. **563**, 597 (2001)
70. A.S. Fruchter, P.M. Vreeswijk, J. Rhoads, I. Burud: GCN 1200 (2001)
71. A.S. Fruchter, S. Pattel, C. Kouveliotou, J. Rhoads, S. Holland, I. Burud, R. Wijers: GCN 1268 (2002)
72. J.U. Fynbo et al.: Astrophys. J. Lett. **542**, L89 (2000)
73. J.U. Fynbo et al.: Astron. Astrophys. **369**, 373 (2001)
74. J.U. Fynbo et al.: Astron. Astrophys. **373**, 796 (2001)
75. J.U. Fynbo, P. Moller: Astron. Astrophys. **388**, 425 (2002)
76. T.J. Galama et al.: Astrophys. J. Lett. **497**, L13 (1998)
77. T.J. Galama et al.: Nature **395**, 670 (1998)

78. T.J. Galama et al.: Nature **398**, 394 (1999)
79. T.J. Galama et al.: Astrophys. J. **536**, 185 (2000)
80. T.J. Galama: In: *Gamma-Ray Bursts: 5th Huntsville Symposium, AIP Conf. Proc, 526*, ed. by M. Kippen, R. Mallozzi, G. Fishman (AIP, New York 2000) p. 303
81. T.J. Galama, R.A.M.J. Wijers: Astrophys. J. Lett. **549**, L209 (2001)
82. G. Gandolfi et al.: In: *Gamma-Ray Bursts: 5th Huntsville Symposium, AIP Conf. Proc. 526*, ed. by M. Kippen, R. Mallozzi, G. Fishman (AIP, New York 2000) p. 23
83. P.M. Garnavich, A. Loeb, K.Z. Stanek: Astrophys. J. Lett. **544**, L11 (2000)
84. B.S. Gaudi, J. Granot, A. Loeb: Astrophys. J. **561**, 178 (2001)
85. G. Ghisellini, D. Lazzati: Mon. Not. R. Astron. Soc. **309**, L7 (1999)
86. G. Ghisellini: astro-ph 0111584 (2001)
87. J. Gorosabel et al.: Astron. Astrophys. **384**, 11 (2002)
88. J. Gorosabel et al.: Astron. Astrophys. **383**, 112 (2002)
89. J. Granot, R. Sari: Astrophys. J. **568**, 820 (2002)
90. J. Granot, A. Panaitescu, P. Kumar, S.E. Woosley: Astrophys. J. Lett. **570**, L61 (2002)
91. P.J. Groot et al.: Astrophys. J. Lett. **493**, L27 (1998)
92. P.J. Groot et al.: Astrophys. J. Lett. **502**, L123 (1998)
93. A. Gruzinov, E. Waxman: Astrophys. J. **511**, 852 (1999)
94. A. Gruzinov: Astrophys. J. Lett. **525**, L29 (1999)
95. A. Gruzinov: Astrophys. J. Lett. **563**, L15 (2001)
96. R. Guzmán, J. Gallego, D.C. Koo, A.C. Phillips, J.D. Lowenthal, S.M. Faber, G.D. Illingworth, N.P. Vogt: Astrophys. J. **489**, 559 (1997)
97. J.P. Halpern, J.R. Thorstensen, D.J. Helfand, E. Costa: Nature **393**, 41 (1998)
98. J.P. Halpern, R. Fesen: GCN 134 (1998)
99. J.P. Halpern, J. Kemp, T. Piran, M.A. Bershady: Astrophys. J. Lett. **517**, L105 (1999)
100. J.P. Halpern et al.: Astrophys. J. **543**, 697 (2000)
101. L. Hanlon et al.: Astron. Astrophys. **359**, 941 (2000)
102. F.A. Harrison et al.: Astrophys. J. Lett. **523**, L121 (1999)
103. F.A. Harrison et al.: Astrophys. J. **559**, 123 ((2001)
104. F.A. Harrison, T.J. Galama, J.S. Bloom, S.M. Castro, S.R. Kulkarni: GCN 1088 (2001)
105. J. Hjorth, M.I. Andersen, H. Pedersen, A.O. Jaunsen: GCN 109 (1998)
106. J. Hjorth, H. Pedersen, A.O. Jaunsen, M.I. Andersen: Astron. Astrophys. Suppl. Ser. **138**, 461 (1999)
107. D.W. Hogg, A.S. Fruchter: Astrophys. J. **520**, 54 (1999)
108. S. Holland et al.: GCN 698 (2000)
109. S. Holland et al.: GCN 726 (2000)
110. S. Holland et al.: GCN 731 (2000)
111. S. Holland et al.: GCN 749 (2000)
112. S. Holland et al.: GCN 778 (2000)
113. S. Holland et al.: Astron. Astrophys. **371**, 52 (2001)
114. S. Holland et al.: Astron. J. **124**, 639 (2002)
115. S. Holland: In: *Relativistic Astrophysics: 20th Texas Symposium, AIP Conf. Proc. 586*, ed. J.C. Wheeler, H. Martel (AIP, New York 2001) p. 593
116. Y.F. Huang, Z.G. Dai, T. Lu: Astron. Astrophys. **355**, L43 (2000)
117. K. Hurley, M.S. Briggs, R.M. Kippen, C. Kouveliotou, C. Meegan, G. Fishman, T. Cline, M. Boer: Astrophys. J. Suppl. **120**, 399 (1997)

118. K. Hurley, M.S. Briggs, R.M. Kippen, C. Kouveliotou, C. Meegan, G. Fishman, T. Cline, M. Boer: Astrophys. J. Suppl. **122**, 497 (1997)
119. K. Hurley et al.: Astrophys. J. **567**, 447 (2002)
120. L. Infante, P.M. Garnavich, K.Z. Stanek, L. Wyrzykowski: GCN 1152 (2001)
121. G.L. Israel et al.: Astron. Astrophys. **348**, L5 (1999)
122. R. Jager et al.: Astron. Astrophys. Suppl. Ser. **125**, 557 (1997)
123. A.O. Jaunsen et al.: Astrophys. J. **546**, 127 (2001)
124. B. Jensen et al.: Astron. Astrophys. **370**, 909 (2001)
125. S. Jha et al.: Astrophys. J. Lett. **554**, L155 (2001)
126. D.L. Kaplan, J.A. Eisner, S.R. Kulkarni, J.S. Bloom: GCN 1069 (2001)
127. R. Kehoe et al.: Astrophys. J. Lett. **554**, L159 (2001)
128. S. Klose: Rev. Mod. Astron. **13**, 129 (2000)
129. S. Klose et al.: Astrophys. J. **545**, 271 (2000)
130. C. Kouveliotou, C.A. Meegan, G.J. Fishman, N.P. Bhat, M.S. Briggs, T.M. Koshut, W.S. Paciesas, G.N. Pendleton: Astrophys. J. Lett. **413**, L101 (1993)
131. M. Krumholz, S. Thorsett, F. Harrison: Astrophys. J. Lett. **506**, L81 (1998)
132. S.R. Kulkarni et al.: Nature **393**, 35 (1998)
133. S.R. Kulkarni et al.: Nature **395**, 663 (1998)
134. S.R. Kulkarni et al.: Nature **398**, 389 (1999)
135. S.R. Kulkarni et al.: In: *Gamma-Ray Bursts: 5th Huntsville Symposium, AIP Conf. Proc. 526*, ed. M. Kippen, R. Mallozzi, G. Fishman (AIP, New York 2000) p. 277
136. P. Kumar, A. Panaitescu: Astrophys. J. Lett. **541**, L9 (2000)
137. D.Q. Lamb, F.J. Castander, D.E. Reichart: Astron. Astrophys. Suppl. Ser. **138,**, 479 (1999)
138. D.Q. Lamb: Astron. Astrophys. Suppl. Ser. **138**, 607 (1999)
139. D.Q. Lamb: Phys. Rep. **333**, 505 (2000)
140. D.Q. Lamb, D.E. Reichart: Astrophys. J. **536**, 1 (2000)
141. D.Q. Lamb: In: *Gamma-Ray Bursts in the Afterglow Era, Proc. International Workshop at Rome, Italy, 17–20 October 2000*, ed. by E. Costa, F. Frontera, J. Hjorth (Springer, Heidelberg 2001) p. 297
142. D.Q. Lamb, D.E. Reichart: In: *Relativistic Astrophysics: 20th Texas Symposium, AIP Conf. Proc. 586*, ed. J.C. Wheeler, H. Martel (AIP, New York 2001) p. 605
143. S.B. Larson: Astrophys. J. **491**, 86 (1997)
144. D. Lazzati, S. Covino, G. Ghisellini: Mon. Not. R. Astron. Soc. **330**, 583 (2002)
145. D. Lazzati, R. Perna, G. Ghisellini: Mon. Not. R. Astron. Soc. **325**, L19 (2001)
146. D. Lazzati et al.: Astron. Astrophys. **378**, 996 (2001)
147. B.C. Lee et al.: Astrophys. J. **561**, 183 (2001)
148. Z.-Y. Li, R.A. Chevalier: Astrophys. J. **551**, 940 (2001)
149. B. Lindgren et al.: GCN 190 (1999)
150. M. Livio, E. Waxman: Astrophys. J. **538**, 187 (2000)
151. A. Loeb, R. Perna: Astrophys. J. **495**, 597 (1998)
152. A.I. MacFadyen, S.E. Woosley: Astrophys. J. **524**, 262 (1999)
153. S. Mao, H.J. Mo: Astron. Astrophys. **339**, L1 (1998)
154. N. Masetti et al.: Astron. Astrophys. **354**, 473 (2000)
155. N. Masetti et al.: Astron. Astrophys. **359**, L23 (2000)
156. N. Masetti et al.: Astron. Astrophys. **374**, 382 (2001)
157. M.V. Medvedev, A. Loeb: Astrophys. J. **526**, 697 (1999)
158. P. Mészáros, M.J. Rees, H. Papathanassiou: Astrophys. J. **432**, 181 (1994)

159. P. Mészáros, M.J. Rees: Astrophys. J. **476**, 232 (1997)
160. P. Mészáros, M.J. Rees, R.A.M.J. Wijers: Astrophys. J. **499**, 301 (1998)
161. P. Mészáros, M.J. Rees: Mon. Not. R. Astron. Soc. **306**, L39 (1999)
162. P. Mészáros: Nucl. Phys. Proc. Suppl. **80**, 63 (2000)
163. M.R. Metzger, S.G. Djorgovski, S.R. Kulkarni, C.C. Steidel, K.L. Adelberger, D.A. Frail, E. Costa, F. Frontera: Nature **387**, 879 (1997)
164. M.R. Metzger, A. Fruchter, N. Masetti, E. Palazzi, E. Pian, S. Klose, B. Stecklum: GCN 733 (2000)
165. R. Moderski, M. Sikora, T. Bulik: Astrophys. J. **529**, 151 (2000)
166. V. Mohan et al.: GCN 1120 (2001)
167. S. Mukherjee, E.D. Feigelson, G. Jogesh Babu, F. Murtagh, C. Fraley, A. Raftery: Astrophys. J. **508**, 314 (1998)
168. S.C. Odewahn et al.: Astrophys. J. Lett. **509**, L5 (1998)
169. B. Paczyński: Astrophys. J. Lett. **494**, L45 (1998)
170. B. Paczyński: astro-ph 0108522 (2001)
171. E. Palazzi et al.: Astron. Astrophys. **336**, L95 (1998)
172. E. Palazzi, E. Pian: Astron. Nach. **322**, 275 (2001)
173. E. Palazzi et al.: in preparation (2003)
174. A. Panaitescu, P. Mészáros: Astrophys. J. **501**, 772 (1998)
175. A. Panaitescu, P. Mészáros, M.J. Rees: Astrophys. J. **503**, 314 (1998)
176. A. Panaitescu, P. Mészáros: Astrophys. J. **526**, 707 (1999)
177. A. Panaitescu, P. Kumar: Astrophys. J. **543**, 66 (2000)
178. A. Panaitescu, P. Kumar: Astrophys. J. **554**, 667 (2001)
179. A. Panaitescu: Astrophys. J. **556**, 1002 (2001)
180. A. Panaitescu, P. Kumar: Astrophys. J. **571**, 779 (2002)
181. H.S. Park et al.: Astrophys. J. Lett. **490**, L21 (1997)
182. H.S. Park et al.: Astron. Astrophys. Suppl. Ser. **138**, 577 (1999)
183. H. Pedersen, M. Andersen: In: *Gamma-Ray Bursts: 2nd Huntsville Symposium, AIP Conf. Proc. 307*, ed. G.J. Fishman (AIP, New York 1994) p. 670
184. H. Pedersen et al.: Astrophys. J. **496**, 311 (1998)
185. H. Pedersen et al.: ESO Messenger **100** 32 (2000)
186. R. Perna, A. Aguirre: Astrophys. J. **543**, 56 (2000)
187. R. Perna, A. Loeb: Astrophys. J. **501**, 467 (1998)
188. E. Pian et al.: Astrophys. J. Lett. **492**, L103 (1998)
189. E. Pian et al.: Astron. Astrophys. **372**, 456 (2001)
190. T. Piran: Phys. Rep. **314**, 575 (1999)
191. L. Piro et al.: Astron. Astrophys. **331**, L41 (1998)
192. L. Piro et al.: Science **290**, 955 (2000)
193. L. Piro et al.: Astrophys. J. **558**, 442 (2001)
194. L. Piro et al.: astro-ph 0201282 (2002)
195. P.A. Price, T.S. Axelrod, B.P. Schmidt: GCN 843 (2000)
196. P.A. Price et al.: Astrophys. J. Lett. **549**, L7 (2001)
197. P.A. Price et al.: Astrophys. J. Lett. **571**, 121 (2002)
198. A.N. Ramaprakash et al.: Nature **393**, 43 (1998)
199. E. Ramirez-Ruiz, E.E. Fenimore, N. Trentham: In: *Proc. of the CAPP2000 Conference on Cosmology and Particle Physics*, ed. J. Garcia-Bellido, R. Durrer, M. Shaposhnikov (AIP, New York 2001) p. 457
200. D.E. Reichart et al.: Astrophys. J. **517**, 692 (1999)
201. D.E. Reichart: Astrophys. J. Lett. **521**, L111 (1999)
202. D.E. Reichart: Astrophys. J. **553**, 235 (2001)

203. D.E. Reichart: Astrophys. J. **554**, 643 (2001)
204. J.E. Rhoads: Astrophys. J. **525**, 737 (1999)
205. J.E. Rhoads, A.S. Fruchter: Astrophys. J. **546**, 117 (2001)
206. J.E. Rhoads: Astrophys. J. **557**, 943 (2001)
207. E. Rol et al.: Astrophys. J. **544**, 707 (2000)
208. E. Rossi, D. Lazzati, M.J. Rees: Mon. Not. R. Astron. Soc. **332**, 945 (2002)
209. K.C. Sahu et al.: Astrophys. J. **540**, 74 (2000)
210. P. Saracco, G. Chincarini, S. Covino, G. Ghisellini, M. Longhetti, F. Zerbi, D. Lazzati, P. Severgnini: GCN 1010 (2001)
211. R. Sari, T. Piran, R. Narayan: Astrophys. J. Lett. **497**, L17 (1998)
212. R. Sari, T. Piran: Astrophys. J. **520**, 641 (1999)
213. R. Sari, T. Piran: Astrophys. J. Lett. **517**, L109 (1999)
214. R. Sari, T. Piran, J.P. Halpern: Astrophys. J. Lett. **519**, L17 (1999)
215. R. Sari, A. Esin: Astrophys. J. **548**, 787 (2001)
216. B.E. Schaefer: In: *Gamma-Ray Bursts: Observations, Analyses, and Theories*, ed. by C. Ho, R.I. Epstein, E.E. Fenimore (Cambridge Univ. Press, Cambridge 1992) p. 107
217. B.E. Schaefer et al.: Astrophys. J. Lett. **524**, L103 (1999)
218. P. Schechter: Astrophys. J. **203**, 297 (1976)
219. D.J. Schlegel, D.P. Finkbeiner, M. Davis: Astrophys. J. **500**, 525 (1998)
220. N. Shaviv, A. Dar: Astrophys. J. **447**, 863 (1995)
221. V. Simon, R. Hudec, G. Pizzichini, N. Masetti: Astron. Astrophys. **377**, 450 (2001)
222. A. Smette et al.: Astrophys. J. **556**, 70 (2001)
223. I.A. Smith et al.: Astron. Astrophys. **347**, 92 (1999)
224. V.V. Sokolov et al.: Astron. Astrophys. **372**, 438 (2001)
225. K.Z. Stanek, P.M. Garnavich, J. Kaluzny, W. Pych, I. Thompson: Astrophys. J. Lett. **522**, L39 (1999)
226. K.Z. Stanek et al.: Astrophys. J. **563**, 592 (2001)
227. B. Stecklum, O. Fischer, S. Klose, R. Mundt, C. Bailer-Jones: In: *Relativistic Astrophysics: 20th Texas Symposium, AIP Conf. Proc. 586*, ed. by J.C. Wheeler, H. Martel (AIP, New York 2001) p. 635
228. G.B. Taylor, J.S. Bloom, D.A. Frail, S.R. Kulkarni, S.G. Djorgovski, B.A. Jacoby: Astrophys. J. Lett. **537**, L17 (2000)
229. C. Tinney, R. Stathakis, R. Cannon, T. Galama: IAUC 6896 (1998)
230. J. van Paradijs et al.: Nature **386**, 686 (1997)
231. J. van Paradijs, C. Kouveliotou, R.A.M.J. Wijers: Astron. Astrophys. **38**, 379 (2000)
232. B.P. Venemans, A.W. Blain: Mon. Not. R. Astron. Soc. **325**, 1477 (2001)
233. M. Vietri, L. Stella: Astrophys. J. Lett. **507**, L45 (1998)
234. P.M. Vreeswijk et al.: Astrophys. J. **523**, 171 (1999)
235. P.M. Vreeswijk et al.: Astrophys. J. **546**, 672 (2001)
236. P.M. Vreeswijk et al.: GCN 496 (1999)
237. P.M. Vreeswijk, A. Fruchter, H. Ferguson, C. Kouveliotou: GCN 751 (2000)
238. X. Wang, A. Loeb: Astrophys. J. **535**, 788 (2000)
239. E. Waxman, B.T. Draine: Astrophys. J. **537**, 796 (2000)
240. N. Weinberg, C. Graziani, D.Q. Lamb, D.E. Reichart: In: *Gamma-Ray Bursts in the Afterglow Era, Proc. International Workshop at Rome, Italy, 17–20 October 2000*, ed. by E. Costa, F. Frontera, J. Hjorth (Springer, Heidelberg 2001) p. 252
241. R.A.M.J. Wijers, T.J. Galama: Astrophys. J. **523**, 177 (1999)

242. R.A.M.J. Wijers et al.: Astrophys. J. Lett. **523**, L33 (1999)
243. J.J.M. in 't Zand et al.: Astrophys. J. **559**, 710 (2001)
244. F.M. Zerbi et al.: Astron. Nach. **322**, 275 (2001)
245. B. Zhang, P. Mészáros: Astrophys. J. **571**, 876 (2002)
246. J. Zhu, H.T. Zhang: GCN 295 (1999)

Radio Observations of γ-Ray Burst Afterglows

Kurt W. Weiler[1], Nino Panagia[2], and Marcos J. Montes[3]

[1] Naval Research Laboratory, Code 7213, Washington, DC 20375-5320, USA
[2] Space Telescope Science Institute, 3700 San Martin Drive, Baltimore, MD 21218, USA and Astrophysics Division, Space Science Department of the European Space Agency
[3] Naval Research Laboratory, Code 7212, Washington, DC 20375-5320, USA

Abstract. Since 1997 the afterglow of γ-ray bursting sources (GRBs) has occasionally been detected in the radio, as well in other wavelengths bands. In particular, the interesting and unusual γ-ray burst GRB980425, thought to be related to the radio supernova SN1998bw, is a possible link between the two classes of objects. Analyzing the extensive radio emission data available for SN1998bw, one can describe its time evolution within the well established framework available for the analysis of radio emission from supernovae. This then allows relatively detailed description of a number of physical properties of the object. The radio emission can best be explained as the interaction of a mildly relativistic ($\Gamma \sim 1.6$) shock with a dense pre-explosion stellar wind-established circumstellar medium that is highly structured both azimuthally, in clumps or filaments, and radially, with observed density enhancements. Because of its unusual characteristics for a type Ib/c supernova, the relation of SN1998bw to GRB980425 is strengthened and suggests that at least some classes of GRBs originate in supernova massive star explosions. Thus, employing the formalism for describing the radio emission from supernovae and following the link through SN1998bw/GRB980425, it is possible to model the gross properties of the radio and optical/IR emission from the half-dozen GRBs with extensive radio observations. From this we conclude that at least some members of the slow-soft class of GRBs can be attributed to the explosion of a massive star in a dense, highly structured CSM that was presumably established by the pre-explosion stellar system.

1 SN1998bw/GRB980425

The suggestion of an association of the type Ib/c supernova (SN) SN1998bw with the γ-ray burster (GRB) GRB980425 may provide evidence for another new phenomenon generated by SNe – at least some types of GRBs may originate in some types of SN explosions. Because SN1998bw/GRB980425 is by far the nearest and best studied of the GRBs, it is worthwhile to examine its radio emission in detail before proceeding to the discussion of the radio emission from other GRBs.

1.1 Background

While generally accepted that most GRBs are extremely distant and energetic (see, e.g., [39,71]), the discovery of GRB980425 [96] on 25.90915 April 1998 and its possible association with a bright supernova, SN1998bw at RA(J2000) =

$19^h35^m03\overset{s}{.}31$, Dec(J2000) $= -52°50'44\overset{''}{.}7$ [103], in the relatively nearby spiral galaxy ESO184-G82 at z=0.0085(distance $\sim$ 40 Mpc for H_0=65 km s^{-1} Mpc^{-1}) [31,32,35,62,83,103,116], has introduced the possibility of a SN origin for at least some types of GRBs. The estimated explosion date of SN1998bw is between 21 – 27 April 1998 [83] corresponds rather well with the time of the GRB980425 outburst. Iwamoto et al. [48] felt that they could restrict the core collapse date for SN1998bw even more from hydrodynamical modeling of exploding C + O stars and, assuming that the SN1998bw optical light curve is energized by ^{56}Ni decay as in type Ia SNe, they then placed the coincidence between the core collapse of SN1998bw to within +0.7/−2 days of the outburst detection of GRB980425.

Classified initially as an SN optical type Ib [83], then type Ic [72], then peculiar type Ic [20,53], then later, at an age of 300 – 400 days, again as a type Ib [73], SN1998bw presents a number of optical spectral peculiarities that strengthen the suspicion that it may be the counterpart of the γ-ray burst.

When the more precise *BeppoSAX Narrow Field Instruments (NFI)* were pointed at the *BeppoSAX* error box 10 hours after the detection of GRB980425, two X-ray sources were present [74]. One of these, named S1 by Pian et al. [74], was coincident with the position of SN1998bw and declined slowly between April 1998 and November 1998. The second X-ray source, S2, that was $\sim$ 4$'$ from the position of SN1998bw, was not (or at best only marginally with a $< 3\sigma$ possible detection six days after the initial detection) detectable again in follow up observations in April, May, and November 1998 [74,75].

However, although concern remains that the Pian et al. [74,75] X-ray source, S2, might have been the brief afterglow from GRB980425 rather than the Pian source S1 associated with SN1998bw, Pian et al. [75] concluded that S1 has a "high probability" of being associated with GRB980425 and that S2 is more likely a variable field source.

1.2 Radio Emission

The radio emission from SN1998bw reached an unusually high 6 cm spectral luminosity at peak of $\sim 6.7 \times 10^{28}$ erg s^{-1} Hz^{-1}, i.e., $\sim$ 3 times higher than either of the well studied, very radio luminous type IIn SN1986J and SN1988Z, and $\sim$ 40 times higher than the average peak 6 cm spectral luminosity of type Ib/c SNe. It also reached this 6 cm peak rather quickly, only $\sim$ 13 days after explosion.

SN1998bw was unusual in its radio emission, but not extreme. For example, the time from explosion to peak 6 cm luminosity for both SN1987A and SN1983N was shorter and, in spite of the fact that SN1998bw has been called "the most luminous radio supernova ever observed," its 6 cm spectral luminosity at peak is exceeded by that of SN1982aa [117]. However, SN1998bw is certainly the most radio luminous type Ib/c radio supernova (RSN) observed so far by a factor of $\sim$ 25 and it reached this higher radio luminosity very early.

1.3 Expansion Velocity

Although unique in neither the speed of radio light curve evolution nor in peak 6 cm radio luminosity, SN1998bw is certainly unusual in the combination of these two factors – very radio luminous very soon after explosion. Kulkarni et al. [58] have used these observed qualities, together with the lack of interstellar scintillation (ISS) at early times, brightness temperature estimates, and physical arguments to conclude that the blastwave from SN1998bw that gives rise to the radio emission must have been expanding relativistically. On the other hand, Waxman and Loeb [109] argued that a sub-relativistic blastwave can generate the observed radio emission. However, both sets of authors agree that a very high expansion velocity ($\gtrsim 0.3c$) is required for the radio emitting region under a spherical geometry.

Simple arguments confirm this high velocity. To avoid the well known Compton Catastrophe, Kellermann and Pauliny-Toth [54] have shown that the brightness temperature $T_{\rm B} < 10^{12}$ K must hold and Readhead [78] has better defined this limit to $T_{\rm B} < 10^{11.5}$ K. From geometrical arguments, such a limit requires the radiosphere of SN1998bw to have expanded at an apparent speed $\gtrsim 230,000$ km s^{-1}, at least during the first few days after explosion. Although such a value is only mildly relativistic ($\Gamma \sim 1.6$; $\Gamma = (1 - \frac{v^2}{c^2})^{-\frac{1}{2}}$), it is still unusually high. However, measurements by Gaensler et al. [30] and Manchester et al. [63] have demonstrated that the radio emitting regions of the type II SN1987A expanded at an average speed of $\sim 35,000$ km s^{-1} ($\sim 0.1c$) over the 3 years from February 1987 to mid-1990 so that, in a very low density environment such as one finds around type Ib/c SNe, very high blastwave velocities appear to be possible.

1.4 Radio Light Curves

An obvious comparison of SN1998bw with other radio supernovae (RSNe) is the evolution of its radio flux density at multiple frequencies and its description by known RSN models (see the chapter by Sramek and Weiler in this volume). The radio data available from the *Australia Telescope Compact Array (ATCA)*[1] are plotted in Fig. 1. SN1998bw shows an early peak that reaches a maximum as early as day 10–12 at 8.64 GHz, a minimum almost simultaneously for the higher frequencies ($\nu \geq 2.5$ GHz) at day $\sim 20 - 24$, then a secondary, somewhat lower peak after the first dip. An interesting characteristic of this "double humped" structure is that it dies out at lower frequencies and is relatively inconspicuous in the 1.38 GHz radio measurements (see Fig. 1).

Such a double humped structure of the radio light curves can be reproduced by a single energy blastwave encountering differing circumstellar medium (CSM) density regimes as it travels rapidly outward. This is a reasonable assumption because complex density structure in the CSM surrounding SNe, giving rise to structure in the radio light curves, is very well known in such objects as SN1979C [68,110,111], SN1980K [67,112], and, particularly, SN1987A [49].

[1] See *http://www.narrabri.atnf.csiro.au/~mwiering/grb/GRB980425/*

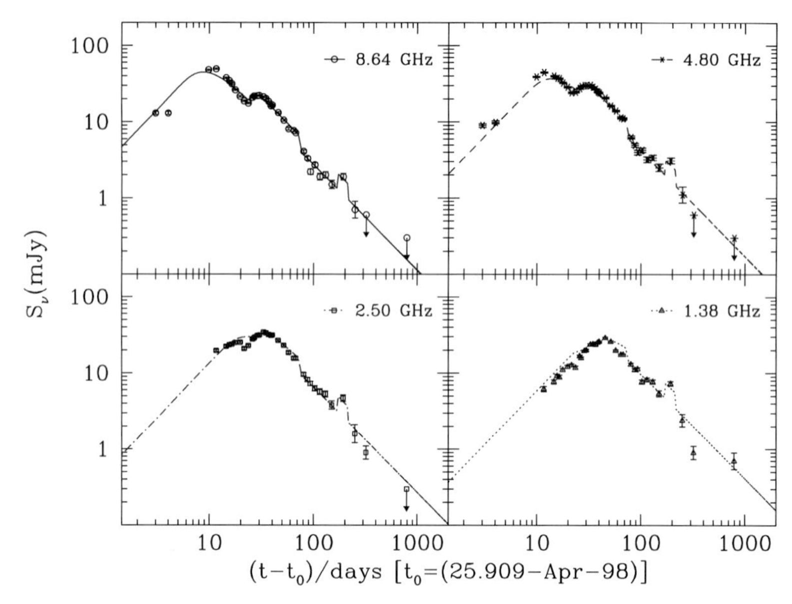

Fig. 1. The radio light curves of SN1998bw at 3.5 cm (8.6 GHz; upper left, *open circles, solid line*), 6.3 cm (4.8 GHz; upper right, *stars, dashed line*), 12 cm (2.5 GHz; lower left, *open squares, dash-dot line*) and 21 cm (1.4 GHz; lower right, *open triangles, dotted line*). The curves are derived from a best fit model described by the Eqs. (1–10) of Sramek and Weiler in this volume and the parameters and assumptions listed in Table 1. During the 50 day intervals from day 25 - 75 and from day 165 - 215 the emission and absorption terms (K_1 and K_3; see the chapter by Sramek and Weiler in, this volume) increase by factors of 1.6 and 2.0, respectively, corresponding to a density increase of 40% with a 6 day boxcar smoothed turn-on and turn-off of the enhanced emission/absorption.

Weiler et al. [114] pointed out what has not been previously recognized, that there is a sharp drop in the radio emission near day $\sim$ 75 and a single measurement epoch at day 192 that is significantly ($\sim$ 60%) higher at all frequencies than expected from the preceding data on day 149 and the following data on day 249.

Weiler et al. [114,115] were able to explain both of these temporary increases in radio emission by the SN blastwave encountering physically similar shells of enhanced density. The first enhancement or "bump" after the initial outburst peak is estimated to start on day 25 and end on day 75, i.e., having a duration of $\sim$ 50 days and turn-on and turnoff times of about 12 days, where the radio emission (K_1) increased by a factor of 1.6 and absorption (K_3) increased by a factor of 2.0 implying a density enhancement of $\sim$ 40% for no change in clump size. Exactly the same density enhancement factor and length of enhancement is compatible with the bump observed in the radio emission at day 192 (i.e., the single measurement within the 100 day gap between measurements on day 149 and day 249), even though the logarithmic time scale of Fig. 1 makes the time interval look much shorter. The decreased sampling interval has only one set of measurements altered by the proposed day 192 enhancement, so Weiler et al. [114] could not determine its length more precisely than < 100 days.

Li and Chevalier [61] proposed an initially synchrotron self-absorbed, rapidly expanding blastwave in a $\rho \propto r^{-2}$ circumstellar wind model to describe the radio light curve for SN1998bw. This is in many ways similar to the Chevalier [13] model for type Ib/c SNe, that also included synchrotron self-absorption (SSA). To produce the first bump in the radio light curves of SN1998bw, Li and Chevalier postulated a boost of blastwave energy by a factor of ~ 2.8 on day $\sim$22 in the observer's time frame. They did not discuss the second bump.

Modeling of the radio data for SN1998bw with the well established formalism for RSNe presented by [114,115] and by Sramek and Weiler in this volume, shows that such an energy boost is not needed. A fast blastwave interacting with a dense, slow, stellar wind-established, ionized CSM, that is modulated in density over time scales similar to those seen for RSNe, can produce a superior fit to the data. No blastwave re-acceleration is required and no SSA at early times is apparent. The parameters of the best fit model are given in Table 1 and shown as the curves in Fig. 1. A visual comparison of the curves in Fig. 1 with those of Li and Chevalier [61] Fig. 9, shows that the purely thermal absorption model with structured CSM provides a superior fit.

One should note that the fit listed in Table 1 and shown as the curves in Fig. 1 requires no "uniform" absorption ($K_2 = 0$) so all of the free-free (f-f) absorption is due to a clumpy medium as described in Eqs. (1) and (5) of Sramek and Weiler in this volume. These results, combined with the estimate of a high blastwave velocity, suggest that the CSM around SN1998bw is highly structured with little, if any, inter-clump gas. The clump filling factor must be high enough to intercept a considerable fraction of the blastwave energy and low enough to let radiation escape from any given clump without being appreciably absorbed by any other clump, which is Case 3 discussed by [114,115] and by Sramek and Weiler in this volume. The blastwave can then easily move at a speed that is a significant fraction of the speed of light, because it is moving in a very low density medium, but still cause strong energy dissipation and relativistic electron acceleration at the clump surfaces facing the SN explosion center.

Weiler et al. [114,115] also noted from the fit given in Table 1 that the presence of a K_4 (see also the chapter by Sramek and Weiler in this volume) factor implies there is a more distant, uniform screen of ionized gas surrounding the exploding system that is too far to be affected by the rapidly expanding blastwave and provides a time independent absorption.

1.5 Physical Parameter Estimates

Using the fitting parameters from Table 1 and Eqs. (11) and (16) from the chapter by Sramek and Weiler in this volume, Weiler et al. [114,115] estimated a mass-loss rate from the pre-explosion star. The proper parameter assumptions are rather uncertain for these enigmatic objects but, for a preliminary estimate, they assumed $t_{\rm i} = 23$ days, $t = (t_{\rm 6cm\ peak} - t_0) = 13.3$ days, $m = -(\alpha - \beta - 3)/3 =$

Table 1. SN1998bw/GRB980425 Modeling Results[a]

Parameter	Value
α (spectral index)	−0.71
β (decline rate)	−1.38
K_1^{b}	2.4×10^3
K_2	0
δ	–
K_3^{b}	1.7×10^3
δ'	−2.80
K_4	1.2×10^{-2}
t_0(Explosion Date)	25.90915 April 1998
$(t_{6\mathrm{cm\ peak}} - t_0)$(days)	13.3
$F_{6\mathrm{cm\ peak}}$(mJy)	37.4
d(Mpc)	38.9
$L_{6\mathrm{cm\ peak}}$ (erg s^{-1} Hz^{-1})	6.7×10^{28}
$\dot{M}$(M$_\odot$ yr^{-1})[c]	2.6×10^{-5}

[a]See the chapter by Sramek and Weiler in this volume for parameter definitions.
[b]Enhanced by a factors of 1.6 (K_1) and 2.0 (K_3), corresponding to a density increase of 40%, over the intervals day 25 – 75 and day 165 – 215, although the latter interval could be as long as 100 days and still be compatible with the available data.
[c]Assuming $t_{\mathrm{i}} = 23$ days, $t = (t_{6\mathrm{cm\ peak}} - t_0) = 13.3$ days, $m = -(\alpha - \beta - 3)/3 = 0.78$, $w_{\mathrm{wind}} = 10$ km s^{-1}, $v_{\mathrm{i}} = v_{\mathrm{blastwave}} = 230,000$ km s^{-1}, $T = 20,000$ K, average number of clumps along the line-of-sight $N = 0.5$, and volume filling factor $\phi = 0.22$ (see the chapter by Sramek and Weiler in this volume, Case 3).

0.78 (Eq. 7 of Sramek and Weiler in this volume), $w_{\mathrm{wind}} = 10$ km s^{-1} (for an assumed red supergiant (RSG) progenitor), $v_{\mathrm{i}} = v_{\mathrm{blastwave}} = 230,000$ km s^{-1}, and $T = 20,000$ K. They also assumed, because the radio emission implies that the CSM is highly clumped (i.e., $K_2 = 0$), that the CSM volume is only sparsely occupied ($N = 0.5$, $\phi = 0.22$; see the chapter by Sramek and Weiler in this volume, Case 3). Within these rather uncertain assumptions, Eqs. (11) and (16) from Sramek and Weiler in this volume yield an estimated mass-loss rate of $\dot{M} \sim 2.6 \times 10^{-5}$ M$_\odot$ yr^{-1} with density enhancements of $\sim 40\%$ during the two known, extended bump periods.

Assuming that the blastwave is traveling at a constant speed of $\sim 230,000$ km s^{-1} the beginning of the first bump on day 25 implies that it starts at $\sim 5.0 \times 10^{16}$ cm and ends on day 75 at $\sim 1.5 \times 10^{17}$ cm from the star. Correspondingly, if it was established by a 10 km s^{-1} RSG wind, the 50 days of enhanced mass-loss ended $\sim 1,600$ yr and started $\sim 4,700$ yr before the explosion. The earlier high mass-loss rate epoch indicated by the enhanced emission on day 192 in the measurement gap between day 149 and day 249 implies, with the same assumptions, that it occurred in the interval between $\sim 9,400$ yr and $\sim 15,700$ yr before explosion. It is interesting to note that the time between the presumed centers of the first and second increased mass-loss episodes of $\sim 9,400$ yr is comparable to the $\sim 12,000$ yr before explosion at which SN1979C had a significant mass-loss rate increase [68] and SN1980K had a significant mass-loss rate decrease [67], thus establishing a possible characteristic time scale of $\sim 10^4$ yr for significant changes in mass-loss rate for pre-explosion massive stars.

1.6 Radio Emission from SN1998bw/GRB980425 and Other GRB Radio Afterglows

Because of the suggestion of a possible relation between SN1998bw and GRB980425, it has remained a tantalizing possibility that the origin of at least some GRBs is in the better known type Ib/c SN phenomenon. First, of course, one must keep in mind that there may be (and probably is) more than one origin for GRBs, a situation that is true for most other classes of objects. For example, SNe, after having been identified as a new phenomenon in the early part of the last century, were quickly split into several subgroups such as Zwicky's types I–V, then coalesced back into just two subgroupings based on Hα absent (type I) or Hα present (type II) in their optical spectra. This simplification has not withstood the test of time, however, and subgroupings of type Ia, type Ib, type Ic, type II, type IIb, type IIpec, type IIn, and others have come into use over the past 20 years (see the chapter by Turatto in this volume).

GRBs, although at a much earlier stage of understanding, have similarly started to split into subgroupings. The two currently accepted groupings are referred to as "fast-hard" and "slow-soft" from the tendency of the γ-ray emission for some to evolve more rapidly (mean duration $\sim$0.2 s) and to have a somewhat harder spectrum than for others that evolve more slowly (mean duration $\sim$20 s) with a somewhat softer spectrum [21].

Because we are only concerned with the radio afterglows of GRBs here, all of our examples fall into the slow-soft classification, at least partly because the fast-hard GRBs fade too quickly for followup observations to obtain the precise positional information needed for identification at longer wavelengths. It is therefore uncertain whether fast-hard GRBs have radio afterglows or even whether the slow-soft GRBs represent a single phenomenon. If, however, we assume that at least some types of slow-soft GRBs have a similar origin and that GRB980425/SN1998bw is a key to this puzzle telling us that such slow-soft GRBs have their origin in at least some types of SNe, we can investigate relations between the two observational phenomena.

2 Gamma-Ray Bursters

2.1 Radio Detections

Beyond GRB980425, there are relatively few GRBs with detected radio afterglows and only six of these as of 31 December 2001 had sufficient radio light curve information to permit approximate model fits. Additionally, because the optical/infrared (OIR) data appear consistent with a synchrotron origin similar to that of the radio emission, we have collected the available OIR data and included it in our model fitting to better constrain the source parameters. Although detailed modeling is beyond the scope of this review, we apply the parameterization of Eqs. (1–10) of Sramek and Weiler in this volume to the available radio and OIR data in an attempt to highlight some of the gross properties of the GRB afterglow processes. Because the OIR data suffer an extinction that is absent in the radio, a zero redshift color excess $E(B-V)$ was also obtained from the fitting by adopting the Galactic law of Savage and Mathis [86].

GRB970508 was discovered by the *BeppoSAX* team on 8.904 May 1997 UT [16]. These results showed detection of an afterglow in all wavelength bands including X-ray [76], optical [5], and radio [22]. Frail and Kulkarni [22] derived a position from their 8.46 GHz *Very Large Array (VLA)*[2] observations of RA(J2000) $= 06^h53^m49\overset{s}{.}45$, Dec(J2000) $= +79°16'19\overset{''}{.}5$ with an error of $\sim 0\overset{''}{.}1$ in both coordinates. Metzger et al. [65] found a redshift of $z = 0.835$, which Bloom et al. [4] confirmed for the host galaxy.

The radio data were obtained from the references [6,26,33,90,94] and the OIR data were obtained from the references [8,12,18,33,34,37,85,87,97]. Representative data for GRB970508 are plotted, along with curves from the best fit model, in Plate 1 and the parameters of the fit are listed in Table 2.

Examination of Plate 1 shows that the parameterization listed in Table 2 describes the data well in spite of the very large frequency and time range. In Plate 1, top panel, the 232 GHz upper limits and the 86.7 GHz limits and detections are in rough agreement with the model fitting; the 8.5 GHz and 4.9 GHz measurements are described very well, and even the 1.5 GHz data are consistent with the parameterization if significant ISS is present (see Sect. 2.3). Waxman et al. [108] have already ascribed the large fluctuations in the flux density at both 8.5 and 4.9 GHz to ISS, and one expects such ISS to also be present at 1.4 GHz.

In Plate 1, bottom panel, although the OIR data show significant scatter at individual frequencies, the data are consistent with a non-thermal, synchrotron origin that has the same decline rate (β) and spectral index (α) as in the radio regime, indicating that no spectral breaks have occurred between the two observing ranges. The color excess of $E(B-V) = 0.08$ mag from Schlegel et al.

[2] The *VLA* telescope of the National Radio Astronomy Observatory is operated by Associated Universities, Inc. under a cooperative agreement with the National Science Foundation.

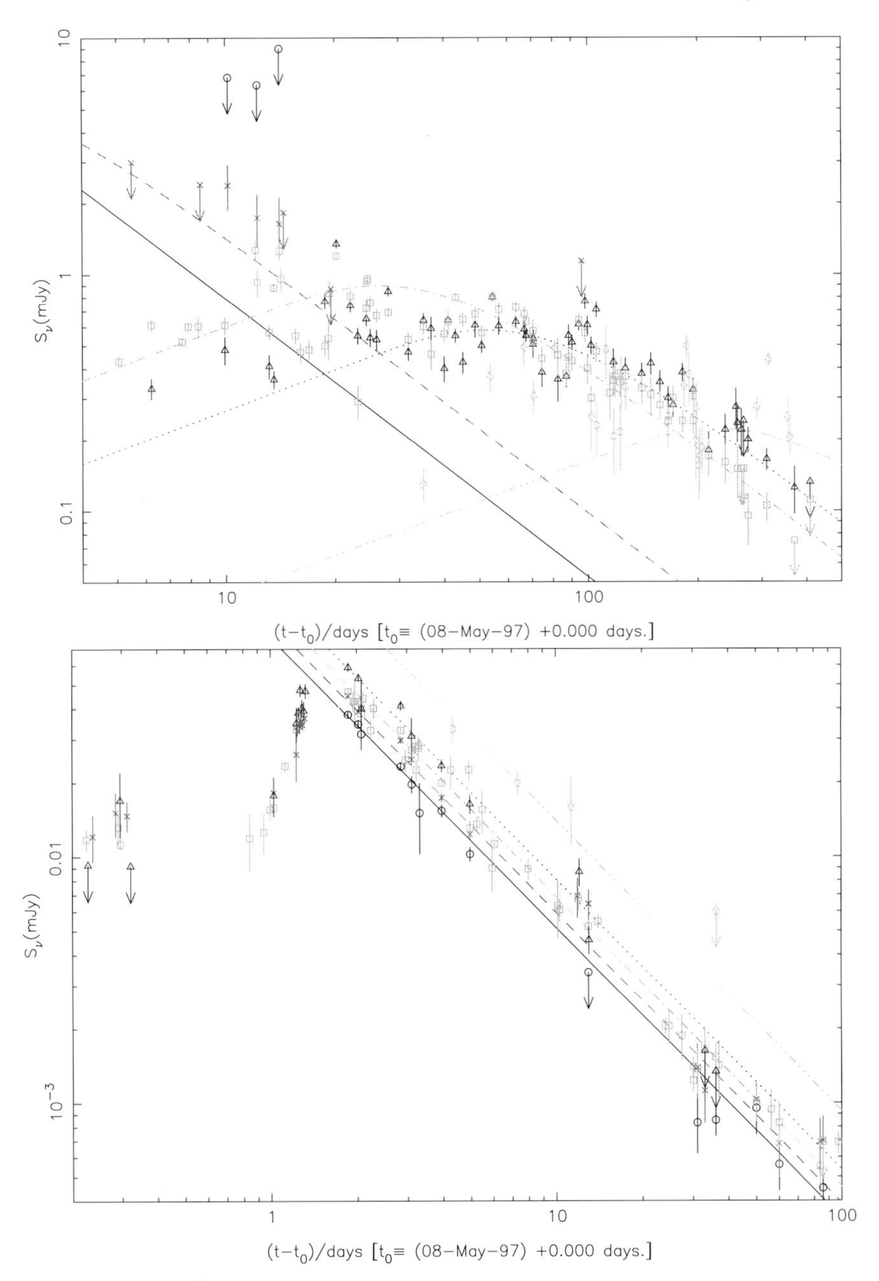

Plate 1. GRB970508 at radio wavelengths (top panel) of 1.3 mm (232 GHz; *open circles, solid line*), 3.5 mm (86.7 GHz; *crosses, dashed line*), 3.5 cm (8.5 GHz; *open squares, dash-dot line*), 6 cm (4.9 GHz; *open triangles, dotted line*), and 20 cm (1.5 GHz; *open diamonds, dash-triple dot line*) and optical/IR bands (bottom panel) of B (681 THz; *circles, solid line*), V (545 THz; *crosses, dashed line*), R (428 THz; *open squares, dash-dot line*), I (333 THz; *open triangles, dotted line*), and K (138 THz; *open diamonds, dash-triple dot line*). (Note that to enhance clarity, the scales are different between the two figures even though the fitting parameters are the same and not all available bands are plotted even though they were used in the fitting.)

Table 2. Fitting Parameters for GRB Radio Afterglows[a]

GRB	$E(B-V)$ Milky Way[b] (mag)	Host Galaxy[c] (mag)	Red-shift[d] (z)	Spectral Index (α)	Decline Rate (β)	K_1^{e}	K_3^{e}	δ'	Peak 6 cm Radio Luminosity[e] ($erg\,s^{-1}Hz^{-1}$)
GRB970508	0.08	0.00	0.835	−0.63	−1.18	1.35×10^{2}	1.81×10^{3}	−1.75	1.39×10^{31}
GRB980329	0.07	0.28	≡ 1.000	−1.33	−1.09	1.54×10^{4}	1.33×10^{5}	−1.16	7.69×10^{30}
GRB980425[f]	-	-	0.0085	−0.71	−1.38	2.37×10^{3}	1.73×10^{3}	−2.80	6.70×10^{28}
GRB980519	0.27	0.00	≡ 1.000	−0.75	−2.08	8.45×10^{1}	1.37×10^{4}	−3.57	7.52×10^{30}
GRB991208	0.02	0.07	0.707	−0.58	−2.27	5.11×10^{2}	3.53×10^{3}	−3.29	2.07×10^{31}
GRB991216	0.63	0.11	1.020	−0.28	−1.38	7.07×10^{0}	2.50×10^{1}	−1.40	1.05×10^{31}
GRB000301C[g]	0.05	0.05	2.034	−0.60	≡ −1.75	2.31×10^{2}	2.72×10^{3}	−2.06	3.77×10^{31}

[a]The fits do not generally require use of $K_2, \delta, K_4, K_5, \delta'', K_6, \delta'''$ (see [114,115] and the chapter by Sramek and Weiler in this volume) so that these parameters are not included.
[b]Galactic extinction in the direction of the γ-ray burster (GRB) was obtained from [43,79,89].
[c]Additional extinction, at the host galaxy redshift (see Sect. 2.1), is needed in some cases to provide a good fit to the OIR data.
[d]Where unknown, the redshift is defined to be $z = 1.000$.
[e]Derived for 6 cm in the rest frame of the observer, not the emitter.
[f]The best fit includes a $K_4 = 1.24 \times 10^{-2}$ term.
[g]The OIR data and radio data appear to have different rates of decline ($\beta_{\rm OIR} \sim -2.0$, $\beta_{\rm radio} \sim -1.5$) implying that there may be a break between the two regimes. For the purposes of this review the average of $\beta \equiv -1.75$ has been adopted.

[89] is consistent with the best fit to the data, implying little extinction in the host galaxy.

The most obvious characteristic of the OIR data is that, in the initial interval between the time of the GRB and day 1.75, the data are not well modeled by the parameterization. This is not surprising, because the modeling contains no parameters to describe any turn-on effects for the synchrotron emission. Thus, the initial OIR data prior to day 1.75 are particularly suitable for constraining explosion models.

GRB980329 was discovered by the *BeppoSAX* team and also by the *Burst and Transient Source Experiment (BATSE)* (*BATSE* Trigger # 6665 [7]) on 29.1559 March 1998 UT [28]. The afterglow was detected in all wavelength bands including X-ray [119], optical [57], and radio [99]. Taylor et al. [99] derived a position from their *VLA* observations of RA(J2000) = $07^h02^m38^s.022$, Dec(J2000) = $+38°50'44''.02$ with an uncertainty of $\pm0''.05$ in each coordinate. Unfortunately, no redshift has been obtained for the optical afterglow of GRB980329 or its inferred parent galaxy so that, except for arguments that is quite distant with, perhaps, z ∼ 5, no reliable distance estimate is available.

The radio data were obtained from [93,94,100] and the OIR data were obtained from [41,79]. Representative data for GRB980329 are plotted, along with curves from the best fit model, in Plate 2 and the parameters of the fit are listed in Table 2.

Examination of Plate 2 shows that the parameterization listed in Table 2 describes the data rather well over the broad parameter space in time and frequency. In Plate 2, top panel, the 352 GHz detections and upper limits are in good agreement with the parameterization; the 8.5 GHz and 4.9 GHz measurements are described very well although there may be some ISS present (see Sect. 2.3); the 1.4 GHz upper limits are consistent with the parameterization.

In Plate 2, bottom panel, the data are mostly upper limits, with only a few detections at K-band (136 THz) and R-band (428 THz). However, these are surprisingly well described by the parameterization if a zero redshift color excess of $E(B-V) = 0.75$ mag is assumed. This value for color excess is much higher than the value of $E(B-V) = 0.073$ mag from Reichart et al. [79] for Galactic color excess in that direction. However, if we assume that the additional color excess arises in the host galaxy and that the color excess scales with redshift as $(1+z)^{-1.25}$, as is appropriate for an adopted *Small Magellanic Cloud (SMC)* extinction law [77], then the host galaxy contributes $E(B-V) = 0.28$ mag (Table 2). Such a color excess is fairly typical for the disk of a late-type galaxy.

Taylor et al. [100] have also modeled the radio data and invoked a somewhat steeper, inverted spectrum with $\alpha = -1.7$ between 4.9 and 8.3 GHz, flattening to $\alpha = -0.8$ between 15 and 90 GHz caused by a SSA component with a turnover frequency near 13 GHz. Extrapolating to higher frequencies, their model predicts a rather low 350 GHz flux density of only $\sim$1.7 mJy that is incompatible with the *James Clerk Maxwell Telescope (JCMT)* measurements.

Such complexity is not be needed, however. Our parameterization listed in Table 2 and shown in Plate 2, top panel, yields a good description of the data and predicts a 350 GHz flux density of $\sim$3.0 mJy, in much better agreement with the observations.

Our model fit to the available first 30 days of radio data indicates that GRB980329 should be detectable with the *VLA* at centimeter wavelengths, with little decline, for an extended period. Although prediction of exact flux densities at later times is not reliable because the decline phase of the model is not well constrained by the few available data, it is interesting to note that GRB980329 has apparently been detected by the *VLA* at 8.46 GHz, 4.86 GHz, and 1.43 GHz for up to 500 days after the outburst [118]. Unfortunately, those data have not been made available to us.

GRB980519 was discovered by the *BeppoSAX* team (*BATSE* Trigger # 6764) on 19.51410 May 1998 UT [69]. The afterglow was detected in all wavelength bands including X-ray [70], optical [50], and radio [23]. Frail et al. [23] derived a position from their *VLA* observations of RA(J2000) = $23^h22^m21\overset{s}{.}49$, Dec(J2000) = $+77°15'43\overset{''}{.}2$ with an uncertainty of $\pm0\overset{''}{.}1$ in each coordinate. Unfortunately, no redshift has been obtained for the optical afterglow so that no distance estimate is available.

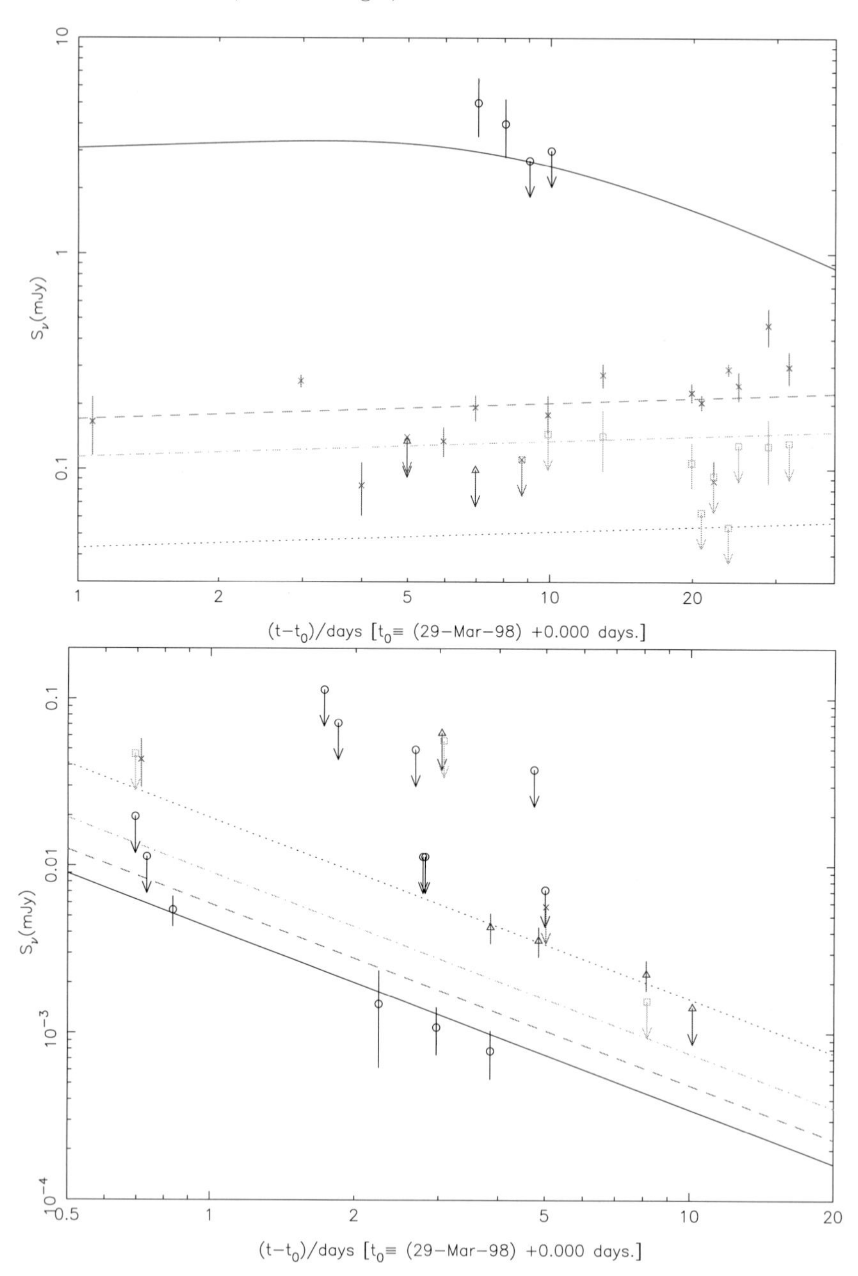

Plate 2. GRB980329 at radio wavelengths (top panel) of 0.9 mm (352 GHz; *open circles, solid line*), 3.5 cm (8.5 GHz; *crosses, dashed line*), 6 cm (4.9 GHz; *open squares, dash-dot line*), and 20 cm (1.4 GHz; *open triangles, dotted line*) and optical/IR bands (bottom panel) of R (428 THz; *open circles, solid line*), I (333 THz; *crosses, dashed line*), J (240 THz; *open squares, dash-dot line*), and K (136 THz; *open triangles, dotted line*). (Note that to enhance clarity, the scales are different between the two figures even though the fitting parameters are the same and not all available bands are plotted even though they were used in the fitting.)

The radio data were obtained from [26] and the OIR data from [42,51,105] Representative data for GRB980519 are plotted, along with curves from the best fit model, in Plate 3 and the parameters of the fit are listed in Table 2.

Examination of Plate 3 shows that the parameterization listed in Table 2 describes the data reasonably well, even though the radio data shown in Plate 3, top panel, are very limited and of relatively poor quality. For example, the reported detections at 4.9 GHz after day 50 are barely more than 3σ and thus of limited reliability, so that the fit is not well constrained and the significance of their deviation from the best fit curve is unknown. The data at both 8.5 GHz and 4.9 GHz have significant fluctuations yielding detections and 3σ upper limits at relatively small time separations so that GRB980519 may be undergoing ISS (see Sect. 2.3). Only upper limits are available at 1.4 GHz, but they are consistent with the best fit model.

In Plate 3, bottom panel, there is good coverage in both frequency and time. Although there is some fluctuation, the data are rather well described by the parameterization if a color excess of $E(B-V) = 0.25$ mag is assumed. This is rather close to the value of $E(B-V) = 0.267$ mag suggested by Schlegel et al. [89] in that direction from their *IRAS* 100 μm maps.

GRB991208 was discovered by the *Ulysses* and *KONUS* and *NEAR* teams on 08.1923 December 1999 UT [47]. The afterglow was detected in the optical [9] and radio [24] wavelength bands. There does not appear to have been an X-ray detection. Frail and Kulkarni [24] derived an position for the radio transient of RA(J2000) = $16^h33^m53\overset{s}{.}50$, Dec(J2000) = $+46°27'20\overset{''}{.}9$ with no error given, but normally *VLA* positional errors are $< 0\overset{''}{.}1$. Dodonov et al. [19] found a redshift for the parent galaxy of $z = 0.707 \pm 0.002$.

The radio data were obtained from Galama et al. [36] and the OIR data from Castro-Tirado et al. [11]. Representative data for GRB991208 are plotted, along with curves from the best fit model, in Plate 4 and the parameters of the fit are listed in Table 2.

Examination of Plate 4 shows that the parameterization listed in Table 2 describes the data reasonably well. In Plate 4, top panel, the radio data are quite limited and have relatively large scatter. For example, the reported detection at 15 GHz on day 7.4 is preceded by a 3σ upper limit of comparable magnitude on day 5.4 and followed by an upper limit on day 12.5, so that the reality of the day 7.4 detection must be called into question. In any case, although the fit is not well constrained by the radio data, the model curves describe its evolution reasonably well. The data are too sparse to judge if ISS is present.

In Plate 4, bottom panel, there is good coverage in both frequency and time. There is some fluctuation, particularly for the four R-band observations on day 4. However, examination of the data reveals that the three high R-band measurements on day 4.026, 4.058, and 4.096 have significantly larger errors (~ 0.4 mag) than the R-band datum on day 4.068 (error = 0.050 mag) that is consistent with the fitted curves. In general, the data are rather well described by the parameterization if a zero redshift color excess of $E(B-V) = 0.15$ mag is

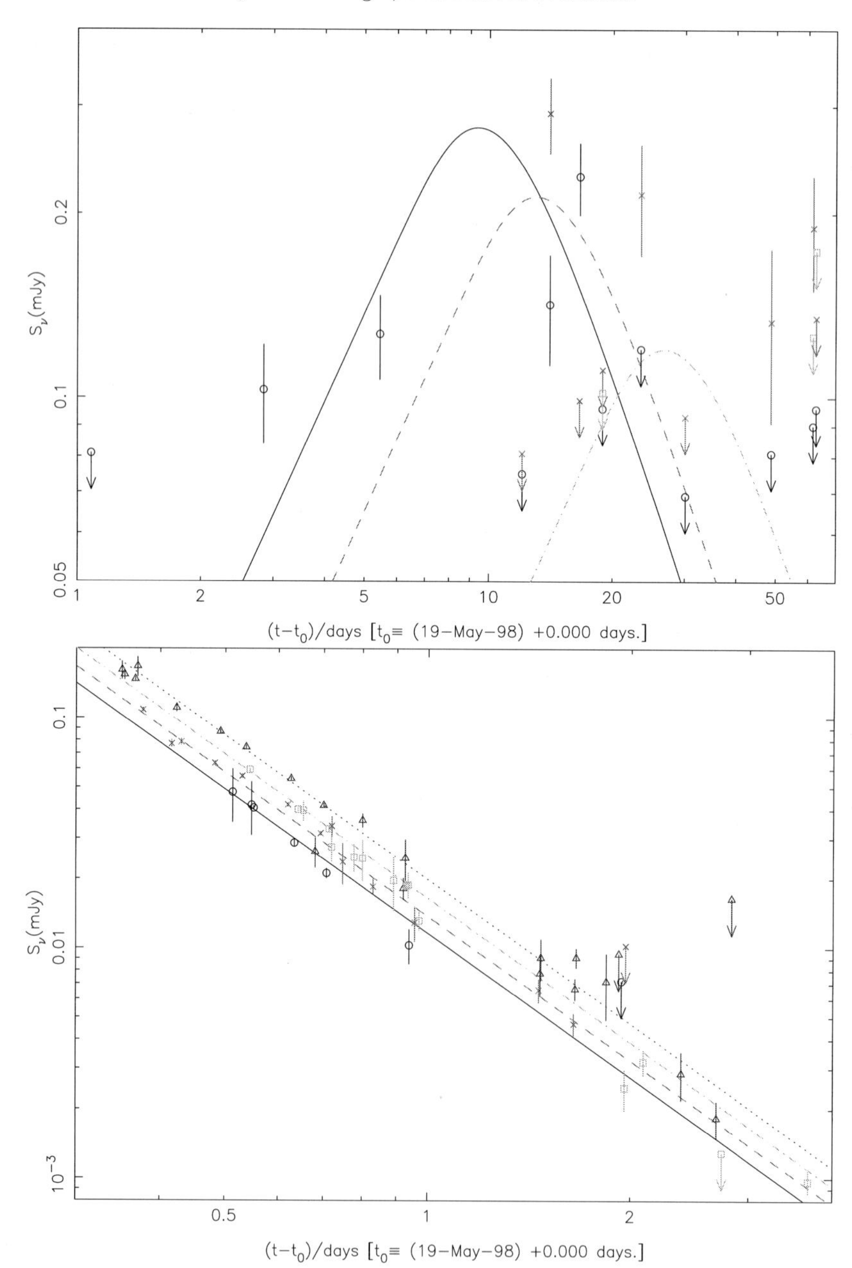

Plate 3. GRB980519 at radio wavelengths (top panel) of 3.5 cm (8.5 GHz; *open circles, solid line*), 6 cm (4.9 GHz; *crosses, dashed line*), and 20 cm (1.4 GHz; *open squares, dash-dot line*) and optical/IR bands (bottom panel) of B (681 THz; *open circles, solid line*), V (545 THz; *crosses, dashed line*), R (428 THz; *open squares, dash-dot line*), and I (333 THz; *open triangles, dotted line*). (Note that to enhance clarity, the scales are different between the two figures even though the fitting parameters are the same and not all available bands are plotted even though they were used in the fitting.)

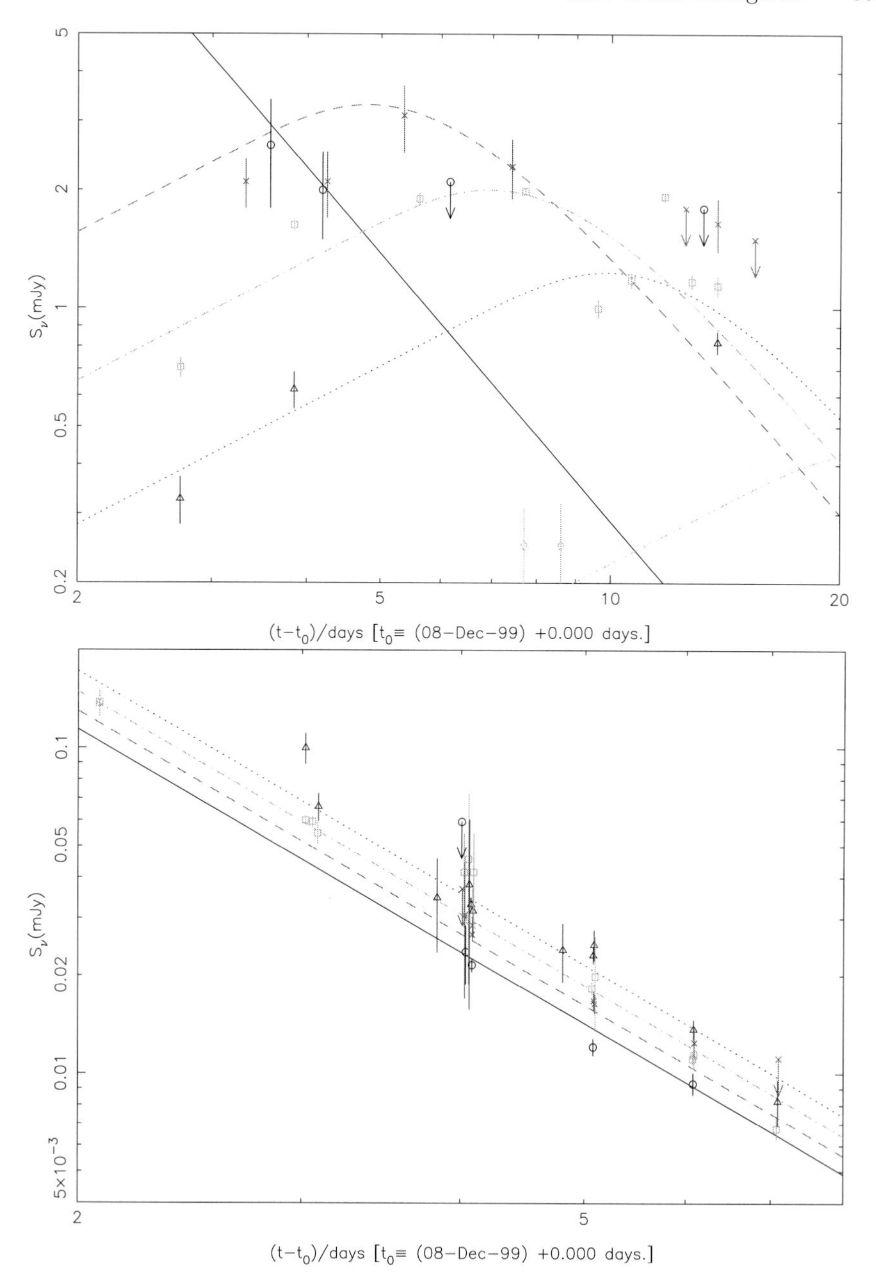

Plate 4. GRB991208 at radio wavelengths (top panel) of 1.2 mm (250 GHz; *open circles, solid line*), 2.0 cm (15.0 GHz; *crosses, dashed line*), 3.5 cm (8.5 GHz; *open squares, dash-dot line*), 6 cm (4.9 GHz; *open triangles, dotted line*), and 20 cm (1.4 GHz; *open diamonds, dash-triple dot line*) and optical/IR bands (bottom panel) of B (681 THz; *open circles, solid line*), V (545 THz; *crosses, dashed line*), R (428 THz; *open squares, dash-dot line*), and I (333 THz; *open triangles, dotted line*). (Note that to enhance clarity, the scales are different between the two figures even though the fitting parameters are the same and not all available bands are plotted even though they were used in the fitting.)

assumed. This is appreciably higher than the value of $E(B-V) = 0.016$ mag suggested by Schlegel et al. [89] in that direction from their *IRAS* 100 μm maps, implying, under the same assumptions as were used for GRB980329 above, a $E(B-V) = 0.07$ mag at the redshift of the host galaxy (Table 2).

GRB991216 was discovered by *BATSE* on 16.671544 December 1999 UT (*BATSE* trigger # 7906) [56]. The afterglow was detected in all wavelength bands including X-ray [98], optical [104], and radio [101]. Taylor and Berger [101] derived a position from their 8.5 GHz *VLA* observations of RA(J2000) = $05^h09^m31\overset{s}{.}297$, Dec(J2000) = $+11°17'07\overset{''}{.}25$ with an error of $\pm0\overset{''}{.}1$ in each coordinate. Vreeswijk et al. [106] suggested a redshift of $z \geq 1.02$ based on the highest redshift of three possible absorption systems seen in the optical afterglow.

The radio data were obtained from Frail et al. [27] and the OIR data from Halpern et al. [43], Frail et al. [27], and Garnavich et al. [38]. Representative data for GRB991216 are plotted, along with curves from the best fit model, in Plate 5 and the parameters of the fit are listed in Table 2.

Examination of Plate 5 shows that the parameterization listed in Table 2 fits the data reasonably well, even with the radio data (Plate 5, top panel) being quite limited and with relatively large scatter. The single detection at 8.5 GHz on day 42.49 surrounded by much lower 3σ upper limits on days 38.28 and 50.51, if correct, is not well described. The data are too sparse to judge if ISS is observed.

In Plate 5, bottom panel there is reasonably good coverage in both frequency and time, particularly at R-band. A satisfactory fit to the data requires a rather high zero redshift color excess of $E(B-V) = 0.90$ mag. This is significantly higher than the value of $E(B-V) = 0.63$ mag suggested by Schlegel et al. [89] in that direction from their *IRAS* 100 μm maps or the value of $E(B-V) = 0.40$ mag derived by Halpern et al. [43] from the Galactic 21 cm column density in that direction. If, as above, we assume that the additional extinction arises in the host galaxy, then with the same assumptions as were used for GRB980329, we obtain a color excess of $E(B-V) = 0.11$ mag at the redshift of the host galaxy (Table 2).

GRB000301C was discovered by the *Ulysses* and *NEAR* teams on 01.4108 March 2000 UT [95]. The afterglow was detected in the optical [29] and radio [2] wavelength bands. There does not appear to have been an X-ray detection. Fynbo et al. [29] derived a position for the optical transient of RA(J2000) = $16^h20^m18\overset{''}{.}6$, Dec(J2000) = $+29°26'36''$ with an error of $\sim 1''$ in both coordinates. Castro et al. [10] (see also [91]) found a redshift for the parent galaxy of $z = 2.0335 \pm 0.003$.

The radio data were obtained from Berger et al. [1] and the OIR data from Sagar et al. [84], Bhargavi and Cowsik [3], Masetti et al. [64], Rhoads and Fruchter [81], and Jensen et al. [52]. Representative data for GRB000301C are plotted, along with curves from the best fit model, in Plate 6 and the parameters of the fit are listed in Table 2.

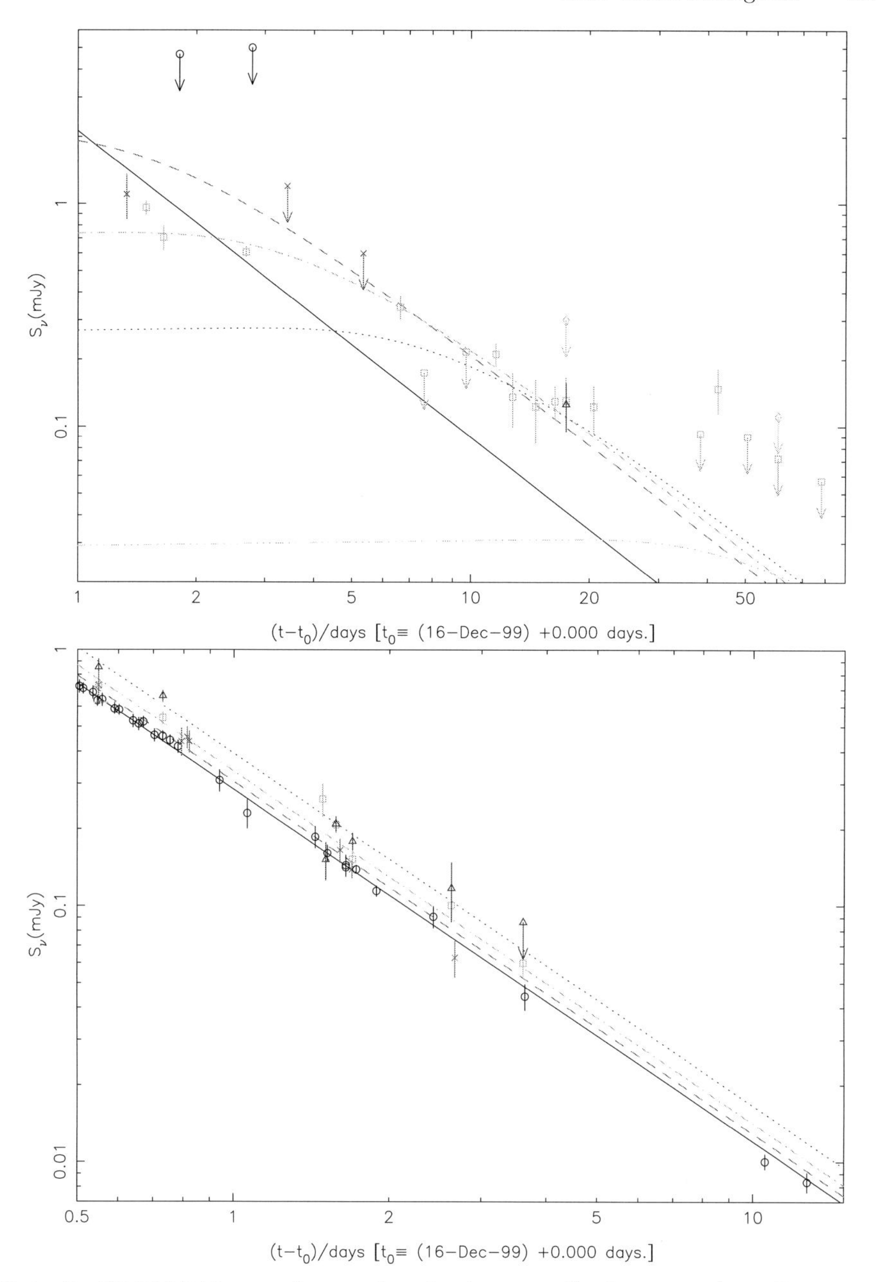

Plate 5. GRB991216 at radio wavelengths (top panel) of 0.9 mm (350 GHz; *open circles, solid line*), 2.0 cm (15.0 GHz; *crosses, dashed line*), 3.5 cm (8.5 GHz; *open squares, dash-dot line*), 6 cm (4.9 GHz; *open triangles, dotted line*),and 20 cm (1.4 GHz; *open diamonds, dash-triple dot line*), and optical/IR bands (bottom panel) of R (428 THz; *open circles, solid line*), I (333 THz; *crosses, dashed line*), J (239 THz; *open squares, dash-dot line*), and K (136 THz; *open triangles, dotted line*). (Note that to enhance clarity, the scales are different between the two figures even though the fitting parameters are the same and not all available bands are plotted even though they were used in the fitting.)

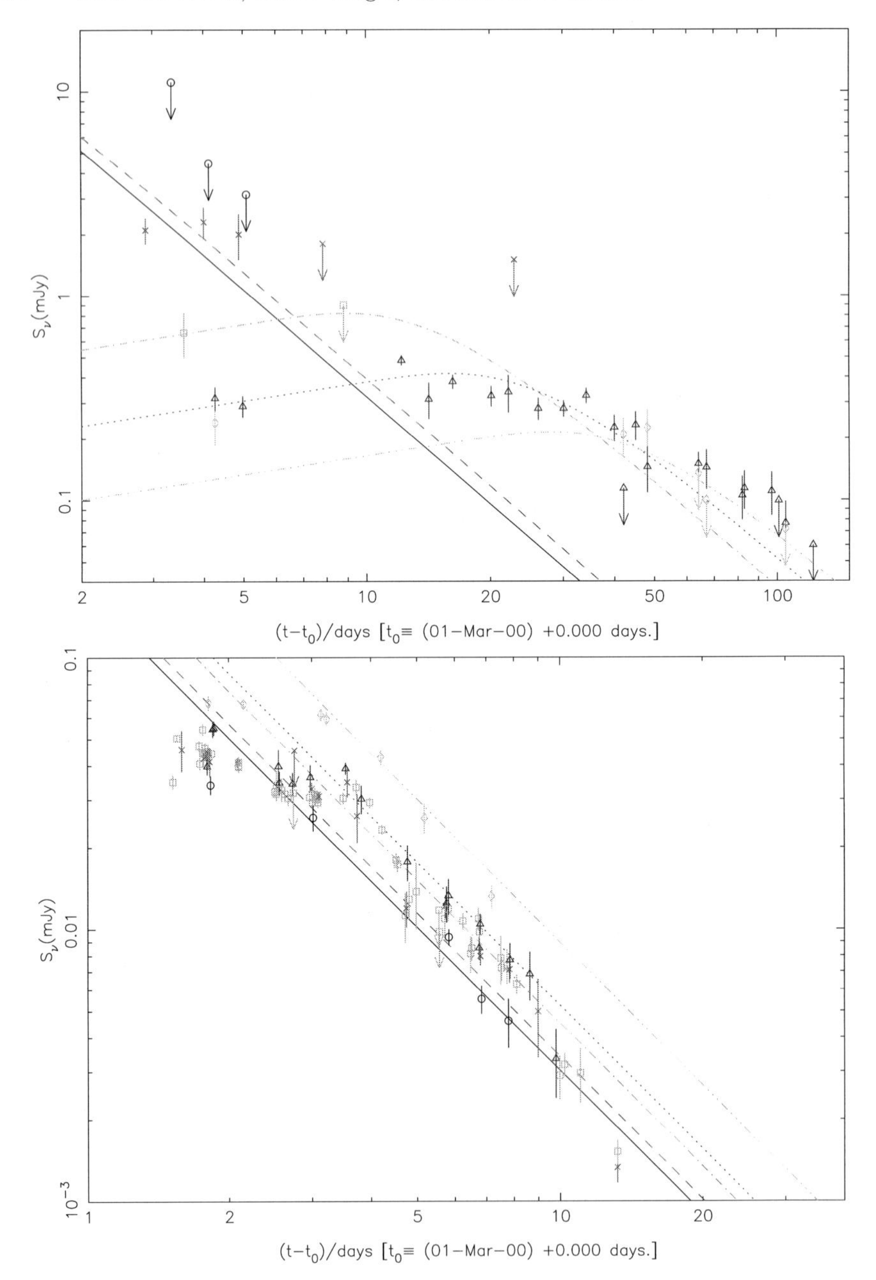

Plate 6. GRB000301C at radio wavelengths (top panel) of 0.9 mm (350 GHz; *open circles, solid line*), 1.2 mm (250 GHz; *crosses, dashed line*), 2.0 cm (15.0 GHz; *open squares, dash-dot line*), 3.5 cm (8.5 GHz; *open triangles, dotted line*), and 6.0 cm (4.9 GHz; *open diamonds, dash-triple dot line*) and optical/IR bands (bottom panel) of U (832 THz; *circles, solid line*), B (681 THz; *crosses, dashed line*), R (428 THz; *open squares, dash-dot line*), I (333 THz; *open triangles, dotted line*), and K (136 THz; *open diamonds, dash-triple dot line*). (Note that to enhance clarity, the scales are different between the two figures even though the fitting parameters are the same and not all available bands are plotted even though they were used in the fitting.)

Examination of Plate 6 shows that the parameterization listed in Table 2 is successful in that it describes the data rather well over the large frequency and time range. In Plate 6, top panel, the 350 GHz upper limits and the 250 GHz detections and limits are in rough agreement with the parameterization; the 15 GHz and 8.5 GHz measurements are also consistent with the modeling, although the model drops off a bit faster at late times than the data and the fit would be improved by a somewhat flatter decline rate of $\beta \sim -1.5$. The 1.4 GHz data are generally consistent with the model, although ISS (see Sect. 2.3) may be present in the first measurement.

In Plate 6, bottom panel, although the OIR data show significant scatter at individual frequencies, they are consistent with a non-thermal, synchrotron origin. A decline rate of $\beta \sim -2.0$, somewhat steeper than in the radio, would provide an improved fit. The fit indicates a best value for the zero redshift color excess of $E(B-V) = 0.25$ mag, significantly higher than the value of $E(B-V) = 0.052$ mag suggested by Schlegel et al. [89] in that direction from their *IRAS* 100 μm maps. As above, this implies a color excess of $E(B-V) = 0.05$ mag at the redshift of the host galaxy (Table 2).

The most obvious characteristic of the OIR data is that the initial interval between the time of the GRB and day 3.5, is not well modeled by the parameterization. This is not surprising, because the modeling contains no parameters to describe any turn-on effects for the synchrotron emission. A similar turn-on effect was seen for the early OIR data for GRB970508, although not for the other GRBs discussed here.

It should be noted that the possible steepening of the flux density decline rate from $\beta \sim -1.5$ (radio) to $\beta \sim -2.0$ (OIR) may indicate a break in the decline rate somewhere between the two observing ranges, possibly owing to synchrotron losses of the high energy electrons.

2.2 Parameterization

Even though the study of the GRB phenomenon is still very much in its early stages and, as has been pointed out above, the GRBs for which we have any afterglow information are all of the slow-soft category with those about which we have significant radio information only a small subset, the parameterization studies (see Table 2) allow us to draw a some tentative conclusions:

- Of the seven relatively well observed radio GRBs (RGRBs) (including GRB980425/SN1998bw), five appear to have similar relativistic electron acceleration processes that generate a spectral index $\alpha \sim -0.6$ to ~ -0.7. The two possible exceptions are GRB980329 ($\alpha \sim -1.3$) and GRB991216 ($\alpha \sim -0.3$) which, interestingly enough, straddle the average spectral index of the other objects. The average spectral index value is very similar to the canonical value for extragalactic radio sources.
- The RGRBs can be described rather well from radio through OIR by the same formalism of Eqs. (1–10) of Sramek and Weiler in this volume that is successful in describing the radio emission from RSNe as arising from

a blastwave/CSM interaction. The obvious conclusion is that at least this subset of GRBs arise from the explosion of a compact, presumably stellar-sized object embedded in a dense CSM.

- Only two of the objects, GRB970508 and GRB000301C, appear to have been observed in a phase for which they show the turn-on of their OIR radiation. Such a turn-on is presumably related to the unmodeled relativistic electron acceleration at very early times and provides a test of explosion models.
- None of the objects requires the inclusion of a uniformly distributed absorbing component to the early time radio absorption (K_2 in Eq. (1) of Sramek and Weiler in this volume) which implies a highly filamentary or clumpy structure in the CSM with which the blastwave is interacting.
- Estimates for the color excess, $E(B-V)$, obtained from the modeling imply that many of the objects suffer appreciable extinction at the redshift of their parent galaxies because fitted zero redshift values significantly exceed estimates of Galactic extinction in their directions. The values for the excess extinction at the redshifts of the host galaxies noted above and in Table 2 are rather typical for late-type galaxies.
- All of the RGRBs except for GRB980425/SN1998bw have similar observed peak 6 cm radio spectral luminosities of $\sim 10^{31}$ erg s^{-1} Hz^{-1}. This implies that GRBs are rare objects for which one must survey very large volumes of space in order to detect examples. Such examples are, therefore, generally at very large distances and we detect only the brightest members of the luminosity distribution. GRB980425/SN1998bw is the only known nearby, lower luminosity exception.
- The prevalence of ISS in the early radio emission from many of these objects implies that the RGRBs are initially very small in angular size although they expand rapidly.

2.3 Discussion

Although many useful conclusions can be drawn from the light curve parameterizations, one should keep in mind that a more detailed study of radio GRBs will have to take into account several physical effects that are not seen in RSNe, and that have not been included in this brief overview, but need further consideration.

Interstellar Scintillation: Because their high radio luminosity and low absorption allows detection at great distance and quite early when they are still of very small angular size, GRB radio afterglows appear to be so compact as to exhibit ISS during the first few days or weeks of detectability. This possibility was proposed by Goodman [40] for GRB970508 based on earlier work by Rickett [80] for pulsars. After consideration of the several regimes of strong, weak, refractive, and diffractive scattering, Goodman [40] concluded that both diffractive and refractive scintillation are possible for radio afterglows and that observation of the effects of scintillation can place limits on their μas (micro-arcsecond) sizes at levels far too small to be resolved with *Very Long Baseline Interferometry (VLBI)*.

GRB970508 shows strong flux density fluctuations at both 8.46 and 4.86 GHz until age ~4 weeks after explosion, that Waxman et al. [108] attributed to diffractive scintillation. After ~1 month, they found that the modulation amplitude decreased, which is consistent with the diffractive scintillation being quenched by the increased size of the radio emitting region. They also took this increasing source size to be consistent with, and supportive of, the "fireball" model predictions (see the chapter by Waxman in this volume). From their 4.86 and 8.46 GHz results, Waxman et al. [108] concluded that the quenching of diffractive scintillation at ~4 weeks implies a size at that epoch of $\sim 10^{17}$ cm and an expansion speed comparable to that of light.

In contrast to the conclusion of Waxman et al. [108] that GRB970508 is undergoing strong diffractive scintillation during the first 30 days after explosion, Smirnova and Shishov [92] concluded that the radio afterglow is, in fact, undergoing only weak scintillation at 4.86 and 8.46 GHz at early times with refractive scintillation dominating at 1.43 GHz.

GRB980329 shows rapid flux density fluctuations at 4.9 and 8.3 GHz that are extinguished by age ~ 3 weeks. Although they did not analyze the scintillation data in detail, Taylor et al. [100] pointed out a similarity to the better studied scintillations of GRB970508 and suggested that the early-time angular size of GRB980329 may be even smaller than the ~ 3 μas inferred for GRB970508 by Goodman [40] and Waxman et al. [108] because GRB980329 is at a lower Galactic latitude, that should more quickly quench ISS.

GRB980519 shows strong modulation of the flux density at 4.86 and 8.46 GHz during the first ~20 days after the GRB burst, that Frail et al. [26] interpreted as being due to diffractive ISS. They derived a resulting maximum radio source size of < 0.4 μas, an extremely compact object. As was seen for GRB970508, the 1.4 GHz emission from GRB980519 seems to be suppressed, in this case below their detectability limit because their three measurements at 1.4 GHz are all upper limits. Frail et al. [26] attributed this suppression to SSA with a turn-over frequency between 1.43 and 4.86 GHz.

Cosmological Effects: Because the GRBs are at cosmological distances ($z \sim 1$), there are two effects that must be taken into account:

- The first is that there is a time dilation that slows the light curve evolution in the observer's frame with respect to the time evolution in the emission frame. This is a straightforward correction and has been elaborated by a number of authors (see, e.g., [15,44,59]). The time dilation results in a true emitted time to 6 cm peak flux density $[(t_{6\text{cm peak}} - t_0)_{\text{emit}}]$ shorter than that actually observed $[(t_{6\text{cm peak}} - t_0)_{\text{obs}}]$. The correction takes the form

$$(t_{6\text{cm peak}} - t_0)_{\text{emit}} = (t_{6\text{cm peak}} - t_0)_{\text{obs}} \, (1+z)^{-1} \tag{1}$$

and must be applied to the measured times from explosion to 6 cm peak flux density to obtain true times.

- The second is that there is a correction of the observed flux density, $(F_{6\text{cm}})_{\text{obs}}$, owing to the redshift making the observed frequency different from that emitted, $(F_{6\text{cm}})_{\text{emit}}$, for sources with non-zero spectral indices. The correction normally takes the form, for $F \propto \nu^{+\alpha}$,

$$(F_{6\text{cm}})_{\text{emit}} = (F_{6\text{cm}})_{\text{obs}} \, (1+z)^{-\alpha} \tag{2}$$

However, Chevalier and Li [14] have proposed an "equality of peaks" on theoretical grounds so that such a correction may be less important than expected.

Relativistic Effects: Essentially all researchers agree that the GRB phenomenon involves relativistic motion of the emitting region. Whether the motion involves a spherical, relativistic fireball or a directed relativistic jet is probably not of concern from the radio observer's standpoint because the emission from the CSM interaction is probably not highly directed.

For relativistic corrections there are two factors – the Lorentz factor Γ $[\Gamma = (1 - \frac{v^2}{c^2})^{-\frac{1}{2}}]$ and, if the motion is directed, the viewing angle θ. If, as expected, the radio emission is not highly directed, then taking $\theta = 0$ leaves only the Lorentz factor Γ to be determined.

The factor Γ is difficult to estimate from observations. Theoretical modeling predicts Γ of several hundred very early in the expansion phase [107] declining to subrelativistic motion after only a few days to a few weeks [17,46]. However, although highly speculative, there is, perhaps, one method for estimating Γ.

Weiler et al. [114] pointed out that by assuming a relation between type Ib/c SN1998bw and GRB980425, they could obtain estimates of $\Gamma \sim 1.6 - 2.0$ for the radio afterglow. They obtained one estimate by postulating that SN1998bw was expanding at $\sim 230,000$ km s^{-1} if its brightness temperature at 6 cm peak flux density was $10^{11.5}$ K, as required by the inverse Compton limit. The second estimate was obtained by assuming that the true 6 cm peak spectral luminosity was the same as that for the average of all known type Ib/c RSNe. Then, if one assumes: 1) all GRB radio afterglows arise in type Ib/c SNe, 2) the postulate of Weiler et al. [113] that all type Ib/c SNe may be approximate standard radio candles with an $L_{6\text{cm peak}} \sim 1.3 \times 10^{27}$ erg s^{-1} Hz^{-1}, and 3) any increase in the observed flux density is solely due to relativistic boosting, then one can estimate a relativistic Γ factor for each GRB radio afterglow.

Applying these assumptions to the peak 6 cm radio luminosity ($L_{6\text{cm peak}}$) values obtained for the GRBs and listed in Table 2, we obtain Γ values ~ 10. Such Γ values are similar to those found for VLBI "superluminal" sources (see e.g., [55,88]) and for the Galactic micro-QSO superluminal sources [45,60,66,102].

3 Summary

We have assembled an overview of the radio emission from GRBs and have shown that there are many similarities to the much better understood radio emission from RSNe. In particular, we have shown that the formalism described by [114,115] and by Sramek and Weiler in this volume can be applied to the radio and OIR emission from a number of GRBs and provides a satisfactory description of their light curves over extremely broad frequency and time ranges. These results then imply that at least some GRBs probably arise in the explosions of massive stars in dense, highly structured circumstellar media. Such media are presumed to have been established by mass-loss from the pre-explosion stellar systems.

Acknowledgments

KWW and MJM thank the *Office of Naval Research (ONR)* for the 6.1 funding supporting this research. Additional information and data on radio emission from SNe and GRBs can be found on *http://rsd-www.nrl.navy.mil/7214/weiler/* and linked pages.

References

1. E. Berger et al.: Astrophys. J. **545**, 56 (2000)
2. F. Bertoldi: GCN 580 (2000)
3. S.G. Bhargavi, R. Cowsik: Astrophys. J. Lett. **545**, L77 (2000)
4. J.S. Bloom, S.G. Djorgovski, S.R. Kulkarni, D.A. Frail: Astrophys. J. Lett. **507**, L25 (1998)
5. H.E. Bond: IAUC 6654 (1997)
6. M. Bremer, T.P. Kirchbaum, T.J. Galama, A.J. Castro-Tirado, F. Frontera, J. van Paradijs, I.F. Mirabel, E. Costa: Astron. Astrophys. **332**, L13 (1998)
7. M.S. Briggs, G. Richardson, R.M. Kippen, P.M. Woods: IAUC 6856 (1998)
8. A.J. Castro-Tirado et al.: Science **279**, 1011 (1998)
9. A.J. Castro-Tirado et al.: GCN 452 (1999)
10. S.M. Castro, A. Diercks, S.G. Djorgovski, S.R. Kulkarni, T.J. Galama, J.S. Bloom, F.A. Harrison, D.A. Frail: GCN 605 (2000)
11. A.J. Castro-Tirado et al.: Astron. Astrophys. **370**, 398 (2001)
12. R. Chary et al.: Astrophys. J. Lett. **498**, L9 (1998)
13. R.A. Chevalier: Astrophys. J. **499**, 810 (1998)
14. R.A. Chevalier, Z.-Y. Li: Astrophys. J. **536**, 195 (2000)
15. S.A. Colgate: Astrophys. J. **232**, 404 (1979)
16. E. Costa et al.: IAUC 6649 (1997)
17. Z.G. Dai, Y.F. Huang, T. Lu: Astrophys. J. **520**, 634 (1999)
18. S.G. Djorgovski et al.: Nature **387**, 876 (1997)
19. S.N. Dodonov, V.L. Afanasiev, V.V. Sokolov, A.V. Moiseev, A.J. Castro-Tirado: GCN 465 (1999)
20. A. Filippenko: IAUC 6969 (1998)
21. G.J. Fishman, C.A. Meegan: Ann. Rev. Astron. Astrophys. **33**, 415 (1995)

22. D.A. Frail, S.R. Kulkarni: IAUC 6662 (1997)
23. D.A. Frail, G.B. Taylor, S.R. Kulkarni: GCN 89 (1998)
24. D.A. Frail, S.R. Kulkarni: GCN 451 (1999)
25. D.A. Frail, G.B. Taylor: GCN 574 (2000)
26. D.A. Frail et al.: Astrophys. J. **534**, 559 (2000)
27. D.A. Frail et al.: Astrophys. J. Lett. **538**, L129 (2000)
28. F. Frontera, E. Costa, L. Piro, P. Soffitta, J. in 't Zand, L. Di Ciolo, A. Tesseri: IAUC 6853 (1998)
29. J.P.U. Fynbo, B.L. Jensen, J. Hjorth, H. Pedersen, J. Gorosabel: GCN 570 (2000)
30. B.M. Gaensler, R.N. Manchester, L. Staveley-Smith, A.K. Tzioumis, J.E. Reynolds, M.J. Kesteven: Astrophys. J. **479**, 845 (1997)
31. T.J. Galama et al.: IAUC 6895 (1998)
32. T.J. Galama et al.: Nature **395**, 670 (1998)
33. T.J. Galama et al.: Astrophys. J. Lett. **500**, L101 (1998)
34. T.J. Galama et al.: Astrophys. J. Lett. **497**, L13 (1998)
35. T.J. Galama et al.: Astron. Astrophys. Suppl. Ser. **138**, 465 (1999)
36. T.J. Galama et al.: Astrophys. J. Lett. **541**, L45 (2000)
37. M.R. Garcia et al.: Astrophys. J. Lett. **500**, L105 (1998)
38. P.M. Garnavich, S. Jha, M.A. Pahre, K.Z. Stanek, R.P. Kirshner, M.R. Garcia, A.H. Szentgyorgyi, J.L. Tonry: Astrophys. J. **543**, 61 (2000)
39. J. Goodman: Astrophys. J. Lett. **308**, L47 (1986)
40. J. Goodman: New Astron. **2**, 449 (1997)
41. J. Gorosabel, A.J. Castro-Tirado, A. Pedrosa, M.R. Zapatero-Osorio, A.J.L. Fernades, M. Feroci, E. Costa, F. Frontera: Astron. Astrophys. **347**, L31 (1999)
42. J.P. Halpern, J. Kemp, T. Piran, M.A. Bershady: Astrophys. J. Lett. **517**, L105 (1999)
43. J.P. Halpern et al.: Astrophys. J. **543**, 697 (2000)
44. M. Hamuy, M.M. Phillips, L.A. Wells, J. Maza: Pub. Astron. Soc. Pacific **105**, 787 (1993)
45. R.M. Hjellming, M.P. Rupen: Nature **375**, 464 (1995)
46. Y.F. Huang, Z.G. Dai, D.M. Wei, T. Lu: Mon. Not. R. Astron. Soc. **298**, 459 (1998)
47. K. Hurley, T. Cline: GCN 450 (1999)
48. K. Iwamoto et al.: Nature **395**, 672 (1998)
49. P. Jakobsen et al.: Astrophys. J. Lett. **369**, L63 (1991)
50. A.O. Jaunsen, J. Hjorth, M.I. Andersen, K. Kjernsmo, H. Pedersen, E. Palazzi: GCN 78 (1998)
51. A.O. Jaunsen et al.: Astrophys. J. **546**, 127 (2001)
52. B.L. Jensen et al.: Astron. Astrophys. **370**, 909 (2001)
53. L.E. Kay, J.P. Halpern, K.M. Leighly: IAUC 6969 (1998)
54. K.I. Kellermann, I.I.K. Pauliny-Toth: Astrophys. J. Lett. **155**, L71 (1969)
55. K.I. Kellermann: Comm. Astrophys. **11**, 69 (1985)
56. M.R. Kippen, R.D. Preece, T. Giblin: GCN 463 (1999)
57. S. Klose, H. Meusinger, H. Lehmann: IAUC 6864 (1998)
58. S.R. Kulkarni, J.S. Bloom, D.A. Frail, R. Ekers, M. Wieringa, R. Wark, J.L. Higdon: IAUC 6903 (1998)
59. B. Leibundgut: Astron. Astrophys. **229**, 1 (1990)
60. A. Levinson, R. Blandford: Astron. Astrophys. Suppl. Ser. **120**, 129 (1996)
61. Z.-Y. Li, R.A. Chevalier: Astrophys. J. **526**, 716 (1999)
62. C. Lidman et al.: IAUC 6895 (1998)

63. R.N. Manchester, B.M. Gaensler, V.C. Wheaton, L. Staveley-Smith, A.K. Tzioumis, M.J. Kesteven, J.E. Reynolds, N.S. Bizunok: Pub. Astron. Soc. Australia **19**, 207 (2002)
64. N. Masetti et al.: Astron. Astrophys. **359**, L23 (2000)
65. M.R. Metzger, S.G. Djorgovski, C.C. Steidel, S.R. Kulkarni, K.L. Adelberger, D.A. Frail: IAUC 6655 (1997)
66. I.F. Mirabel, L.F. Rodriguez: Nature **371**, 46 (1994)
67. M.J. Montes, S.D. Van Dyk, K.W. Weiler, R.A. Sramek, N. Panagia: Astrophys. J. **506**, 874 (1998)
68. M.J. Montes, K.W. Weiler, S.D. Van Dyk, R.A. Sramek, N. Panagia, R. Park: Astrophys. J. **532**, 1124 (2000)
69. J.M. Muller, J. Heise, C. Butler, F. Frontera, L. Di Ciolo, G. Gandolfi, A. Coletta, P. Soffitta: IAUC 6910 (1998)
70. L. Nicastro, L.A. Antonelli, G. Celidonio, M.R. Daniele, C. De Libero, G. Spoliti, L. Piro, E. Pian: IAUC 6912 (1998)
71. B. Paczynski: Astrophys. J. Lett. **308**, L43 (1986)
72. F. Patat, A. Piemonte: IAUC 6918 (1998)
73. F. Patat, E. Cappellaro, L. Rizzi, M. Turatto, S. Benetti: IAUC 7215 (1999)
74. E. Pian et al.: Astron. Astrophys. Suppl. Ser. **138**, 463 (1999)
75. E. Pian et al.: Astrophys. J. **536**, 778 (2000)
76. L. Piro et al.: IAUC 6656 (1997)
77. M.L. Prevot, J. Lequeux, L. Prevot, E. Maurice, B. Rocca-Volmerange: Astron. Astrophys. **132**, 389 (1984)
78. A.C.S. Readhead: Astrophys. J. **426**, 51 (1994)
79. D.E. Reichart et al.: Astrophys. J. **517**, 692 (1999)
80. B.J. Rickett: Mon. Not. R. Astron. Soc. **150**, 67 (1970)
81. J.E. Rhoads, A.S. Fruchter: Astrophys. J. **546**, 117 (2001)
82. M. Ruderman: NY Acad. Sci. Ann. **262**, 164 (1975)
83. E.M. Sadler, R.A. Stathakis, B.J. Boyle, R.D. Ekers: IAUC 6901 (1998)
84. R. Sagar, V. Mohan, S.B. Pandey, A.K. Pandey, C.S. Stalin, A.J. Castro-Tirado: Bull. Astron. Soc. India **28**, 499 (2000)
85. K.C. Sahu et al.: Astrophys. J. Lett. **489**, L127 (1997)
86. B.D. Savage, J.S. Mathis: Ann. Rev. Astron. Astrophys. **17**, 73 (1979)
87. B. Schaefer et al.: IAUC 6658 (1997)
88. P.A.G. Scheuer: IAU Symp. 110: VLBI & Compact Radio Sources **110**, 197 (1984)
89. D.J. Schlegel, D.P. Finkbeiner, M. Davis: Astrophys. J. **500**, 525 (1998)
90. D.S. Shepherd, D.A. Frail, S.R. Kulkarni, M.R. Metzger: Astrophys. J. **497**, 859 (1998)
91. A. Smette, A. Fruchter, T. Gull, K. Sahu, H. Ferguson, L. Petro, D. Lindler: GCN 603 (2000)
92. T.V. Smirnova, V.I. Shishov: Astron. Rep. **44**, 421 (2000)
93. I.A. Smith, R.P.J. Tilanus: GCN 50 (1998)
94. I.A. Smith et al.: Astron. Astrophys. **347**, 92 (1999)
95. D.A. Smith, K. Hurley, T. Cline: GCN 568 (2000)
96. P. Soffitta et al.: IAUC 6884 (1998)
97. V.V. Sokolov, A.I. Kopylov, S.V. Zharikov, M. Feroci, L. Nicastro, E. Palazzi: Astron. Astrophys. **334**, 117 (1998)
98. T. Takeshima, C. Markwardt, F. Marshall, T. Giblin, R.M. Kippen: GCN 478 (1999)

99. G.B. Taylor, D.A. Frail, S.R. Kulkarni: GCN 40 (1998)
100. G.B. Taylor, D.A. Frail, S.R. Kulkarni, D.S. Shepherd, M. Feroci, F. Frontera: Astrophys. J. Lett. **502**, L115 (1998)
101. G.B. Taylor, E. Berger: GCN 483 (1999)
102. S.J. Tingay et al.: Nature **374**, 141 (1995)
103. C. Tinney, R. Stathakis, R. Cannon, T. Galama: IAUC 6896 (1998)
104. R. Uglesich, N. Mirabal, J. Halpern, S. Kassin, S. Novati: GCN 472 (1999)
105. F.J. Vrba et al.: Astrophys. J. **528**, 254 (2000)
106. P.M. Vreeswijk et al.: GCN 496 (1999)
107. E. Waxman: Astrophys. J. Lett. **489**, L33 (1997)
108. E. Waxman, S.R. Kulkarni, D.A. Frail: Astrophys. J. **497**, 288 (1998)
109. E. Waxman, A. Loeb: Astrophys. J. **515**, 721 (1999)
110. K. Weiler, S. Van Dyk, N. Panagia, R. Sramek, J. Discenna: Astrophys. J. **380**, 161 (1991)
111. K. Weiler, S. Van Dyk, J. Pringle, N. Panagia: Astrophys. J. **399**, 672 (1992)
112. K. Weiler, S. Van Dyk, N. Panagia, R. Sramek: Astrophys. J. **398**, 248 (1992)
113. K.W. Weiler, S.D. Van Dyk, M.J. Montes, N. Panagia, R.A. Sramek: Astrophys. J. **500**, 51 (1998)
114. K.W. Weiler, N. Panagia, M.J. Montes: Astrophys. J. **562**, 670 (2001)
115. K.W. Weiler, N. Panagia, M.J. Montes, R.A. Sramek: Ann. Rev. Astron. Astrophys. **40**, 387 (2002)
116. S.E. Woosley, R.G. Eastman, B.P. Schmidt: Astrophys. J. **516**, 788 (1999)
117. Q.F. Yin: Astrophys. J. **420**, 152 (1994)
118. C.H. Young, D.A. Frail, S.R. Kulkarni: Bull. Am. Astron. Soc. **195**, 71.05 (1999)
119. J. in 't Zand et al.: IAUC 6854 (1998)

Gamma-Ray Bursts: The Underlying Model

Eli Waxman

Weizmann Institute of Science, Rehovot 76100, Israel

Abstract. A pedagogical derivation is presented of the "fireball" model of γ-ray bursts, according to which the observable effects are due to the dissipation of the kinetic energy of a relativistically expanding wind, a "fireball." The main open questions are emphasized, and key afterglow observations, that provide support for this model, are briefly discussed. The relativistic outflow is, most likely, driven by the accretion of a fraction of a solar mass onto a newly born (few) solar mass black hole. The observed radiation is produced once the plasma has expanded to a scale much larger than that of the underlying "engine," and is therefore largely independent of the details of the progenitor, whose gravitational collapse leads to fireball formation. Several progenitor scenarios, and the prospects for discrimination among them using future observations, are discussed. The production in γ-ray burst fireballs of high energy protons and neutrinos, and the implications of burst neutrino detection by kilometer-scale telescopes under construction, are briefly discussed.

1 Introduction

The widely accepted interpretation of the phenomenology of γ-ray bursts (GRBs), of $0.1-1$ MeV photons lasting for a few seconds (see [39] for a review), is that the observable effects are due to the dissipation of the kinetic energy of a relativistically expanding wind, a "fireball," whose primal cause is not yet known. The recent detection of "afterglows," delayed low energy (X-ray to radio) emission of GRBs (see [69] for a review), confirmed the cosmological origin of the bursts through the redshift determination of several GRB host-galaxies, and supported standard model predictions of afterglows that result from the collision of an expanding fireball with its surrounding medium (see [78,92] for reviews).

The fireball model is described in Sects. 2, 3, and 4 of this chapter. The phenomenological arguments suggesting that fireball formation is likely regardless of the underlying progenitor are presented, and fireball hydrodynamics and radiative processes are discussed in detail in Sects. 2 and 3, respectively. The main open questions related to fireball physics are discussed in Sect. 3.4. Since both the theory and the implications of afterglow observations are extensively discussed in other chapters of this volume, we include in Sect. 4 of this chapter only a brief discussion of several key afterglow implications. We also limit the theoretical discussion of fireball evolution to the GRB production phase that precedes the afterglow phase during which evolution is dominated by the interaction of the fireball with its surrounding medium. We do discuss, however, the initial non-self-similar onset of this interaction, which marks the onset of the afterglow phase.

The GRB progenitors are not yet known. We present in Sect. 5 the constraints imposed by observations on possible progenitors, and discuss the (presently) leading candidates. Hints provided by afterglow observations, which are extensively discussed in separate chapters of this volume, are briefly reviewed.

The association of GRBs with ultra-high energy cosmic-rays (UHECR), the evidence for which is strengthened by recent afterglow observations, is based on two key points [115]: 1) he constraints that a dissipative ultra-relativistic wind must satisfy in order to allow acceleration of protons to energy $\sim 10^{20}$ eV, the highest observed cosmic-ray energy, are remarkably similar to the constraints imposed on the fireball wind by γ-ray observations, and 2) the inferred local ($z = 0$) GRB energy generation rate of γ-rays is remarkably similar to the local generation rate of UHECR implied by cosmic-ray observations. We briefly discuss in Sect. 6 production of high energy protons and neutrinos in GRB fireballs (see [120,121] for more detailed reviews). The GRB model for UHECR production makes unique predictions that may be tested with operating and planned large area UHECR detectors [27,29,106,114][1]. In this review we focus, however, on more recent predictions of neutrino emission, which may be tested with planned high energy neutrino telescopes [57]. Detection of the predicted neutrino signal will confirm the GRB fireball model for the production of UHECR and may allow discrimination between different fireball progenitor scenarios. Moreover, a detection of even a handful of neutrino events correlated with GRBs will allow a test for neutrino properties, e.g., flavor oscillation and coupling to gravity, with accuracy many orders of magnitude better than currently possible.

2 The Fireball Model: Hydrodynamics

2.1 Relativistic Expansion

General phenomenological considerations, based on γ-ray observations, indicate that, regardless of the nature of the underlying sources, GRBs are produced by the dissipation of the kinetic energy of a relativistically expanding fireball. The rapid rise time and short duration, ~ 1 ms, observed in some bursts [13,38] imply that the sources are compact, with a linear scale comparable to a light-ms, $r_0 \sim 10^7$ cm. The high γ-ray luminosity implied by cosmological distances, $L_\gamma \sim 10^{52}$ erg s^{-1}, then results in a very high optical depth to pair creation since the energy of observed γ-ray photons is above the threshold for pair production. The number density of photons at the source n_γ is approximately given by

$$L_\gamma = 4\pi r_0^2 c n_\gamma \epsilon, \tag{1}$$

[1] See also *http://www.physics.adelaide.edu.au/astrophysics/FlysEye.html*, *http://www-ta.icrr.u-tokyo.ac.jp/*, and *http://www.auger.org/*

where $\epsilon \simeq 1$MeV is the characteristic photon energy. Using $r_0 \sim 10^7$cm, the optical depth for pair production at the source is

$$\tau_{\gamma\gamma} \sim r_0 n_\gamma \sigma_T \sim \frac{\sigma_T L_\gamma}{4\pi r_0 c\epsilon} \sim 10^{15}, \tag{2}$$

where σ_T is the Thomson cross section.

The high optical depth implies that a thermal plasma of photons, electrons, and positrons is created, a fireball which then expands and accelerates to relativistic velocities [50,87]. The optical depth is reduced by relativistic expansion of the source. If the source expands with a Lorentz factor Γ, the energy of photons in the source frame is smaller by a factor Γ compared to that in the observer frame, and most photons may therefore be below the pair production threshold.

A lower limit for Γ may be obtained in the following way [10,67]. The GRB photon spectrum is well fitted in the *Burst and Transient Source Experiment (BATSE)* detectors range, 20 keV to 2 MeV [39], by a combination of two power-laws, $dn_\gamma/d\epsilon_\gamma \propto \epsilon_\gamma^{(\alpha-1)}$ (α is the flux density spectral index, $F_\nu \propto \nu^{+\alpha}$) with different values of α at low and high energy [9]. Here, $dn_\gamma/d\epsilon_\gamma$ is the number of photons per unit photon energy. The break energy (where α changes) in the observer frame is typically $\epsilon_{\gamma b} \sim 1$ MeV, with $\alpha \simeq 0$ at energies below the break and $\alpha \simeq -1$ above the break. In several cases, the spectrum has been observed to extend to energies > 100 MeV [39,103]. Consider then a high energy test photon, with observed energy ϵ_t, trying to escape the relativistically expanding source. Assuming that, in the source rest frame, the photon distribution is isotropic, and that the spectrum of high energy photons follows $dn_\gamma/d\epsilon_\gamma \propto \epsilon_\gamma^{-2}$, the mean free path for pair production (in the source rest frame) for a photon of energy $\epsilon_t' = \epsilon_t/\gamma$ (in the source rest frame) is

$$l_{\gamma\gamma}^{-1}(\epsilon_t') = \frac{1}{2}\frac{3}{16}\sigma_T \int d\cos\theta(1-\cos\theta) \int_{\epsilon_{\rm th}(\epsilon_t',\theta)}^{\infty} d\epsilon \frac{U_\gamma}{2\epsilon^2} = \frac{1}{16}\sigma_T \frac{U_\gamma \epsilon_{\rm t}'}{(m_e c^2)^2} \,. \tag{3}$$

Here, $\epsilon_{\rm th}(\epsilon_t', \theta)$ is the minimum energy of photons that may produce pairs interacting with the test photon, given by $\epsilon_{\rm th}\epsilon_t'(1-\cos\theta) \geq 2(m_e c^2)^2$ (θ is the angle between the photons' momentum vectors). U_γ is the photon energy density (in the range corresponding to the observed *BATSE* range) in the source rest-frame, given by

$$L_\gamma = 4\pi r^2 \gamma^2 c U_\gamma. \tag{4}$$

(Note that we have used a constant cross section, $3\sigma_T/16$, above the threshold $\epsilon_{\rm th}$.) The cross section drops as $\log(\epsilon)/\epsilon$ for $\epsilon \gg \epsilon_{\rm th}$; however, since the number density of photons drops rapidly with energy, this does not introduce a large correction to $l_{\gamma\gamma}$.

The source size constraint implied by the variability time is modified for a relativistically expanding source. Since in the observer frame almost all photons propagate at a direction making an angle $< 1/\Gamma$ with respect to the expansion direction, radiation seen by a distant observer originates from a conical section of the source around the source-observer line of sight, with opening angle $\sim 1/\Gamma$.

Photons which are emitted from the edge of the cone are delayed, compared to those emitted on the line of sight, by $r/2\Gamma^2 c$. Thus, the constraint on source size implied by variability on time scale Δt is

$$r \sim 2\Gamma^2 c\Delta t. \tag{5}$$

The time r/c required for significant source expansion corresponds to comoving time (measured in the source frame) $t_{\rm co.} \approx r/\Gamma c$. The two-photon collision rate at the source frame is $t_{\gamma\gamma}^{-1} = c/l_{\gamma\gamma}$. Thus, the source optical depth to pair production is $\tau_{\gamma\gamma} = t_{\rm co.}/t_{\gamma\gamma} \approx r/\Gamma l_{\gamma\gamma}$. Using Eqs. (3) and (5) we have

$$\tau_{\gamma\gamma} = \frac{1}{128\pi} \frac{\sigma_T L_\gamma \epsilon_t}{c^2 (m_e c^2)^2 \Gamma^6 \Delta t}. \tag{6}$$

Requiring $\tau_{\gamma\gamma} < 1$ at ϵ_t we obtain a lower limit for Γ,

$$\Gamma \geq 250 \left[L_{\gamma,52} \left(\frac{\epsilon_t}{100\mathrm{MeV}} \right) \Delta t_{-2}^{-1} \right]^{1/6}, \tag{7}$$

where $L_\gamma = 10^{52} L_{\gamma,52}$ erg s^{-1}and $\Delta t = 10^{-2} \Delta t_{-2}$ s.

2.2 Fireball Evolution

As the fireball expands it cools, the photon temperature T_γ in the fireball frame decreases, and most pairs annihilate. Once the pair density is sufficiently low, photons may escape. However, if the observed radiation is due to photons escaping the fireball as it becomes optically thin, two problems arise. First, the photon spectrum is quasi-thermal, in conflict with observations. Second, the source size, $r_0 \sim 10^7$ cm, and the total energy emitted in γ-rays, $\sim 10^{53}$ erg, suggest that the underlying energy source is related to the gravitational collapse of a ~ 1 $M_\odot$ object. Thus, the plasma is expected to be loaded with baryons which may be injected with the radiation or present in the atmosphere surrounding the source. A small baryonic load, $\geq 10^{-8}$ $M_\odot$, increases the optical depth due to Thomson scattering on electrons associated with the loading protons, so that most of the radiation energy is converted to kinetic energy of the relativistically expanding baryons before the plasma becomes optically thin [88,102]. To overcome both problems it was proposed [95] that the observed burst is produced once the kinetic energy of the ultra-relativistic ejecta is re-randomized by some dissipation process at large radius, beyond the Thomson photosphere, and then radiated as γ-rays. Collision of the relativistic baryons with the interstellar medium (ISM) [95], and internal collisions within the ejecta itself [79,86,91], were proposed as possible dissipation processes. Most GRBs show variability on time scales much shorter (typically 10^{-2} times) than the total GRB duration. Such variability is hard to explain in models where the energy dissipation is due to external shocks [98,127]. Thus, it is believed that internal collisions are responsible for the emission of γ-rays.

Let us first consider the case where the energy release from the source is "instantaneous," i.e., on a time scale of r_0/c. We assume that most of the energy is released in the form of photons, i.e., that the fraction of energy carried by baryon rest mass M satisfies $\eta^{-1} \equiv Mc^2/E \ll 1$. The initial thickness of the fireball shell is r_0. Since the plasma accelerates to relativistic velocity, all fluid elements move with velocity close to c, and the shell thickness remains constant at r_0 (this changes at very late time, as discussed below). We are interested in the stage where the optical depth (due to pairs and/or electrons associated with baryons) is high, but only a small fraction of the energy is carried by pairs.

The entropy of a fluid component with zero chemical potential is $S = V(e + p)/T$, where e, p and V are the (rest frame) energy density, pressure and volume. For the photons $p = e/3 \propto T_\gamma^4$. Since initially both the rest mass and thermal energy of baryons is negligible, the entropy is provided by the photons. Conservation of entropy implies

$$r^2 \Gamma(r) r_0 T_\gamma^3(r) = \text{const}, \tag{8}$$

and conservation of energy implies

$$r^2 \Gamma(r) r_0 \Gamma(r) T_\gamma^4(r) = \text{const}\,. \tag{9}$$

Here $\Gamma(r)$ is the shell Lorentz factor. Combining (8) and (9) we find

$$\Gamma(r) \propto r, \quad T_\gamma \propto r^{-1}, \quad n \propto r^{-3}, \tag{10}$$

where n is the rest frame (comoving) baryon number density.

As the shell accelerates the baryon kinetic energy, ΓMc^2, increases. It becomes comparable to the total fireball energy when $\Gamma \sim \eta$, at radius $r_f = \eta r_0$. At this radius most of the energy of the fireball is carried by the baryon kinetic energy, and the shell does not accelerate further. Eq. (9) describing energy conservation is replaced with $\Gamma = \text{const}$. Eq. (8), however, still holds. Eq. (8) may be written as $T_\gamma^4/nT_\gamma = \text{const}$ (constant entropy per baryon). This implies that the ratio of radiation energy density to thermal energy density associated with the baryons is r independent. Thus, the thermal energy associated with the baryons may be neglected at all times, and Eq. (8) holds also for the stage where most of the fireball energy is carried by the baryon kinetic energy. Thus, for $r > r_f$ we have

$$\Gamma(r) = \Gamma \approx \eta, \quad T \propto r^{-2/3}, \quad n \propto r^{-2}. \tag{11}$$

Let us consider now the case of extended emission from the source, on time scale $\gg r_0/c$. In this case, the source continuously emits energy at a rate L, and the energy emission is accompanied by mass-loss rate

$$\dot{M} = L/\eta c^2. \tag{12}$$

For $r < r_f$ the fluid energy density is relativistic, $aT_\gamma^4/nm_pc^2 = \eta r_0/r$, and the speed of sound is $\sim c$. The time it takes the shell at radius r to expand significantly is r/c in the observer frame, corresponding to $t_{\text{co.}} \sim r/\Gamma c$ in the shell frame. During this time sound waves can travel a distance $cr/\Gamma c$ in the shell frame, corresponding to $r/\Gamma^2 = r/(r/r_0)^2 = (r_0/r)r_0$ in the observer frame. This implies that at the early stages of evolution, $r \sim r_0$, sound waves had enough time to smooth out spatial fluctuations in the fireball over a scale r_0, but that regions separated by $\Delta r > r_0$ cannot interact with each other. Thus, if the emission extends over a time $t_{\rm GRB} \gg r_0/c$, a fireball of thickness $ct_{\rm GRB} \gg r_0$ would be formed, which would expand as a collection of independent, roughly uniform, sub-shells of thickness r_0. Each sub-shell would reach a final Lorentz factor Γ_f, which may vary between sub-shells. This implies that different sub-shells may have velocities differing by $\Delta v \sim c/2\eta^2$, where η is some typical value representative of the entire fireball. Different shells emitted at times differing by Δt, $r_0/c < \Delta t < t_{\rm GRB}$, may therefore collide with each other after a time $t_c \sim c\Delta t/\Delta v$, i.e., at a radius

$$r_i \approx 2\Gamma^2 c\Delta t = 6 \times 10^{13} \Gamma_{2.5}^2 \Delta t_{-2}\ {\rm cm}, \tag{13}$$

where $\Gamma = 10^{2.5}\Gamma_{2.5}$. The minimum internal shock radius, $r \sim \Gamma^2 r_0$, is also the radius at which an individual sub-shell may experience significant change in its width r_0, due to Lorentz factor variation across the shell.

2.3 The Allowed Range of Lorentz Factors and Baryon Loading

The acceleration, $\Gamma \propto r$, of fireball plasma is driven by radiation pressure. Fireball protons are accelerated through their coupling to the electrons, which are coupled to fireball photons. We have assumed in the analysis presented above, that photons and electrons are coupled throughout the acceleration phase. However, if the baryon loading is too low, radiation may decouple from fireball electrons already at $r < r_f$. The fireball Thomson optical depth is given by the product of comoving expansion time, $r/\Gamma(r)c$, and the photon Thomson scattering rate, $n_e c\sigma_T$. The electron and proton comoving number densities are equal, $n_e = n_p$, and are determined by equating the r independent mass flux carried by the wind, $4\pi r^2 c\Gamma(r) n_p m_p$, to the mass-loss rate from the underlying source, which is related to the rate L at which energy is emitted through $\dot{M} = L/(\eta c^2)$. Thus, during the acceleration phase, $\Gamma(r) = r/r_0$ and the Thomson optical depth, τ_T, is $\tau_T \propto r^{-3}$. The Thomson optical depth drops below unity at a radius $r < r_f = \eta r_0$ if $\eta > \eta_*$, where

$$\eta_* = \left(\frac{\sigma_T L}{4\pi r_0 m_p c^3}\right)^{1/4} = 1.0 \times 10^3 L_{52}^{1/4} r_{0,7}^{-1/4}. \tag{14}$$

Here $r_0 = 10^7 r_{0,7}$ cm.

If $\eta > \eta_*$ radiation decouples from the fireball plasma at $\Gamma = r/r_0 = \eta_*^{4/3}\eta^{-1/3}$. If $\eta \gg \eta_*$, then most of the radiation energy is not converted to kinetic energy prior to radiation decoupling, and most of the fireball energy escapes in the form of thermal radiation. Thus, the baryon load of fireball shells, and the corresponding final Lorentz factors, must be within the range $10^2 \le \Gamma \approx \eta \le \eta_* \approx 10^3$ in order to allow the production of the observed non-thermal γ-ray spectrum.

2.4 Fireball Interaction with the Surrounding Medium

As the fireball expands, it drives a relativistic shock (blastwave) into the surrounding gas, e.g., into the ISM gas if the explosion occurs within a galaxy. In what follows, we refer to the surrounding gas as ISM gas, although the gas need not necessarily be interstellar. At early times, the fireball is little affected by the interaction with the ISM. At late times, most of the fireball energy is transferred to the ISM, and the flow approaches the self-similar blastwave solution of Blandford and McKee [17]. At this stage a single shock propagates into the ISM, behind which the gas expands with Lorentz factor

$$\Gamma_{BM}(r) = \left(\frac{17E}{16\pi n m_p c^2}\right)^{1/2} r^{-3/2} = 150 \left(\frac{E_{53}}{n_0}\right)^{1/2} r_{17}^{-3/2}, \quad (15)$$

where $E = 10^{53} E_{53}$ erg is the fireball energy, $n = 1 n_0$ cm^{-3} is the ISM number density, and $r = 10^{17} r_{17}$ cm is the shell radius. The characteristic time at which radiation emitted by shocked plasma at radius r is observed by a distant observer is $t \approx r/4\Gamma_{BM}^2 c$ [119].

The transition to self-similar expansion occurs on a time scale $t_{\rm SS}$ (measured in the observer frame) comparable to the longer of the two time scales set by the initial conditions: the (observer) GRB duration $t_{\rm GRB}$ and the (observer) time t_Γ at which the self-similar Lorentz factor equals the original ejecta Lorentz factor Γ, $\Gamma_{BM}(t = t_\Gamma) = \Gamma$. Since $t = r/4\Gamma_{BM}^2 c$,

$$t = \max\left[t_{\rm GRB}, 5\left(\frac{E_{53}}{n_0}\right)^{1/3} \Gamma_{2.5}^{-8/3}\,{\rm s}\right]. \quad (16)$$

During the transition, plasma shocked by the reverse shocks expands with Lorentz factor close to that given by the self-similar solution,

$$\Gamma_{\rm tr.} \simeq \Gamma_{BM}(t = t_{\rm SS}) \simeq 245 \left(\frac{E_{53}}{n_0}\right)^{1/8} t_1^{-3/8}, \quad (17)$$

where $t = 10 t_1$ s. The unshocked fireball ejecta propagate at the original expansion Lorentz factor, Γ, and the Lorentz factor of plasma shocked by the reverse shock in the rest frame of the unshocked ejecta is $\simeq \Gamma/\Gamma_{\rm tr.}$. If $t \simeq t_{\rm GRB} \gg t_\Gamma$ then $\Gamma/\Gamma_{\rm tr.} \gg 1$, the reverse shock is relativistic, and the Lorentz factor associated with the random motion of protons in the reverse shock is $\Gamma_p^R \simeq \Gamma/\Gamma_{\rm tr.}$.

If, on the other hand, $t \simeq t_\Gamma \gg t_{\rm GRB}$ then $\Gamma/\Gamma_{\rm tr.} \sim 1$, and the reverse shock is not relativistic. Nevertheless, the following argument suggests that the reverse shock speed is not far below c, and that the protons are therefore heated to relativistic energy, $\Gamma_p^R - 1 \simeq 1$. The comoving time, measured in the fireball ejecta frame prior to deceleration, is $t_{\rm co.} \simeq r/\Gamma c$. The expansion Lorentz factor is expected to vary across the ejecta, $\Delta\Gamma/\Gamma \sim 1$, due to variability of the underlying GRB source over the duration of its energy release. Such variation would lead to expansion of the ejecta, in the comoving frame, at relativistic speed. Thus, at the deceleration radius, $t_{\rm co.} \simeq \Gamma t$, the ejecta width exceeds $\simeq c t_{\rm co.} \simeq \Gamma c t$.

Since the reverse shock should cross the ejecta over a deceleration time scale, $\simeq \Gamma t$, the reverse shock speed must be close to c. We therefore conclude that the Lorentz factor associated with the random motion of protons in the reverse shock is approximately given by $\Gamma_p^R - 1 \simeq \Gamma/\Gamma_{\rm tr.}$ for both $\Gamma/\Gamma_{\rm tr.} \sim 1$ and $\Gamma/\Gamma_{\rm tr.} \gg 1$.

Since $t_{\rm GRB} \sim 10$ s is typically comparable to t_Γ, the reverse shocks are typically expected to be mildly relativistic.

2.5 Fireball Geometry

We have assumed in the discussion so far that the fireball is spherically symmetric. However, a jet-like fireball behaves as if it were a conical section of a spherical fireball as long as the jet opening angle is larger than Γ^{-1}. This is due to the fact that the linear size of causally connected regions, $ct_{\rm co.} \sim r/\Gamma$ in the fireball frame, corresponds to an angular size $ct_{\rm co.}/r \sim \Gamma^{-1}$. Moreover, due to the relativistic beaming of radiation, a distant observer cannot distinguish between a spherical fireball and a jet-like fireball, as long as the jet opening angle $\theta > \Gamma^{-1}$. Thus, as long as we are discussing processes that occur when the wind is ultra-relativistic, $\Gamma \sim 300$ (prior to significant fireball deceleration by the surrounding medium), our results apply for both a spherical and a jet-like fireball. In the latter case, L (E) in our equations should be understood as the luminosity (energy) the fireball would have carried had it been spherically symmetric.

3 The Fireball Model: Radiative Processes

3.1 Gamma-Ray Emission

If the Lorentz factor variability within the wind is significant, internal shocks will reconvert a substantial part of the kinetic energy to internal energy. The internal energy may then be radiated as γ-rays by synchrotron and inverse Compton emission of shock-accelerated electrons. The internal shocks are expected to be mildly relativistic in the fireball rest frame, i.e., characterized by Lorentz factor $\Gamma_i - 1 \sim$ a few. This is due to the fact that the allowed range of shell Lorentz factors is $\sim 10^2$ to $\sim 10^3$ (see Sect. 2.3), implying that the Lorentz factors associated with the relative velocities are not very large. Since internal shocks are mildly relativistic, we expect results related to particle acceleration in sub-relativistic shocks (see [16] for a review) to be valid for acceleration in internal shocks. In particular, electrons are expected to be accelerated to a power law energy distribution, $dn_e/d\Gamma_e \propto \Gamma_e^{-p}$ for $\Gamma_e > \Gamma_m$, with $p \simeq 2$ [5,12,18].

The minimum Lorentz factor Γ_m is determined by the following consideration. Protons are heated in internal shocks to random velocities (in the wind frame) $\Gamma_p^R - 1 \approx \Gamma_i - 1 \approx 1$. If electrons carry a fraction ξ_e of the shock internal energy, then $\Gamma_m \approx \xi_e(m_p/m_e)$. The characteristic frequency of synchrotron emission is determined by Γ_m and by the strength of the magnetic field. Assuming that a fraction ξ_B of the internal energy is carried by the magnetic field,

$4\pi r_i^2 c\Gamma^2 B^2/8\pi = \xi_B L_{\rm int.}$, the characteristic observed energy of synchrotron photons, $\epsilon_{\gamma b} = \Gamma\hbar\Gamma_m^2 eB/m_e c$, is

$$\epsilon_{\gamma b} \approx 1\xi_B^{1/2}\xi_e^{3/2}\frac{L_{\gamma,52}^{1/2}}{\Gamma_{2.5}^2 \Delta t_{-2}}\ \text{MeV}. \tag{18}$$

In deriving Eq. (18) we have assumed that the wind luminosity carried by internal plasma energy, $L_{\rm int.}$, is related to the observed γ-ray luminosity through $L_{\rm int.} = L_\gamma/\xi_e$. This assumption is justified since the electron synchrotron cooling time is short compared to the wind expansion time (unless the equipartition fraction ξ_B is many orders of magnitude smaller than unity), and hence electrons lose all their energy radiatively. Fast electron cooling also results in a synchrotron spectrum $dn_\gamma/d\epsilon_\gamma \propto \epsilon_\gamma^{-1-p/2} = \epsilon_\gamma^{-2}$ at $\epsilon_\gamma > \epsilon_{\gamma b}$, consistent with observed GRB spectra [9].

At present, there is no theory that allows the determination of the values of the equipartition fractions ξ_e and ξ_B. Eq. (18) implies that fractions not far below unity are required to account for the observed γ-ray emission. We note that build up of magnetic field to near equipartition by electromagnetic instabilities is expected to be a generic characteristic of collisionless shocks (see the discussion in [16] and references therein), and is inferred to occur in other systems such as in supernova remnant shocks (see, e.g., [24,60]).

3.2 Break Energy Distribution

The γ-ray break energy $\epsilon_{\gamma b}$ of most GRBs observed by *BATSE* detectors is in the range of 100 keV to 300 keV [22]. It may appear from Eq. (18) that the clustering of break energies in this narrow energy range requires fine tuning of fireball model parameters, which should naturally produce a much wider range of break energies. This is, however, not the case [55]. Consider the dependence of $\epsilon_{\gamma b}$ on Γ. The strong Γ dependence of the pair production optical depth, Eq. (6), implies that if the value of Γ is smaller than the minimum value allowed by Eq. (7), for which $\tau_{\gamma\gamma}(\epsilon_\gamma = 100\text{MeV}) \approx 1$, most of the high energy photons in the power-law distribution produced by synchrotron emission, $dn_\gamma/d\epsilon_\gamma \propto \epsilon_\gamma^{-2}$, would be converted to pairs. This would lead to high optical depth due to Thomson scattering on $e^\pm$, and hence to strong suppression of the emitted flux [55]. For fireball parameters such that $\tau_{\gamma\gamma}(\epsilon_\gamma = 100\text{MeV}) \approx 1$, the break energy implied by Eqs. (18) and (7) is

$$\epsilon_{\gamma b} \approx 1\xi_B^{1/2}\xi_e^{3/2}\frac{L_{\gamma,52}^{1/6}}{\Delta t_{-2}^{2/3}}\ \text{MeV}. \tag{19}$$

As explained in Sect. 2.3, shell Lorentz factors cannot exceed $\eta_* \simeq 10^3$, for which break energies in the X-ray range, $\epsilon_{\gamma b} \sim 10$ keV, may be obtained. We note, however, that the radiative flux would be strongly suppressed in this case too [55]. If the typical Γ of radiation emitting shells is close to η_*, then the range of Lorentz factors of wind shells is narrow, which implies that only a small fraction of wind kinetic energy would be converted to internal energy which can be radiated from the fireball.

Thus, the clustering of break energies at ~ 1 MeV is naturally accounted for, provided that the variability time scale satisfies $\Delta t \leq 10^{-2}$ s, which implies an upper limit on the source size, since $\Delta t \geq r_0/c$ (see [36,49] for alternative explanations). We note, that a large fraction of bursts detected by *BATSE* show variability on the shortest resolved time scale, ~ 10 ms [127]. In addition, a natural consequence of the model is the existence of low luminosity bursts with low, $1-10$ keV, break energies [55]. Such "X-ray bursts" may have recently been identified [64].

For internal collisions, the observed γ-ray variability time, $\sim r_i/\Gamma^2 c \approx \Delta t$, reflects the variability time of the underlying source, and the GRB duration reflects the duration over which energy is emitted from the source. Since the wind Lorentz factor is expected to fluctuate on time scales ranging from the shortest variability time r_0/c to the wind duration $t_{\rm GRB}$, internal collisions will take place over a range of radii, $r \sim \Gamma^2 r_0$ to $r \sim \Gamma^2 c t_{\rm GRB}$.

3.3 Afterglow Emission

Let us consider the radiation emitted from the reverse shocks during the transition to self-similar expansion. The characteristic electron Lorentz factor (in the plasma rest frame) is $\Gamma_m \simeq \xi_e(\Gamma/\Gamma_{\rm tr.})m_p/m_e$, where the internal energy per proton in the shocked ejecta is $\simeq (\Gamma/\Gamma_{\rm tr.})m_p c^2$. The energy density U is $E \approx 4\pi r^2 ct\Gamma_{\rm tr.}^2 U$, and the number of radiating electrons is $N_e \approx E/\Gamma m_p c^2$. Using Eq. (17) and $r = 4\Gamma_{\rm tr.}^2 ct$, the characteristic (or peak) energy of synchrotron photons (in the observer frame) is [124]

$$\epsilon_{\gamma m} \approx \hbar \Gamma_{\rm tr.}\Gamma_m^2 \frac{eB}{m_e c} = 2\xi_{e,-1}^2 \xi_{B,-1}^{1/2} n_0^{1/2} \Gamma_{2.5}^2 \ \mathrm{eV}, \tag{20}$$

and the specific luminosity, $L_\epsilon = dL/d\epsilon_\gamma$, at $\epsilon_{\gamma m}$ is

$$L_m \approx (2\pi\hbar)^{-1}\Gamma_{\rm tr.}\frac{e^3 B}{m_e c^2} N_e \approx 10^{61} \xi_{B,-1}^{1/2} E_{53}^{5/4} t_1^{-3/4} \Gamma_{2.5}^{-1} n_0^{1/4} \ \mathrm{s}^{-1}, \tag{21}$$

where $\xi_e = 0.1\xi_{e,-1}$, and $\xi_B = 0.1\xi_{B,-1}$.

Here too, we expect a power law energy distribution, $dN_e/d\Gamma_e \propto \Gamma_e^{-p}$ for $\Gamma_e > \Gamma_m$, with $p \simeq 2$. Since the radiative cooling time of electrons in the reverse shock is long compared to the ejecta expansion time, the specific luminosity extends in this case to energy $\epsilon_\gamma > \epsilon_{\gamma m}$ as $L_\epsilon = L_m(\epsilon_\gamma/\epsilon_{\gamma m})^{-1/2}$, up to photon energy $\epsilon_{\gamma c}$. Here $\epsilon_{\gamma c}$ is the characteristic synchrotron frequency of electrons for which the synchrotron cooling time, $6\pi m_e c/\sigma_T \gamma_e B^2$, is comparable to the ejecta (rest frame) expansion time, $\sim \Gamma_{\rm tr.} t$. At energy $\epsilon_\gamma > \epsilon_{\gamma c}$,

$$\epsilon_{\gamma c} \approx 0.1 \xi_{B,-1}^{-3/2} n_0^{-1} E_{53}^{-1/2} t_1^{-1/2} \ \mathrm{keV}, \tag{22}$$

the spectrum steepens to $L_\epsilon \propto \epsilon_\gamma^{-1}$.

The shock driven into the ISM continuously heats new gas, and produces relativistic electrons that may produce the delayed afterglow radiation observed on time scales $t \gg t_{SS}$, typically of order days to months. As the shock wave decelerates, the emission shifts to lower frequency with time. Since afterglow emission on such long time scale is extensively discussed in other chapters of this volume, we do not discuss in detail the theory of late-time afterglow emission.

3.4 Open Questions: Magnetic Field and Electron Coupling

The emission of radiation in both the GRB and afterglow phases is assumed to arise from synchrotron emission of shock accelerated electrons. To match observations, the magnetic field behind the shocks must be close to equipartition and a significant fraction of the internal shock energy must be carried by electrons, that is, ξ_B and ξ_e should be close to unity, of order 10%. During the afterglow phase, shock compression of the existing ISM field yields a field many orders of magnitude smaller than needed. Thus, the magnetic field is most likely generated in, and by, the shock wave. A similar process is likely necessary to generate the field required for synchrotron emission during the GRB phase, i.e., in the internal fireball shocks. Although a magnetic field close to equipartition at the base of the wind frozen into the fireball plasma may not be many orders of magnitude below equipartition during the internal shock phase, significant amplification is nevertheless required. It is well known that near equipartition magnetic fields may be generated in collisionless shocks through the Weibel instability (see, e.g., [63,97]). However, the field is generated on microscopic, skin-depth, scale and is therefore expected to rapidly decay, unless its coherence length grows to a macroscopic scale [53,54]. The process by which such scale increase is achieved is not understood, and probably related to the process of particle acceleration [54].

In order to produce the observed spectrum during both afterglow and GRB phases, electrons must be accelerated in the collisionless shocks to a power-law distribution, $dn_e/d\Gamma_e \propto \Gamma_e^{-p}$ with $p \simeq 2$. As mentioned in Sect. 3.1, such distribution is expected in the internal shocks, which are mildly relativistic. Recent numeric and analytic calculations of particle acceleration via the first order Fermi mechanism in relativistic shocks show that similar spectral indices, $p \approx 2.2$, are obtained for highly-relativistic shocks as well [11,65]. The derivation of electron spectral indices is based, in both the non-relativistic and relativistic cases, on a phenomenological description of electron scattering and, therefore, does not provide a complete basic principle description of the process. In particular, these calculations do not allow one to determine the fraction of energy ξ_e carried by electrons.

4 The Fireball Model: Key Afterglow Implications

The following point should be clarified in the context of afterglow observations – the distribution of GRB durations is bimodal, with broad peaks at $t_{GRB} \sim 0.2$ s and $t_{\rm GRB} \sim 20$ s [39]. The majority of bursts belong to the long duration, $t_{\rm GRB} \sim 20$ s, class. The detection of afterglow emission was made possible thanks to the accurate GRB positions provided on hour time scale by the *BeppoSAX* satellite [28]. Since the detectors on board this satellite trigger only on long bursts, afterglow observations are not available for the smaller population of short, $t_{\rm GRB} \sim 0.2$ s, bursts. Thus, while the discussion of the fireball model presented in Sects. 2 and 3, based on γ-ray observations and on simple phenomenological arguments, applies to both long and short duration bursts, the discussion below of afterglow observations applies to long duration bursts only. It should, therefore, be kept in mind that short duration bursts may constitute a different class of GRBs which, for example, may be produced by a different class of progenitors and may have a different redshift distribution than the long duration bursts.

Afterglow observations led to the confirmation, as mentioned in Sect. 1, of the cosmological origin of GRBs [69], and supported [117,126] standard model predictions [62,80,90,112] of afterglows that result from synchrotron emission by electrons accelerated to high energy in the highly relativistic shock driven by the fireball into its surrounding gas. As discussed in separate chapters of this volume, both the spectral and temporal behavior of afterglow emission are in general agreement with model predictions.

Since afterglow emission results from the interaction of the fireball with ambient medium, it does not provide direct information on the evolution of the fireball at the earlier stage during which the GRB is produced. Nevertheless, afterglow observations may be used to indirectly test underlying model assumptions and constrain model parameters relevant for this earlier stage. We describe below several afterglow observations which have important implications for the GRB phase of the model.

4.1 Fireball Size and Relativistic Expansion

Radio observations of GRB970508 allowed a direct determination of the fireball size and a direct confirmation of its relativistic expansion. As explained in Sect. 2.1, radiation seen by a distant observer originates from a conical section of the fireball around the source-observer line of sight, with opening angle $\sim 1/\Gamma$, and photons which are emitted from the edge of the cone are delayed compared to those emitted along the line of sight, by $r/2\Gamma^2c$. Thus, the apparent radius of the emitting cone is $R = r/\Gamma(r) = 2\Gamma(r)ct$ where r and t are related by $t = r/2\Gamma(r)^2c$. (A detailed calculation of fireball emission introduces only a small correction, $R = 1.9\Gamma(r)ct$ [119]. Using Eq. 15, we find that the apparent size of the fireball during its self-similar expansion into the surrounding medium is given by

$$R = 0.8 \times 10^{17} \left(\frac{1+z}{2}\right)^{-5/8} \left(\frac{E_{52}}{n_0}\right)^{1/8} \left(\frac{t}{1\rm week}\right)^{5/8} \rm cm. \tag{23}$$

We have chosen here the normalization $E = 10^{52}E_{52}$ erg since this is the energy inferred for GRB970508 [118,125] (note, however, the very weak dependence on

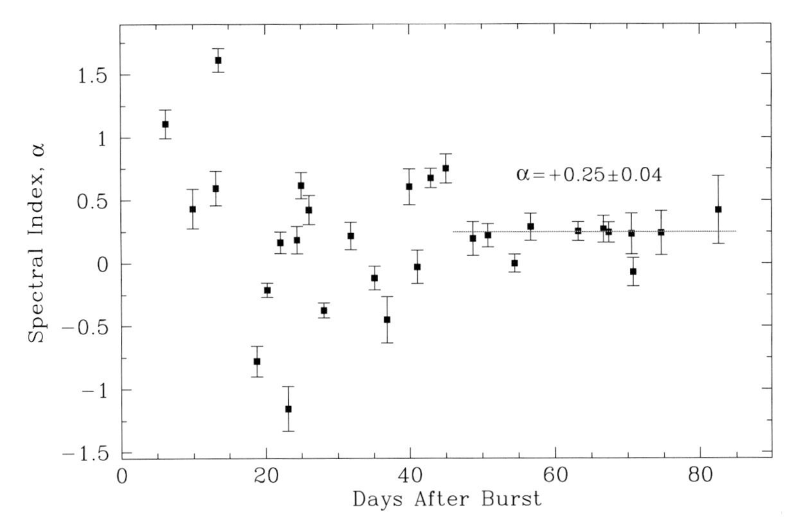

Fig. 1. The ratio between the GRB970508 afterglow radio fluxes at 4.86 GHz and 8.46 GHz, $\alpha \equiv \log[f_\nu(4.86\mathrm{GHz})/f_\nu(8.46\mathrm{GHz})]/\log(4.86/8.46)$, is shown as a function of time following the burst (modified from [41]). The rapid variations at early times are due to narrow band diffractive scintillation, and their quenching at late times is due to the expansion of the source beyond a critical size given by Eq. (24). This is a direct confirmation of model predictions, according to which highly relativistic plasma ejection is responsible for the observed radiation.

E and n). The factor $1+z$ appears due to the redshift between source and observer time intervals.

Scattering by irregularities in the local interstellar medium ISM may modulate the observed fireball radio flux [51]. If scattering produces multiple images of the source, interference between the multiple images may produce a diffraction pattern, leading to strong variations of the flux as the observer moves through the pattern. In order for the diffraction patterns produced by different points on the source to be similar, so that the pattern is not smoothed out due to large source size, the apparent size of a source a redshift $z=1$ must satisfy [125]

$$R < R_{\mathrm{sc.}} \approx 10^{17} \frac{\nu_{10}^{6/5}}{h_{75}} \left(\frac{SM}{10^{-3.5}\mathrm{m}^{-20/3}\mathrm{kpc}} \right)^{-3/5} \mathrm{cm}, \tag{24}$$

where $\nu = 10\nu_{10}$ GHz, h_{75} is the Hubble Constant in units of 75 km s^{-1} Mpc^{-1}, and the scattering measure, SM, a measure of the strength of the electron density fluctuations, is normalized to its characteristic Galactic value.

Comparing Eqs. (23) and (24) we find that, on time scale of weeks, the apparent fireball size is comparable to the maximum size for which diffractive scintillation is possible. On shorter time scales, therefore, strong modulation of the radio flux is expected, while on longer time scales we expect diffractive scintillation to be quenched. This is exactly what had been observed for GRB970508, as demonstrated in Fig. 1. Observations are therefore in agreement with fireball model predictions: They imply a source size consistent with model prediction, Eq. (23) which, in particular, imply expansion at a speed comparable to that of light.

4.2 The Nature of the Fireball Plasma

Due to present technical limitations of the experiments, afterglow radiation is observed in most cases only on time scale $\gg 10$ s. In one case, however, that of GRB990123, optical emission was detected on ~ 10 s time scale [1]. The most natural explanation of the observed optical radiation is synchrotron emission from electrons accelerated to high energy in the reverse shocks driven into fireball ejecta at the onset of interaction with the surrounding medium [81,99], as explained in Sect. 3.3. This observation provides, therefore, direct constraints on the fireball ejecta plasma. First, it provides strong support for one of the underlying assumptions of the dissipative fireball scenario described in Sect. 2.2, that the energy is carried from the underlying source in the form of proton kinetic energy. This is due to the fact that the observed radiation is well accounted for in a model where a shock propagates into fireball plasma composed of protons and electrons (rather than, e.g., a pair plasma). Second, comparison of the observed flux with model predictions from Eqs. (20) and (21) implies $\xi_e \sim \xi_B \sim 10^{-1}$.

4.3 Gamma-Ray Energy and GRB Rate

Following the determination of GRB redshifts, it is now clear that most GRB sources lie within the redshift range $z \sim 0.5$ to $z \sim 2$, with some bursts observed at $z > 3$. For the average GRB γ-ray fluence, 1.2×10^{-5} erg cm^{-2} in the 20 keV to 2 MeV band, this implies characteristic isotropic γ-ray energy and luminosity $E_\gamma \sim 10^{53}$ erg and $L_\gamma \sim 10^{52}$ erg s^{-1} (Here, we assume a flat universe with $\Omega = 0.3$, $\Lambda = 0.7$, and $H_0 = 65$ km s^{-1} Mpc^{-1}. These estimates are consistent with more detailed analyses of the GRB luminosity function and redshift distribution. Mao and Mo [77], e.g., find, for the cosmological parameters we use, a median GRB energy of $\approx 0.6 \times 10^{53}$ erg in the $50 - 300$ keV band, corresponding to a median GRB energy of $\approx 2 \times 10^{53}$ erg in the 20 keV to 2 MeV band.

Since most observed GRB sources lie within the redshift range $z \sim 0.5 - 2$, observations essentially determine the GRB rate per unit volume at $z \sim 1$. The observed rate of 10^3 yr^{-1} implies $R_{\rm GRB}(z=1) \approx 3$ Gpc^{-3} yr^{-1} (for $\Omega = 0.3$, $\Lambda = 0.7$, and $H_0 = 65$ km s^{-1} Mpc^{-1}. The present, $z = 0$, rate is less well constrained since available data are consistent with both no evolution of GRB rate with redshift and with strong evolution (following, e.g., the star formation rate), in which $R_{\rm GRB}(z=1)/R_{\rm GRB}(z=0) \sim 10$ [61,68]. A detailed analysis by, e.g., Schmidt [101] leads, assuming $R_{\rm GRB}$ is proportional to the star formation rate, to $R_{\rm GRB}(z=0) \sim 0.5$ Gpc^{-3} yr^{-1}.

If fireballs are conical jets of solid angle $\Delta\Omega$ then, clearly, the total γ-ray energy is smaller by a factor $\Delta\Omega/4\pi$ than the isotropic energy, and the GRB rate is larger by the same factor.

4.4 Fireball Geometry

Afterglow observations suggest that at least some GRBs are conical jets, of opening angle $\theta \sim 10^{-1}$ corresponding to a solid angle $\Delta\Omega \sim 10^{-2}$ [26,70,100]. As explained in Sect. 2.5, the discussion in Sect. 2 and Sect. 3 is limited to the stage where the wind is ultra-relativistic, $\Gamma \sim 300$, prior to significant fireball deceleration by the surrounding medium, and is hence equally valid for both a spherical and a jet-like fireball. A jet like geometry has, of course, profound implications for the underlying progenitor (see Sect. 5): it implies that the underlying source must produce a collimated outflow and, if $\theta \sim 10^{-1}$ is indeed typical, it implies that the characteristic γ-ray energy emitted by the source is $\approx 10^{51}$ erg rather than $\approx 10^{53}$ erg implied by the assumption of isotropy.

4.5 Fireball γ-Ray Efficiency

Afterglow observations imply that a significant fraction of the energy initially carried by the fireball is converted into γ-rays, i.e., that the observed γ-ray energy provides a rough estimate of the total fireball energy. This has been demonstrated for one case, that of GRB970508, by a comparison of the total fireball energy derived from long term radio observations with the energy emitted in γ-rays [41,125], and for a large number of bursts by a comparison of observed γ-ray energy with the total fireball energy estimate based on X-ray afterglow data [42]. Freedman and Waxman [42] demonstrated that a single measurement of the X-ray afterglow flux on the time scale of a day provides a robust estimate of the fireball energy per unit solid angle, ε, averaged over a conical section of the fireball of opening angle $\theta \sim 0.1$. Applying their analysis to *BeppoSAX* afterglow data they demonstrated that the ratio of observed γ-ray to total fireball energy per unit solid angle, $\varepsilon_\gamma/\varepsilon$, is of order unity.

The inferred high radiative efficiency implies that a significant fraction of the wind kinetic energy must be converted to internal energy in internal shocks, and that electrons must carry a significant fraction of the internal energy, i.e., that ξ_e should be close to unity. We have already shown in Sect. 3.1 and Sect. 3.2, following [55], that efficient conversion of kinetic to internal energy and ξ_e values close to unity may naturally lead to ~ 1 MeV spectral break energies, in accordance with observations.

5 Progenitors Clues

The observational constraints that provide the main hints regarding the nature of the underlying GRB progenitors are the rapid, ~ 1 ms, γ-ray signal variability, and the total γ-ray energy emitted by the source, $\approx 10^{53}(\Delta\Omega/4\pi)$ erg. The rapid variability implies a compact object, of size smaller than a light millisecond, $\sim 10^7$ cm, and mass smaller than $\sim 30\ \mathrm{M}_\odot$ (the mass of a black hole with $\sim 10^7$ cm Schwarzschild radius). The energy released in γ-rays corresponds to a $0.1(\Delta\Omega/4\pi)\ \mathrm{M}_\odot$ rest mass energy. The most natural way for triggering the

GRB event is therefore the accretion of a fraction of a solar mass onto a (several) solar mass black hole (or possibly a neutron star). The dynamical time of such a source, comparable to its light crossing time, is sufficiently short to account for the observed rapid variability and, if significant fraction of the gravitational binding energy release is converted to γ-rays, the source will meet the observed energy requirements.

Most cosmological GRB models therefore have at their basis gravitational collapse of a several solar-mass progenitor to a black hole. Within the context of the fireball model, the observed variability is determined by the dynamical time of the source, which determines the variability in the ejected wind properties, while the GRB duration, ≥ 10 s for long bursts, reflects the wind duration, i.e., the duration over which energy is extracted from the source. The characteristic time for the gravitational collapse is of the order of the dynamical time, i.e., much shorter than the wind duration. Most models therefore assume that, following collapse and black hole formation, some fraction of the progenitor mass forms an accretion disk which powers the wind through gradual mass accretion. The characteristic time scale for accretion is set by the disk viscosity, which is uncertain and assumed to correspond to the observed GRB duration.

Progenitor models differ in the scenario for black hole formation, and in the process assumed to convert disk energy into relativistic outflow. The two leading progenitor scenarios are, at present, collapses of massive stars [89,128], and mergers of compact objects [50,87]. In the former case, the progenitor is a massive rotating star, e.g., a $\sim 15\ M_{\odot}$ helium star evolved (by mass-loss) from a $\sim 30\ M_{\odot}$ main sequence star. The collapse of the progenitor's $\sim 2\ M_{\odot}$ iron core leads to the formation of a black hole surrounded by an accretion disk composed of mantle plasma [128]. In the latter case, the merger of two neutron stars leads to the formation of a black hole surrounded by a disk produced by neutron star disruption during the merger process (see, e.g., [37]). A similar scenario involves neutron star disruption during a neutron star–black hole merger (see, e.g., [71]).

Two types of processes are widely considered for the extraction of disk energy: neutrino emission and magneto-hydrodynamic (MHD) processes. The viscous dissipation of energy, driving mass accretion, heats the dense disk plasma leading to the emission of thermal neutrinos of all flavors. Neutrino annihilation along the rotation axis in the vicinity of the black hole may then produce an electron-positron pair plasma fireball. The fraction of rest mass which is dissipated during accretion is typically of order 10% (the specific energy at the last stable orbit of a non-rotating black-hole, at three Schwarzschild radii, is $c^2/6$). Accretion of $0.1\ M_{\odot}$ over a second may therefore lead to a neutrino luminosity of $\approx 10^{52.5}$ erg s^{-1}, of which $\approx 1\%$ would be deposited by annihilation to drive a fireball. The resulting wind luminosity, $\sim 10^{50}$ erg s^{-1}, may be too low to drive a spherical fireball, but may be sufficient if the fireball is collimated into $\Delta\Omega/4\pi \sim 10^{-2}$.

If equipartition magnetic fields, $\sim 10^{15}$ G, are built in the disk (e.g., by convective motion) the dissipated energy may be extracted electromagnetically from the disk. Although the process by which such a strong field is generated, as well as the details of the energy extraction mechanism in the presence of such field are not understood, there is evidence from various astrophysical systems (active galactic nuclei and micro-quasars, see, e.g., [74]), for the formation of MHD driven jets which carry $\sim 10\%$ of the disk binding energy. The presence of equipartition fields may also allow the extraction of energy directly, e.g., via the Blandford-Znajek mechanism [19], from the rotating black hole [72]. For rapid rotation, the available energy in this case is comparable to the collapsed rest mass. Thus, MHD processes are often invoked to drive a relativistic wind with efficiency much higher than that estimated for a neutrino driven wind.

We note in this context that a possible alternative to the above models may be the formation from stellar collapse of a fast rotating neutron star with ultra-high magnetic field [107,109]. If a fast rotating, millisecond period, neutron star is produced by the collapse with $\sim 10^{15}$ G field, then the resulting electromagnetic energy luminosity is sufficient to drive a GRB wind.

One problem which all models are facing is the baryon loading. In order to allow acceleration to the high Lorentz factors implied by observations, $\Gamma \sim 10^3$, the total mass entrained within the expanding plasma must be smaller than $\sim 10^{-4}$ $M_\odot$, i.e., the mass-loss rate should be smaller than $\sim 10^{-6}$ $M_\odot$ s^{-1}. The neutrino luminosity is expected to drive mass-loss at a much higher rate. It is generally assumed that mass flow towards the rotation axis is inhibited (e.g., by high pressure of fireball plasma along the rotation axis), thus allowing the formation of a sufficiently baryon free fireball collimated along the rotation axis. In the case of a massive star progenitor, the fireball jet is assumed to form along the rotation axis of the star, where rapid rotation leads to lower mantle and envelope density. The collimation of the fireball may be due in this case to the presence of a low density funnel along the rotation axis, and the resulting jet must penetrate through the stellar mantle and envelope in order to allow the production of an observable GRB. Recent numerical and analytical calculations of the propagation of high entropy jets through stellar progenitors indicate that such a scenario may be viable [2,44,83].

Afterglow observations provide several hints which indicate that long duration GRBs, at least those for which afterglows have been detected, are associated with massive star progenitors. The location of GRBs within host galaxies, the presences of iron lines, and the evidence for a supernova association all imply massive star progenitors. Since this issue is discussed in detail in other chapters in this volume, we address it only briefly here.

Most GRB afterglows are localized within the optical image of a host galaxy [20]. This is in disagreement with simple analyses of the neutron star merger scenario which predict that the high velocity of such binaries should carry many of them outside of the host prior to merger. This result is, however, uncertain since it depends on model parameters which are only poorly constrained, such as the distribution of initial binary separations. Evidence for the presence of iron lines has been found in X-ray data for two bursts [4,93]. While the presence of iron lines strongly suggests a massive stellar progenitor, as it indicates the presence of iron enriched environment, the confidence level of their detection is moderate. There is evidence in three cases that a supernova may be associated with the GRB [47,110]. The evidence is, however, not yet conclusive (see, e.g., [124]). Finally, the synchrotron emission produced by a shock driven by the fireball into its surrounding medium depends on the density of the ambient medium. Thus, the temporal and spectral dependence of this afterglow emission may distinguish between the high density environment characteristic of a massive stellar wind, expected to exist in the case of a massive stellar progenitor, and the low density ISM expected, e.g., in merger scenarios. Present observations are not yet conclusive, since data on time scales much shorter than one day are required to distinguish between the two cases [75].

It is clear from the above discussion, that future afterglow observations providing more detailed information on the burst environment and location will play a crucial role in placing stringent constraints on progenitor models. In addition to the points discussed in the previous paragraph, the destruction of dust [34,43,96,124] and time dependence of atomic, ionic [21,76] and molecular H_2 [34,124] lines due to photoionization may be detectable for a burst in a molecular cloud environment characteristic of star forming regions.

6 High Energy Protons and Neutrinos from GRB Fireballs

6.1 Fermi Acceleration in GRBs

In the fireball model, the observed radiation is produced, both during the GRB and the afterglow, by synchrotron emission from shock accelerated electrons. In the region where electrons are accelerated, protons are also expected to be shock accelerated. This is similar to what is thought to occur in supernovae remnant shocks, where synchrotron radiation of accelerated electrons is the likely source of non-thermal X-rays (recent *Advanced Satellite for Cosmology and Astrophysics (ASCA)* observations give evidence for acceleration of electrons in the remnant of SN1006 to 10^{14} eV [66]), and where shock acceleration of protons is believed to produce cosmic rays with energy extending to $\sim 10^{15}$eV (see, e.g., [16] for a review). Thus, it is likely that protons, as well as electrons, are accelerated to high energy within GRB fireballs.

We consider proton Fermi acceleration in fireball internal and reverse shocks (see Sect. 2.2 and Sect. 2.4 respectively). Since these shocks are mildly relativistic, with Lorentz factors $\Gamma_i - 1 \sim 1$ in the wind frame (see Sects. 2.4 and 3.1), the predicted energy distribution of accelerated protons is [12,16] $dn_p/d\epsilon_p \propto \epsilon_p^{-2}$, similar to the electron energy spectrum inferred from the observed photon spectrum.

Two constraints must be satisfied by fireball wind parameters in order to allow proton acceleration to $\epsilon_p > 10^{20}$ eV in internal and reverse shocks. First, the proton acceleration time, $t_a \sim R_L/c$ where R_L is the proton Larmor radius, must be smaller than the wind expansion time [85,111,115], $t_d \sim r_i/\Gamma c$ (in the wind frame). This constraint sets a lower limit to the magnetic field carried by the wind, which may be expressed as [115]:

$$\xi_B/\xi_e > 0.02\Gamma_{2.5}^2\epsilon_{p,20}^2 L_{\gamma,52}^{-1}. \tag{25}$$

Here, $\epsilon_p = 10^{20}\epsilon_{p,20}$ eV. Recall that ξ_B is the fraction of the wind energy density which is carried by magnetic field, $4\pi r^2 c\Gamma^2(B^2/8\pi) = \xi_B L$, and ξ_e is the fraction of wind energy carried by shock accelerated electrons. Since the electron energy is lost radiatively, $L_\gamma \approx \xi_e L$.

The second constraint that should be satisfied is that the proton synchrotron loss time must exceed t_a, setting an upper limit to the magnetic field. The latter constraint may be satisfied simultaneously with the lower limit to the magnetic field, Eq. (25), provided [115]

$$\Gamma > 130\epsilon_{p,20}^{3/4}\Delta t_{-2}^{-1/4}. \tag{26}$$

The constraints that must be satisfied to allow acceleration of protons to energy $> 10^{20}$ eV are remarkably similar to those inferred from γ-ray observations: $\Gamma > 100$ is implied by observed γ-ray spectra (see Sects. 2.1 and 2.3), and magnetic field close to equipartition, $\xi_B \sim 1$, is required in order for electron synchrotron emission to account for the observed radiation (see Sect. 3.1).

It has recently been claimed [48] that the conditions at the external shock driven by the fireball into the ambient gas are not likely to allow proton acceleration to ultra-high energy. Regardless of the validity of this claim, it is irrelevant for the acceleration in internal shocks, the scenario considered for UHECR production in GRBs [111,115]. Moreover, it is not at all clear that UHECRs cannot be produced at the external shock, since the magnetic field may be amplified ahead of the shock by the streaming of high energy particles. For discussion of high energy proton production in the external shock and its possible implications see Dermer [33].

6.2 UHECR Flux and Spectrum

The local ($z = 0$) energy production rate in γ-rays by GRBs is roughly given by the product of the characteristic GRB γ-ray energy, $E \approx 2 \times 10^{53}$ erg, and the local GRB rate. Under the assumption that the GRB rate evolution is similar to the star-formation rate evolution, the local GRB rate is $\sim 0.5\mathrm{Gpc}^{-3}\ \mathrm{yr}^{-1}$ [101] (see Sect. 4.3), implying a local γ-ray energy generation rate of $\approx 10^{44}\mathrm{erg}\mathrm{Mpc}^{-3}\mathrm{yr}^{-1}$. The energy observed in γ-rays reflects the fireball energy in accelerated electrons. Thus, if accelerated electrons and protons carry similar energy (as indicated by afterglow observations [42] (see, however, [108]) then the GRB production rate of high energy protons is

$$\epsilon_p^2(d\dot{n}_p/d\epsilon_p)_{z=0} \approx 10^{44}\ \mathrm{erg\ Mpc^{-3}\ yr^{-1}}. \tag{27}$$

In Fig. 2 we compare the observed UHECR spectrum with that predicted by the GRB model. The generation rate (Eq. 27) of high energy protons is

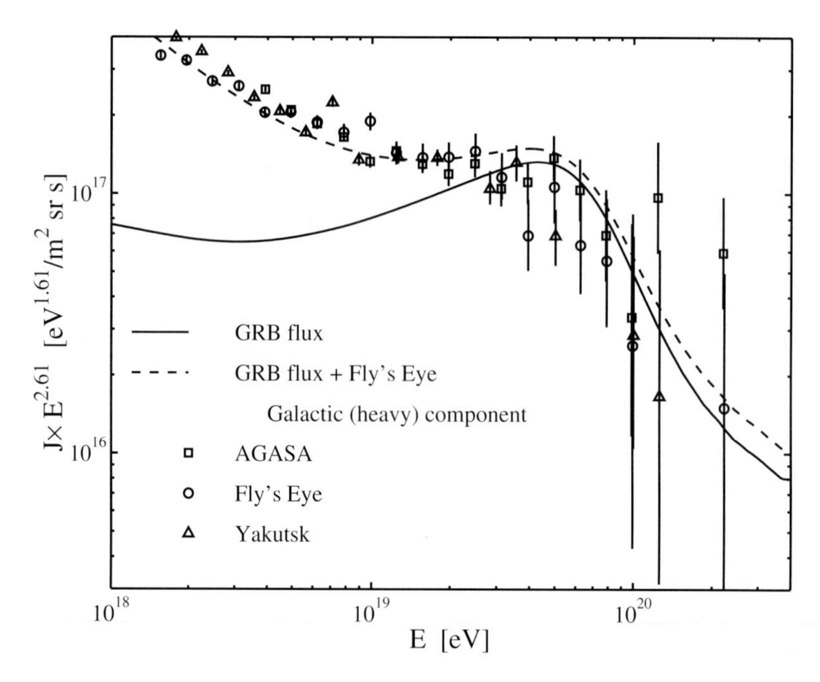

Fig. 2. The UHECR flux expected in a cosmological model where high-energy protons are produced at a rate $(\epsilon_p^2 d\dot{n}_p/d\epsilon_p)_{z=0} = 0.8 \times 10^{44}$ erg Mpc^{-3} yr^{-1} as predicted in the GRB model (Eq. 27, solid line), compared to the *High Resolution Fly's Eye (HiRes)* [15], *Yakutsk Extensive Air Shower Array (Yakutsk)* [35] and *Akeno Giant Air Shower Array (AGASA)* [105] data. 1σ flux error bars are shown. The dashed line is the sum of the GRB model flux and the *HiRes* fit (with normalization increased by 25%) to the Galactic heavy nuclei component [15], $J_G \propto E^{-3.5}$, which dominates below $\sim 10^{19}$ eV.

remarkably similar to that required to account for the flux of $> 10^{19}$ eV cosmic-rays. The flux at lower energies is most likely dominated by heavy nuclei of Galactic origin [15], as indicated by the flattening of the spectrum at $\approx 10^{19}$ eV and by the evidence for a change in composition at this energy [14,15,31,46,113].

The suppression of model flux above $10^{19.7}$ eV is due to energy loss of high energy protons in interaction with the microwave background, i.e., to the "GZK cutoff" [52,129]. The available data do not allow determination of the existence (or absence) of the "cutoff" with high confidence. The *Akeno Giant Air Shower Array (AGASA)* results show an excess ($\sim 2.5\sigma$ confidence level) of events compared to model predictions above 10^{20} eV. This excess is not confirmed, however, by other experiments. Moreover, since the 10^{20} eV flux is dominated by sources at distances < 100 Mpc, over which the distribution of known astrophysical systems (e.g., galaxies and clusters of galaxies) is inhomogeneous, significant deviations from model predictions presented in Fig. 1 for a uniform source distribution are expected [116]. Clustering of cosmic-ray sources leads to a standard deviation, σ, in the expected number, N, of events above 10^{20} eV given by [8]

$$\sigma/N = 0.9(d_0/10\text{Mpc})^{0.9}, \tag{28}$$

where d_0 is the unknown scale length of the source correlation function and $d_0 \sim 10$ Mpc for field galaxies.

Although the rate of GRBs out to a distance of 100 Mpc from Earth, the maximum distance traveled by $> 10^{20}$ eV protons, is in the range of 10^{-2} to 10^{-3} yr, the number of different GRBs contributing to the flux of $> 10^{20}$ eV protons at any given time may be large. This is due to the dispersion, Δt, in proton arrival time, which is expected due to deflection by intergalactic magnetic fields and may be as large as 10^7 yr. This implies that the number of sources contributing to the flux at any given time may be as large as [115] $\sim \Delta t \times 10^{-3}$ yr $\sim 10^4$.

6.3 Neutrino Production

A burst of $\sim 10^{14}$ eV neutrinos accompanying observed γ-rays is a natural consequence of the conventional fireball scenario [122]. The neutrinos are produced by π^+ created in interactions between fireball γ-rays and accelerated protons. The key relation is between the observed photon energy, ϵ_γ, and the accelerated proton's energy, ϵ_p, at the photo-meson threshold of the Δ-resonance. In the observer frame,

$$\epsilon_\gamma \, \epsilon_p = 0.2 \, \Gamma^2 \text{ GeV}^2. \tag{29}$$

For $\Gamma \approx 300$ and $\epsilon_\gamma = 1$ MeV, we see that characteristic proton energies $\sim 10^{16}$ eV are required to produce pions. The pion typically carries $\approx 20\%$ of the interacting proton energy, and this energy is roughly equally distributed between the leptons in the decay $\pi^+ \to \mu^+ + \nu_\mu \to e^+ + \nu_e + \overline{\nu}_\mu + \nu_\mu$. Thus, proton interaction with fireball γ-rays is expected to produce $\sim 10^{14}$ eV neutrinos.

The fraction $f_\pi(\epsilon_p)$ of proton energy lost to pion production is determined by the number density of photons in the dissipation region and is $\approx 20\%$ at high proton energy for fireball wind parameters implied by γ-ray observations [56,122]. Assuming that GRBs produce high energy protons at a rate given by Eq. (27), the intensity of high energy neutrinos is [122]

$$\epsilon_\nu^2 \Phi_{\nu_x} \approx 10^{-9} \left(\frac{f_\pi}{0.2}\right) \min\left(1, \frac{\epsilon_\nu}{10^{14}\text{eV}}\right) \text{ GeV cm}^{-2} \text{ sr}^{-1} \text{ s}^{-1}. \tag{30}$$

Here, ν_x stands for ν_μ, $\bar{\nu}_\mu$ or ν_e. The neutrino flux of Eq. (30) is suppressed at high energy, $> 10^{16}$ eV, due to synchrotron energy loss of pions and muons [94,122].

During the transition to self-similarity, high energy protons accelerated in the reverse shock may interact with the 10 eV to 1 keV photons radiated by the accelerated electrons to produce, through pion decay, a burst of duration $\sim t_\pi$ of ultra-high energy, $10^{17} - 10^{19}$ eV, neutrinos [123] as indicated by Eq. (29). The flux of these neutrinos depends on the density of gas surrounding the fireball.

It is weak, and undetectable by experiments under construction, if the density is $n \sim 1\ \mathrm{cm}^{-3}$, a value typical for the ISM. If GRBs result, however, from the collapse of massive stars, then the fireball is expected to expand into a pre-existing wind and the transition to self-similar behavior takes place at a radius where the wind density is $n \approx 10^4\ \mathrm{cm}^{-3} \gg 1\ \mathrm{cm}^{-3}$. In this case, a typical GRB at $z \sim 1$ is expected to produce a neutrino fluence [30,123]

$$\epsilon_\nu^2 \Phi_{\nu_x} \approx 10^{-2.5} \left(\frac{\epsilon_\nu}{10^{17}\mathrm{eV}}\right)^\alpha \ \mathrm{GeV\ cm}^{-2}, \tag{31}$$

where $\alpha = 0$ for $\epsilon_\nu > 10^{17}$eV and $\alpha = 1$ for $\epsilon_\nu < 10^{17}$eV. The neutrino flux is expected to be strongly suppressed at energy $> 10^{19}$ eV, since protons are not expected to be accelerated to energy $\gg 10^{20}$ eV.

6.4 Implications of Neutrino Emission

The predicted intensity of 10^{14} eV neutrinos produced by photo-meson interactions with observed 1 MeV photons, Eq. (30), implies a detection of ~ 10 neutrino induced muons per year in planned 1 km^3 Čerenkov neutrino detectors, correlated in time and direction with GRBs [3,58,122]. The predicted intensity of 10^{17} eV neutrinos, produced by photo-meson interactions during the onset of fireball interaction with its surrounding medium in the case of fireball expansion into a pre-existing wind, Eq. (31), implies a detection of several neutrino induced muons per year in a 1 km^3 detector. In this case, the predicted flux of 10^{19} eV neutrinos may also be detectable by planned large air-shower detectors [23,73,104].

Detection of high energy neutrinos will test the shock acceleration mechanism and the suggestion that GRBs are the sources of ultra-high energy protons, since $\geq 10^{14}$ eV ($\geq 10^{18}$ eV) neutrino production requires protons of energy $\geq 10^{16}$ eV ($\geq 10^{19}$ eV). The dependence of the $\sim 10^{17}$ eV neutrino flux on fireball environment implies that the detection of high energy neutrinos will also provide constraints on the GRB progenitors. Furthermore, it has recently been pointed out [84] that if GRBs originate from core-collapse of massive stars, then a burst of ≥ 5 TeV neutrinos may be produced by photo-meson interaction while the jet propagates through the envelope, with TeV fluence implying $0.1 - 10$ neutrino events per individual collapse in a 1 km^3 neutrino telescope. (The neutrino flux which may result from nuclear collisions in the expanding jet is more difficult to detect due to the low energy of neutrinos, ~ 10 GeV, produced by this process [7,32,82]).

Detection of neutrinos from GRBs could be used to test the simultaneity of neutrino and photon arrival to an accuracy of ~ 1 s (~ 1 ms for short bursts), checking the assumption of special relativity that photons and neutrinos have the same limiting speed. These observations would also test the weak equivalence principle, according to which photons and neutrinos should suffer the same time delay as they pass through a gravitational potential. With 1 s accuracy, a burst at 100 Mpc would reveal a fractional difference in limiting speed of 10^{-16}, and a fractional difference in gravitational time delay of order 10^{-6} (considering the

Galactic potential alone). Previous applications of these ideas to SN1987A, where simultaneity could be checked only to an accuracy of order several hours, yielded much weaker upper limits, of order 10^{-8} and 10^{-2} for fractional differences in the limiting speed and time delay respectively [6].

The model discussed above predicts the production of high energy muon and electron neutrinos. However, if the atmospheric neutrino anomaly has the explanation usually given [25,40,45], oscillation to ν_τ's with mass $\sim$ 0.1 eV, then one should detect equal numbers of ν_μ's and ν_τ's. Up-going τ's, rather than μ's, would be a distinctive signature of such oscillations. Since ν_τ's are not expected to be produced in the fireball, looking for τ's would be an "appearance experiment." To allow flavor change, the difference in squared neutrino masses, Δm^2, should exceed a minimum value proportional to the ratio of source distance and neutrino energy [6]. A burst at 100 Mpc producing 10^{14} eV neutrinos can test for $\Delta m^2 \geq 10^{-16}$ eV^2, 5 orders of magnitude more sensitive than solar neutrinos.

Acknowledgments

This work was supported in part by grants from the Israel-US BSF (BSF-9800343), MINERVA, and AEC (AEC-38/99). EW is the Incumbent of the Beracha foundation career development chair.

References

1. C.W. Akerlof et al.: Nature **398**, 400 (1999)
2. M. Aloy et al.: Astrophys. J. Lett. **531**, L119 (2000)
3. J. Alvarez-Muniz, F. Halzen, D.W. Hooper: Phys. Rev. **D62**, 093015 (2000)
4. L. Amati et al.: Science **290**, 953 (2000)
5. W.I. Axford, E. Leer, E., G. Skadron: In: *Proc. 15th Internat. Cosmic Ray Conference at Plovdiv, Bulgaria, August 13–26, 1977* (B'lgarska Akademia na Naukite, Sofia 1978) p. 132
6. J.N. Bahcall: In: *N*eutrino Astrophysics (Cambridge University Press, Cambridge 1989) p. 438
7. J.N. Bahcall, P. Mészáros: Phys. Rev. Lett. **85**, 1362 (2000)
8. J.N. Bahcall, E. Waxman: Astrophys. J. **542**, 542 (2000)
9. D. Band et al.: Astrophys. J. **413**, 281 (1993)
10. M. Baring: Astrophys. J. **418**, 391 (1993)
11. J. Bednarz, M. Ostrowski: Phys. Rev. Lett. **80**, 3911 (1998)
12. A.R. Bell: Mon. Not. R. Astron. Soc. **182**, 147 (1978)
13. P.N. Bhat et al.: Nature **359**, 217 (1992)
14. D.J. Bird et al.: Phys. Rev. Lett. **71**, 3401 (1993)
15. D.J. Bird et al.: Astrophys. J. **424**, 491 (1994)
16. R. Blandford, D. Eichler: Phys. Rep. **154**, 1 (1987)
17. R.D. Blandford, C.F. Mckee: Phys. Fluids **19**, 1130 (1976)
18. R.D. Blandford, J.P. Ostriker: Astrophys. J. Lett. **221**, L229 (1978)
19. R.D. Blandford, R.L. Znajek: Mon. Not. R. Astron. Soc. **179**, 433 (1977)
20. J.S. Bloom, S.R. Kulkarni, S.G. Djorgovski: Astron. J. **123**, 1111 (2002)

21. M. Böttcher, C.D. Dermer, E.P. Liang: Astron. Astrophys. Suppl. Ser. **138**, 343 (1999)
22. J.J. Brainerd et al.: In: *Proc. 19th Texas Symposium on Relativistic Astrophysics and Cosmology at Paris, France, December 14–18, 1998*, ed. J. Paul, T. Montmerle, E. Aubourg (CEA, Saclay 1998)
23. K.S. Capelle, J.W. Cronin, G. Parente, E. Zas: Astropart. Phys. **8**, 321 (1998)
24. P.J. Cargill, K. Papadopoulos: Astrophys. J. Lett. **329**, L29 (1988)
25. D. Casper et al.: Phys. Rev. Lett. **66**, 2561 (1991)
26. A.J. Castro-Tirado et al.: Science **283**, 2069 (1999)
27. S.C. Corbató et al.: Nucl. Phys. B **28B**, 36 (1992)
28. E. Costa et al.: Nature **387**, 783 (1997)
29. J.W. Cronin: Nucl. Phys. B **28B**, 213 (1992)
30. Z.G. Dai, T. Lu: Astrophys. J. **551**, 249 (2001)
31. B.R. Dawson, R. Meyhandan, K.M. Simpson: Astropart. Phys. **9**, 331 (1998)
32. E.V. Derishev, V.V. Kocharovsky, Vl.V. Kocharovsky: Astrophys. J. **521**, 640 (1999)
33. C.D. Dermer: Astrophys. J. **574**, 65 (2002)
34. T.D. Draine, L. Hao: Astrophys. J. **569**, 780 (2002)
35. N.N. Efimov et al.: In: *Proc. of the Internat. Symp. on Astrophys. Aspects of the Most Energetic Cosmic-Rays*, ed. by M. Nagano, F. Takahara (World Scientific, Singapore 1991) p. 20
36. D. Eichler, A. Levinson: Astrophys. J. **529**, 146 (2000)
37. D. Eichler, M. Livio, D.N. Schramm: Nature **340**, 126 (1989)
38. G.J. Fishman et al.: Astrophys. J. Suppl. **92**, 229 (1994)
39. G.J. Fishman, C.A. Meegan: Ann. Rev. Astron. Astrophys. **33**, 415 (1995)
40. G.L. Fogli, E. Lisi: Phys. Rev. **D52**, 2775 (1995)
41. D. Frail, E. Waxman, S. Kulkarni: Astrophys. J. **537**, 191 (2000)
42. D.L. Freedman, E. Waxman: Astrophys. J. **547**, 922 (2001)
43. A.S. Fruchter, J.H. Krolik, J.E. Rhoads: Astrophys. J. **563**, 597 (2001)
44. C.L. Fryer, S.E. Woosley: Astrophys. J. Lett. **502**, L9 (1998)
45. Y. Fukuda et al.: Mod. Phys. Lett. B **B335**, 237 (1994)
46. T.K. Gaisser et al.: Phys. Rev. **D47**, 1919 (1993)
47. T. Galama et al.: Nature **395**, 670 (1998)
48. Y.A. Gallant, A. Achterberg: Mon. Not. R. Astron. Soc. **305**, L6 (1999)
49. G. Ghisellini, A. Celotti: Astrophys. J. Lett. **511**, L93 (1999)
50. J. Goodman: Astrophys. J. Lett. **308**, L47 (1986)
51. J. Goodman: New Astron. **2**, 449 (1997)
52. K. Greisen: Phys. Rev. Lett. **16**, 748 (1966)
53. A. Gruzinov, E. Waxman: Astrophys. J. **511**, 852 (1999)
54. A. Gruzinov: Astrophys. J. **563**, 15 (2001)
55. D. Guetta, M. Spada, E. Waxman: Astrophys. J. **557**, 399 (2001)
56. D. Guetta, M. Spada, E. Waxman: Astrophys. J. **559**, 101 (2001)
57. F. Halzen: In: *Weak Interactions and Neutrinos, Proc. of the 17th Internat. Workshop at Cape Town, South Africa, January 24–30, 1999*, ed. by C.A. Dominguez, R.D. Viollier (World Scientific Publishers, Singapore 2000) p. 123
58. F. Halzen, D.W. Hooper: Astrophys. J. Lett. **527**, L93 (1999)
59. N. Hayashida et al.: Astrophys. J. **522**, 225 (1999)
60. D.J. Helfand, R.H. Becker: Astrophys. J. **314**, 203 (1987)
61. D.W. Hogg, A.S. Fruchter: Astrophys. J. **520**, 54 (1999)

62. J.I. Katz: Astrophys. J. Lett. **432**, L107 (1994)
63. Y. Kazimura, J.I. Sakai, T. Neubert, S.V. Bulanov: Astrophys. J. Lett. **498**, L183 (1998)
64. R.M. Kippen: In: *Gamma-Ray Bursts in the Afterglow Era, Proc. International Workshop at Rome, Italy, 17–20 October 2000*, ed. by E. Costa, F. Frontera, J. Hjorth (Springer, Heidelberg 2001)
65. J.G. Kirk, A.W. Guthmann, Y.A. Gallant, A. Achterberg: Astrophys. J. **542**, 235 (2000)
66. K. Koyama et al.: Nature **378**, 255 (1995)
67. J.H. Krolik, E.A. Pier: Astrophys. J. **373**, 277 (1991)
68. M. Krumholtz, S.E. Thorsett, F.A. Harrison: Astrophys. J. Lett. **506**, L81 (1998)
69. S.R. Kulkarni et al.: SPIE **4005**, 9 (2000)
70. S.R. Kulkarni et al.: Nature **398**, 389 (1999)
71. J.M. Lattimer, D.N. Schramm: Astrophys. J. Lett. **192**, L145 (1974)
72. A. Levinson, D. Eichler: Astrophys. J. **418**, 386 (1993)
73. J. Linsley: In: *Proc. 19th Internat. Cosmic Ray Conference at La Jolla, CA, USA, August 11-23, 1985*, ed. by F.C. Jones et al.(NASA Conf. Publ. No. 2376, 1985) p. 438
74. M. Livio: Phys. Rep. **311**, 225 (1999)
75. M. Livio, E. Waxman: Astrophys. J. **538**, 187 (2000)
76. A. Loeb, R. Perna: Astrophys. J. Lett. **503**, L35 (1998)
77. S. Mao, H.J. Mo: Astron. Astrophys. **339**, L1 (1998)
78. P. Mészáros: Astron. Astrophys. Suppl. Ser. **138**, 533 (1999)
79. P. Mészáros, M. Rees: Mon. Not. R. Astron. Soc. **269**, 41P (1994)
80. P. Mészáros, M. Rees: Astrophys. J. **476**, 232 (1997)
81. P. Mészáros, M. Rees: Mon. Not. R. Astron. Soc. **306**, L39 (1999)
82. P. Mészáros, M. Rees: Astrophys. J. Lett. **541**, L5 (2000)
83. P. Mészáros, M. Rees: Astrophys. J. **556**, 37 (2001)
84. P. Mészáros, E. Waxman: Phys. Rev. Lett. **87**, 1101 (2001)
85. M. Milgrom, V. Usov: Astrophys. J. Lett. **449**, L37 (1995)
86. R. Narayan, B. Paczyńsk, T. Piran: Astrophys. J. Lett. **395**, L83 (1992)
87. B. Paczyński: Astrophys. J. Lett. **308**, L43 (1986)
88. B. Paczyński: Astrophys. J. **363**, 218 (1990)
89. B. Paczyński: In: *Gamma-Ray Bursts, 4th Huntsville Symposium at Huntsville, AL, USA, September 15–20, 1997*, ed. by C. Meegan, R. Preece, T. Koshut (AIP Press, New York 1998), p.783
90. B. Paczyński, J. Rhoads: Astrophys. J. Lett. **418**, L5 (1993)
91. B. Paczyński, G. Xu: Astrophys. J. **427**, 708 (1994)
92. T. Piran: Phys. Rep. **333**, 529 (2000)
93. L. Piro et al.: Science **290**, 955 (2000)
94. J.P. Rachen, P. Mészáros: Phys. Rev. D **58**, 123005 (1998)
95. M. Rees, P. Mészáros: Mon. Not. R. Astron. Soc. **258**, 41P (1992)
96. D.E. Reichart: astro-ph 0107546 (2001)
97. R.Z. Sagdeev: Rev. Plasma Phys. **4**, 23 (1966)
98. R. Sari, T. Piran: Astrophys. J. **485**, 270 (1997)
99. R. Sari, T. Piran: Astrophys. J. Lett. **517**, L109 (1999)
100. R. Sari, T. Piran, J. Halpern: Astrophys. J. Lett. **519**, L17 (1999)
101. M. Schmidt: Astrophys. J. **552**, 36 (2001)
102. A. Shemi, T. Piran: Astrophys. J. Lett. **365**, L55 (1990)
103. M. Sommer et al.: Astrophys. J. Lett. **422**, L63 (1994)

104. Y. Takahashi: In: *Proc. 24th Internat. Cosmic-Ray Conference at Rome, Italy, August 28–September 8, 1995*, ed. by N. Iucci, E. Lamanna (XXIV ICRC, Rome 1995) p. 595
105. M. Takeda et al.: Phys. Rev. Lett. **81**, 1163 (1998)
106. M. Teshima et al.: Nucl. Phys. B **28B**, 169 (1992)
107. C. Thompson: Mon. Not. R. Astron. Soc. **270**, 480 (1994)
108. T. Totani: Astron. Astrophys. Suppl. Ser. **142**, 443 (2000)
109. V.V. Usov: Mon. Not. R. Astron. Soc. **267**, 1035 (1994)
110. J. van Paradijs et al.: Science **286**, 693 (1999)
111. M. Vietri: Astrophys. J. **453**, 883 (1995)
112. M. Vietri: Astrophys. J. Lett. **478**, L9 (1997)
113. A.A. Watson: Nucl. Phys. B **22B**, 116 (1991)
114. A.A. Watson: In: *Nuclear and Particle Physics, Inst. of Physics Conf. Series*, ed. I.J.D. MacGregor, A.T. Doyle (IoP, Bristol 1993) p. 135
115. E. Waxman: Phys. Rev. Lett. **75**, 386 (1995)
116. E. Waxman: Astrophys. J. Lett. **452**, L1 (1995)
117. E. Waxman: Astrophys. J. Lett. **485**, L5 (1997)
118. E. Waxman: Astrophys. J. Lett. **489**, L33 (1997)
119. E. Waxman: Astrophys. J. Lett. **491**, L19 (1997)
120. E. Waxman: Nucl. Phys. B **87**, 345 (2000)
121. E. Waxman: In: *ICTP Summer School at ICTP, Italy, June 2000 and VI Gleb Wataghin School at UNICAMP, Brazil, July 2000*, Lecture Notes (astro-ph/0103186)
122. E. Waxman, J.N. Bahcall: Phys. Rev. Lett. **78**, 2292 (1997)
123. E. Waxman, N.J. Bahcall: Astrophys. J. **541**, 707 (2000)
124. E. Waxman, B.T. Draine: Astrophys. J. **537**, 796 (2000)
125. E. Waxman, S. Kulkarni, D. Frail: Astrophys. J. **497**, 288 (1998)
126. A.M.J. Wijers, M.J. Rees, P. Mészáros: Mon. Not. R. Astron. Soc. **288**, L51 (1997)
127. E. Woods, A. Loeb: Astrophys. J. **453**, 583 (1995)
128. S.E. Woosley: Astrophys. J. **405**, 273 (1993)
129. G.T. Zatsepin, V.A. Kuzmin: JETP Lett. **4**, 78 (1966)

Ambient Interaction Models for γ-Ray Burst Afterglows

Zhi-Yun Li and Roger A. Chevalier

Department of Astronomy, University of Virginia, Charlottesville, VA 22903, USA

Abstract. We review ambient interaction models for γ-ray burst afterglows, with an eye on constraining the nature of the progenitors and the geometry of the explosion. Evidence is presented for two types of progenitors. The radio afterglow of GRB980425/SN1998bw and the multi-frequency observations of GRB970508 can be fitted by a blastwave expanding into a wind-type medium, pointing to a massive star progenitor. The broadband afterglow data of GRB990123 and GRB990510 are better modeled by a jet expanding into a constant density medium, implying a compact star merger origin. Among other well observed γ-ray bursts, the jet model appears to be more widely applicable than the wind model, although some cases are ambiguous. The model fits often require a deviation of the energy distribution of the radiating electrons from the commonly assumed single power-law form, particularly for sources with a rapid decline and/or a pronounced steepening in the optical light curves. Transition to non-relativistic evolution has been suggested as an alternative explanation for the light curve steepening, although to produce steepening on the order of days or less would require very high ambient densities which are generally difficult to reconcile with radio observations. Major open issues include the hydrodynamics of jet-ambient medium interaction, a self-consistent determination of the electron energy distribution, and the effects of pair production on the early afterglows which are expected to be particularly large for the wind interaction model.

1 Introduction

The simplest model of the afterglows of γ-ray bursts (GRBs) involves a spherical relativistic blastwave expanding into a constant density, presumably interstellar, medium [56,68]. The afterglows are emitted by nonthermal electrons accelerated at the shock front to an energy distribution usually assumed to be a power law above some cutoff determined by the shock velocity. The predicted power-law decay with time of the afterglow emission was subsequently observed at X-ray [19] and optical [110] wavelengths for GRB970228, giving basic confirmation to this now "standard" picture [113,123].

To date, nearly 50 GRB afterglows have been observed at more than one frequency (for reviews of afterglow observations, see the chapters by Pian, by Frontera, and by Weiler et al. in this volume). A dozen or so of these are well observed at multiple frequencies to allow for detailed modeling. Some features of these afterglows turn out to be difficult to accommodate in the standard model. The most noticeable is the relatively steep decline of the optical light curves, sometimes preceded by flatter evolution. The steep decline is usually attributed to a collimated or jet-like, rather than spherical, initial energy injection [93,102]. Alternatively, it could be due to the transition of blastwave evolution to the non-relativistic regime [25], a non-standard electron energy distribution [63,75], and/or an expansion into a wind-type ambient medium [17]. The question of a stellar wind ($\rho \propto r^{-s}$ with $s \sim 2$) established circumstellar medium (CSM) versus a constant density ($s = 0$) interstellar medium (ISM) surrounding the explosion is a crucial one for the progenitors of GRBs, since massive stars, one of the leading candidates for GRB progenitors [74,125], should be surrounded by a wind while GRBs resulting from compact star mergers, the other leading candidate, are expected to be surrounded by the general ISM (see the chapter by Waxman, in this volume).

The plan of this chapter is as follows: in Sect. 2 we discuss the likely environments of GRBs relevant to afterglow evolution; in Sect. 3, we outline the basics of various ambient interaction models for GRB afterglows; and in Sect. 4 these models are applied to individual GRBs with relatively well observed afterglows. In Sect. 5 we find that some of the sources are probably wind interactors, while others are better modeled as interacting with a constant density ISM.

2 The Ambient Medium

Models for the GRB afterglows indicate that the emission comes from a region $\sim 10^{16} - 10^{18}$ cm from the source of the explosion. The nature of the material in this region depends on the GRB progenitors, which are presently not known. There are two main possibilities. In one model, the progenitors are massive stars at the end of their lives [74,125]. In this case, the interaction is expected to be primarily with a radially decreasing density wind established by mass-loss from the progenitor prior to the explosion. In the other model, the progenitors are mergers of two compact objects for which there are a number of possibilities [37]. In this case, there may be some debris from the merger, but the interaction is expected to be primarily with a constant density ISM. We consider both of these possibilities.

If the progenitors are massive stars, there is an analogy to the explosions of core-collapse supernovae, for which there is abundant evidence that they are interacting with the winds from the progenitor stars (see the chapters by Chevalier and Fransson and by Sramek and Weiler in this volume). In general, the interaction appears to be with the free wind from the progenitor star, with density

$$\rho_{\rm wind} = \dot{M}/(4\pi r^2 w_{\rm wind}), \tag{1}$$

where $\dot{M}$ is the mass-loss rate from the star and $w_{\rm wind}$ is the wind velocity. In most of the supernova cases, the radial range that is observed is out to a few $\times 10^{17}$ cm, so that the mass-loss characteristics have not substantially changed during the time that mass is supplied to the wind and $\rho_{\rm wind} \propto r^{-2}$, although deviations from this simple structure have been reported (see [119,120] and the chapter by Sramek and Weiler in volume).

The density in the wind depends on the type of progenitor. Red supergiant stars, which are thought to be the progenitors of most type II supernovae, have slow (~ 10 km s^{-1}) dense winds. Wolf-Rayet (W-R) stars, which are thought to be the progenitors of type Ib and type Ic supernovae, have faster ($\sim 1,000$ km s^{-1}), lower density winds. SN1987A is a special case in which the star was a RSG, but became a B3 blue supergiant (BSG) about 10^4 years before the explosion. The result for SN1987A was a complex circumstellar medium, including a dense ring at a radius of $\sim 6 \times 10^{17}$ cm.

If GRBs do have massive stellar progenitors, there are a number of arguments suggesting that W-R stars are the most likely progenitors:

- SN1998bw, the best case of a SN-GRB association (with GRB980425), was of type Ic, with a probable W-R progenitor.
- The high energy of GRBs suggests that a moderately massive (> 3 M$_\odot$) black hole is involved which, in turn, requires a massive, $\gtrsim 20 - 25$ M$_\odot$ progenitor [28]. These stars are likely to be W-R stars at the end of their lives [43].
- The relativistic flow from a central object may be able to penetrate a relatively compact W-R star, but probably cannot penetrate an extended RSG star [65].
- If a rapidly rotating black hole is required for the GRB explosion, a merger at the center of a massive star is one way to obtain a rapidly rotating core with the merger process giving rise to a W-R star.

None of these arguments is definitive, but they do point toward the most plausible progenitors.

The circumstellar medium created by a 35 M$_\odot$ initial mass star that passes through a W-R phase and a prior RSG phase has been simulated by Garcia-Segura et al. [43]. They find that during the W-R phase a broken shell is created at a radius of a few pc or more. This type of shell has been observed around a number of W-R stars. Inside of the shell is a region of shocked wind with an approximately constant density [15]. The size of the shocked region is determined by what is required to decelerate the wind and is typically less than half of the shell radius. In general, the wind ram pressure $p = \rho_{\rm wind} w_{\rm wind}^2$ determines the size of the wind bubble that is created. The pressure is

$$p/k = 3.6 \times 10^9 \; \dot{M}_{-5} w_8 r_{17}^{-2} \; \mathrm{cm}^{-3} \;\; \mathrm{K}, \tag{2}$$

where

$$\dot{M}_{-5} = \dot{M}/10^{-5} \; \mathrm{M}_\odot \; \mathrm{yr}^{-1} \tag{3}$$

$$w_8 = w_{\rm wind}/10^8 \; \mathrm{cm \; s}^{-1} \tag{4}$$

$$r_{17} = r/10^{17} \mathrm{cm}. \tag{5}$$

For comparison, the interstellar pressure in the solar neighborhood is ~ 3000 cm^{-3} K. The region out to the wind termination shock can be described by a density

$$\rho_{\rm wind} = 5 \times 10^{11} A_* r^{-2} \; \mathrm{g \; cm}^{-3}, \tag{6}$$

where

$$A_* = 1\dot{M}_{-5} w_8^{-1}. \tag{7}$$

Considering the radial range relevant to GRB afterglows, we expect this to apply to most cases, unless the wind is especially weak or the surrounding pressure is extraordinarily high. Ramirez-Ruiz et al. [89] have followed the circumstellar mass loss from W-R stars and found that the free wind does not extend to large radii in their treatment.

For bursts that interact directly with the ISM, the size scale of the afterglows suggests that the medium has a roughly constant density, although the ISM is known to have considerable density inhomogeneity on a range of scales. The range of interstellar densities in a galaxy like our own extends from $\sim 10^{-3}$ cm^{-3} in the hot ISM to $\gtrsim 10^6$ cm^{-3} in the compact cores in molecular clouds. The high densities are found in a very small volume fraction so that a typical event is unlikely to occur in such a region. The situation may change in a starburst region where a significant fraction of the gas may have a density $\gtrsim 10^3$ cm^{-3} [104]. The radio bright, compact supernova remnants in starburst regions such as M82 may be interacting with this dense interstellar component [14]. The finding that GRB010222 may have occurred within a very active starburst [35] indicates that the interstellar environments of GRBs may have unusual properties.

Although the interstellar and wind models are the two main types of environments considered for afterglows, there is a different scenario involving a massive star, motivated by the observation of possible Fe lines in the X-ray spectra of afterglows [2,85,86,126]. The observations appear to require a substantial mass of Fe at $r \sim 10^{16}$ cm, so one suggestion is that a supernova occurred before a GRB and that the ejecta have had time to expand to this radius [111]. This requires that the GRB be delayed by months while the supernova expands into the progenitor wind, creating a complex circumburst region in the inner part of the wind.

3 Ambient Interaction Models

3.1 Standard ISM Interaction Model

We start with the simplest model of GRB afterglows involving synchrotron emission from a spherical relativistic blastwave propagating into a constant density interstellar medium. The model was worked out in some detail by Mészáros and Rees [68] in advance of any afterglow detections. It has since been elaborated upon by others (see, e.g., [113,114,122]) and, in particular, [103]). Here, we will follow the formalism of Sari et al. [103], which has been widely used to interpret the spectra and light curves of GRB afterglows. This standard model serves as a benchmark against which other models will be compared.

The standard model assumes that the synchrotron emitting electrons are accelerated at the shock front to a power-law distribution of Lorentz factor

$$N(\Gamma_{\rm e})\, d\Gamma_{\rm e} \propto \Gamma_{\rm e}^{-p}\, d\Gamma_{\rm e} \tag{8}$$

(with $p > 2$) above some minimum cutoff $\Gamma_{\rm m}$, which is determined by the shock velocity. This distribution is further modified by synchrotron cooling in the downstream flow. Denoting by $\Gamma_{\rm c}$ the Lorentz factor of an electron which cools in a blastwave expansion time, in the "fast cooling" case with $\Gamma_{\rm c} < \Gamma_{\rm m}$, the electron distribution is given by $\Gamma_{\rm e}^{-2}$ between $\Gamma_{\rm c}$ and $\Gamma_{\rm m}$ and by $\Gamma_{\rm e}^{-p-1}$ above $\Gamma_{\rm m}$. This electron distribution produces a synchrotron spectrum of the flux

$$F_\nu \propto \nu^{-1/2}, \ \nu_{\rm c} \le \nu \le \nu_{\rm m}, \tag{9}$$

where $\nu_{\rm c}$ and $\nu_{\rm m}$ are the characteristic frequencies of the photons produced by electrons with $\Gamma_{\rm c}$ and $\Gamma_{\rm m}$ respectively, and

$$F_\nu \propto \nu^{-p/2}, \ \nu > \nu_{\rm m}, \tag{10}$$

$$F_\nu \propto \nu^{1/3}, \ \nu < \nu_{\rm c}. \tag{11}$$

The spectrum is modified by synchrotron self-absorption (SSA), which typically occurs at relatively low frequencies. Assuming that the SSA frequency $\nu_{\rm SSA}$ is less than the cooling frequency $\nu_{\rm c}$ and that the absorbing electrons that have cooled for different lengths of time are well mixed spatially [47], we have

$$F_\nu \propto \nu^{2}, \ \nu < \nu_{\rm SSA}. \tag{12}$$

In the opposite, slow cooling case with $\Gamma_{\rm c} > \Gamma_{\rm m}$, the electron distribution steepens to $\Gamma_{\rm e}^{-p-1}$ above $\Gamma_{\rm c}$. Assuming that the SSA frequency, $\nu_{\rm SSA}$, is less than the characteristic frequency $\nu_{\rm m}$, we have the following spectrum:

$$F_\nu \propto \nu^{2}, \ \nu < \nu_{\rm SSA}, \tag{13}$$

$$F_\nu \propto \nu^{1/3}, \ \nu_{\rm SSA} < \nu < \nu_{\rm m}, \tag{14}$$

$$F_\nu \propto \nu^{-(p-1)/2}, \ \nu_{\rm m} < \nu < \nu_{\rm c}, \tag{15}$$

$$F_\nu \propto \nu^{-p/2}, \ \nu > \nu_{\rm c}. \tag{16}$$

These broken power-law expressions are often used to interpret the instantaneous spectra of GRB afterglows. They tend to agree with the more accurately determined spectra to within a factor of a few (see, e.g., [46]).

To obtain light curves, one must first determine the time evolution of the characteristic frequencies ($\nu_{\rm SSA}$, $\nu_{\rm m}$ and $\nu_{\rm c}$) and the peak flux $F_{\nu,\rm max}$. The evolution depends on the dynamics of the blastwave. For the spherical blastwave propagating into a constant density medium envisioned in the standard model, the dynamics are described by the self-similar solutions of Blandford and McKee [7]: the blastwave Lorentz factor Γ varies with the distance R from the explosion center as

$$\Gamma \propto R^{-3/2} \tag{17}$$

in the adiabatic regime and

$$\Gamma \propto R^{-3} \tag{18}$$

in the fully radiative regime. The Lorentz factor and distance are related to the detector time, t, as

$$t \propto R/\Gamma^2. \tag{19}$$

Making the standard assumption that a constant fraction ϵ_e (ϵ_B) of the blastwave energy goes into the electrons (magnetic fields) and making use of the jump conditions for relativistic shocks, one obtains the following scalings:

$$\nu_{SSA} \propto t^0; \tag{20}$$
$$\nu_m \propto t^{-3/2}; \tag{21}$$
$$\nu_c \propto t^{-1/2}; \tag{22}$$
$$F_{\nu,\max} \propto t^0 \tag{23}$$

for the slow cooling case, which is always in the adiabatic regime. In the fast cooling case, if the electron energy fraction ϵ_e is close to unity, then the blastwave is fully radiative, and the scalings become:

$$\nu_{SSA} \propto t^{-4/5}; \tag{24}$$
$$\nu_c \propto t^{-2/7}; \tag{25}$$
$$\nu_m \propto t^{-12/7}; \tag{26}$$
$$F_{\nu,\max} \propto t^{-3/7}. \tag{27}$$

If on the other hand for $\epsilon_e \ll 1$, the evolution remains adiabatic and the scalings are the same as in the slow cooling case, except for the SSA frequency which now decreases with time as

$$\nu_{SSA} \propto t^{-1/2}. \tag{28}$$

These scalings are combined with the instantaneous spectra listed above to obtain light curves.

The light curve at a given observing frequency ν can be described by a broken power-law with breaks at various characteristic times. These include the times t_{SSA}, t_m and t_c when the characteristic frequencies ν_{SSA}, ν_m and ν_c pass, respectively, the observing frequency ν. Ignoring SSA, Sari et al. [103] obtained two types of light curves in two frequency regimes, separated by a critical frequency ν_0 at which the characteristic times t_m and t_c become equal, i.e.,

$$t_m(\nu_0) = t_c(\nu_0) \equiv t_0. \tag{29}$$

The critical time t_0 divides the early, fast cooling part of the light curve from the later, slow cooling part. In the high frequency regime where $\nu > \nu_0$, the ordering of the characteristic times is

$$t_c < t_m < t_0 \tag{30}$$

which yields a light curve of

$$F_\nu \propto t^{1/6},\ t < t_c, \tag{31}$$
$$F_\nu \propto t^{-1/4},\ t_c < t < t_m, \tag{32}$$
$$F_\nu \propto t^{(2-3p)/4},\ t > t_m \tag{33}$$

in the case of adiabatic evolution. In the opposite case of fully radiative evolution, which is possible only before the critical time t_0 when the electrons are fast cooling, one has instead

$$F_\nu \propto t^{-1/3},\ t < t_c, \tag{34}$$
$$F_\nu \propto t^{-4/7},\ t_c < t < t_m, \tag{35}$$
$$F_\nu \propto t^{(2-6p)/7},\ t_m < t < t_0. \tag{36}$$

In the low frequency regime where $\nu < \nu_0$, the ordering of the characteristic times becomes $t_0 < t_m < t_c$. The light curve is then

$$F_\nu \propto t^{1/6},\ t < t_0, \tag{37}$$
$$F_\nu \propto t^{1/2},\ t_0 < t < t_m, \tag{38}$$
$$F_\nu \propto t^{3(1-p)/4},\ t_m < t < t_c, \tag{39}$$
$$F_\nu \propto t^{(2-3p)/4},\ t > t_c \tag{40}$$

in the adiabatic case. In the fully radiative case, one has

$$F_\nu \propto t^{-1/3},\ t < t_0 \tag{41}$$

instead.

SSA modifies the light curves. We will concentrate on the case of adiabatic evolution here and below; it is more likely applicable than the fully radiative case during most of the afterglow phase of a GRB evolution (see, e.g., [68]). The SSA is characterized by the absorption frequency ν_{SSA}, which decreases with time as

$$\nu_{SSA} \propto t^{-1/2} \tag{42}$$

during the fast cooling period $t < t_0$. It has a constant value, denoted by

$$\nu_{SSA,0},\ t_0 < t < t_{SSAm}, \tag{43}$$

where t_{SSAm} is another critical time when the characteristic frequencies ν_{SSA} and ν_m become equal. In the frequency regime $\nu > \nu_{SSA,0}$, both the high and low frequency light curves discussed above have an additional power-law segment

$$F_\nu \propto t \tag{44}$$

before the characteristic time t_{SSA}, which is smaller than any other characteristic time. Collecting all time and frequency dependences, we finally have

$$F_\nu \propto t\nu^2,\ t < t_{SSA}, \tag{45}$$
$$F_\nu \propto t^{1/6}\nu^{1/3},\ t_{SSA} < t < t_c, \tag{46}$$
$$F_\nu \propto t^{-1/4}\nu^{-1/2},\ t_c < t < t_m, \tag{47}$$
$$F_\nu \propto t^{(2-3p)/4}\nu^{-p/2},\ t > t_m \tag{48}$$

for the high frequency light curve, and

$$F_\nu \propto t\nu^2,\ t < t_{SSA}, \tag{49}$$
$$F_\nu \propto t^{1/6}\nu^{1/3},\ t_{SSA} < t < t_0, \tag{50}$$
$$F_\nu \propto t^{1/2}\nu^{1/3},\ t_0 < t < t_m, \tag{51}$$
$$F_\nu \propto t^{3(1-p)/4}\nu^{-(p-1)/2},\ t_m < t < t_c, \tag{52}$$
$$F_\nu \propto t^{(2-3p)/4}\nu^{-p/2},\ t > t_c \tag{53}$$

for the low frequency light curve. The time and frequency dependences of the flux are sometimes parameterized using

$$F_\nu \propto \nu^\alpha t^\beta . \tag{54}$$

The relations between the spectral index, α, and the time decay index, β,

$$\beta = 3(1-p)/4 = 3\alpha/2,\ t_{\rm m} < t < t_{\rm c} \text{ or } \nu_{\rm m} < \nu < \nu_{\rm c}, \tag{55}$$

$$\beta = (2-3p)/4 = (1+3\alpha)/2,\ t > t_{\rm c} \text{ or } \nu > \nu_{\rm c} > \nu_{\rm m} \tag{56}$$

in the most relevant, slow cooling case, are often used to interpret optical and X-ray afterglow observations at relatively late times (of order one day or more). In the frequency regime $\nu < \nu_{\rm SSA,0}$, more relevant to radio afterglows, one has the ordering $t_0 < t_{\rm m} < t_{\rm SSA} < t_{\rm c}$, which yields

$$F_\nu \propto t\nu^2,\ t < t_0, \tag{57}$$

$$F_\nu \propto t^{1/2}\nu^2,\ t_0 < t < t_{\rm m}, \tag{58}$$

$$F_\nu \propto t^{5/4}\nu^{5/2},\ t_{\rm m} < t < t_{\rm SSA}, \tag{59}$$

$$F_\nu \propto t^{3(1-p)/4}\nu^{-(p-1)/2},\ t_{\rm SSA} < t < t_{\rm c}, \tag{60}$$

$$F_\nu \propto t^{(2-3p)/4}\nu^{-p/2},\ t > t_{\rm c}. \tag{61}$$

We reiterate that the scalings involving SSA in the fast cooling case with $t < t_0$ are derived assuming that the energy distribution of the cooling electrons is spatially homogeneous in the emission region behind the shock front. The opposite situation where layers of cooling electrons at different distances from the shock front remain unmixed has been considered in detail by Granot et al. [47] and Granot and Sari [48].

3.2 Wind Interaction Model

The first modification of the standard model we consider is the density distribution of the ambient medium. Mészáros et al. [71] studied the general case of a power-law ambient density distribution $n \propto r^{-s}$, with an arbitrary power index s. Chevalier and Li [16,17] examined the specific case of $s = 2$, corresponding to a constant mass-loss rate, constant velocity, circumstellar wind, possibly of a W-R origin. Some features of the wind interaction model are also described in Dai and Lu [24] and Panaitescu et al. [81]. Here, we follow the formalism of Chevalier and Li [17].

The wind interaction model has the same instantaneous afterglow spectra as in the standard ISM interaction model, but different light curves. The differences in light curves come from the blastwave dynamics. For a relativistic blastwave propagating in an $s = 2$ medium, Blandford and McKee [7] showed that its Lorentz factor is $\Gamma \propto R^{-1/2}$ in the adiabatic regime and $\Gamma \propto R^{-1}$ in the fully radiative regime. Concentrating on the adiabatic evolution as before and making the standard assumption about the fractions of the blastwave energy going into the electrons and magnetic fields, one finds that the characteristic frequencies and the peak flux scale with time as

$$\nu_{\rm SSA} \propto t^{-3/5}; \nu_{\rm m} \propto t^{-3/2}; \nu_{\rm c} \propto t^{1/2}; F_{\nu,\rm max} \propto t^{-1/2} \tag{62}$$

in the slow cooling case. In the fast cooling case, the scalings are the same except for the SSA frequency, which now decreases with time more rapidly as $\nu_{\rm SSA} \propto t^{-8/5}$.

The above scalings of the characteristic frequencies define four critical times, with four corresponding critical frequencies which divide the light curves into five distinct frequency regimes. For the typical parameters adopted by Chevalier and Li [17], the light curve that is most relevant to the optical and X-ray afterglows has the ordering $t_{\rm SSA} < t_{\rm m} < t_0 < t_{\rm c}$, which yields

$$F_\nu \propto t^{7/4}\nu^{5/2}, \ t < t_{\rm SSA}, \tag{63}$$
$$F_\nu \propto t^{-1/4}\nu^{-1/2}, \ t_{\rm SSA} < t < t_{\rm m}, \tag{64}$$
$$F_\nu \propto t^{(2-3p)/4}\nu^{-p/2}, \ t_{\rm m} < t < t_0, \tag{65}$$
$$F_\nu \propto t^{(2-3p)/4}\nu^{-p/2}, \ t_0 < t < t_{\rm c}, \tag{66}$$
$$F_\nu \propto t^{(1-3p)/4}\nu^{-(p-1)/2}, \ t > t_{\rm c}. \tag{67}$$

Note that as the cooling frequency $\nu_{\rm c}$ moves across the observing frequency ν from below, the light curve steepens from $\beta = (2-3p)/4$ to $\beta = (1-3p)/4$ by a modest amount $\Delta\beta = 1/4$. The same amount of steepening, from $\beta = 3(1-p)/4$ to $\beta = (2-3p)/4$, also occurs in the standard ISM interaction model, as $\nu_{\rm c}$ moves across ν from above. A signature of the wind interaction model is the relatively fast decline of the light curve in the non-cooling spectral region $\nu_{\rm m} < \nu < \nu_{\rm c}$, where $\beta = (1-3p)/4$ compared with $\beta = 3(1-p)/4$ for the ISM case. If the cooling frequency falls between optical and X-ray wavelengths, the wind model would predict the X-ray light curve declining less steeply than the optical light curve, whereas the opposite would be true for the ISM model.

At the much lower radio frequencies, the characteristic times typically have the following ordering: $t_{\rm c} < t_0 < t_{\rm SSA} < t_{\rm m}$, which leads to a light curve with

$$F_\nu \propto t^{7/4}\nu^{5/2}, \ t < t_{\rm c}, \tag{68}$$
$$F_\nu \propto t^2\nu^2, \ t_{\rm c} < t < t_0, \tag{69}$$
$$F_\nu \propto t\nu^2, \ t_0 < t < t_{\rm SSA}, \tag{70}$$
$$F_\nu \propto t^0\nu^{1/3}, \ t_{\rm SSA} < t < t_{\rm m}, \tag{71}$$
$$F_\nu \propto t^{(1-3p)/4}\nu^{-(p-1)/2}, \ t > t_{\rm m}. \tag{72}$$

In the slow cooling case with $t > t_0$ most relevant to radio observations, the fairly steep rise of radio flux $F_\nu \propto t$ at the self-absorbed frequencies could, in principle, distinguish the wind model from the standard ISM model where the rise is slower ($F_\nu \propto t^{1/2}$). In practice, the difference is masked to a large extent by interstellar scintillation (ISS).

3.3 Transition to Non-Relativistic Evolution

After a GRB blastwave sweeps up a mass equivalent to the rest mass of the explosion, its evolution becomes non-relativistic. In a medium of constant density n, this happens at a radius of order

$$R_{\rm nr} = 1.2 \times 10^{18} E_{52}^{1/3} n^{-1/3} \text{ cm} \tag{73}$$

where E_{52} is the explosion energy in units of 10^{52} erg and n the ambient number density in units of cm^{-3}. The transition occurs on a time scale

$$t_{\rm nr} = R_{\rm nr}/c = 1.2(1+z) E_{52}^{1/3} n^{-1/3} \text{ yr} \tag{74}$$

where z is the cosmological redshift. This is longer than the typical duration of GRB afterglow observations, unless the explosion energy is much lower than 10^{52} erg [32] and/or the density is much higher than 1 cm^{-3} [25]. Provided that the minimum electron Lorentz factor $\Gamma_{\rm m} > 1$, the temporal and frequency dependences of the flux in the non-relativistic regime can be obtained in a way similar to that in the relativistic regime [123]. Here, we follow the detailed treatment of Frail et al. [32]. The electrons are typically slow cooling, with

$$\nu_{\rm m} \propto t^{-3}; \nu_{\rm c} \propto t^{-1/5} \tag{75}$$

and the peak flux at $\nu_{\rm m}$ is given by

$$F_{\nu,\max} \propto t^{3/5}. \tag{76}$$

The SSA frequency could either be greater or smaller than $\nu_{\rm m}$. There are two relevant orderings of characteristic frequencies: $\nu_{\rm SSA} < \nu_{\rm m} < \nu_{\rm c}$ and $\nu_{\rm m} < \nu_{\rm SSA} < \nu_{\rm c}$. In the former case,

$$\nu_{\rm SSA} \propto t^{6/5}, \tag{77}$$

one has

$$F_\nu \propto \nu^2 t^{-2/5}, \quad \nu < \nu_{\rm SSA}, \tag{78}$$

$$F_\nu \propto \nu^{1/3} t^{8/5}, \quad \nu_{\rm SSA} < \nu < \nu_{\rm m}, \tag{79}$$

$$F_\nu \propto \nu^{-(p-1)/2} t^{3(7-5p)/10}, \quad \nu_{\rm m} < \nu < \nu_{\rm c}, \tag{80}$$

$$F_\nu \propto \nu^{-p/2} t^{(4-3p)/2}, \quad \nu > \nu_{\rm c}. \tag{81}$$

In the latter case,

$$\nu_{\rm SSA} \propto t^{(2-3p)/(p+4)}, \tag{82}$$

and

$$F_\nu \propto \nu^2 t^{13/5}, \quad \nu < \nu_{\rm m}, \tag{83}$$

$$F_\nu \propto \nu^{5/2} t^{11/10}, \quad \nu_{\rm m} < \nu < \nu_{\rm SSA}, \tag{84}$$

$$F_\nu \propto \nu^{-(p-1)/2} t^{3(7-5p)/10}, \quad \nu_{\rm SSA} < \nu < \nu_{\rm c}, \tag{85}$$

$$F_\nu \propto \nu^{-p/2} t^{(4-3p)/2}, \quad \nu > \nu_{\rm c}. \tag{86}$$

Note that the light curve decreases with time faster in the non-relativistic regime (where $\beta = 3[7-5p]/10$ and $\beta = [4-3p]/2$ before and after the cooling break t_c, or $\beta = -1.65$ and $\beta = -1.75$ for $p = 2.5$) than in the relativistic regime (where the corresponding $\beta = 3[1-p]/4$ and $\beta = [2-3p]/4$ or $\beta = -1.125$ and $\beta = -1.375$ for $p = 2.5$).

In the wind interaction model with an ambient density $\rho = Ar^{-2}$, the transition to non-relativistic evolution occurs at a distance

$$R_{\rm nr} = 1.8 \times 10^{18} E_{52}/A_* \ {\rm cm}, \tag{87}$$

where A_* is the coefficient A in units of 5×10^{11} g cm^{-1} [17], corresponding to a time

$$t_{\rm nr} = 1.9(1+z)E_{52}/A_* \ {\rm yr}, \tag{88}$$

again longer than most afterglow observations for typical parameters. In the non-relativistic regime, one finds

$$\nu_{\rm m} \propto t^{-7/3}; \nu_{\rm c} \propto t, \tag{89}$$

and the peak flux

$$F_{\nu,\rm max} \propto t^{-1/3} \ {\rm at}\ \nu_{\rm m}. \tag{90}$$

The relevant orderings are $\nu_{\rm SSA} < \nu_{\rm m} < \nu_{\rm c}$ and $\nu_{\rm m} < \nu_{\rm SSA} < \nu_{\rm c}$. In the former case,

$$\nu_{\rm SSA} \propto t^{13/15}, \tag{91}$$

and

$$F_\nu \propto \nu^2 t^{-1}, \ \nu < \nu_{\rm SSA}, \tag{92}$$
$$F_\nu \propto \nu^{1/3} t^{4/9}, \ \nu_{\rm SSA} < \nu < \nu_{\rm m}, \tag{93}$$
$$F_\nu \propto \nu^{-(p-1)/2} t^{(5-7p)/6}, \ \nu_{\rm m} < \nu < \nu_{\rm c}, \tag{94}$$
$$F_\nu \propto \nu^{-p/2} t^{(8-7p)/6}, \ \nu > \nu_{\rm c}. \tag{95}$$

In the latter case,

$$\nu_{\rm SSA} \propto t^{(4-7p)/[3(p+4)]}, \tag{96}$$

and

$$F_\nu \propto \nu^2 t^{-1}, \ \nu < \nu_{\rm m}, \tag{97}$$
$$F_\nu \propto \nu^{5/2} t^{1/6}, \ \nu_{\rm m} < \nu < \nu_{\rm SSA}, \tag{98}$$
$$F_\nu \propto \nu^{-(p-1)/2} t^{(5-7p)/6}, \ \nu_{\rm SSA} < \nu < \nu_{\rm c}, \tag{99}$$
$$F_\nu \propto \nu^{-p/2} t^{(8-7p)/6}, \ \nu > \nu_{\rm c}. \tag{100}$$

As in ISM interaction, the light curve decreases with time faster in the non-relativistic regime (where $\beta = [8-7p]/6$ and $\beta = [5-7p]/6$ before and after the cooling break t_c, or $\beta = -1.58$ and $\beta = -2.08$ for $p = 2.5$) than in the relativistic regime (where the corresponding $\beta = [2-3p]/4$ and $\beta = [1-3p]/4$ or $\beta = -1.375$ and $\beta = -1.625$ for $p = 2.5$).

3.4 Jet Model

Starting with Rhoads [93], jet models of afterglows have been widely discussed in connection with the steepening of light curves. We will first outline the asymptotic analysis of Sari et al. [102] and then comment on possible complications. Recent reviews of the subject include [83,95].

Let θ_0 be the initial angular width of the jet. When the Lorentz factor Γ drops below θ_0^{-1}, the jet starts to spread sideways, changing its dynamics. The spreading occurs around a time

$$t_{\rm jet} = 3[(1+z)/2](E_{52}/n)^{1/3}(\theta_0/0.2)^{8/3} \text{ days} \tag{101}$$

in a constant density medium and

$$t_{\rm jet} = 2[(1+z)/2](E_{52}/A_*)(\theta_0/0.2)^4 \text{ days} \tag{102}$$

in a radially decreasing density wind [17]. Before $t_{\rm jet}$, the standard spherical results apply. After $t_{\rm jet}$, the jet Lorentz factor decreases with distance exponentially [93], yielding

$$\Gamma \propto t^{-1/2}; \nu_{\rm m} \propto t^{-2}; \nu_{\rm c} \propto t^0, \tag{103}$$

and the peak flux

$$F_{\nu,\rm max} \propto t^{-1} \text{ at } \nu_{\rm m}. \tag{104}$$

In the case $\nu_{\rm SSA} < \nu_{\rm m} < \nu_{\rm c}$, one finds that

$$\nu_{\rm SSA} \propto t^{-1/5}, \tag{105}$$
$$F_\nu \propto \nu^2 t^0,\ \nu < \nu_{\rm SSA}, \tag{106}$$
$$F_\nu \propto \nu^{1/3} t^{-1/3},\ \nu_{\rm SSA} < \nu < \nu_{\rm m}, \tag{107}$$
$$F_\nu \propto \nu^{-(p-1)/2} t^{-p},\ \nu_{\rm m} < \nu < \nu_{\rm c}, \tag{108}$$
$$F_\nu \propto \nu^{-p/2} t^{-p},\ \nu > \nu_{\rm c}. \tag{109}$$

In the case $\nu_{\rm m} < \nu_{\rm SSA} < \nu_{\rm c}$, we have instead

$$\nu_{\rm SSA} \propto t^{-2(p+1)/(p+4)}, \tag{110}$$
$$F_\nu \propto \nu^2 t^0,\ \nu < \nu_{\rm m}, \tag{111}$$
$$F_\nu \propto \nu^{5/2} t,\ \nu_{\rm m} < \nu < \nu_{\rm SSA}, \tag{112}$$
$$F_\nu \propto \nu^{-(p-1)/2} t^{-p},\ \nu_{\rm SSA} < \nu < \nu_{\rm c}, \tag{113}$$
$$F_\nu \propto \nu^{-p/2} t^{-p},\ \nu > \nu_{\rm c}. \tag{114}$$

These scalings apply to both ISM and wind interaction models.

The steepening of light curves to an asymptotic scaling of $F_\nu \propto t^{-p}$ is a signature of the jet models. How sharply the change from one temporal slope to another actually happens remains controversial. Mészáros and Rees [69] pointed out that seeing the edge of a jet when its Lorentz factor drops below θ_0^{-1} would also steepen the light curve by a factor of $\Delta\beta = 3/4$ in a constant density medium and $\Delta\beta = 1/2$ in a wind. Semi-analytic calculations taking into account both the sideways spreading and edge effect find that the transition from one slope to another tends to be continuous, spanning one decade or more in the observer's time [59,73,117], especially in a wind-type ambient medium. These calculations adopted a simplified set of 1D equations for the jet dynamics, which need to be checked against numerical simulations. Preliminary 2D calculations of Granot et al. [49] show that the blastwave has an egg-like shape, differing considerably from that predicted by analytic or semi-analytic models. A steepening of light curves does occur around the time $\Gamma \propto \theta_0^{-1}$, as expected from simple arguments. Whether the sharpness and amount of steepening match the predictions of the semi-analytic models remains to be seen.

3.5 Other Modifications

Many additional effects have been considered in the afterglow literature on top of the basic interaction models outlined above. We will limit ourselves to three of the more commonly discussed ones: 1) inverse Compton scattering, 2) pair production, and 3) a non-standard electron energy distribution.

Inverse Comptonization of synchrotron photons by relativistic electrons could not only produce a high energy component of afterglow emission that is potentially observable, but could also dominate the cooling of electrons and affect the blastwave dynamics [18,76,80,116]. Sari and Esin [100] examined these effects in detail, and concluded that the inverse Compton spectra broadly resemble the primary synchrotron spectra in shape, although significant differences do exist, especially at the high frequency end where a broken power-law description is no longer adequate. They showed that, as long as the fraction of the blastwave energy in electrons ϵ_e exceeds that in magnetic fields ϵ_B, the inverse Compton emission dominates the synchrotron emission in cooling the electrons in the fast cooling regime. Depending on the ratio ϵ_e/ϵ_B, the domination can extend well into the slow cooling regime, changing the value of the cooling frequency ν_c and thus the light curves of afterglow emission.

Thompson and Madau [109] considered pair formation as the γ-ray photons from the GRB proper propagate ahead of the external shock front and interact with the seed photons back scattered by the ambient medium. A simplified discussion of the process is presented in Mészáros et al. [67]. Beloborodov [3] went one step further and solved for the dynamics of the pair-loaded medium. These studies concluded that pair production can dramatically increase the radiative efficiency of the blastwave and potentially broaden the original pulses of γ-rays if the ambient density is high enough. The effects on afterglows are expected to be large at early times, especially for the wind interaction model, although details are yet to be worked out.

Most afterglow models assume that the radiating electrons have a power-law energy distribution at injection above some cutoff. The power-law indexes inferred from afterglow observations under this assumption span a wide range, from $p \sim 2$ (or below) to $p \sim 3$ (see, e.g., [17,77,78]). The lack of a universal value for p calls into question the assumption that the shock front accelerates electrons to a power-law that is constant with energy and with time. Indeed, in the Crab Nebula, arguably the best studied astrophysical synchrotron source, a break in the injection spectrum is required [1]. This motivated Li and Chevalier [63] to consider a broken power-law distribution of the electron energy that steepens at high energies in the context of the spherical wind interaction model. Similar non-standard energy distributions have been adopted in other models [23,75,118]. A major attraction of the invoked steepening of the electron energy distribution is that it can lead to a steepening of the afterglow light curves. The light curve steepening is chromatic, and should be distinguishable from the achromatic steepening due to jet effects if a wide wavelength coverage is available.

4 Application to Individual Sources

We now apply the interaction models to the dozen or so GRBs whose afterglows are relatively well observed. It turns out that for most of the sources the models are not unique. We shall start the discussion with the cases that, in our view, are less controversial.

4.1 Probable Collimated ISM Interactors: GRB990123 and GRB990510

The first two GRBs that show breaks in their optical afterglows are GRB990123 [58] and GRB990510 [51]. The breaks in both sources are achromatic and are interpreted as due to jets. In the case of GRB990123, the break occurs at a time $t_b = 1.68 \pm 0.19$ days, depending somewhat on the fitting function adopted [53]. Before the break, the temporal decay index at optical wavelengths is $\beta = -1.12 \pm 0.08$. The optical spectral index is found to be $\alpha = -0.750 \pm 0.068$. These two indexes can be fitted by an ISM interaction model with $p = 2.5$ in the adiabatic regime (where $\beta = -3[p-1]/4$ and $\alpha = -[p-1]/2$ for $\nu_{\rm m} < \nu < \nu_{\rm c}$). The decay index $\beta = -1.44 \pm 0.07$ at X-ray wavelengths is also consistent with the model, provided that the X-rays are in the cooling regime (where $\beta = [2-3p]/4$ for $\nu > \nu_{\rm c}$). The observed steeper decline in X-rays than in optical is expected for ISM interaction but not for wind interaction [16]. After the break, the optical light curves steepen quickly to $\beta = -1.69 \pm 0.06$ while the spectral index α remains approximately the same [53]. The observed amount of steepening, $\Delta\beta = 0.57 \pm 0.10$, agrees marginally with that expected of a jet with a fixed opening angle (where $\Delta\beta = 3/4$ for ISM interaction). It is less than half of the asymptotic value $\Delta\beta = (p+3)/4 = 1.38$ expected from the sideways expansion of a jet. The relatively small amount of steepening was taken as evidence for seeing the edge of a non-spreading jet [53,58,69], although it does not rule out jet spreading as the cause of steepening, since it takes time for the light curves to reach the asymptotic slope. Indeed, Panaitescu and Kumar [77] obtained a reasonable fit to the multi-frequency data of GRB990123 using a semi-analytic jet model taking into account the possibility of lateral expansion.

The breaks in the optical light curves of GRB990510 appear to be smoother than those in GRB990123. Fitting a continuous function, Harrison et al. [51] obtained a break time of $t_b = 1.20 \pm 0.08$ days. Before the break, the decay index $\beta = -0.82 \pm 0.02$, which implies $p = 2.1$ in the adiabatic regime for ISM interaction. The expected spectral index $\alpha = -0.55$ is consistent with $\alpha = -0.61 \pm 0.12$ determined by Stanek et al. [107] or $\alpha = -0.531 \pm 0.019$ by Holland et al. [53]. The latter authors find some evidence that α decreases with time which, if real, is not explained. The relatively flat decay rate of $\beta > -1$ is difficult to accommodate in the wind interaction model; it requires a very flat electron energy distribution with $p < 2$ [17]. Even then, the predicted spectral index would not match that observed [77]. After the break, the measured $\beta = -2.18 \pm 0.05$, which is consistent with the expected asymptotic value $\beta = -p$ [102]. The ISM interacting jet model is further supported by radio data, which are consistent with the expected $F_\nu \propto t^{-1/3}$ evolution [51]. X-ray data are available between 0.3 and 2 days, bracketing the break in the optical [61]. The light curve can be fitted either by a single power-law with $\beta = -1.42 \pm 0.07$ or a broken power-law, and is well modeled using an ISM interacting jet model with the cooling frequency ν_c between the optical and X-rays [77,82]. The detection of polarization from this event [20,124] is also consistent with the jet interpretation [44,99], although the observed polarization is small and does not require a jet.

4.2 Probable Spherical Wind Interactors: GRB980425 and GRB970508

The probable association of GRB980425 with the radio supernova SN1998bw at $z = 0.0085$ is discussed by Galama, by Iwamoto et al., and by others in this volume. Here, we are concerned with the radio emission from this source, which is extraordinary [58](see also the chapter by Weiler et al. in this volume). Kulkarni et al. [58] argued that the shock responsible for the radio emission was relativistic, based on the high brightness temperature, on a SSA interpretation of the early evolution, and on scintillation results. Using the standard synchrotron theory, Li and Chevalier [62] were able to deduce from the radio data that the shock was expanding into a medium with an approximately r^{-2} density profile at a speed comparable to the speed of light. A more detailed model, taking into account of the shock dynamics and relativistic effects and assuming the standard power-law electron distribution above some cutoff, reproduces the radio light curves after about day 10 reasonably well, provided that the shock energy is increased by a factor of ~ 2.5 at $\sim 10^2$ days in the rest frame of the explosion to explain the rise in the radio fluxes observed between days $20 - 40$. On the other hand, Weiler et al. (see [119,120] and their chapter in this volume) proposed a density enhancement of $\sim 40\%$ in the CSM to explain the rise in flux density between days $5 - 75$ with a similar enhancement occurring around day 192.

Based on the radio spectra at day 12 and day 15, and adopting a mono-energetic distribution for the synchrotron emitting electrons, Waxman and Loeb [115] inferred a shock speed of ~ 0.3c, which is a factor of ~ 2 smaller than that inferred by Li and Chevalier [62] around the same time. The longer term evolution of the radio source implies a power law particle energy distribution, as generally observed in GRB afterglows. Weiler et al. [119] were able to fit a parameterized model to the radio data, and came to a similar conclusion that the shock is mildly relativistic and is interacting with a clumpy progenitor wind.

Chevalier and Li [17] applied the spherical wind interaction model to GRB970508 which, unlike GRB980425, is at a cosmological distance of $z = 0.835$ [72]. Extensive optical data up to hundreds of days are available for this source (see the chapter by Pian in this volume). They follow a power-law after about day 2 with $\beta = -1.141 \pm 0.014$ and $\alpha = -1.11 \pm 0.06$ [39]. The X-rays have a similar decay index of $\beta = -1.1 \pm 0.1$ [84] and the optical/X-ray spectral index is consistent with $\alpha = -1.1$ [39]. Based on the radio to X-ray spectrum on day 12.1, Galama et al. [40] deduced that the cooling frequency is just below optical frequencies. At the optical and X-ray frequencies above the cooling frequency $\nu > \nu_c$, the observed value of β implies $p = 2.2$, which yields a spectral index α consistent with that observed for both the ISM and wind interaction models. Radio observations are crucial for distinguishing between these two models. In Fig. 1a, we show the predictions of a wind interaction model at three wavelengths. They fit the radio data reasonably well, especially after about day 100 when the interstellar scintillation dies down [32]. In Fig. 1, we present an analytic fit to the R-band data, which is used in conjunction with the radio fit to obtain the best model parameters: $\epsilon_e = 0.2$, $\epsilon_B = 0.1$, $E_{52} = 0.3$, and $A_* = 0.3$. The inferred mass-loss rate of 3×10^{-6} $M_\odot$ yr^{-1}(for a wind speed of 10^3 km s^{-1}) is in the expected range of a W-R star. The optical afterglow brightens unexpectedly around day 2. Its temporal and spectral behaviors before day 2 are not explained by the wind model.

The peak flux F_{ν_m} and SSA frequency ν_{SSA} of GRB970508 are inferred to decrease with time, which is consistent with the wind model but not with the standard spherical ISM model in the relativistic regime [39]. Frail et al. [32] discussed in depth these problems and suggested jet effects and a transition to non-relativistic expansion as possible solutions. They proposed three phases of evolution in a constant density medium: 1) a relativistic jet phase, 2) a jet spreading phase, and 3) a non-relativistic, spherical expansion phase. The jet spreading reduces the peak flux F_{ν_m}, and the emission expected from the non-relativistic phase is shown to be consistent with the radio data, although the complete model has yet to be calculated. The relatively slow decline of the optical afterglow for over 100 days [36] was unusual compared to other well studied afterglows.

4.3 Jets, Winds, or Non-relativistic Evolution?

Besides the four sources discussed above, there are a half dozen or so GRBs with reasonable multi-frequency coverage of afterglows, particularly at radio frequencies (see also the chapter by Weiler et al. in this volume), which allow for detailed modeling. These include GRB980519, GRB991208, GRB991216, GRB000301C, GRB000418, GRB000926, and GRB010222 as of August 2001. The majority of these sources show clear steepening in the optical light curves

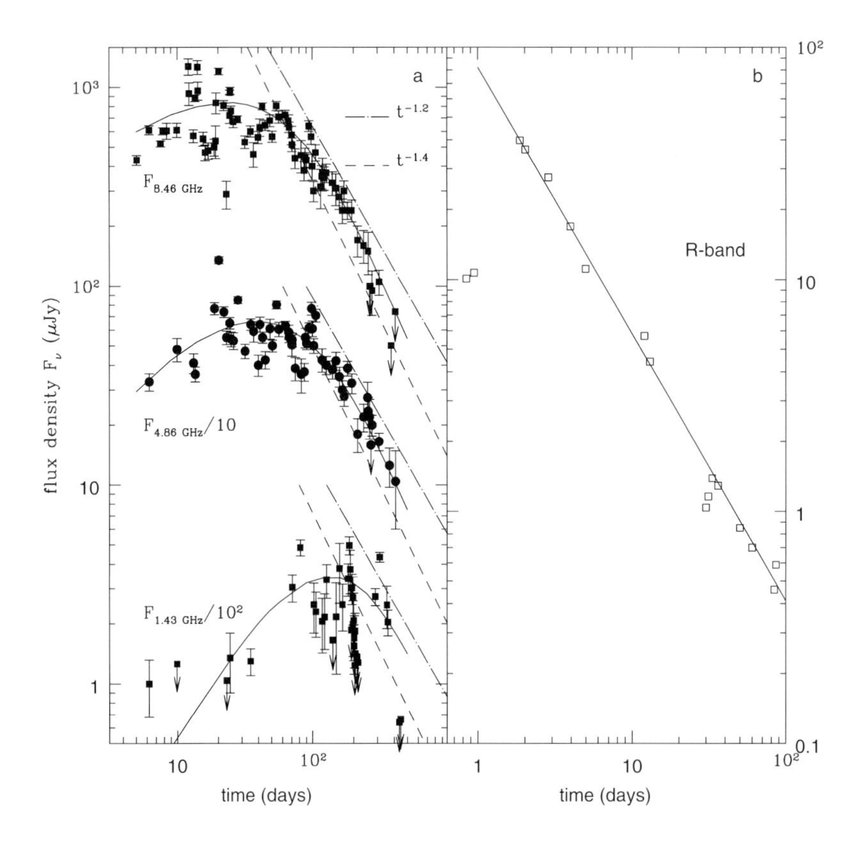

Fig. 1. Wind interaction model for the afterglow of GRB970508 (adopted from Chevalier and Li [17]). Radio data are taken from Frail et al. [30,32] and R-band data from Sokolov et al. [106].

(GRB991216, GRB000301C, GRB000926, and GRB010222), while in others the evidence for light curve steepening is weaker (GRB980519 and GRB991208) or absent (GRB000418). A variety of models has been proposed for these sources. One common conclusion is that the simplest, standard model involving a relativistic spherical blastwave expanding into a constant density medium is inadequate, and further refinements are necessary. GRB000418 was modeled by Berger et al. [6], who found that both a spherical wind model and an ISM interacting jet model fit the available radio to optical data reasonably well. For the remaining six sources, Panaitescu and Kumar [77–79] constructed semi-analytic jet models for both constant density and wind-type ambient media. They found that ISM interacting jet models can plausibly explain the broadband emission of all sources, and that wind interacting jet models are compatible with the emission of GRB991208 and GRB991216 and can marginally accommodate the afterglows of GRB000301C and GRB010222 but cannot explain the observations of GRB980519 and GRB000926. Other authors have modeled these sources individually and, in some cases, come to different conclusions. We discuss these sources in turn.

GRB980519 is one of three sources that show unusually steep light curve decay (with $\beta < -2$) but weak or no evidence for breaks (the other two being GRB980326 and GRB991208). Chevalier and Li [16] proposed that the steep decay can be explained in a spherical wind model, provided that the electron energy index p is close to 3, a value that is higher than normally found in GRB afterglows but is within the range found in radio supernovae [12]. Radio data provide some support to this interpretation [33] but do not rule out the jet model, first advocated for this source by Sari et al. [102]. Indeed, Jaunsen et al. [55] have recently found some evidence for a sharp break in the R-band light curve, which they interpreted as due to a jet expanding into a wind-type medium. The relative sparseness of the data available around and after the break makes its identification less secure. If true, the sharpness of the break would be difficult to understand in a jet model, especially if the ambient medium is wind-like [59]. The spherical wind model of Chevalier and Li [16] did not take into account inverse Compton scattering, which is important for the parameters they adopted (see, e.g., [100]). Inverse Compton scattering lowers the cooling frequency, and poses a problem for the wind model in fitting the X-ray afterglow.

GRB991208 is unique in that the radio data are sampled well enough to allow for a determination of the evolution of the characteristic frequencies $\nu_{\rm SSA}$ and $\nu_{\rm m}$ and the peak flux $F_{\nu,\rm max}$ at $\nu_{\rm m}$. The inferred scalings [42]

$$\nu_{\rm SSA} \propto t^{-0.15\pm0.23}; \nu_{\rm m} \propto t^{-1.7\pm0.7}; F_{\nu,\rm max} \propto t^{-0.47\pm0.20} \tag{115}$$

are compatible with the predictions of either the ISM interacting jet model

$$\nu_{\rm SSA} \propto t^{-1/5}; \nu_{\rm m} \propto t^{-2}; F_{\nu,\rm max} \propto t^{-1} \tag{116}$$

or the spherical wind model

$$\nu_{\rm SSA} \propto t^{-3/5}; \nu_{\rm m} \propto t^{-3/2}; F_{\nu,\rm max} \propto t^{-1/2} \tag{117}$$

but not with those of the spherical ISM model

$$\nu_{\rm SSA} \propto t^{0}; \nu_{\rm m} \propto t^{-3/2}; F_{\nu,\rm max} \propto t^{0}. \tag{118}$$

Galama et al. [42] pointed out a problem with the wind model: the rate of optical light curve decline implies $p \geq 3.3$ and such a large value of p would be inconsistent with the radio to optical spectrum at day 7.3. The inconsistency motivated Li and Chevalier [63] to seek a model with a non-standard energy distribution of electrons. They found that a spherical wind model with a broken power-law electron energy distribution can reproduce all data as well as, if not better than, the jet model. Interestingly, in the jet model of Panaitescu and Kumar [78] for this source, where the jet dynamics is followed semi-analytically, the steep decay in the optical is mostly attributed to a steepening in the electron energy distribution, as in the wind model for this source, rather than jet effects.

GRB000301C was modeled by Li and Chevalier [63] using the same spherical wind model with an electron energy distribution that steepens at a certain (high) energy as the one applied to GRB991208. The quality of the overall fit to the broad band radio to optical data is again comparable to that of an ISM interacting jet model [4]. The radio data at 8.48 GHz are particularly well sampled for this source. The fact that its temporal decay (with $\beta \approx -1.4$ [4]) is much shallower than that observed in R-band (with an asymptotic value of $\beta \approx -2.7$) provides some evidence for a steepening of the electron energy distribution at high energies (see also [78]). The steepening in energy distribution produces a steepening in the optical light curves, which could mimic the jet effects. A difference is that the optical spectrum should steepen with time in the former case but not in the latter. Rhoads and Fruchter [96] presented IR, optical, and UV data at several epochs, showing some evidence for spectral steepening with time. These data are fitted reasonably well by the wind model. The spectral steepening would be more pronounced in the optical to X-ray regime. Unfortunately, X-ray observations are not available for this source. Kumar and Panaitescu [60] attributed the steepening of the R-band light curve to a sudden large drop in the density of the ambient medium and Dai and Lu [26] to transition to non-relativistic evolution. The predictions of these models on radio emission remain to be worked out and compared with observations.

GRB991216 is a source well observed in radio, optical and X-rays. The decay indexes at these wavelengths ($\beta = -0.82 \pm 0.02$ at 8.46 GHz, -1.33 ± 0.01 at R-band and -1.61 ± 0.06 at $2 - 10$ keV) are all different [34]. There is some evidence for light curve steepening at R-band, although the data at late times are too sparse to tightly constrain the time of transition from one power-law decay to another and the decay index after the transition [50]. The optical to X-ray data are consistent with an ISM interacting jet model [50], but the radio data are not [34]. Frail et al. [34] attributed the bulk of the radio emission to either the reverse shock as in GRB991023 [101] or a second forward shock. Panaitescu and Kumar [77] examined the multi-frequency data, and concluded that all of the data can be fitted reasonably well using an ISM interacting jet model, provided that a large curvature exists in the electron energy distribution, with p changing from ~ 1.2 to ~ 2.1. GRB991216 is therefore the third source (after GRB991208 and GRB000301C discussed above; see also [78]) for which a steepening in the electron energy distribution is proposed, although in all three cases the interpretation is model dependent.

GRB000926 stands apart from other sources in that its X-ray afterglow is observed nearly two weeks after the explosion [87]. There is clear evidence for steepening in the optical light curves, with the decay index changing from $\beta = -1.46 \pm 0.11$ to $\beta = -2.38 \pm 0.07$ around 1.8 ± 0.1 days [38,52,88,97], pointing to a jet explanation. However, the decay after the break is significantly steeper than that in X-rays, estimated to be $\beta = -1.89^{+0.19}_{-0.16}$ [87], which is difficult to explain in the standard jet model. Piro et al. [87] proposed an alternative model involving a mildly collimated blastwave expanding into a dense uniform medium ($n \sim 3 \times 10^4$ cm^{-3}) with the transition to non-relativistic evolution occurring ~ 5 days after the explosion. Harrison et al. [52] showed that such a high density appears to have difficulties reproducing the radio data. They suggested that the standard jet model, either in a constant density or wind-like medium, can fit the broad band data reasonably well, provided that the inverse Compton emission contributes significantly to the X-ray emission (see also [79,127]).

GRB010222 was well observed in X-rays, with a decay index $\beta = -1.33 \pm 0.04$ [54] that is nearly identical to that in the optical after a break around ~ 0.5 days (see, e.g., [66]). Before the break, the optical light curve is significantly flatter, with $\beta \sim -0.6$ to -0.8. The steepening of optical light curves can be interpreted as due to jet effects [21,98,108], provided that the electron energy distribution is very flat (with $p \sim 1.5$) as required by the asymptotic relation $p = -\beta$ after the break [102]. This relation, originally derived for $p > 2$, may not be applicable to the case of $p < 2$ [3,23]. Masetti et al. [66] and in 't Zand et al. [54] favored an alternative model, in which the transition to non-relativistic evolution occurs rapidly in a very dense medium ($n \sim 10^6$ cm^{-3}). The inferred $p \approx 2.2$ is more in line with those inferred for other GRBs (see however [78]). It is not clear, however, whether this high density model can reproduce the early radio detection of the source [5].

5 Discussion and Conclusions

No single model explains all of the dozen or so relatively well observed GRB afterglows (see, however, [22] for a different opinion). The model that comes closest appears to be the ISM interacting jet model at the present time. It fits particularly well the multi-frequency data of GRB990123 and GRB990510. To reproduce the afterglow observations of GRB991208, GRB991216 and GRB000301C, a relatively flat electron energy distribution with $p < 2$ is required, but the flat distribution must steepen at high energies to $p > 2$ for the total electron energy to remain bounded. The model also appears capable of describing the data of GRB000418 [6] and GRB000926, provided that inverse Compton scattering contributes significantly to the X-ray emission of the latter [52]. In addition, the steep decay of the optical light curves of GRB980326 and GRB980519 has been attributed to jet effects [102], and GRB970508 has been modeled as a mildly collimated jet making a transition to a spherical, non-relativistic evolution [32].

However, a number of the above sources can be modeled equally well or perhaps better by the wind interacting spherical model. This is particularly true for GRB970508. In the case of GRB991208 and GRB000301C, a non-standard electron energy distribution is required [63], as is the case for GRB991216 (Li and Chevalier, in preparation). In addition, the wind model can fit the extensive radio data of GRB980425/SN1998bw, provided that a late energy injection occurs or the clumpy CSM is radially structured (see [119,120] and the chapter by Weiler et al. in this volume). Furthermore, the optical and X-ray afterglow data of GRB970228 are compatible with the wind model, after the subtraction of plausible supernova emission from the optical light curve [41,90]. Frail et al. [31] searched for this source in the radio for the first year, and did not detect any afterglow. The upper limits are of order 100 μJy or less at 8.46 GHz where the monitoring is most frequent. This absence of a radio afterglow is not easy to explain in the standard, ISM interacting spherical model, but is compatible with either a spherical wind model or an ISM interacting jet model [17], since in both models the peak flux decreases with frequency. The jet model may have difficulty with the relatively flat decay of the optical light curve ($\beta \sim -1.6$) unless the electron energy distribution is very flat and/or the asymptotic slope after the jet break has yet to be reached. A relatively complete optical/IR data set also exists for GRB980703 [8,10,53]. The inferred large extinction intrinsic to its bright host galaxy, coupled with the sparseness of the published X-ray [112] and radio [8] data, makes it impossible to constrain models firmly.

The presence or absence of supernova emission in the optical afterglow may provide a powerful means for corroborating the ISM or wind model. The ISM interacting sources are not expected to be accompanied by supernovae, and one indeed finds no evidence for supernovae in the most probable ISM interactors: GRB990123 and GRB990510. The wind interacting sources, on the other hand, are expected to be associated with supernovae, and the best wind interactors GRB980425 and GRB970508 are probably associated with supernovae. The evidence of supernova association is strong for GRB980425, as reviewed in the chapters by Galama and by Iwamoto et al. in this volume. The evidence is weaker for GRB970508, based mainly on a "shoulder" in the late time light curve at the I_c-band [105]; supernova emission is not clearly seen in the R-band [36]. In the case of another possible wind interactor, GRB970228, evidence for a supernova has been marshaled by Reichart [90] and Galama et al. [41]; the alternative explanation involving dust echoes [29] appears less likely [91], especially if the GRB progenitors are W-R stars instead of RSG stars [13]. Association with a supernova has also been proposed for other sources, including GRB980326 [9], GRB991208 [11], and possibly GRB000418 [27,57]. Their afterglow observations appear to be compatible with both the jet and wind models. In no source can the presence of a supernova be excluded definitively [22,38].

The expansion into a wind-type medium and the association with a supernova would point to a massive star origin for GRBs. The massive star connection is indirectly supported by the analysis of optically "dark" bursts by Reichart [92], which indicates that the majority of GRBs may be tied to giant molecular clouds, the sites of massive star formation. The connection may be strengthened by the presence of possible Fe line emission in the X-ray afterglows of GRB970508 [85], GRB970828 [126] 991216 [86], and 000214 [2], which implies a substantial amount of circumburst material. We have discussed GRB970508 as a wind interactor. GRB970828 and GRB000214 are optically undetected, which is compatible with the idea that they are tied to giant molecular clouds and thus (indirectly) to massive stars. Their afterglow data are insufficient to constrain ambient interaction models. However, some models for the Fe lines require a large Fe mass and densities $\gtrsim 10^8$ cm^{-3} at $r \approx 10^{16}$ cm [111,121], much larger than the densities inferred in afterglow models and than expected in the wind from a massive progenitor star. The distribution of the dense gas is asymmetric and it is not present along the line of sight to the GRB, which is the region probed by the afterglow emission. The strong angular variation of the density is surprising; in the "supranova" model [111], the supernova would be expected to affect the density along the line of sight. The required Fe mass would be greatly reduced if the X-ray lines come from a much denser region closer to the explosion center, as in the collapsar-bubble model of Mészáros and Rees [70].

The absence of supernova emission in the afterglows of GRB990123 and GRB990510 is consistent with their being ISM interactors with compact star merger progenitors. It, therefore, appears that there are two types of burst progenitors, with GRB970508 and GRB980425 representing the best examples of one type, and GRB990123 and GRB990510 the other ([17], see also [64]). The possible presence of supernova emission in the optical afterglows of GRB970228, GRB980326, GRB991208, and GRB000418 coupled with the fact that their afterglows can be fitted by the wind model, plausibly puts them in the massive star category. This assignment is weakened, however, by the fact that the afterglows of these sources can be fitted by the ISM interacting jet model as well. For the other GRBs with relatively well observed afterglows, the progenitor types are even less certain.

The interpretation of GRB afterglows in terms of ambient interaction models is complicated by several major uncertainties. For the jet model, the angular distribution of matter and energy inside the jet is not clear and the hydrodynamics of the jet-ambient medium interaction remains uncertain. Numerical simulations are beginning to address these issues [49]. An open issue common to both the jet and wind models is the shape of the energy distribution of the radiative electrons. Fitting the afterglow data of several GRBs requires a curvature in the energy distribution. How the required curvature comes about is not understood. This problem is particularly severe for the cases that demand a flat distribution with $p < 2$. In such cases, the afterglow emission depends sensitively on the way in which the energy distribution steepens at high energies, and the steepening has not been treated self-consistently. Another open issue is how the postshock energy is distributed among electrons, protons, and magnetic field. In addition, the spherical wind model faces an interesting dilemma: the wind model implies a massive star progenitor, and a collimated flow is expected if it must escape from the center of a star. To some extent, the spherical model can be justified by the slow apparent evolution of a jet in a wind [45,59]. It may also be possible that, upon passing through the star, the jet becomes uncollimated due to a sudden weakening of the lateral confinement.

We conclude that there is evidence for two types of GRB afterglows in different environments: 1) a constant density ISM and 2) a wind of possibly W-R star origin. The types are not immediately distinguishable, partly because, at an age of a few days, the preshock wind density is comparable to an interstellar density. At an age of seconds to minutes, the preshock density is much higher for the wind case, which could make the types more distinguishable. Pair production in the ambient medium could substantially modify the afterglow emission. Its effects are expected to be far greater in the wind case than in the ISM case [3,67,109]. In addition, the large difference in density for the two cases at small distances from the center of explosion affects the dynamics of blastwave energization, and thus the prompt emission [17,101]. While some theoretical work has been done on the prompt and early afterglow emission, much more is needed. Hopefully, rapid follow-up observations of GRB afterglows, to be enabled by the *High Energy Transient Explorer (HETE-2)* and *Swift Gamma-Ray Burst Explorer (Swift)* satellites, will put tighter constraints on the ambient interaction models and the nature of GRB progenitors.

Acknowledgments

This work was supported in part by NASA grant NAG5-8130.

References

1. E. Amato, M. Salvati, R. Bandiera, F. Pacini, L. Woltjer: Astron. Astrophys. **359**, 1107 (2000)
2. L.A. Antonelli et al.: Astrophys. J. Lett. **545**, L39 (2000)
3. A.M. Beloborodov: Astrophys. J. **565**, 808 (2002)
4. E. Berger et al.: Astrophys. J. **545**, 56 (2000)
5. E. Berger et al.: GCN 968 (2001)
6. E. Berger et al.: Astrophys. J. **556**, 556 (2001)
7. R.D. Blandford, C.F. McKee: Phys. Fluids **19**, 1130 (1976)
8. J.S. Bloom et al.: Astrophys. J. Lett. **508**, L21 (1998)
9. J.S. Bloom et al.: Nature **401**, 453 (1999)
10. A.J. Castro-Tirado et al.: Astrophys. J. Lett. **511**, L85 (1999)
11. A.J. Castro-Tirado et al.: Astron. Astrophys. **370**, 398 (2001)
12. R.A. Chevalier: Astrophys. J. **499**, 810 (1998)
13. R.A. Chevalier: In: *Gamma-Ray Bursts in the Afterglow Era, Proc. International Workshop at Rome, Italy, 17–20 October 2000*, ed. by E. Costa, F. Frontera, J. Hjorth (Springer, Heidelberg 2001) p. 142
14. R.A. Chevalier, C. Fransson: Astrophys. J. Lett. **558**, L27 (2001)
15. R.A. Chevalier, J.N. Imamura: Astrophys. J. **270**, 554 (1983)
16. R.A. Chevalier, Z.-Y. Li: Astrophys. J. Lett. **520**, L29 (1999)
17. R.A. Chevalier, Z.-Y. Li: Astrophys. J. **536**, 195 (2000)
18. J. Chiang, C.D. Dermer: Astrophys. J. **512**, 699 (1999)
19. E. Costa et al.: Nature **387**, 783 (1997)
20. S. Covino et al.: Astron. Astrophys. **348**, L1 (1999)
21. R. Cowsik, T.P. Prabhu, G.C. Anupama, B.C. Bhatt, D.K. Sahu, S. Ambika, Padmakar, S.G. Bhargavi: Bull. Astron. Soc. India **29**, 157 (2001)
22. S. Dado, A. Dar, A. De Rujula: Astron. Astrophys. **388**, 1079 (2002)
23. Z.G. Dai, K.S. Cheng: Astrophys. J. Lett. **558**, L109 (2001)
24. Z.G. Dai, T. Lu: Mon. Not. R. Astron. Soc. **298**, 87 (1998)
25. Z.G. Dai, T. Lu: Astrophys. J. Lett. **519**, L155 (1999)
26. Z.G. Dai, T. Lu: Astron. Astrophys. **367**, 501 (2001)

27. A. Dar, A. De Rujula: astro-ph 0008474 (2000)
28. E. Ergma, E.P.J. van den Heuvel: Astron. Astrophys. **331**, L29 (1998)
29. A.A. Esin, R.D. Blandford: Astrophys. J. Lett. **534**, L151 (2000)
30. D.A. Frail, S.R. Kulkarni, L. Nicastro, M. Feroci, G.B. Taylor: Nature **389**, 261 (1997)
31. D.A. Frail, S.R. Kulkarni, D.S. Shepherd, E. Waxman: Astrophys. J. Lett. **502**, L119 (1998)
32. D.A. Frail, E. Waxman, S.R. Kulkarni: Astrophys. J. **537**, 191 (2000)
33. D.A. Frail et al.: Astrophys. J. **534**, 559 (2000)
34. D.A. Frail et al.: Astrophys. J. Lett. **538**, L129 (2000)
35. D.A. Frail et al.: Astrophys. J. **565**, 829 (2002)
36. A.S. Fruchter et al.: Astrophys. J. **545**, 664 (2000)
37. C.L. Fryer, S.E. Woosley, D.H. Hartmann: Astrophys. J. **526**, 152 (1999)
38. J.U. Fynbo et al.: Astron. Astrophys. **373**, 796 (2001)
39. T.J. Galama et al.: Astrophys. J. Lett. **497**, L13 (1998)
40. T.J. Galama, R.A.M.J. Wijers, M. Bremer, P.J. Groot, R.G. Strom, C. Kouveliotou, J. van Paradijs: Astrophys. J. Lett. **500**, L97 (1998)
41. T.J. Galama et al.: Astrophys. J. **536**, 185 (2000)
42. T.J. Galama et al.: Astrophys. J. Lett. **541**, L45 (2000)
43. G. García-Segura, N. Langer, M.-M. MacLow: Astron. Astrophys. **316**, 133 (1996)
44. G. Ghisellini, D. Lazzati: Mon. Not. R. Astron. Soc. **309**, L7 (1999)
45. L.-J. Gou, Z.G. Dai, Y.F. Huang, T. Lu: Astron. Astrophys. **368**, 464 (2001)
46. J. Granot, T. Piran, R. Sari: Astrophys. J. **527**, 236 (1999)
47. J. Granot, T. Piran, R. Sari: Astrophys. J. Lett. **534**, L163 (2000)
48. J. Granot, R. Sari: Astrophys. J. **568**, 820 (2002)
49. J. Granot, M. Miller, T. Piran, W.-M. Suen, P.A. Hughes: In: *Gamma-Ray Bursts in the Afterglow Era, Proc. International Workshop at Rome, Italy, 17–20 October 2000*, ed. by E. Costa, F. Frontera, J. Hjorth (Springer, Heidelberg 2001) p. 312
50. J.P. Halpern et al.: Astrophys. J. **543**, 697 (2000)
51. F.A. Harrison et al.: Astrophys. J. Lett. **523**, L121 (1999)
52. F.A. Harrison et al.: Astrophys. J. Lett. **559**, 123 (2001)
53. S. Holland et al.: Astron. Astrophys. **371**, 371 (2001) 0703
54. J.J.M. in 't Zand et al.: Astrophys. J. **559**, 710 (2001)
55. A.O. Jaunsen et al.: Astrophys. J. **546**, 127 (2001)
56. J.I. Katz: Astrophys. J. **422**, 248 (1994)
57. S. Klose et al.: Astrophys. J. **545**, 271 (2000)
58. S.R. Kulkarni et al.: Nature **398**, 389 (1999)
59. P. Kumar, A. Panaitescu: Astrophys. J. Lett. **541**, L9 (2000)
60. P. Kumar, A. Panaitescu: Astrophys. J. Lett. **541**, L51 (2000)
61. E. Kuulkers et al.: Astrophys. J. **538**, 638 (2000)
62. Z.-Y. Li, R.A. Chevalier: Astrophys. J. **526**, 716 (1999)
63. Z.-Y. Li, R.A. Chevalier: Astrophys. J. **551**, 940 (2001)
64. M. Livio, E. Waxman: Astrophys. J. **538**, 187 (2000)
65. A.I. MacFadyen, S.E. Woosley, A. Heger: Astrophys. J. **550**, 410 (2001)
66. N. Masetti et al.: Astron. Astrophys. **374**, 382 (2001)
67. P. Mészáros, E. Ramirez-Ruiz, M.J. Rees: Astrophys. J. **554**, 660 (2001)
68. P. Mészáros, M.J. Rees: Astrophys. J. **476**, 232 (1997)
69. P. Mészáros, M.J. Rees: Mon. Not. R. Astron. Soc. **306**, L39 (1999)

70. P. Mészáros, M.J. Rees: Astrophys. J. Lett. **556**, L37 (2001)
71. P. Mészáros, M.J. Rees, R.A.M.J. Wijers: Astrophys. J. **499**, 301 (1998)
72. M.R. Metzger et al.: Nature **387**, 878 (1997)
73. R. Moderski, M. Sikora, T. Bulik: Astrophys. J. **529**, 151 (2000)
74. B. Paczyński: Astrophys. J. Lett. **494**, L45 (1998)
75. A. Panaitescu: Astrophys. J. **556**, 1002 (2001)
76. A. Panaitescu, P. Kumar: Astrophys. J. **543**, 66 (2000)
77. A. Panaitescu, P. Kumar: Astrophys. J. **554**, 667 (2001)
78. A. Panaitescu, P. Kumar: Astrophys. J. Lett. **560**, L49 (2001)
79. A. Panaitescu, P. Kumar: Astrophys. J. **571**, 779 (2002)
80. A. Panaitescu, P. Mészáros: Astrophys. J. **501**, 772 (1998)
81. A. Panaitescu, P. Mészáros, M.J. Rees: Astrophys. J. **503**, 315 (1998)
82. E. Pian et al.: Astron. Astrophys. **372**, 456 (2001)
83. T. Piran, J. Granot: In: *Gamma-Ray Bursts in the Afterglow Era, Proc. International Workshop at Rome, Italy, 17–20 October 2000*, ed. by E. Costa, F. Frontera, J. Hjorth (Springer, Heidelberg 2001) p. 300
84. L. Piro et al.: Astron. Astrophys. **331**, L41 (1998)
85. L. Piro et al.: Astrophys. J. Lett. **514**, L73 (1999)
86. L. Piro et al.: Science **290**, 955 (2000)
87. L. Piro et al.: Astrophys. J. **558**, 442 (2001)
88. P.A. Price et al.: Astrophys. J. Lett. **549**, L7 (2001)
89. E. Ramirez-Ruiz, L.M. Dray, P. Madau, C.A. Tout: Mon. Not. R. Astron. Soc. **327**, 829 (2001)
90. D.E. Reichart: Astrophys. J. Lett. **521**, L111 (1999)
91. D.E. Reichart: Astrophys. J. **554**, 643 (2001)
92. D.E. Reichart: astro-ph 0107546 (2001)
93. J.E. Rhoads: Astrophys. J. Lett. **487**, L1 (1997)
94. J.E. Rhoads: Astrophys. J. **525**, 737 (1999)
95. J.E. Rhoads: astro-ph 0103028 (2001)
96. J.E. Rhoads, A.S. Fruchter: Astrophys. J. **546**, 117 (2001)
97. R. Sagar, S.B. Pandey, V. Mohan, D. Bhattacharya, A.J. Castro-Tirado: Bull. Astron. Soc. India **29**, 1 (2001)
98. R. Sagar et al.: Bull. Astron. Soc. India **29**, 91 (2001)
99. R. Sari: Astrophys. J. Lett. **524**, L43 (1999)
100. R. Sari, A.A. Esin: Astrophys. J. **548**, 787 (2001)
101. R. Sari, T. Piran: Astrophys. J. Lett. **517**, L109 (1999)
102. R. Sari, T. Piran, J.P. Halpern: Astrophys. J. Lett. **519**, L17 (1999)
103. R. Sari, T. Piran, R. Narayan: Astrophys. J. Lett. **497**, L17 (1998)
104. P.M. Solomon: In: *Starburst Galaxies: Near and Far*, ed. by L. Tacconi, D. Lutz (Springer, Heidelberg 2001) p. 173
105. V.V. Sokolov: In: *Gamma-Ray Bursts in the Afterglow Era, Proc. International Workshop at Rome, Italy, 17–20 October 2000*, ed. by E. Costa, F. Frontera, J. Hjorth (Springer, Heidelberg 2001) p. 136
106. V.V. Sokolov, A.I. Kopylov, S.V. Zharikov, M. Feroci, L. Nicastro, E. Palazzi: Astron. Astrophys. **334**, 117 (1998)
107. K.Z. Stanek, P.M. Garnavich, J. Kaluzny, W. Pych, I. Thompson: Astrophys. J. Lett. **522**, L39 (1999)
108. K.Z. Stanek et al.: Astrophys. J. **563**, 592 (2001)
109. C. Thompson, P. Madau: Astrophys. J. **538**, 105 (2000)
110. J. van Paradijs et al.: Nature **386**, 686 (1997)

111. M. Vietri, G.Ghisellini, D. Lazzati, F. Fiore, L. Stella: Astrophys. J. Lett. **550**, L43 (2001)
112. P.M. Vreeswijk et al.: Astrophys. J. **523**, 171 (1999)
113. E. Waxman: Astrophys. J. Lett. **485**, L5 (1997)
114. E. Waxman: Astrophys. J. Lett. **489**, L33 (1997)
115. E. Waxman, A. Loeb: Astrophys. J. **515**, 721 (1999)
116. D.M. Wei, T. Lu: Astrophys. J. **505**, 252 (1998)
117. D.M. Wei, T. Lu: Astrophys. J. **541**, 203 (2000)
118. D.M. Wei, T. Lu: Astron. Astrophys. **381**, 731 (2002)
119. K. Weiler, N. Panagia, M. Montes: Astrophys. J. **562**, 670 (2001)
120. K.W. Weiler, N. Panagia, M.J. Montes, R.A. Sramek: Ann. Rev. Astron. Astrophys. **40**, 387 (2002)
121. C. Weth, P. Mészáros, T. Kallman, M.J. Rees: Astrophys. J. **534**, 581 (2000)
122. R.A.M.J. Wijers, T.J. Galama: Astrophys. J. **523**, 177 (1999)
123. R.A.M.J. Wijers, M.J. Rees, P. Mészáros: Mon. Not. R. Astron. Soc. **288**, L51 (1997)
124. R.A.M.J. Wijers et al.: Astrophys. J. Lett. **523**, L33 (1999)
125. S.E. Woosley: Astrophys. J. **405**, 273 (1993)
126. A. Yoshida et al.: Astron. Astrophys. Suppl. Ser. **138**, 433 (1999)
127. S.A. Yost et al.: In: *Gamma-Ray Bursts in the Afterglow Era, Proc. Internat. Workshop in Rome, Italy, 17–20 October 2000*, ed. by E. Costa, F. Frontera, J. Hjorth (Springer, Heidelberg 2001) p. 204

Cosmological Studies with γ-Ray Bursts

Abraham Loeb

Astronomy Department, Harvard University, Cambridge, MA 02138, USA

Abstract. Gamma-Ray Burst explosions from the first generation of stars offer an exciting opportunity to probe the epoch of reionization. Clues about how and when the intergalactic medium was ionized can be read off their UV emission spectrum. Roughly one percent of all GRBs should be strongly gravitationally lensed by intervening stars. A microlensed light curve can be inverted to reconstruct the surface brightness profile of the GRB image on the sky with micro-arcsecond resolution.

1 Introduction

Since their discovery four decades ago, quasars have been used as powerful lighthouses which probe the intervening Universe out to high redshifts, $z \sim 6$ [9,28]. The spectra of almost all quasars show strong emission lines of metals, indicating super-solar enrichment of the emitting gas [27]. This implies that, at least in the cores of galaxies, formation of massive stars and their evolution to supernovae preceded the observed quasar activity. If γ-ray bursts (GRBs) originate from the remnants of massive stars (such as neutron stars or black holes), as seems likely based on recent estimates of their energy output [10,11,48], then they should exist at least out to the same redshift as quasars. Although GRBs are transient events, their peak optical-UV flux can be as bright as that of quasars. Hence, GRBs promise to be as useful as quasars in probing the high-redshift Universe.

Not much is known observationally about the Universe in the redshift interval $z = 6 - 30$, when the first generation of galaxies condensed out of the primordial gas left over from the Big Bang (see reviews [3,33]). Observations of the *Cosmic Microwave Background (CMB)* anisotropies indicate that the cosmic gas became neutral at $z \sim 1000$ and remained so at least down to $z \sim 30$ (see, e.g., [46]). On the other hand, the existence of transmitted flux short ward of the Lyα resonance in the spectrum of the highest-redshift quasars and galaxies (see, e.g., Fig. 1), indicates that the intergalactic medium was reionized to a level better than 99.9999% by a redshift $z \sim 6$. This follows from the fact that the Lyα optical depth of the intergalactic medium at high-redshifts ($z \gg 1$) is [23],

$$\tau_\alpha = \frac{\pi e^2 f_\alpha \lambda_\alpha n_{HI}(z)}{m_e c H(z)} \approx 6.45 \times 10^5 x_{HI} \left(\frac{\Omega_b h}{0.03}\right) \left(\frac{\Omega_m}{0.3}\right)^{-1/2} \left(\frac{1+z}{10}\right)^{3/2}, \quad (1)$$

where $H \approx 100h$ kms^{-1} Mpc$^{-1}\Omega_m^{1/2}(1+z)^{3/2}$ is the Hubble parameter at the source redshift z, $f_\alpha = 0.4162$ and $\lambda_\alpha = 1216$ Å are the oscillator strength and

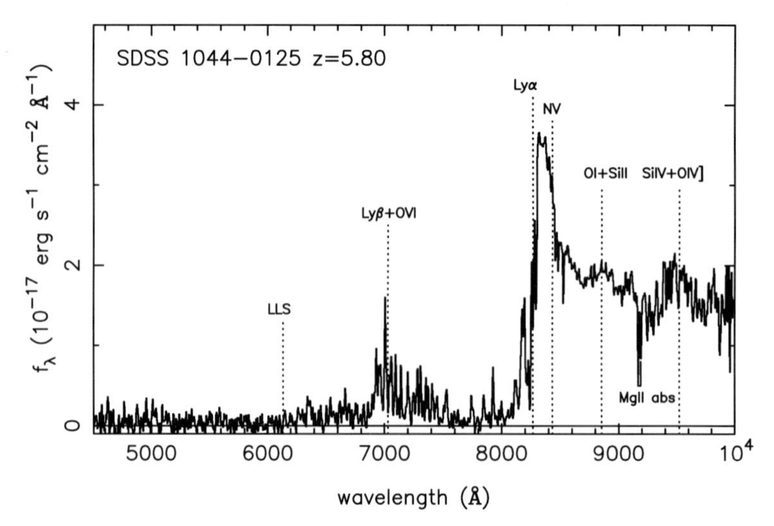

Fig. 1. Optical spectrum of a $z = 5.8$ quasar, discovered by the Sloan Digital Sky Survey [9].

the wavelength of the Lyα transition; $n_{HI}(z)$ is the average intergalactic density of neutral hydrogen at the source redshift (assuming primordial abundances); Ω_m and Ω_b are the present-day density parameters of all matter and of baryons, respectively; and x_{HI} is the average fraction of neutral hydrogen. Modeling [9] of the transmitted flux in Fig. 1 implies $\tau_\alpha < 0.5$ or $x_{HI} < 10^{-6}$, i.e., the low-density gas throughout the Universe is fully ionized at $z \sim 6$.

The question of how and when the Universe was reionized defines a new frontier in observational cosmology [33]. The UV spectrum of GRB afterglows can be used to probe the ionization and thermal state of the intergalactic gas during the epoch of reionization, at redshifts $z \sim 7 - 10$ [36]. The stretching of the temporal evolution of GRB light curves by the cosmological redshift factor $(1 + z)$, makes it easier for an observer to react in time and take a spectrum of their optical-UV emission at its peak.

Energy arguments suggest that reionization resulted from photoionization rather than from collisional ionization [14,45]. The corresponding sources of UV photons were either stars or quasars. Recent simulations of the first generation of stars that formed out of the primordial metal-free gas indicate that these stars were likely massive [1,6]. If GRBs result from the formation of compact stellar remnants, such as black holes or neutron stars, then the fraction of all stars that lead to GRBs may have been higher at early cosmic times. This, however, is true only if the GRB phenomenon is triggered on a time scale much shorter than the age of the Universe at the corresponding redshift, which for $z \gg 1$ is $\sim 5.4 \times 10^8$ yr $(h/0.7)^{-1}(\Omega_m/0.3)^{-1/2}[(1 + z)/10]^{-3/2}$. This condition may not hold, for example, for neutron star binaries with an excessively long coalescence time.

In Sect. 2 we briefly discuss the expected properties of GRB afterglows at high redshift. We then describe the use of distant GRBs for two studies: 1) probing the intergalactic medium (IGM) during the epoch of reionization (Sect. 3); and 2) finding intervening stars at cosmological distances through their gravitational lensing effect (Sect. 4).

2 Properties of High-Redshift GRB Afterglows

Young (days to weeks old) GRBs outshine their host galaxies in the optical regime. In the standard hierarchical model of galaxy formation, the characteristic mass, and hence optical luminosity, of galaxies and quasars declines with increasing redshift [3,24–26,44]. Hence, GRBs should become easier to observe than galaxies or quasars at increasing redshift. Similarly to quasars, GRB afterglows possess broad-band spectra which extend into the rest-frame UV and can probe the ionization state and metallicity of the IGM out to the epoch when it was reionized at a redshift $z \sim 7 - 10$ [33]. Lamb and Reichart [32] have extrapolated the observed γ-ray and afterglow spectra of known GRBs to high redshifts and emphasized the important role that their detection might play in probing the IGM. Simple scaling of the long wavelength spectra and temporal evolution of afterglows with redshift implies that, at a fixed time lag after the GRB trigger in the observer's frame, there is only a mild change in the observed flux at IR or radio wavelengths as the GRB redshift increases. Ciardi and Loeb [7] demonstrated this behavior using a detailed extrapolation of the GRB fireball solution into the non-relativistic regime (see the 2 μmcurves in Fig. 2). Despite the strong increase of the luminosity distance with redshift, the observed flux for a given observed age is almost independent of redshift, in part because of the special spectrum of GRB afterglows (see Fig. 4), but mainly because afterglows are brighter at earlier times and a given observed time refers to an earlier intrinsic time in the source frame as the source redshift increases. The mild dependence of the long-wavelength ($\lambda_{\rm obs} > 1$ μm) flux on redshift differs from other high-redshift sources such as galaxies or quasars, which fade rapidly with increasing redshift [3,24–26,44]. Hence, GRBs provide exceptional lighthouses for probing the Universe at $z = 6 - 30$, at the epoch when the first stars had formed.

Assuming that the GRB rate is proportional to the star formation rate and that the characteristic energy output of GRBs is $\sim 10^{52}$ erg, Ciardi and Loeb [7] predicted that there are always ~ 15 GRBs from redshifts $z > 5$ across the sky which are brighter than ~ 100 nJy at an observed wavelength of ~ 2 μm. The infrared spectrum of these sources could be taken with future telescopes such as the *Next Generation Space Telescope (NGST)* (planned for launch in 2009[1]), as a follow-up on their early X-ray localization with the *Swift Gamma-Ray Burst Explorer (Swift)* satellite (planned for launch in 2003[2].

[1] See *http://ngst.gsfc.nasa.gov/*

[2] See *http://swift.sonoma.edu/*

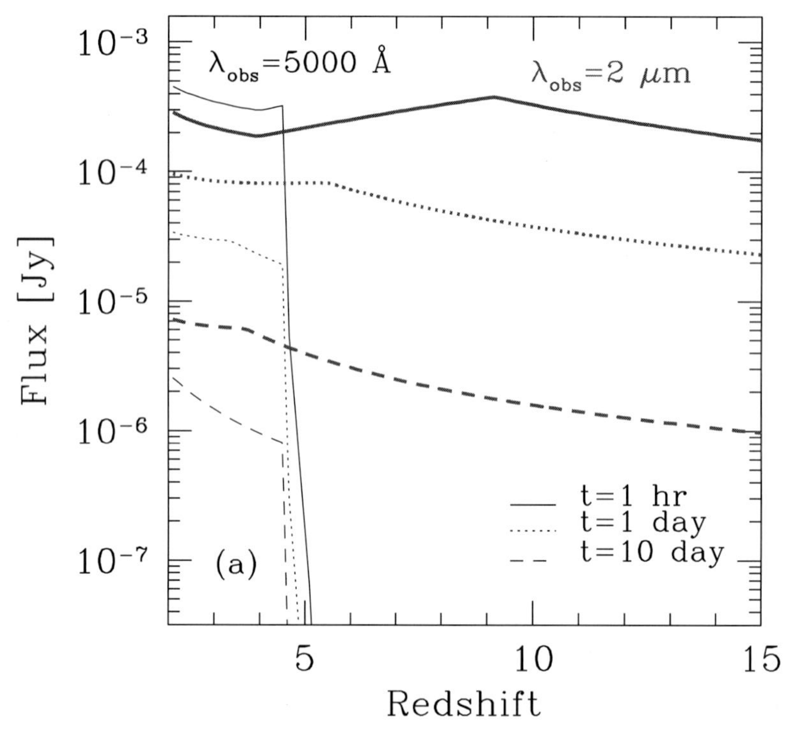

Fig. 2. Theoretical expectation for the observed afterglow flux of a GRB as a function of its redshift (from [7]). The curves refer to an observed wavelength of 5000 Å (thin lines) and 2 μm (thick lines). Different line types refer to different observed times after the GRB trigger, namely 1 hour (solid line), 1 day (dotted line) and 10 days (dashed line). The 5000 Å flux is strongly absorbed at $z > 4.5$ by intergalactic hydrogen. However, at infrared and radio wavelengths the observed afterglow flux shows only a mild dependence on the source redshift.

The redshifts of GRB afterglows can be estimated photometrically from either the Lyman limit or Lyα troughs in their spectra. At low redshifts, the question of whether the Lyman limit or Lyα trough interpretation applies depends on the absorption properties of the host galaxy. If the GRB originates from within the disk of a star-forming galaxy, then the afterglow spectrum will likely show a damped Lyα trough. At $z > 6$ the Lyα trough would inevitably exist since the intergalactic Lyα opacity is $> 90\%$ (see Fig. 13 in [44]). Interestingly, an absorption feature in the afterglow spectrum which is due to the neutral hydrogen within a molecular cloud or the disk of the host galaxy, is likely to be time dependent due to the ionization caused by the UV illumination of the afterglow itself along the line of sight [40].

So far, there have been two claims for high-redshift GRBs. Fruchter [12] argued that the photometry of GRB980329 is consistent with a Lyα trough due to IGM absorption at $z \sim 5$. Anderson et al. [2] inferred a redshift of $z = 4.5$ for GRB000131 based on a crude optical spectrum that was taken by

DETERMINING THE REIONIZATION REDSHIFT

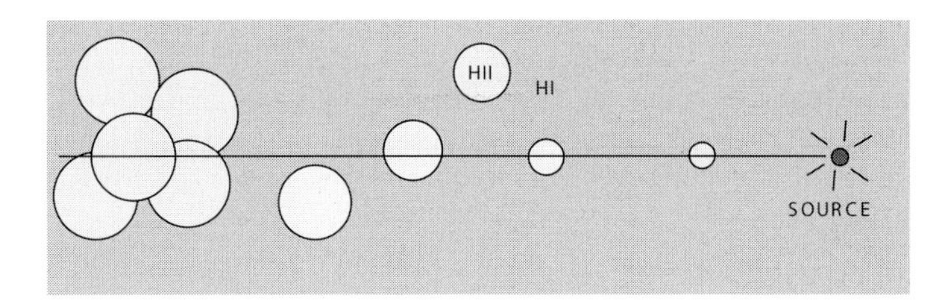

$$1 < \frac{1+z_s}{1+z_{reion}} < \frac{\lambda_\alpha}{\lambda_\beta} = 1.18$$

Fig. 3. Sketch of the expected spectrum of a source at a redshift z_s slightly above the reionization redshift $z_{\rm reion}$. The transmitted flux due to HII bubbles in the pre-reionization era, and the Lyα forest in the post-reionization era, are exaggerated for illustration.

the *European Southern Observatory (ESO) Very Large Telescope (VLT)* a few days after the GRB trigger. (These cases emphasize the need for a coordinated observing program that will alert 10 m class telescopes to take spectra of an afterglow about a day after the GRB trigger, based on a photometric assessment obtained with a smaller telescope using the Lyman limit or Lyα troughs that the GRB may originate at a high redshift.)

In the following two sections, we use two examples to illustrate the usefulness of GRB afterglows for cosmological studies.

3 Probing the Intergalactic Medium

The UV spectra of GRB afterglows can be used to measure the evolution of the neutral intergalactic hydrogen with redshift. Figure 3 illustrates schematically the expected absorption just beyond the reionization redshift. Resonant scattering suppresses the spectrum at all wavelengths corresponding to the Lyα

resonance prior to reionization. Since the Lyα cross-section is very large, any transmitted flux prior to reionization reflects a large volume of ionized hydrogen along the line of sight. If the GRB is located at a redshift larger by $> 18\%$ than the reionization redshift, then the Lyα and the Lyβ troughs will overlap. Unlike quasars, GRBs do not ionize the IGM around them; their limited energy supply of $\sim 10^{52}$ erg [10,11,48] can ionize only $\sim 4 \times 10^5$ $M_\odot$ of neutral hydrogen within their host galaxy.

Quasar spectra indicate the existence of an IGM metallicity which is a fraction of a percent of the solar value [8]. The metals were likely dispersed into the IGM through outflows from galaxies, driven by either supernova or quasar winds [3,14]. Detection of metal absorption lines in the spectrum of GRB afterglows, produced either in the IGM or the host galaxy of the GRB, can help unravel the evolution of the IGM metallicity with redshift and its link to the evolution of galaxies.

4 Cosmological Microlensing of γ-Ray Bursts

Loeb and Perna [34] noted the coincidence between the angular size of a solar mass lens at a cosmological distance and the micro-arcsecond (μas) size of the image of a GRB afterglow. They therefore suggested that microlensing by stars can be used to resolve the photospheres of GRB fireballs at cosmological distances. (Alternative methods, such as radio scintillations, provide a constraint on the radio afterglow image size [18,48] but do not reveal its detailed surface brightness distribution because of uncertainties in the scattering properties of the Galactic interstellar medium (ISM).)

The fireball of a GRB afterglow is predicted to appear on the sky as a ring (in the optical band) or a disk (at low radio frequencies) which expands laterally at a superluminal speed, $\sim \Gamma c$, where $\Gamma \gg 1$ is the Lorentz factor of the relativistic blastwave which emits the afterglow radiation [20,21,39,42,47]. For a spherical explosion into a constant density medium (such as the ISM), the physical radius of the afterglow image is of order the fireball radius over Γ, or more precisely [20,21]

$$R_{\mathrm{s}} = 3.9 \times 10^{16} \left(\frac{E_{52}}{n_1}\right)^{1/8} \left(\frac{t_{\mathrm{days}}}{1+z}\right)^{5/8} \mathrm{cm}, \tag{2}$$

where E_{52} is the hydrodynamic energy output of the GRB explosion in units of 10^{52} ergs, n_1 is the ambient medium density in units of 1 cm^{-3}, and t_{days} is the observed time in days. At a cosmological redshift z, this radius of the GRB image occupies an angle $\theta_{\mathrm{s}} = R_{\mathrm{s}}/D_A$, where $D_A(z)$ is the angular diameter distance at the GRB redshift, z. For the typical cosmological distance, $D_A \sim 10^{28}$ cm, the angular size is of order a μas. Coincidentally, this image size is comparable to the Einstein angle of a solar mass lens at a cosmological distance,

$$\theta_{\mathrm{E}} = \left(\frac{4GM_{\mathrm{lens}}}{c^2 D}\right)^{1/2} = 1.6 \left(\frac{M_{\mathrm{lens}}}{1M_\odot}\right)^{1/2} \left(\frac{D}{10^{28}\ \mathrm{cm}}\right)^{-1/2} \mu\mathrm{as}, \tag{3}$$

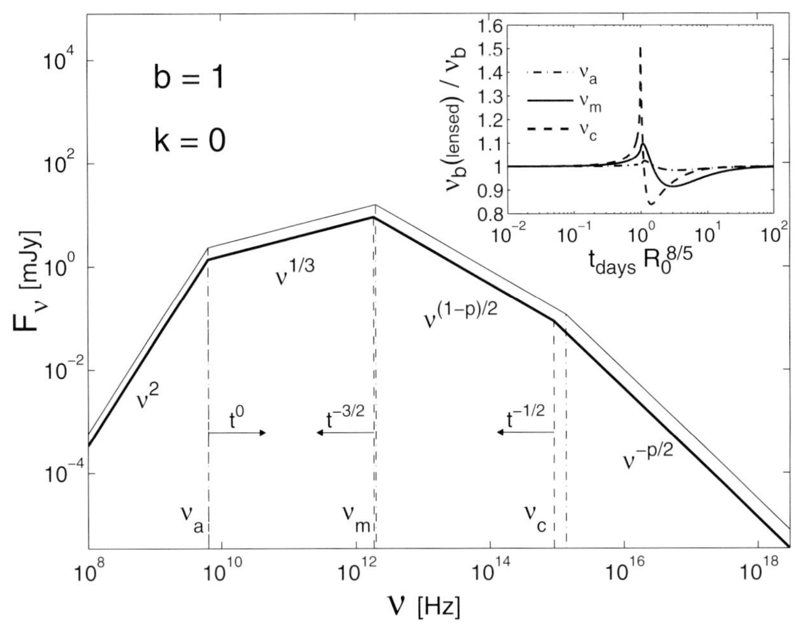

Fig. 4. A typical broken power-law spectrum of a GRB afterglow at a redshift $z = 1$ (from [19]). The observed flux density, F_ν, as a function of frequency, ν, is shown by the boldface solid line at an observed time $t_{\rm days} = 1$ for an explosion with a total energy output of 10^{52} erg in a uniform ISM ($k = 0$) with a hydrogen density of 1 cm^{-3}, and post-shock energy fractions in accelerated electrons and magnetic field of $\epsilon_e = 0.1$ and $\epsilon_B = 0.03$, respectively. The thin solid line shows the same spectrum when it is microlensed by an intervening star with an impact parameter equal to the Einstein angle and $R_0 \equiv [\theta_s(1 \text{ day})/\theta_{\rm E}] = 1$. The insert shows the excess evolution of the break frequencies $\nu_{\rm b} = \nu_a$, ν_m and ν_c (normalized by their unlensed values) due to microlensing.

where $M_{\rm lens}$ is the lens mass, and $D \equiv D_{\rm os}D_{\rm ol}/D_{\rm ls}$ is the ratio of the angular diameter distances between the observer and the source, the observer and the lens, and the lens and the source [43]. Loeb and Perna [34] showed that, because the ring expands laterally faster than the speed of light, the duration of the microlensing event is only a few days rather than tens of years as is the case for more typical astrophysical sources which move at a few hundred km s^{-1} or $\sim 10^{-3}c$.

The microlensing light curve goes through three phases: 1) constant magnification at early times, when the source is much smaller than the source-lens angular separation, 2) peak magnification when the ring-like image of the GRB first intersects the lens center on the sky, and (3) fading magnification as the source expands to larger radii.

Granot and Loeb [19] calculated the radial surface brightness profile (SBP) of the image of a GRB afterglow as a function of frequency and ambient medium properties, and inferred the corresponding magnification light curves due to mi-

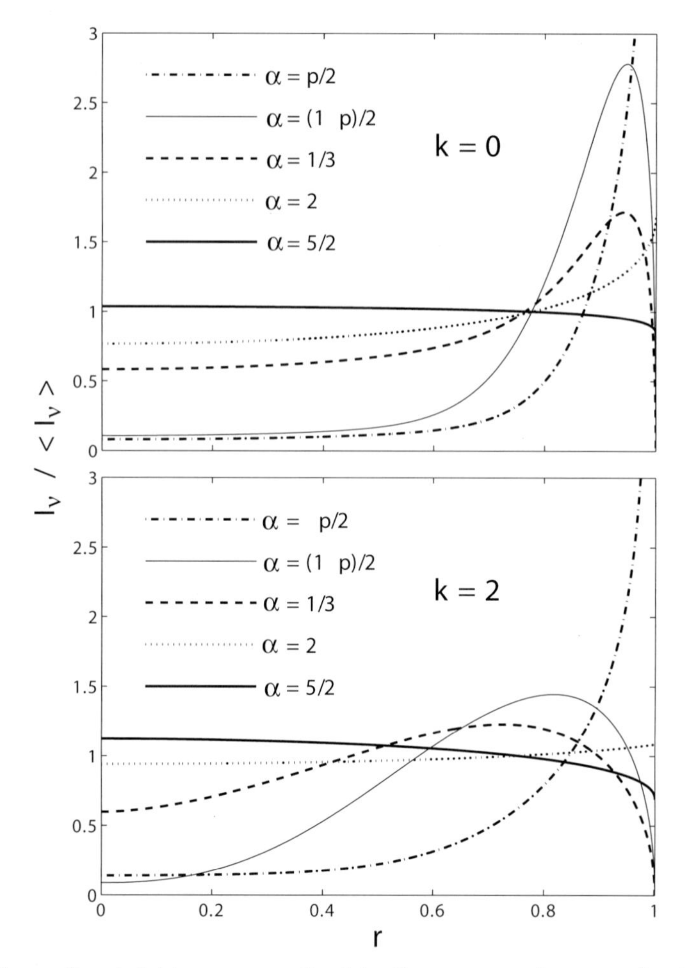

Fig. 5. The surface brightness, normalized by its average value, as a function of the normalized radius, r, from the center of a GRB afterglow image (where $r = 0$ at the center and $r = 1$ at the outer edge). The image profile changes considerably between different power-law segments of the afterglow spectrum, $F_\nu \propto \nu^\alpha$ (see Fig. 4). There is also a strong dependence on the power-law index of the radial density profile of the external medium around the source, $\rho \propto R^{-k}$ (taken from [19]).

crolensing by an intervening star. The afterglow spectrum consists of several power-law segments separated by breaks, as illustrated by Fig. 4. The image profile changes considerably across each of the spectral breaks, as shown in Fig. 5. It also depends on the power-law index, k, of the radial density profile of the ambient medium into which the GRB fireball propagates. Gaudi and Loeb [16] have shown that intensive monitoring of a microlensed afterglow light curve can be used to reconstruct the parameters of the fireball and its environment. The dependence of the afterglow image on frequency offers a fingerprint that can be used to identify a microlensing event and distinguish it from alternative interpre-

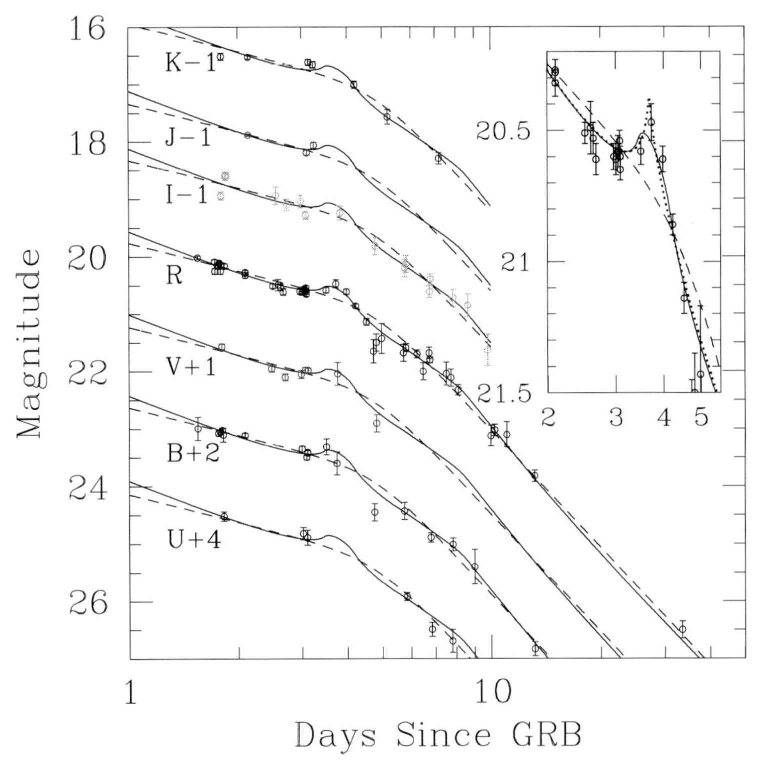

Fig. 6. *UBVRIJK* photometry of GRB000301C as a function of time in days from the GRB trigger (from [15,17]). Data points have been offset by the indicated amount for clarity. The dashed line is the best-fit smooth, double power-law light curve (with no lensing), while the solid line is the overall best-fit microlensing model, where the SBP has been determined from direct inversion. The inset shows the *R*-band data only. The dotted line is the best-fit microlensing model with theoretically calculated SBP, for $k = 0$ and $\nu > \nu_c$.

tations. It can also be used to constrain the relativistic dynamics of the fireball and the properties of its gaseous environment. At the highest frequencies, the divergence of the surface brightness near the edge of the afterglow image ($r = 1$ in Fig. 5) depends on the thickness of the emitting layer behind the relativistic shock front, which is affected by the length scale required for particle acceleration and magnetic field amplification behind the shock [22,35].

Ioka and Nakamura [29] considered the more complicated case where the explosion is collimated and centered around the viewing axis. More general orientations that violate circular symmetry need to be considered in the future.

4.1 GRB000301C

Garnavich et al. [15] have reported the possible detection of a microlensing magnification feature in the optical/infrared (OIR) light curve of GRB000301C (see

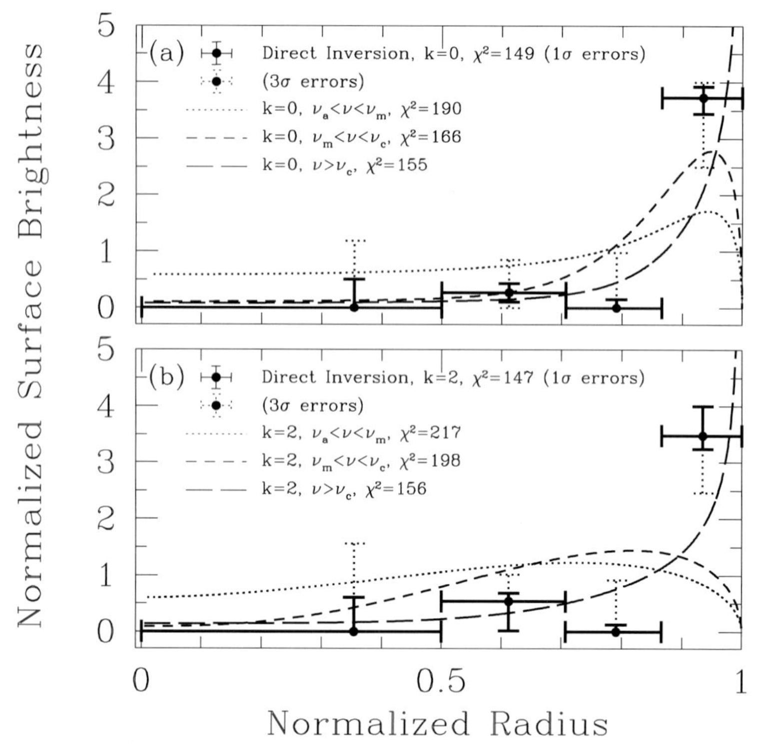

Fig. 7. Fitting GRB000301C with different SBPs as a function of normalized radius (from [17]). The points are the SBPs determined from direct inversion, with 1σ errors (solid line) and 3σ errors (dotted line). The curves are theoretically calculated SBPs for various frequency regimes (see Fig. 5): 1) Uniform external medium, $k = 0$ and 2) stellar wind environment, $k = 2$. The number of degrees of freedom is 92 for the direct inversion points and 89 for the curves.

Fig. 6). The achromatic transient feature is well fitted by a microlensing event of a 0.5 $M_\odot$ lens separated by an Einstein angle from the source center, and resembles the prediction of Loeb and Perna [34] for a ring-like source image with a narrow fractional width ($\sim$ 10%). Alternative interpretations relate the transient achromatic brightening to a higher density clump into which the fireball propagates [4], or to a refreshment of the decelerating shock either by a shell which catches up with it from behind or by continuous energy injection from the source [49]. However, the microlensing model has a smaller number of free parameters. If, with better data, a future event shows the generic temporal and spectral characteristics of a microlensing event, then these alternative interpretations will be much less viable. A galaxy $2''$ from GRB000301C might be the host of the stellar lens, but current data provide only an upper limit on its surface brightness at the GRB position. The existence of an intervening galaxy increases the probability for microlensing over that of a random line of sight.

Gaudi et al. [17] have shown that direct inversion of the observed light curve for GRB000301C yields a surface brightness profile (SBP) of the afterglow image which is strongly limb brightened, as expected theoretically (see Fig. 7).

Obviously, realistic lens systems could be more complicated due to the external shear of the host galaxy or a binary companion. Mao and Loeb [31] calculated the magnification light curves in these cases, and found that binary lenses may produce multiple peaks of magnification. They also demonstrated that all afterglows are likely to show variability at the level of a few percent about a year following the explosion, due to stars which are separated by tens of Einstein angles from the line of sight.

It is useful to determine the probability for microlensing. If the lenses are not strongly clustered so that their cross-sections overlap on the sky, then the probability for having an intervening lens star at a projected angular separation θ from a source at a redshift $z \sim 2$ is $P \sim 0.3\Omega_{\star}(\theta/\theta_E)^2$ [5,37,38,41], where $\Omega_{\star}$ is the cosmological density parameter of stars. The value of $\Omega_{\star}$ is bounded between the density of the luminous stars in galaxies and the total baryonic density as inferred from Big Bang nucleosynthesis, $7 \times 10^{-3} < \Omega_{\star} < 5 \times 10^{-2}$ [13]. Hence, all GRB afterglows should show evidence for events with $\theta < 30\theta_E$, for which microlensing provides a small perturbation to the light curve [31]. (This crude estimate ignores the need to subtract those stars which are embedded in the dense central regions of galaxies, where macrolensing dominates and the microlensing optical depth is of order unity.) However, only one out of roughly a hundred afterglows is expected to be strongly microlensed with an impact parameter smaller than the Einstein angle. Indeed, Koopmans and Wambsganss [30] have found that the *a posteriori* probability for the observed microlensing event of GRB000301C along a random line of sight is between $0.7 - 2.7\%$ if $20 - 100\%$ of the dark matter is in compact objects.

Microlensing events are rare but precious. Detailed monitoring of a few strong microlensing events among the hundreds of afterglows detected per year by the forthcoming *Swift* satellite, could be used to constrain the environment and the dynamics of relativistic GRB fireballs, as well as their magnetic structure and particle acceleration process.

Acknowledgments

I thank all my collaborators on this subject: the young theorists with whom I studied the above topics (Rennan Barkana, Benedetta Ciardi, Scott Gaudi, Jonathan Granot, Zoltan Haiman, Shude Mao, Misha Medvedev, and Rosalba Perna) and the observers who introduced me to the data on GRB000301C (Peter Garnavich and Kris Stanek).

References

1. T. Abel, G. Bryan, M.L. Norman: Astrophys. J. **540**, 39 (2000)
2. M.I. Andersen et al.: Astron. Astrophys. **364**, L54 (2000)

3. R. Barkana, A. Loeb: Phys. Rep. **349**, 125 (2001)
4. E. Berger et al.: Astrophys. J. **545**, 56 (2000)
5. O.M. Blaes, R.L. Webster: Astrophys. J. Lett. **391**, L63 (1992)
6. V. Bromm, P.S. Coppi, R.B. Larson: Astrophys. J. **527**, 5 (1999)
7. B. Ciardi, A. Loeb: Astrophys. J. **540**, 687 (2000)
8. S.L. Ellison, A. Songaila, J. Schaye, M. Pettini: Astron. J. **120**, 1175 (2000)
9. X. Fan et al.: Astron. J. **120**, 1167 (2000)
10. D.A. Frail et al.: Astrophys. J. Lett. **562**, L55 (2001)
11. D.L. Freedman, E. Waxman: Astrophys. J. **547**, 922 (2001)
12. A. Fruchter: Astrophys. J. Lett. **512**, L1 (1999)
13. M. Fukugita, C.J. Hogan, P.J.E. Peebles: Nature **366**, 309 (1998)
14. S.R. Furlanetto, A. Loeb: Astrophys. J. **556**, 619 (2001)
15. P.M. Garnavich, A. Loeb, K.Z. Stanek: Astrophys. J. Lett. **544**, L11 (2000)
16. B.S. Gaudi, A. Loeb: Astrophys. J. **558**, 643 (2001)
17. B.S. Gaudi, J. Granot, A. Loeb: Astrophys. J. **561**, 178 (2001)
18. J. Goodman: New Astron. **2**, 449 (1997)
19. J. Granot, A. Loeb: Astrophys. J. Lett. **551**, L63 (2001)
20. J. Granot, T. Piran, R. Sari: Astrophys. J. **513**, 679 (1999)
21. J. Granot, T. Piran, R. Sari: Astrophys. J. **527**, 236 (1999)
22. A. Gruzinov, E. Waxman: Astrophys. J. **511**, 852 (1999)
23. J.E. Gunn, B.A. Peterson: Astrophys. J. **142**, 1633, (1965)
24. Z. Haiman, A. Loeb: Astrophys. J. **483**, 21 (1997)
25. Z. Haiman, A. Loeb: Astrophys. J. **503**, 505 (1998)
26. Z. Haiman, A. Loeb: Astrophys. J. **552**, 459 (2001)
27. F. Hamann, G. Ferland: Ann. Rev. Astron. Astrophys. **37**, 487 (1999)
28. F.D.A. Hartwick, D. Schade: Ann. Rev. Astron. Astrophys. **28**, 437 (1990)
29. K. Ioka, T. Nakamura: Astrophys. J. **561**, 703 (2001)
30. L.V.E. Koopmans, J. Wambsganss: Mon. Not. R. Astron. Soc. **325**, 131 (2001)
31. S. Mao, A. Loeb: Astrophys. J. **547**, L97 (2001)
32. D.Q. Lamb, D.E. Reichart: Astrophys. J. **536**, 1 (2000)
33. A. Loeb, R. Barkana: Ann. Rev. Astron. Astrophys. **39**, 19 (2001)
34. A. Loeb, R. Perna: Astrophys. J. **495**, 597 (1998)
35. M.V. Medvedev, A. Loeb: Astrophys. J. **526**, 697 (1999)
36. J. Miralda-Escudé: Astrophys. J. **501**, 15 (1998)
37. R. Nemiroff: Astrophys. Space Sci. **259**, 309 (1998)
38. R. Nemiroff, J.P. Norris, J.T. Bonnell, G.F. Marani: Astrophys. J. Lett. **494**, L173 (1998)
39. A. Panaitescu, P. Mészáros: Astrophys. J. Lett. **493**, L31 (1998)
40. R. Perna, A. Loeb: Astrophys. J. **501**, 467 (1998)
41. W.H. Press, J.E. Gunn: Astrophys. J. **185**, 397 (1973)
42. R. Sari: Astrophys. J. Lett. **494**, L49 (1998)
43. P. Schneider, J. Ehlers, E.E. Falco: In: *Gravitational Lenses* (Springer, Heidelberg 1992) p. 25
44. D. Stern, H. Spinrad: Pub. Astron. Soc. Pacific **111**, 1475 (1999)
45. M. Tegmark, J. Silk, A. Evrard: Astrophys. J. **417**, 54 (1993)
46. X. Wang, M. Tegmark, M. Zaldarriaga: astro-ph 0105091 (2001)
47. E. Waxman: Astrophys. J. Lett. **491**, L19 (1997)
48. E. Waxman, S.R. Kulkarni, D.A. Frail: Astrophys. J. **497**, 288 (1998)
49. Z. Zhang, P. Meszaros: Astrophys. J. Lett. **552**, L35 (2001)

Subject Index

Druck: Strauss Offsetdruck, Mörlenbach
Verarbeitung: Schäffer, Grünstadt